Central and Peripheral Mechanisms of Cardiovascular Regulation

NATO ASI Series

Advanced Science Institutes Series

A series presenting the results of activities sponsored by the NATO Science Committee, which aims at the dissemination of advanced scientific and technological knowledge, with a view to strengthening links between scientific communities.

The series is published by an international board of publishers in conjunction with the NATO Scientific Affairs Division

A	**Life Sciences**	Plenum Publishing Corporation
B	**Physics**	New York and London
C	**Mathematical and Physical Sciences**	D. Reidel Publishing Company Dordrecht, Boston, and Lancaster
D	**Behavioral and Social Sciences**	Martinus Nijhoff Publishers
E	**Engineering and Materials Sciences**	The Hague, Boston, and Lancaster
F	**Computer and Systems Sciences**	Springer-Verlag
G	**Ecological Sciences**	Berlin, Heidelberg, New York, and Tokyo

Recent Volumes in this Series

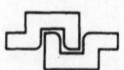

Series A: Life Sciences

Central and Peripheral Mechanisms of Cardiovascular Regulation

Edited by

A. Magro

Wadsworth Center for Laboratories and Research
Albany, New York

W. Osswald

University of Porto
Porto, Portugal

D. Reis

Cornell University Medical College
New York, New York

and

P. Vanhoutte

Mayo Clinic and Mayo Foundation
Rochester, Minnesota

Plenum Press
New York and London
Published in cooperation with NATO Scientific Affairs Division

Proceedings of a NATO Advanced Study Institute on
the Molecular Basis for the Central and Peripheral Regulation
of Vascular Resistance,
held April 29–May 10, 1985,
in Altavilla Milicia, Sicily, Italy

Library of Congress Cataloging in Publication Data

NATO Advanced Study Institute on the Molecular Basis for the Central and Peripheral Regulation of Vascular Resistance (1985: Altavilla Milicia, Italy)
Central and peripheral mechanisms of cardiovascular regulation.

(NATO ASI series. Series A, Life sciences; v. 109)
"Proceedings of a NATO Advanced Study Institute on the Molecular Basis for
the Central and Peripheral Regulation of Vascular Resistance, held April 29–May
10, 1985, in Altavilla Milicia, Sicily, Italy"—T.p. verso.
"Published in cooperation with NATO Scientific Affairs Division."
Includes bibliographies and index.
1. Blood pressure—Regulation—Congresses. 2. Nervous system, Vasomo-
tor—Congresses. 3. Cardiovascular system—Effect of drugs on—Congresses. 3.
Cardiovascular system—Effect of drugs on—Congresses. I. Magro, Albert M. II.
North Atlantic Treaty Organization. Scientific Affairs Division. III. Title. IV. Series.
[DNLM: 1. Blood Pressure—drug effects—congresses. 2. Muscle, Smooth, Vas-
cular—drug effects—congresses. 3. Vascular Resistance—drug ef-
fects—congresses. WG 106 N279c 1985]
QP109.N38   1985   615'.71   86-15101
ISBN-13: 978-1-4615-9473-4   e-ISBN-13: 978-1-4615-9471-0
DOI: 10.1007/978-1-4615-9471-0

© 1986 Plenum Press, New York
Softcover reprint of the hardcover 1st edition 1986
A Division of Plenum Publishing Corporation
233 Spring Street, New York, N.Y. 10013

PREFACE

There is increasing awareness that the autonomic nervous system, through its central and peripheral pathways, plays a critical role in the regulation of the circulation. Peripherally, the autonomic representation, largely that of sympathetic nerves, innervate virtually all segments of the vascular tree as well as the adrenal medulla. Through the interaction of nerve terminals, their transmitters, receptors and intracellular mediators in smooth muscle, sympathetic neurons control vascular tone as well as the basal performance of the heart.

In turn, the performance of the autonomic nervous system is highly controlled by the brain. Once viewed as a black box with only a vague influence on cardiovascular performance, the introduction of concepts and techniques of neuroscience into the field of cardiovascular medicine has led to the realization of the critical role of this organ in cardiovascular control. It is now well recognized that within the brain, the represenation of cardiovascular function is highly restricted anatomically, engages a number of specific transmitters for its actions, and has highly selective and topographically restricted functions to influence circulatory performance.

This volume represents a selected overview of some mechanisms through which the autonomic nervous system contributes to overall cardiovascular control. The contributors touch upon the most critical control points now recognized to be of relevance in control of the circulation. These range through studies of: vascular reactivity, cellular mechanisms mediating smooth muscle contraction, the action of drugs upon blood vessels and nerve terminals and, finally, central neural pathway transmitters and their actions. In selecting the contributors, the editors have not only sought expertise but also for representation of the major areas presently at the frontier in understanding the role of the autonomic nervous system in circulatory control. It is hoped that the reader will confirm the editors' belief that this monograph will be of interest not only to basic scientists but also to cardiologists and other practitioners.

A. Magro
W. Osswald
D. Reis
P. Vanhoutte

ACKNOWLEDGMENT

We are grateful to Dr. Craig Sinclair, Director of the NATO ASI Program, for his assistance and advice in arranging the Institute, where lectures from the contributors preceded and stimulated the Reviews of this Volume. We are also grateful to the Scientific Affairs Division of the North Atlantic Treaty Organization, especially to the Chairman Dr. Henry Durand and the Science Committee, for granting the financial support which made the Institute possible.

The Editors and Contributors would like to express their sincere thanks to Alice Magro for her special administrative qualities and for the invaluable assistance she has given in the proofing, layout and indexing of this Volume.

CONTENTS

INACTIVATION OF CATECHOLAMINES IN THE BLOOD VESSEL WALL

Walter Osswald

Laboratorio de Farmacologia
Faculdade de Medicina do Porto
4200 Porto, Portugal

INTRODUCTION

It is well known that effector organ responses to catecholamines are evanescent in character, disappearing in a few minutes after bolus administration of these substances or short electrical stimulation of the adrenergic innervation of the effector organ. Since experiments on isolated vessels uniformly have shown that after termination of electric stimulation (leading to release of endogenous transmitter) or washout of the added exogenous catecholamine the motor effects quickly disappear, the conclusion is inescapable that the catecholamines are rapidly removed from the biophase and/or metabolized to inactive or at least much less active substances. Thus, the blood vessel wall itself must be endowed with mechanisms which are able to inactivate the catecholamines. In fact, a large amount of evidence has accumulated, showing that the mechanisms presently known to dispose of catecholamines in nonvascular structures are all present and operant in the blood vessel wall. An appreciable number of comprehensive reviews has dealt with these vascular mechanisms (Somlyo and Somlyo, 1970; Iversen, 1971; Spector, Tarver and Berkowitz, 1972; Trendelenburg, 1972, 1980; Bevan and Su, 1973; Langer, 1974; Osswald, 1976, 1978, 1979, 1984; Kalsner, 1979; Bevan, Bevan and Duckles, 1980; Bevan et al, 1980; De la Lande, 1981; Vanhoutte, Verbeuren and Webb, 1981; Osswald and Guimarães, 1983).

Elucidation of the processes leading to inactivation of catecholamines in the blood vessel wall is of particular importance for a better understanding of a series of problems concerning cardiovascular regulation, especially the maintenance of vascular tone, both in health and in disease. An eventual physiological role of blood vessels in regulating plasma levels of catecholamines of sympathoadrenal origin strongly suggests itself, and may well be altered in some pathological states, e.g. hypertensive disease, as discussed later on. On the other hand, a number of drugs, some of which are of appreciable therapeutic relevance, interfere with inactivation of catecholamines and thus may exert effects of therapeutic and/or toxicological consequence. For all these reasons, in this chapter an attempt will be made to review these interrelated topics.

MECHANISMS OF INACTIVATION

As stated above, the blood vessels so far studied appear to possess all enzymatic and non-enzymatic mechanisms which play relevant roles in the inactivation of catecholamines. Detailed description of these mechanisms is beyond the scope of this short review, but may be found in the reviews quoted in the Introduction. In short, the following mechanisms must be considered:

1. Diffusion. By definition a passive, gradient-dependent mechanism, diffusion occurs in every vascular tissue. Its importance depends on the structural characteristics of the vessel under study and on the experimental setup. In incubation experiments, the highly active sites of loss present in the tissue generate concentration gradients, i.e. the amine concentration throughout the extracellular space is not homogeneous (Trendelenburg, 1984). In the rabbit aorta (a much used experimental model), for example, the tissue characteristics of a large average neuromuscular distance and a strictly adventio-medial innervation create for this relatively thick vessel conditions which favour much higher diffusion constants in the adventitia than in the media (Török and Bevan, 1971, Henseling, 1983). Thus, it is understandable that if the rabbit aorta is incubated for short periods with low concentrations of amine, neuronal mechanisms of inactivation will be overestimated and extraneuronal mechanisms underestimated, because of the rapid and easy equilibration of the neuronal sites with the medium offered to the adjacent adventitia, whereas access to the media is diffusion limited (Trendelenburg, 1984).

2. Uptake processes. Active transmembraneous transport mechanisms have been described which are able to translocate the highly polar catecholamines from the extracellular space to the interior of neuronal and nonneuronal cellular elements. These mechanisms have been described as neuronal (uptake 1) and extraneuronal (uptake 2) processes (Paton, 1960, Iversen, 1967, 1971; reviews by Paton 1976 and Trendelenburg, 1980). These mechanisms must play a prominent role in inactivation of catecholamines, since they precede metabolism, which occurs exclusively in the cells themselves (see below, under 4). These mechanisms are saturable at quite different amine concentrations, the K_m of the extraneuronal uptake being about 100 times higher than that of the neuronal uptake; relatively specific inhibitors of these uptakes have been extensively studied (e.g. cocaine and tricyclic antidepressant drugs for neuronal uptake, corticosteroid Δ, 0-methylated derivatives of catecholamines and oestradiol for extraneuronal uptake). Uptake represents in itself an inactivation mechanism, because it contributed to clear the amines from the biophase (where receptors are present); however, if the enzymatic machinery behind the uptake process is not functioning, (i.e. is inhibited by suitable drugs) uptake may function as a temporary loss, because the intact amine is again able to cross the membrane and reach the biophase. This has been demonstrated to occur in the extraneuronal (but not in the neuronal) system, when its enzymes are blocked (Trendelenburg 1974, Guimarães and Paiva 1977, Henseling 1980).

3. Binding to tissue constituents. First described to occur by Avakian and Gillespie (1968), binding of intact molecules to collagen, elastin, endothelial cells, fibroblasts and nuclei may occur in a variety of tissues, including the vascular ones (review by Gillespie, 1976). Very little is known about the importance of this mechanism, but the fact that the total volume of these tissue constituents is much higher than that of nerve terminals and smooth muscle cells taken together (Coimbra, Ribeiro-Silva and Osswald, 1974, Branco et al. 1981) should encourage more detailed investigations into the role played by these mechanisms. Since the initial studies were made with the fluorescence technique, which requires the use of high concentrations of amines, it was thought that the process only played a role for these high, unphysiologic concentrations. However, resorting to autoradiography allowed Branco et al (1981) to conclude that this "binding" occurs even for relatively low concentrations of isoprenaline and may represent the structural counterpart of the corticosteroid-resistant, non 0-methylating site of loss described in the rabbit aorta (Henseling 1980). This corticosteroid-resistant site of loss

and its metabolizing capacity appear to present wide interorgan variations. It is practically nonexistent in the rat heart (Bönisch and Trendelenburg, 1974) and increases in importance in the following order: canine mesenteric arteries (Osswald, Garrett and Guimarães, 1975), dog saphenous vein (Azevedo and Osswald, 1976, Paiva and Guimarães, 1978, Branco and Osswald, 1980), rabbit aorta (Henseling, 1980, Branco et al 1981, Sweet and Levin, 1983, Levin and Wilson, 1983, Magill and Levin, 1984) and rabbit ear artery, where it predominates (Head et al, 1980). A recent critical appraisal of this corticosteroid-resistant site of loss is to be found in the review by Trendelenburg (1984).

4. <u>Enzymatic breakdown</u>. Oxidative deamination through monoamine oxidase (MAO) and O-methylation through catechol O-methyl transferase (COMT) result in the formation of metabolites which are much less active than the parent compounds (in the order of 100 to 1000 times less: Holtz, Osswald and Stock, 1960, Renson, Weissbach and Udenfriend, 1964). These enzymatic mechanisms can only operate after entry of the amines into the cell (by active uptake - see section 2 - or by passive entry, due to pinocytosis - Azevedo, Teixeira and Sarmento, 1984 - or diffusion, depending on the relatively low lipophilicity of the catecholamines - Mack and Bönisch, 1979). Since the active uptake processes play the major role in most tissues studied so far, the sequence of uptake and intracellular metabolism represents, in fact, a "metabolizing system" or a "pump and enzyme system" (Kurahashi, Rawlow and Trendelenburg, 1980, Trendelenburg, 1984). It is evident that the intact system can not be described in terms of kinetic constants, since the constants of both the uptake process and the enzyme influence the function of the entire system. It is also evident that blockade of either the uptake or the enzyme may have very different consequences on the inactivation of amines. For example, for the extraneuronal O-methylating system it has been demonstrated that it functions as an "irreversible site of loss" when the enzyme is intact or not saturated, but that it functions as a "reversible site of loss" when the enzyme is saturated or inhibited (Bönisch, 1978). Thus, the system is transformed from a "pump and enzyme" one into a "pump and leak" system; a mathematical model has been developed (Kurahashi et al., 1980) which well agree with experimental data (Bryan, 1984, Kurahashi, et al., 1984).

The concept of metabolizing systems is not only helpful in our progressive understanding of the function of these sites of loss in the inactivation of catecholamines but also allows us to understand why it is so difficult to compare results obtained in intact tissues with those gained in experiments in which diffusion and uptake processes are excluded (homogenates, purified enzyme preparations). Another important point which must not be forgotten in the attempts to define relative roles of mechanisms of inactivation is that of the relationship between the two enzymes, MAO and COMT. Firstly, the metabolites resulting from the action of MAO on catecholamines (dihydroxyphenylglycol - DOPEG and dihydroxymandelic acid - DOMA, which result from reduction and oxidation, respectively, of the unstable aldehyde formed by oxidative deamination of adrenaline and noradrenaline; dihydroxyphenylethanol - DOPET and dihydroxyphenylacetic acid - DOPAC, which stem in a parallel way from dopamine) are substrates of COMT and give rise, through O-methylation, to the corresponding derivatives (methoxyhydroxyphenylglycol - MOPEG, methoxyhydroxymandelic or "vanillylmandelic" acid - VMA, methoxyhydroxyphenylethanol - MOPET and homovanillic acid - HVA). On the other hand, the O-methylated compounds resulting from the action of COMT (normetanephrine - NMN, metanephrine - MN, methoxytyramine - MT) are substrates of MAO and thus give rise to MOPEG, VMA, MOPET and HVA. The O-methylated and deaminated metabolites (OMDA) thus represent the final metabolites of the catecholamines and should be the only ones found if the tissue contains both MAO and COMT, the enzymes are not saturated and transport mechanisms would allow for an even distribution of the amines and metabolites to the intracellular enzymatic sites. As discussed further on, this does not happen under most experimental conditions; the efflux of metabolites from the cells depend very much on their physicochemical properties (review by Trendelenburg et al., 1979).

A further complication lies in the fact that two different types of MAO have been demonstrated to coexist in blood vessels (Caramona, 1983, 1983). As reviewed by Mantle and Tipton (1982), the evidence accumulated since Johnston (1968) first suggested the presence in tissues of more than one form of MAO has led to the present definition of two forms of MAO, namely A and B, which may be studied with the help of specific inhibitors and of preferential substrates (Kinemuchi, Fowler and Tipton, 1984) and differ in structure and amino acid sequence (Denney and Abell, 1984). In the lateral saphenous vein of the dog, both types of MAO are represented, the ratio of MAO-A/MAO-B (as indicated by the ratio of V_{max} for 5-hydroxytryptamine/V_{max} for B-phenylethylamine) being of 0.51. Tyramine, which is a substrate devoid of specificity, is deaminated, in saphenous vein homogenates, by both forms of MAO, although the activity of MAO-B predominates (Caramona, 1982). Since MAO is present both in nerve terminals and extraneuronal cells (review by Fowler, Magnusson and Ross 1984) it is of interest to pinpoint the localization of the two types of

MAO. Recent studies from this laboratory (Caramona, 1983; Caramona and Soares-da-Silva, 1984) on several normal and denervated venous and arterial vessels allowed to conclude that there is a very good correlation between adrenergic innervation and MAO-A content (but not MAO-B content) and that MAO-A is located both inside and outside of the nerve terminals, whereas MAO-B is entirely extraneuronal. Thus, the suggestion that neuronal MAO belongs to the A type (Goridis and Neff, 1971, Vanhoutte, et al., 1977) could be confirmed. However, it should be stressed that this does not mean that a "transmitter-specific" form of MAO (Goridis and Neff, 1971) exists, since intra- and extraneuronal MAO-A did not differ in their sensitivity to inhibitors (Osswald and Caramona, 1984).

Another point of interest in the inter-relationship of the two enzymes (COMT and MAO) is that pertaining to its localization in cells. For neuronal elements, all existing evidence points to the presence of MAO (type A - see above) but not of COMT (review by Trendelenburg, 1980). For extraneuronal cells, however, the situation is quite different, since, as aptly discussed by Trendelenburg (1984), the results obtained in a number of experiments are consistent with the co-existence of COMT and MAO (type A and B - see above) in the same cells. However, we do not know if all extraneuronal cells contain both types of enzymes or if the distribution of enzyme activity in different cell types is homogeneous, although it is evident that the smooth muscle cells of the rabbit aorta are endowed with both COMT and MAO (Levin and Wilson, 1977).

5. Nonenzymatic formation of condensation products. Catecholamines may easily form aldehyde condensation products, the tetrahydroisoquinolines (TIQs). Thus, the condensation of parent amines with unstable and highly reactive aldehydes originated from them by oxidative deamination (by MAO) represents at least a theoretical possibility. Holtz, Stock and Westermann (1963) identified such a product (tetrahydropapaveroline - THP) as resulting from the reaction of dopamine with the corresponding aldehyde, due to the action of MAO. More recently, Teixeira and Macedo (1981) found that the relaxant effect of dopamine on vascular smooth muscle contracted by prostaglandin $F_2\alpha$ in the presence of an α-adrenoceptor blocking agent appeared to be due to activation of dopamine and β-adrenoceptors by a substance formed through the action of MAO. The use of inhibitors of MAO and of the formation of THP, as well as incubation experiments (Teixeira, Almeida and Figueiredo, 1983) led the authors to suggest that nonenzymatic formation of THP is an important pathway in vascular tissue (for dopamine, but not for adrenaline).

The question of topographical localization of inactivation mechanisms has already been raised in the preceding section. However, it is felt that a very short review of the sites of inactivation and of their interrelations should be attempted at this place. It is by no means clearly established where and in which proportion the inactivation proceeds, since most studies conducted on this subject have dealt with entire segments of blood vessels or with perfused vascular areas and only global inactivation can be measured in most of these experimental setups. Nevertheless, some indications allow the summarization of this question in the following manner :

Most arterial and venous vessels are characterized by the existence of an adventiomedial nervous ground plexus, although the adrenergic innervation pattern may differ and take the form of medial innervation, i.e. of a distribution of nerve terminals throughout the media. This also means that mean neuromuscular distance may vary markedly (from 60 to 4000 nm - see review by Osswald and Guimarães, 1983) from vessels showing close apposition (as some arterioles and veins) to large arteries where no true synaptic cleft exists. These facts are of obvious importance for the fate of the transmitter, after its release from the nerve terminals. The type and density of innervation is also of importance for the disposition of exogenous or blood-borne catecholamines. In the lateral saphenous vein of the dog, nervous tissue represents 7.9% of the wall (Coimbra et al., 1974), whereas in the rabbit thoracic aorta nervous elements account only for 0.5% of the volume (Branco et al., 1981); it is evident that in the former vessel neuronal systems of inactivation (uptake, metabolism by MAO-A, storage in dense-core vesicles) must play a more important role than in the rabbit aorta, and this is actually what happens, as discussed below.

Smooth muscle cells are endowed with both the O-methylating and the deaminating system; in the rabbit aorta the adventitia (containing all nervous elements) can be separated from the media (the endothelium can easily be eliminated by rubbing) as described by Maxwell, Eckhardt and Wastila (1968). This preparation of the isolated (noninnervated) media has been used by several authors (Levin 1974, Levin and Wilson, 1977, 1983, Henseling and Trendelenburg 1978, Henseling 1980, Branco et al., 1981, Sweet and Levin 1983, Magill and Levin 1984) and shown to be a good model for the extraneuronal metabolizing systems, i.e. of the "pump and

7

enzyme" systems described in the previous section.

As far as the endothelium is concerned, there is some evidence pointing in the sense of the presence of uptake and metabolic processes in at least some endothelial cells, e.g. those of the cerebral microcirculation (Owman, Edvinsson and Hardebo, 1980) or of the lung vessels (Gillis, 1980). However, a number of observations suggest that these properties are not restricted to the pulmonary and cerebral endothelium (Bevan and Török, 1970, Török and Bevan, 1971, Osswald, Guimarães and Coimbra, 1971). Azevedo and Osswald (1976) described an active uptake mechanism (blocked by cortexone) in the endothelium of the saphenous vein; on the other hand, Lowe and Creveling (1979) and Lowe (1979), in immunocytochemical and ultrastructural autoradiography studies, described the presence of COMT and MAO in endothelial cells of the rat aorta and of capillaries of the heart. Other cellular elements present in blood vessels (fibroblasts, Schwann cells, fat cells), although capable of accumulating and metabolizing isoprenaline (Branco et al., 1981) probably play only a minor role in the inactivation of catecholamines (Osswald and Guimarães, 1983).

In this respect, it is important to state that we are just beginning to realize that the spatial relationship (i.e. the relative vicinity) of some sites of loss with the source of the amine or the receptors of the effector cell may be of paramount significance for the relevance of these sites of loss in the inactivation of that amine. For example, although fibroblasts to not appear to play a quantitatively important role in metabolism of noradrenaline, their close vicinity to nerve endings in several vessels (Azevedo and Soares-da-Silva, 1981, Soares-da-Silva and Azevedo, 1985) makes them suitable candidates for a role in regulation of transmitter concentration in the juxtaneuronal biophase. On the other hand, it has been unequivocally proven that α- and β-adrenoceptors show a differential localization in relation to nerve terminals and COMT activity, α-adrenoceptors being more closely related to neuronal elements and β-adrenoceptors to sites of COMT activity (Guimarães, 1975, Guimarães et al., 1975, Belfrage, Fredholm and Rosell, 1977, Guimarães and Paiva, 1977, Bevan et al., 1978, Winquist and Bevan, 1979). These observations led to the concept of two different biophases for α- and β-adrenoceptors (Guimarães, 1982, Guimarães and Paiva, 1981 a and b). Thus it is clear that the blockade of one site of loss or enzymatic system may alter the effects of a catecholamine in a more or less pronounced way than the measure of the actual changes in overall concentration of that catecholamine could make plausible. A typical example is that a certain degree of inhibition of COMT potentiates many times more the relaxing than the contraction response to the same concentration of exogenous adrenaline; on the other hand, blockade of neuronal uptake (by example by a tricyclic antidepressant drug) may enhance much more the vasoconstrictive response than the relaxing one.

INACTIVATION OF THE TRANSMITTER

It is well known that the biological half-life of noradrenaline released from nerve terminals located in vascular tissue is short, and in fact this a prerequisite for the physiological role of any transmitter substance, if a fine control of the function of the effector organ is the ultimate goal of transmitter release. The experimental data accumulated over the years confirm this assumption. The transmitter may be released by electrical stimulation (at frequencies which are thought to mimic physiologic conditions of nerve impluses) or by indirectly acting amines (like tyramine) and since these mechanisms of release are different, it is not surprising that inactivation mechanisms may also differ, as shown in the following subsections.

a) Transmitter released by electrical stimulation. The fate of the transmitter depends on the structure of the vessel under study. In blood vessels with medial innervation and/or small neuromuscular distance, the neuronal metabolizing system plays, not unexpectedly, a major role. The released transmitter is re-uptaken into the nerve terminals and subsequently deaminated (to DOPEG) or reincorporated into storage vesicles; metabolism and reincorporation represent secondary events, since removal of the transmitter from the biophase already represents an inactivation, by definition. DOPEG formation represents, in these experimental conditions, a good index of intraneuronal deamination. A recent report (Caramona, Araujo and Brandão, 1985) shows that the formation of DOPEG induced by electrical stimulation is abolished by clorgyline (a selective MAO-A inhibitor) and unchanged by deprenyl (a selective MAO-B inhibitor). These results are in good agreement with the characterization of neuronal MAO as belonging to the A type (Caramona, 1982, 1983).

This is the general picture observed to occur in vessels like the canine lateral saphenous vein (Brandão and Guimarães, 1974, Brandão, 1977; 1979, Muldoon, et al., 1979, Verbeuren and Vanhoutte, 1982), the rabbit portal vein (Hughes, 1972), the rat tail artery (Wyse, 1976) and the gracilis muscle artery (Kahan, Hjemdahl and Dahlöf, 1984). The fraction which escapes re-uptake is subject to O-methylation and deamination, after extraneuronal uptake, giving origin to NMN and to OMDA, or appears in the outflow as unchanged noradrenaline (diffusion). Although perfusion or superfusion methods must favour the process of diffusion, it is clear that even under in vivo conditions diffusion to the circulating blood does occur (Sollevi, Hjemdahl and Fredholm, 1981, Bradley and Hjemdahl, 1984, Kahan et al., 1984).

In vessels with a wide neuromuscular distance and/or with adventitio-medial type of innervation, diffusion and extraneuronal handling play a more important role. This has been shown to be the case for vessels like the pulmonary artery of the rabbit (Endo, et al., 1977, the rabbit aorta (Nedergaard and Schrold, 1980, Schrold and Nedergaard, 1981, Henseling, 1981) and the central artery of the rabbit ear (Bevan, et al., 1972, Allen, Rand and Story, 1973). In the rabbit aorta, however, oxidative deamination appears to play only a limited role. The results obtained in the presence of adequate blockers of uptake or metabolism clearly show that extraneuronal handling plays a role in the final disposition of the transmitter and functions as an alternative inactivation pathway when neuronal mechanisms are blocked (Brandão and Guimarães, 1974, Brandão, 1977; 1979). The results presented above concern the physiological transmitter, noradrenaline. However, if adrenaline is used to label the neuronal stores and afterwards released by electrical stimulation (which causes a release of the false transmitter which does not differ from that of noradrenaline, as shown by almost identical fractional releases and total overflow), the disposition of the amine shows significant differences from that of noradrenaline. Brandão, Paiva and Guimarães (1980) demonstrated that, under these circumstances, DOPEG formation represents, for adrenaline, only a small fraction of that which is formed when noradrenaline is used, and that O-methylation plays a much more important role for adrenaline than for noradrenaline. This observation reflects the much higher degree of neuronal uptake for noradrenaline and of extraneuronal uptake for adrenaline, as discussed in the following section. Thus, not only structural aspects but also the nature of the released amine play a role in the disposition of catecholamines released by electrical stimulation.

b) <u>Transmitter released by tyramine</u>. In contrast with what happens with electrical stimulation, indirectly acting amines (like tyramine) release the transmitter by a nonexocytotic, calcium - independent mechanism. The metabolism of the transmitter released by tyramine was studied in the dog saphenous vein by Brandao, Monteiro and Osswald, 1978, Brandao, et al., 1980, 1981 and by Rapoport, Takimoto and Cho (1981). In contrast with what happens after electrical stimulation (see above), the transmitter released by tyramine is deaminated to DOPEG before it reaches the synaptic gap; therefore, for low tyramine concentrations, DOPEG made the major contribution to the released material (noradrenaline and metabolites). Increasing the concentration of tyramine resulted in a parallel increase in outflow of unchanged noradrenaline and of O-methylated and deaminated metabolites and in a sharp decrease of the DOPEG contribution. Furthermore, it was found that tyramine is able to mobilize all the noradrenaline stored in the neurones, i.e. that there is no bound fraction refractory to tyramine action, whereas such a bound fraction resists electrical stimulation. Thus, the mode of release plays a role in the disposition of the released transmitter.

INACTIVATION OF EXOGENOUS CATECHOLAMINES

The difficulties in surveying the reports dealing with inactivation of exogenous catecholamines in blood vessels arise from their large number as well as from the fact that very differing methodologies have been resorted to and that concentration of amines and vessels used also vary widely. Some critical remarks on these aspects may be found in the review by Osswald and Guimarães (1983). An attempt will be made to give a short overview of the main aspects of interest.

a) <u>Noradrenaline</u>. Early reports dealing with perfusion experiments (Celander and Mellander 1955, Gryglewski and Vane 1970) showed that noradrenaline infused into the blood supply of perfused vascular beds is rapidly and efficiently removed. Osswald and Branco (1973) confirmed this, in the perfused hind-limb of the dog, and demonstrated that both neuronal and extraneuronal mechanisms are involved in removal of noradrenaline. The accumulation of intact noradrenaline in vascular tissue levels off while increasing the concentration of infused noradrenaline, but removal is not saturated, thus showing that metabolism continues to operate when storage capacity is exhausted. Blockade of neuronal uptake by cocaine leads to an increase in importance of

11

extraneuronal mechanisms; sympathetic denervation had the most marked effect on removal capacity, depressing it to the same level as that observed after blockade of neuronal and extraneuronal uptake. This observation suggests that denervation impairs extraneuronal mechanisms, as was later confirmed (see following section).

In incubation experiments with the saphenous vein, Paiva and Guimaraes (1978) found, in good agreement with these results, that noradrenaline was subject mainly to neuronal uptake, accumulating in the terminals in unchanged form and being metabolized to DOPEG; extraneuronal events were of less importance, unless neuronal uptake was blocked. Increasing the concentration of the amine (to values above the apparent K_m of the extraneuronal system) resulted in an increase in extraneuronal metabolism, deamination predominating over O-methylation. Incubation of a vessel with different structural characteristics, the rabbit aorta, led, as was to be expected, to different results. The work of Maxwell, et al. (1968), Levin (1974), Levin and Wilson (1977), Takimoto, Cho and Shaeffer (1977), Henseling (1980), Levin and Wilson (1983), was conducted on the intact aorta or on its isolated constituents (media and adventitia). In this vessel, O-methylation is predominant and due to extraneuronal metabolism (in the media), whereas DOPEG formation occurs primarily in the innervated adventitia. However, in the media blockade of O-methylation increases accumulation of noradrenaline and its deamination and blockade of MAO results in augmentation of the formation of the O-methylated metabolite. Thus, it is apparent that increase in activity of the metabolizing system compensates (at least partially) for the inhibition of the competing system. The same type of shifting was reported to occur in the canine mesenteric arteries (Garrett and Branco, 1977). In this vessel, with asymmetric and dense adrenergic innervation, neuronal uptake and subsequent deamination to DOPEG represent the major pathway of inactivation.

Since these results have been obtained with ^{3}H-7,8 labelled noradrenaline, some caution must be used in their interpretation, since the tritium label in position 8 hinders the deamination by MAO (Starke et al., 1980). However, very recent experiments conducted with noradrenaline exclusively labelled in position 7 (Grohmann et al., 1985) confirm the results obtained by Paiva and Guimaraes (1978), although they also show that the importance of deamination may have been underestimated in the past, due to this methodological pitfall.

Since, as described above, MAO-A and MAO-B coexist in vascular tissue, it was of interest to find out the contribution of each type of MAO to the inactivation of noradrenaline. Using the canine saphenous vein, Caramona, Araujo and Brandao (1985) studied this question. This investigation clearly showed that clorgyline, but not deprenyl, markedly depressed the formation of DOPEG, either in the absence or in the presence of cocaine (which reduced by itself DOPEG formation - see also Paiva and Guimaraes, 1978). Thus, it was concluded that noradrenaline is deaminated by MAO-A, but not by MAO-B, and that its deamination may occur intra- and extraneuronally, according to experimental conditions (higher concentrations of the amine or blockade of neuronal uptake favouring extraneuronal deamination). This should however not be interpreted as demonstrating substrate specificity of noradrenaline for MAO-A. In fact, Osswald and Caramona (1984) and Caramona and Osswald (1985) demonstrated that in homogenates of the same vascular tissue (saphenous vein) noradrenaline is deaminated by both MAO-A and MAO-B. The interpretation of this apparent discrepancy with the results of Caramona et al. (1985) is that in the intact tissue noradrenaline has no access to the sites where MAO-B is located, perhaps because this type of MAO is placed in series with COMT. Since NMN is a good substrate for MAO-A and MAO-B (Caramona and Osswald, 1985), the latter enzyme may play a role in the final disposition of noradrenaline (but not on its inactivation, since NMN is almost inactive).

b) <u>Adrenaline</u>. Studies comparing disposition of adrenaline <u>vs</u>. that of noradrenaline show that quantitative differences exist. Teixeira (1977) found that, in the perfused dog hind-limb, removal of adrenaline is significantly greater that that of noradrenaline and that a substantial part of the removed adrenaline is subject to metabolization and less prone to accumulate in unchanged form than noradrenaline. Extraneuronal mechansims appeared, therefore, to play a more important role for adrenaline than for noradrenaline. The same conclusion was drawn by Paiva and Guimaraes (1978), using incubated saphenous vein strips. In their experiments, neuronal accumulation of noradrenaline was double in magnitude than that of adrenaline. Furthermore, they found that O-methylation played a more important role than deamination for adrenaline, the opposite happening with noradrenaline. Abrahamsen and Nedergaard (1985), using the isolated rabbit aorta, studied the neuronal accumulation of adrenaline and the influence of various drugs on this parameter. Of particular interest in the context of this topic was the fact that at all concentrations used, the accumulation of noradrenaline was significantly higher than that of adrenaline; at low concentrations the difference was more marked than at higher concentrations. Very recently, Grohmann et al. (1985) described a preferential neuronal uptake (and metabolism) for noradrenaline, and a preferential extraneuronal uptake (and metabolism) for adrenaline. Thus, all pertinent evidence points in the sense of a predominant role of the extraneuronal metabolizing system for adrenaline, and this may be of physiological significance, as discussed below.

c) <u>Dopamine</u>. Although there is a growing evidence for an independent role for dopamine in peripheral neurotransmission, including blood vessels (for a review, see Symposium, 1983), there is scant information on the handling of this physiological catecholamine by vascular tissues. In the case of the dog saphenous vein, dopamine was found to be an excellent substrate for neuronal uptake, deamination by MAO and extraneuronal metabolism (Paiva and Osswald, 1980). For the human mesenteric and renal arteries (two vascular areas for which the presence of dopaminergic mechanism is highly probable), Lacković, Relja and Neff (1982) described the presence of dopamine and of its metabolites DOPAC and HVA, the latter predominanting over the former. Since dopamine is a very good substrate for neuronal uptake, even better than noradrenaline (Grohmann et al., 1985), as well as a substrate for MAO and COMT, there is an urgent need for further studies on the disposition of dopamine in blood vesels.

d) <u>Isoprenaline</u>. This nonphysiological catecholamine has been widely used as a model for the study of the extraneuronal metabolizing system, since it is only subject to a minor degree of neuronal uptake (Azevedo and Osswald 1976, Branco and Osswald 1980, Head et al., 1980, Branco et al., 1981) and is not deaminated by MAO. Thus, it is very helpful to define the characteristics of the corticosteroid-sensitive extraneuronal O-methylating system, which also functions as an accumulating site of loss if COMT is inhibited or saturated by a high concentration of isoprenaline (review by Trendelenburg, 1980). A mathematical model (Kurahashi et al., 1980) has been shown to represent the O-methylating system with appreciable accuracy (Trendelenburg, 1984, Bryan, 1984, Kurahashi et al., 1984). Recent work by Guimaraes and Paiva (1984) and Paiva and Guimaraes (1985) with the dog saphenous vein demonstrated that the half-saturating outside concentrations of adrenaline, noradrenaline and isoprenaline (which correspond to the apparent K_m of the system) are low and in the same order of magnitude as those described for the rabbit aorta and ear artery, as well as several nonvascular tisues. Adrenaline and isoprenaline had almost identical half-saturating outside concentrations, whereas noradrenaline had a significantly higher one. All three amines competed for the same O-methylating system; isoprenaline shifted the metabolism of adrenaline from O-methylation to deamination, a fact which well agrees with the view that MAO and COMT coexist in the same extraneuronal cells, as discussed above.

INTERPLAY OF INACTIVATION MECHANISMS AND THEIR RELATIVE IMPORTANCE IN DISPOSITION OF CATECHOLAMINES

It is clear that it is not possible to draw generalizations as to the relative importance of the mechanisms described above in the disposition of catecholamines; and, on the other hand, that these mechanisms are not independent, but rather interrelated in such a way that over-or underactivity of one of them does influence the role played by the other ones. The relevance of a given mechanism depends on at least several factors, the more important of them appearing to be:

a) The chemical nature of the amine - adrenaline, noradrenaline, dopamine are characterized by different affinities for uptake mechanisms and metabolizing enzymes;

b) The source of the amine - transmitter noradrenaline is inactivated in a way, which is quantitatively dissimilar to that of exogenous noradrenaline; for the transmitter, it is of consequence how it was released, i.e. by an exocytotic (nerve impulse, K^+, etc.) or nonexocytotic (indirect amines) mechanism;

c) The structural and biochemical characteristics of the particular vessel - type and density of innervation, amount of smooth muscle, presence of other cellular and noncellular elements constitute the structural basis for different uptake, binding and metabolizing systems, as well as for more or less facilitated diffusion;

d) Concentration in the biophases and mode of entry of the amine - high, nonphysiologic concentrations may easily override uptake and/or metabolic K_ms. On the other hand, the intimal route of access (which is the usual mode for circulating catecholamines) favours intimal and medial mechanisms of inactivation, whereas adventitial application or simultaneous application to both intimal and advential surfaces (as often seen under experimental conditions) emphasizes, in most arterial vessels, neuronal mechanisms (due to the predominant type of adventitio-medial innervation).

As stated above, there is an interrelationship between different mechanisms of inactivation, suppression of one of them leading to compensation by increased activity of another one, in what could be called vicarious activity. As described before, inhibition of one enzymatic system does therefore not lead forcefully to decreased total formation of metabolites, the other system being able to compensate (partially or totally) for the inhibited one. In a similar way, acute blockade of neuronal uptake leads to a more marked expression of the activity of the extraneuronal mechanisms. However, chronic suppression of neuronal mechanisms (by denervation) has opposite effects, causing a marked impairment of the corticosteroid-sensitive extraneuronal mechanisms, as shown to occur in the lateral saphenous vein of the dog (and the rabbit ear artery) by Branco et al. (1984): O-methylation and accumulation of isoprenaline and noradrenaline were markedly reduced, and morphological changes of both smooth muscle cells and fibroblasts were apparent. The existence of a "trophic" role of the adrenergic innervation on extraneuronal events is thus clear, in good agreement with observations of a significant decrease of arterial tricarboxilic acid cycle enzymes (Zemplenyi and Fronek, 1981) and increased susceptibility to dietary atherosclerosis (Fronek, 1983) after chemical denervation by 6-hydroxydopamine.

In summary, it may be concluded that for the rapid inactivation of the transmitter released by nerve impulses, neuronal reuptake and subsequent intraneuronal deamination by MAO-A, as well as re-incorporation in storage vesicles, represent the main events in vessels with medial innervation and/or narrow synaptic clefts; in vessels with adventitio-medial innervation and large mean neuromuscular distances, extraneuronal mechanisms (especially O-methylation) and diffusion appear to play a significantly more important role. For the inactivation of catecholamines carried in the blood (either of vascular source and reaching the circulation by diffusion or of adrenal origin), extraneuronal (intimal and medial) mechanisms appear to play a predominant, although not an exclusive, role.

ROLE OF BLOOD VESSELS IN THE REGULATION OF PLASMA LEVELS OF CATECHOLAMINES

The enormous area of endothelial lining and the appreciable mass of vascular tissue endowed with inactivation mechanisms can not be ignored in discussing the regulation of plasma levels of catecholamines. The vasculature is indeed the most likely candidate for the role of a buffer system regulating the plasma concentrations of the three physiological catecholamines, adrenaline, noradrenaline and dopamine. The extraneuronal mechanisms (in the media and, most probably, in the intima) appear as playing a physiological role in this regulation by disposing of the circulating catecholamines (Osswald and Branco, 1973, Bönisch, Uhlig and Trendelenburg, 1974, De la Lande, Harvey and Holt, 1974, Nicholas et al., 1974, Osswald, et al., 1975, Paiva and Guimaraes, 1978, Trendelenburg, 1980, De la Lande, 1981). It is as yet unclear what, if any, pathophysiological implications this buffer mechanism may have, but there are indications that disturbed mechanisms of vascular inactivation may exist in pathological conditions, of which hypertensive disease is an outstanding example. The results obtained by Garrett, Castro-Tavares and Branco (1981) in blood vessels of perinephritic hypertensive dogs, as well as observations in another hypertensive model, the neurogenic hypertensive rat (Langer et al., 1975) point in that direction. Moreover, there is evidence that an increase in the release of noradrenaline from blood vessels of hypertensive animals, probably due to alterations in presynaptic receptors, may contribute to the development and/or maintenance of hypertension (review by Westfall and Meldrum, 1985).

INTERFERENCE OF THERAPEUTICALLY USED DRUGS ON INACTIVATION OF CATECHOLAMINES

Drugs which interfere with some of the inactivation mechanisms referred to above as playing a role in the disposition of physiological catecholamines have been known for a long period and some of them are used on a large scale for therapeutic reasons, although the goals to be achieved may have nothing to do with the interference with cardiovascular mechanisms. Thus, tricyclic antidepressants (which are potent inhibitors of neuronal uptake) are used in the treatment of depressive illness, but little is known about possible changes observed in the vascular inactivation mechanisms of patients subject to this treatment. This is surprising, since cardiovascular side effects (arrhythmias, decrease in myocardial contractility, cardiomyopathy, hypertension, antagonism of the blood pressure lowering effect of anti-hypertensive drugs, potentiation of the effects of exogenous catecholamines) have been observed in many cases treated with these drugs (review by Davies, 1981). It is tempting to establish a speculative relationship between the blockade of neuronal uptake (and therefore inhibition of reuptake of the transmitter), caused by these antidepressive drugs, and the cardiovascular side effects referred to above. It would not be surprising if a similar link would be established between the hypertension observed in Cushing's disease or after treatment with corticotrophin or adrenal corticosteroids and the blocking action exerted by the latter on extraneuronal mechanisms, specially for the corticosteroids with less salt-retaining capacity.

Some other classes of drugs have been used for the treatment of vascular pathological conditions and may owe their therapeutic usefullness to actions exerted on vascular inactivation mechanisms. Flavonoids and other so called venotropic drugs have a very widespread clinical use in the symptomatic management of chronic venous insufficiency (prevaricose stage) and, at least for some of them, the reduction in compliance of capacity vessels caused by these drugs and the clinical benefit obtained are undebatable. Therefore, an effect on the adrenergic mechanisms of the vein wall would be a candidate for the explanation of the effects observed. In fact, for three drugs (diosmin, clobenoside, hydroxyethylrutosides) it has been shown that they are able to depress the activity of the extraneuronal, O-methylating system, the hypothesis having been advanced that the drugs may increase the concentration of adrenaline (carried in the circulation) in the venous biophase, by hindering its O-methylation (Heusser and Osswald, 1977, Araújo and Osswald, 1984, Araújo, Gulati and Osswald, 1985).

The use of the newer reversible and specific inhibitors of MAO-A may be of special interest in some conditions like idiopathic or drug-induced postural hypotension. In fact, it has been shown that in blood vessels both types of MAO (A and B) are present and operant, that MAO-A is located both neuronally and extraneuronally and, finally, that only MAO-A plays a role in the deamination of the transmitter (see the subsection on Mechanisms of Inactivation). It is clear that reasonably selective (and reversible) inhibition of MAO-A must result in an enhanced availability of the transmitter; if the drug exerts, simultaneously, a blocking effect on neuronal uptake (cocaine-like effect), this profile of action leads to an increase in sympathetic tone. Such a combination of effects has been described for amezinium (Traut, Brode and Hoffmann, 1981, Araujo, Caramona and Osswald, 1983, Starke, 1983, Caramona, Araujo and Osswald, 1984), a

drug which has been used with success in hypotensive states (review by
Oloff, 1981). Recently, desmethylamiflamine (FLA 668 (+)) has been shown
to exert very similar effects, specifically inhibiting MAO-A and (at
somewhat higher concentrations) exerting a cocaine-like effect and may
therefore be found to share the same clinical indications as amezinium
(Caramona, Esplugues and Osswald, 1985).

Thus, it is clear that drugs which interfere with vascular
inactivation mechanisms represent entities with potential therapeutic
interest and that the advances made in elucidation of sites and mechanisms
of action constitute the basis for the rational design of new therapeutic
approaches.

CONCLUDING REMARKS

Blood vessels are equipped with the enzymatic and nonenzymatic
mechanisms known to dispose of catecholamines. In the present chapter, an
attempt was made to briefly review: the mechanisms of inactivation present
and operant in the blood vessel wall, the morphological aspects of
particular relevance for the topography of these mechanisms, the mode of
inactivation of the sympathetic transmitter (when released by nervous
activity or by indirectly acting amines), the inactivation of exogenous
catecholamines, the relative role played by the different inactivation
mechanisms and the interrelationships existing between them. Furthermore,
the role of blood vessels in the regulation of plasma levels of
catecholamines is dealt with; in the last section the changes in activity
of the inactivation mechanisms due to commonly used drugs are discussed
(independently of the fact that they represent side-effects of drugs used
for other purposes or goals of the therapy itself).

It is neither possible nor desirable to try to summarize, in a few
sentences, the wealth of information gleaned from the very large number of
papers dealing with the problems under study. Instead, the biased author
would like to underline some general assessments which appear to be of
more marked relevance, from the heuristic point of view:

1. The fine control of sympathetic tone in blood vessels is guaranteed by
the efficient and rapid inactivation of the released transmitter in the
blood vessel wall itself. Neuronal reuptake and subsequent events play a
major role in this inactivation.

2. The blood vessels appear to represent the main buffer system in regulation of the levels of circulating catecholamines, the extraneuronal system being credited with an important physiological role.

3. A fine interplay between different inactivation mechanisms allows for mutual compensation of hypoactivity or exclusion of a single one.

4. Adrenergic innervation regulates function and morphology of extraneuronal systems.

5. There is growing evidence for the existence of altered inactivation patterns in pathological states as well as under the influence of therapeutically used drugs.

REFERENCES

Abrahamsen, J., and Nedergaard, O. A., 1985, Accumulation of ^{3}H-adrenaline by rabbit aorta, Blood Vessels 22:32-46.

Allen, G. S., Rand, M. J., and Story, D. F., 1973, Techniques for studying adrenergic transmitter release in an isolated perfused artery, Cardiovasc. Res. 7:423-428.

Araújo, D., Caramona, M. M., and Osswald, W., 1983, On the mechanism of action of amezinium methylsulphate on the dog saphenous vein, Eur. J. Pharmacol. 90:203-214.

Araújo, D., and Osswald, W., 1984, A study of the effects of two venotropic drugs on the O-methylating extraneuronal system of the dog saphenous vein, Blood Vessels 21:132.

Araújo, D., Gulati, O., and Osswald, W., 1985, Effects of two venotropic drugs on inactivation and O-methylation of catecholamines in an isolated canine vein, Arch. Int. Pharmacodyn., in print.

Avakian, O. S., and Gillespie, J. S., 1968, Uptake of noradrenaline by adrenergic nerves, smooth muscle and connective tissue in isolated perfused arteries and its correlation with the vasoconstrictor response, Br. J. Pharmacol. 32:168-184.

Azevedo, I., and Osswald, W., 1976, Uptake, distribution and metabolism of isoprenaline in the dog saphenous vein, Naunyn-Schmiedeberg's Arch. Pharmacol. 295:141-147.

Azevedo, I., and Soares-da-Silva, P., 1981, Are fibroblasts adrenergically innervated cells?, Blood Vessels 18:330-332.

Azevedo, I., Teixeira, A. A., and Sarmento, A., 1984, Effects of catecholamines on the pinocytotic activity of endothelium, smooth muscle cells and fibroblasts, Blood Vessels 21:132-133.

Belfrage, E., Fredholm, B. B., and Rosell, S., 1977, Effect of catechol-O-methyltransferase (COMT) inhibition on the vascular and metabolic responses to noradrenaline, isoprenaline and sympathetic nerve stimulation in canine subcutaneous adipose tissue, Naunyn-Schmiedeberg's Arch. Pharmacol. 300:11-17.

Bevan, J. A., Bevan, R. D., Purdy, R. E., Robinson, C. P., Su, C., and Waterson, J. G., 1972, Comparison of adrenergic mechanisms in an elastic and a muscular artery of the rabbit, Circ. Res. 30:541-548.

Bevan, J. A., Pegram, B. L., Prehn, J. L., and Winquist, R. J., 1978, Beta-adrenergic receptor mediated vasodilatation, in: "Mechanisms of vasodilatation," P. M. Vanhoutte and I. Leusen, ed., pp 258-265, S. Karger, Basel.

Bevan, J. A., Bevan, R. D., and Duckles, S. P., 1980, Adrenergic regulation of vascular smooth muscle, in: "Vascular smooth muscle," Hdbk of Physiology, Vol 11, Sect 2, D. F. Bohr, A. P. Somlyo, and H. V. Sparks Jr., eds, pp 515-566, American Physiological Society, Bethesda.

Bevan, J. A., Godfraind, T., Maxwell, R. A., and Vanhoutte, P. M., 1980, Chapter 12, Disposition of norephinephrine in the blood wall, in: "Vascular neuroeffector mechanisms," pp 147-180, Raven, New York.

Bevan, J. A., and Su, C., 1973, Sympathetic mechanisms in blood vessels: nerve and muscle relationships, Ann. Rev. Pharmacol. 13:269-285.

Bevan, J. A., and Torok, J., 1970, Movement of norepinephrine through the media of rabbit aorta, Circ. Res. 27:325-331.

Bönisch, H., 1978, Further studies on the extraneuronal uptake and metabolism of isoprenaline in the perfused rat heart, Naunyn-Schmiedeberg's Arch. Pharmacol. 303:121-131.

Bönisch, H., and Trendelenburg, U., 1974, Extraneuronal removal, accumulation and O-methylation of isoprenaline in the perfused heart, Naunyn-Schmiedeberg's Arch. Pharmacol. 283:191-218.

Bönisch, H., Uhlig, W., and Trendelenburg, U., 1974, Analysis of the compartments involved in the extraneuronal storage and metabolism of isoprenaline in the perfused heart, Naunyn-Schmiedeberg's Arch Pharmacol. 283:233-244.

Bradley, T., and Hjemdahl, P., 1984, Further studies on renal nerve stimulation induced release of noradrenaline and dopamine from the canine kidney in situ, Acta Physiol. Scand. 122:369-379.

Branco, D., Azevedo, I., Sarmento, A., and Osswald, W., 1981, The fate of isoprenaline in the isolated rabbit aorta. Radiochemical and morphological observations, Naunyn-Schmiedeberg's Arch. Pharmacol. 316:120-125.

Branco, D., and Osswald, W., 1980, The fate of isoprenaline after inhibition of O-methylation and of extraneuronal uptake, Eur. J. Pharmacol. 67:247-253.

Branco, D., Teixeira, A. A., Azevedo, I., and Osswald, W., 1984, Structural and functional alterations caused at the extraneuronal level by sympathetic denervation of blood vessels, Naunyn-Schmiedeberg's Arch. Pharmacol. 326:302-312.

Brandão, F., 1977, Inactivation of norepineprine in an isolated vein, J. Pharmacol. Exper. Therp. 203:23-29.

Brandão, F., 1979, "Inactivation of the sympathetic transmitter in a vascular structure" (in Portuguese), Ph.D. Thesis, University of Porto.

Brandão, F., and Guimarães, S., 1974, Inactivation of endogenous noradrenaline released by electrical stimulation in vitro of dog saphenous vein, Blood Vessels 11:45-54.

Brandão, F., Monteiro, J. G., Osswald, W., 1978, Differences in the metabolic fate of noradrenaline released by electrical stimulation or by tyramine, Naunyn-Schmiedeberg's Arch. Pharmacol. 305:37-40.

Brandão, F., Paiva, M. Q., and Guimarães, S., 1980, The role of neuronal and extraneuronal systems in the metabolism of adrenaline and noradrenaline released from nerve terminals by electrical stimulation, Naunyn-Schmiedeberg's Arch. Pharmacol. 311:1-7.

Brandão, F., Rodrigues-Pereira, E., Monteiro, J. G., and Davidson, R., 1981, A kinetic study of the release of noradrenaline by tyramine, Naunyn-Schmiedeberg's Arch. Pharmacol. 318:83-87.

Brandão, F., Rodrigues-Pereira, E., Monteiro, J. G., and Osswald, W., 1980, Characteristics of tyramine induced release of noradrenaline: mode of action of tyramine and metabolic fate of the transmitter, Naunyn-Schmiedeberg's Arch. Pharmacol. 311:9-15.

Bryan, L. J., 1984, The rat heart: a model extraneuronal O-methylating system, in: "Neuronal and extraneuronal events in autonomic pharmacology," W. W. Fleming, S. Z. Langer, K.-H. Graefe, N. Weiner, eds., pp 111-120, Raven, New York.

Caramona, M. M., 1982, Monoamine oxidase of types A and B in the saphenous vein and mesenteric artery of the dog, Naunyn-Schmiedeberg's Arch. Pharmacol. 319:121-124.

Caramona, M. M., 1983, Localization of monoamine oxidase of type A and B in blood vesels with different innervation patterns, Naunyn-Schmiedeberg's Arch. Pharmacol. 324:185-189.

Caramona, M. M., Araújo, D., and Brandão, F., 1985, Influence of MAO A and MAO B on the inactivation of noradrenaline in the saphenous vein of the dog, Naunyn-Schmiedeberg's Arch. Pharmacol. 328:401-406.

Caramona, M. M., Araújo, D., and Osswald, W., 1984, Amezinium: and "indirect amine" which inhibits MAO A, in: "Monoamine oxidase and disease," K. F. Tipton, P. Dostert and M. S. Benedetti, eds., pp 567-568, Academic Press, London.

Caramona, M. M., Esplugues, J. V., and Osswald, W., 1985, Effects of a monoamine oxidase A inhibitor, FLA 688(+) on the adrenergic mechanisms of the dog saphenous vein, Naunyn-Schmiedeberg's Arch. Pharmacol., in print.

Caramona, M. M., and Osswald, W., 1985, Effects of clorgyline and (-) deprenyl on the deamination of normetanephrine and noradrenaline in strips and homogenates of the canine saphenous vein, Naunyn-Schmiedeberg's Arch. Pharmacol. 328:396-400.

Caramona, M. M., and Soares-da-Silva, P., 1984, Effects of pretreatments with 6-hydroxydopamine (6-OHDA) on MAO A and MAO B activities, Blood Vessels 21:136.

Celander, O., and Mellander, S., 1955, Elimination of adrenaline and noradrenaline from the circulating blood, Nature (Lond.) 176:973-974.

Coimbra, A., Ribeiro-Silva, A., and Osswald, W., 1974, Fine structural and autoradiographic study of the adrenergic innervation of the dog lateral saphenous vein, Blood Vessels 11:128-144.

Davies, D. M., 1981, "Textbook of adverse drug reactions", 2nd ed., Oxford University Press, Oxford.

De la Lande, I. S., 1981, Inactivation and metabolism of catecholamines, Proc. 4th Meeting on Adrenergic Mechanisms, pp 75-91, University of Porto, Porto.

De la Lande, I. S., Harvey, J. A., and Holt, S., 1974, Responses of the rabbit coronary arteries to autonomic agents, Blood Vessels 11:319-337.

Denney, R. M., and Abell, C. W., 1984, The genetics of MAO, in: "Monoamine oxidase and disease,", K. F. Tipton, P. Dostert and M. S. Benedetti, eds., pp 243-252, Academic Press, London.

Endo, T., Starke, K., Bangerter, A., and Taube, H. D., 1977, Presynaptic receptor system on the noradrenergic neurons of the rabbit pulmonary artery, Naunyn-Schmiedeberg's Arch. Pharmacol. 296:229-247.

Fowler, C. J., Magnusson, O., and Ross, S. B., 1984, Intra- and extraneuronal monoamine oxidase, Blood Vessels 21:126-131.

Fronek, K., 1983, Trophic effect of the sympathetic nervous system on vascular smooth muscle, Ann. Biomed. Eng. 11:607-615.

Garrett, J., Branco, D., 1977, Uptake and metabolism of noradrenaline by the mesenteric arteries of the dog, Blood Vessels 14:43-54.

Garrett, J., Castro-Tavares., and Branco, D., 1981, Uptake and metabolism of noradrenaline by blood vessels of perinephritic hypertensive dogs, Blood Vessels 18:100-109.

Gillespie, J. S., 1976, Extraneuronal uptake of catecholamines in smooth muscle and connective tissue, in: "The mechanisms of neuronal and extraneuronal transport of catecholamines," D. M. Paton, ed., pp 325-354, Raven, New York.

Gillis, C. N., 1980, Metabolism of vasoactive hormones by pulmonar vascular endothelium: possible functional significance, in: "Vascular neuroeffector mechanism," J. A. Bevan, T. Godfraind, R. A. Maxwell, and P. M. Vonhoutte eds., pp 304-34, Raven, New York.

Goridis, C., and Neff, N. H., 1971, Monoamine oxidase in sympathetic nerves: a transmitter specific enzyme type, Br. J. Pharmacol. 43:814-818.

Grohmann, M., Henseling, M., Cassis, L., and Trendelenburg, U., 1985, The errors introduced by a tritium label in position 8 of catecholamines, Naunyn-Schmiedeberg's Arch. Pharmacol. (in press).

Gryglewski, R., and Vane, J. R., 1970, The inactivation of noradrenaline and isoprenaline in dogs, Br. J. Pharmacol. 39:573-584.

Guimarães, S., 1975, Further study of the adrenoceptors of the saphenous vein of the dog: influence of factors which interfere with the concentration of agonists at the receptor level, Eur. J. Pharmacol. 34:9-19.

Guimarães, S., 1982, Two adrenergic biophases in blood vessels, Trends Pharmacol. Sc. 3:159-161.

Guimarães, S., and Paiva, M. Q., 1977, The role played by extraneuronal system in the disposition of noradrenaline and adrenaline in vessles, Naunyn-Schmiedeberg's Arch. Pharmacol. 296:279-287.

Guimarães, S., and Paiva, M. Q., 1981a, Two distinct adrenoceptor-biophases in the vasculature: one for α- and the other for β-agonists, Naunyn-Schmiedeberg's Arch. Pharmacol. 316:195-199.

Guimarães, S., and Paiva, M. Q., 1981b, Two different biophases for adrenaline released by electrical stimulation or tyramine from the sympathetic nerve endings of the dog saphenous vein, Naunyn-Schmiedeberg's Arch. Pharmacol. 316:200-204.

Guimarães, S., and Paiva, M. Q., 1984, Kinetics of the extraneuronal O-methylating system of the dog saphenous vein, in: "Neuronal and extraneuronal events in autonomic pharmacology," W. W. Fleming, S. Z. Langer, K.-H. Graefe and N. Weiner, eds., pp 131-138, Raven, New York.

Guimarães, S., Azevedo, I., Cardoso, W., and Oliveira, M. C., 1975, Relation between the amount of smooth muscle in venous tissue and the degree of supersensitivity to isoprenaline caused by inhibition of catechol-O-methyl transferase, Naunyn-Schmiedeberg's Arch. Pharmacol. 286:401-412.

Head, R. J., De la Lande, I. S., Irvine, R. J., and Johnson, S. M., 1980, Uptake and O-methylation of isoprenaline in the rabbit ear artery, Blood 17:229-245.

Henseling, M., 1980, Distribution and metabolic fate of ^{3}H-noradrenaline in rabbit aorta and their influence on muscle contraction, in: "Vascular neuroeffector mechanisms," J. A. Bevan, T. Godfraind, R., Maxwell, and P. M. Vanhoutte, eds., pp 160-170, Raven, New York.

Henseling, M., 1981, The effect of inhibition of uptake 2 on the stimulation-evoked overflow of noradrenaline in the rabbit aorta, Naunyn-Schmiedeberg's Arch. Pharmacol. 316:R54.

Henseling, M., 1983, Kinetic constants for uptake and metabolism of ^{3}H-(-)-noradrenaline in rabbit aorta, Naunyn-Schmiedeberg's Arch Pharmacol. 323:12-23.

Henseling, M., and Trendelenburg, U., 1978, Stereoselectivity of the accumulation and metabolism of noradrenaline in rabbit aortic strips, Naunyn-Schmiedeberg's Arch. Pharmacol. 302:195-206.

Heusser, J., and Osswald, W., 1977, Toxicological properties of diosmin and its action on the isolated venous tissue of the dog, Arch. Farmacol. Toxicol. 3:33-40.

Holtz, P., Osswald, W., and Stock, K., 1960, Über die pharmakologische Wirksamkeit von 3-0-Methyladrenalin und 3-0-Methylnoradrenalin und ihre Beeinflussung durch Cocain, Naunyn-Schmiedeberg's Arch Exp. Pathol. Pharmak 239:1-13.

Holtz, P., Stock, K., and Westermann, E., 1963, Über die Blutdruckwirkung des Dopamin, Naunyn-Schmiedeberg's Arch. Exp. Pathol. Pharmak. 246:133-146.

Hughes, J., 1972, Evaluation of mechansims controlling the release and inactivation of the adrenergic transmitter in the rabbit portal vein and vas deferens, Br. J. Pharmacol. 44:472-491.

Iversen, L. L., 1967, "The uptake and storage of noradrenaline in sympathetic nerves," Cambridge University Press, Cambridge.

Iversen, L. L., 1971, Role of transmitter uptake in synaptic neurotransmission, Br. J. Pharmacol. 41:571-591.

Johnston, J. P., 1968, Some observations upon a new inhibitor of monoamine oxidase in brain tissue, Biochem. Pharmacol. 17:1285-1297.

Kahan, T., Hjemdahl, P., and Dahlof, C., 1984, Relationship between the overflow of endogenous and radiolabelled noradrenaline from canine blood perfused gracilis muscle, Acta Physiol. Scand. 122:571-582.

Kalsner, S., 1979, Termination of agonist action on autonomic receptor, in: "Trends in autonomic pharmacology," S. Kalsner, ed., Vol 1, pp 265-287, Urban and Schwarzenberg, Baltimore.

Kinemuchi, H., Fowler, C. J., and Tipton, K. F., 1984, Substrate specificities of the two forms of monoamine oxidase, in: "Monoamine oxidase and disease," K. F. Tipton, P. Dostert, and M. S. Benedetti, eds., pp 53-62, Academic Press, London.

Kurahashi, K., Rawlow, A., and Trendelenburg, U., 1980, A mathematical model representing the extraneuronal O-methylating system of the perfused rat heart, Naunyn-Schmiedeberg's Arch Pharmacol. 311:17-32.

Kurahashi, K., Magaribuchi, T., Sono, K., Fujiwara, M., 1984, Extraneuronal O-methylating system of isoprenaline in the perfused rat heart: mathematical consideration on time course, in: "Neuronal and extraneuronal events in autonomic pharmacology," W. W. Fleming, S. Z. Langer, K.-H. Graefe, and N. Weiner, eds., pp 121-129, Raven, New York.

Lackovic, Z., Relja, M., and Neff, N. N., 1982, Catabolism of endogenous dopamine in peripheral tissues: is there an independent role for dopamine in peripheral neurotransmission?, J. Neurochem. 38:1453-1458.

Langer, S. Z., 1974, Selective metabolic pathways for noradrenaline in the peripheral and in the central nervous system, Med. Biol. 57:372-383.

Langer, S. Z., Granata, A. R., Enero, M. A., and Krieger, E. M., 1975, Overflow of labelled transmitter elicited by nerve stimulation in the perfused mesenteric arteries of rats after the development of neurogenic hypertension, Blood Vessels 12:368-369.

Levin, J. A., 1974, The uptake and metabolism of ^{3}H-l-and ^{3}H-dl-norepinephrine by intact rabbit aorta and by isolated adventitia and media, J. Pharmacol. Exp. Ther. 190:210-226.

Levin, J. A., and Wilson, S. E., 1977, The effect of monoamine oxidase and catechol-O-methyl transferase inhibitors on the accumulation and metabolism of l-^{3}H-norepinephrine by the adventitia and media of rabbit aorta, J. Pharmacol. Exp. Ther. 203:598-609.

Levin, J. A., and Wilson, S. E., 1983, Effects of inhibitors of neuronal and extraneuronal uptake on the accumulation and metabolism of ^{3}H-l-norepinephrine in rabbit aorta, Blood Vessels 20:234-244.

Lowe, M. C., 1979, Autoradiographic localization of monoamine oxidase in cardiovascular tissues, in: "Catecholamines: basic and clinical frontiers," E. Usdin, I. J. Kopin, and J. Barchas, eds., pp 1351–1353, Pergamon, New York.

Lowe, M. C., and Creveling, C. R., 1979, Immunocytochemical localization of catechol-O-methyl transferase in cardiovascular tissues, in: "Catecholamines: basic and clinical frontiers," E. Usdin, I. J. Kopin, and J. Barchas, eds., pp 219–221, Pergamon, New York.

Mack, F., and Bönisch, H., 1979, Dissociation constants and lipophilicity of catecholamines and related compounds, Naunyn-Schmiedeberg's Arch Pharmacol. 310:1–9.

Magill, G. T., and Levin, J. A., 1984, Selectivity of inhibitors of neuronal and extraneuronal ^{3}H-1-norepinephrine uptake in rabbit aorta, Blood Vessles 21:175–179.

Mantle, T. J., and Tipton, K. F., 1982, in: "Trends in autonomic pharmacology," S. Kalsner, ed., pp 523–542, Urban and Schwarzenberg, Baltimore.

Maxwell, R. A., Eckhardt, S. B., Wastila, W. B., 1968, Concerning the distribution of endogenous norepinephrine in the adventitial and medial-intimal layers of the rabbit aorta and the capacity of these layers to bind tritiated norepinephrine, J. Pharmacol. Exp. Ther. 161:34–39.

Muldoon, S. M., Tyce, G. M., Mayer, T. P., and Rorie, D. K., 1979, High pressure liquid chromatographic analysis of endogenous norepinephrine (NE) released from canine saphenous veins, in: "Catecholamines: basic and clinical frontiers," E. Usdin, E. J. Kopin, and J. Barchas, eds., pp 1488–1490, Pergamon, New York.

Nedergaard, O., and Schrold, J., 1980, Effect of neuronal and extraneuronal uptake inhibitors on the fate of ^{3}H-noradrenaline released spontaneously and by electrical field stimulation of rabbit isolated aorta and adventitia, in: "Vascular neuroeffector mechanisms," J. A. Bevan, T. Godfraind, R. A. Maxwell, and P. M. Vanhoutte, eds., pp 170–176, Raven, New York.

Nicholas, T. E., Strum, J. M., Angelo, L. S., and Junod, A. F., 1974, Site and mechanism of uptake of ^{3}H-1-norepinephrine by isolated perfused rat lungs, Circ. Res. 35:670–680.

Oloff, S. H., 1981, A synposis of the therapeutic trials of amezinium, Arzneim.-Forsch. (Drug Res.) 31:1657–1671.

Osswald, W., 1976, Transmitter disposition mechanisms, in "Vascular neuroeffector mechanisms," J. A. Bevan, G. Burnstock, B. Johansson, R. A. Maxwell, and O. Nedergaard, eds., pp 123–130, Karger, Basel.

Osswald, W., 1978, Disposition of vasoconstrictor agonists, in: "Mechanisms of vasodilatation," P. M. Vanhoutte, and I. Leusen, eds., pp 89-97, Karger, Basel.

Osswald, W., 1979, Inactivation of the sympathetic neurotransmitter, Bol. Soc. Port. Cardiol. 17 (Suppl 2):399-407.

Osswald, W., 1984, Enzymatic inactivation of catecholamines in the venous wall, Inter. Angiol. 3:49-52.

Osswald, W., and Branco, D., 1973, The effects of drugs and denervation on removal and accumulation of noradrenaline in the perfused hind-limb of the dog, Naunyn-Schmiedeberg's Arch. Pharmacol. 277:175-190.

Osswald, W., and Caramona, M. M., 1984, Role of MAO A and of MAO B in the metabolism of noradrenaline, in: "Neuronal and extraneuronal events in autonomic pharmacology," W. W. Fleming, S. Z. Langer, K.-H. Graefe, and N. Weiner, eds., pp 23-36, Raven, New York.

Osswald, W., and Guimarães, S., 1983, Adrenergic mechanisms in blood vessels: morphological and pharmacological aspects, Rev. Physiol. Biochem. Pharmacol. 96:53-122.

Osswald, W., Garrett, J., Guimarães, S., 1975, Extraneuronal uptake and metabolism in dog vascular structures, in: "Proceedings of the 6th International Congress of Pharmacology," J. Tuomisto, and K. Paasonen, eds., Vol 2, pp 149-158, Helsinki.

Osswald, W., Guimarães, S., Coimbra, A., 1971, The termination of action of catecholamines in the isolated venous tissue of the dog, Naunyn-Schmiedeberg's Arch. Pharmacol. 269:15-31.

Owman, C., Edvinsson, L., and Hardebo, J. E., 1980, Amine mechanisms and contractile properties of the cerebral microvascular endothelium, in: "Vascular neuroeffector mechanisms," J. A. Bevan, T. Godfraind, R. A. Maxwell, and P. M. Vanhoutte, eds., pp 277-299, Raven, New York.

Paiva, M. Q., and Guimaraes, S., 1978, A comparative study of the uptake and metabolism of noradrenaline and adrenaline by the isolated saphenous vein of the dog, Naunyn-Schmiedeberg's Arch. Pharmacol. 303:221-228.

Paiva, M. Q., and Guimaraes, S., 1985, The kinetic characteristics of the extraneuronal O-methylating system of the dog saphenous vein and the supersensitivity to catecholamines caused by its inhibition, Naunyn-Schmiedeberg's Arch. Pharmacol. 327:48-55.

Paiva, M. Q., and Osswald, W., 1980, Inactivation of some vasoconstrictor agonists by saphenous vein strips of the dog, J. Pharm. Pharmacol. 32:227-228.

Paton, D. M., 1976, Characteristics of uptake of noradrenaline by adrenergic neurons, in: "The mechanism of neuronal and extraneurotransport of catecholamines," D. M. Paton, ed., pp 49-66, Raven, New York.

Paton, W. D. M., 1960, Discussion, in: Ciba Foundation Symposium on catecholamines," J. R. Vane, G. E. W. Wolstenholme, and M. O'Connor, eds., pp 124-127, Churchill, London.

Rapoport, R. M., Takimoto, G. S., and Cho, A. K., 1981, Compartmental analysis of tyramine-induced norepinephrine depletion, Pharmacol. 222:235-242.

Renson, J., Weissbach, H., and Udenfriend, S., 1964, Studies on the biological activities of the aldehydes derived from norepinephrine, serotonin, tryptamine and histamine, J. Pharmcol. Exp. Ther. 143:326-331.

Schrold, J., and Nedergaard, O., 1981, Effect of cocaine and corticosterone on the metabolism of ^{3}H-noradrenaline released from rabbit isolated aorta and adventitia, Acta Pharmacol. Toxicol. 48:233-241.

Soares-da-Silva, P., and Azevedo, I., 1985, Fibroblasts and sympathetic innervation of blood vessels, Blood Vessels (submitted for publication).

Sollevi, K., Hjemdahl, P., and Fredholm, B. B., 1981, Endogenous adenosine inhibits lipolysis induced by nerve stimulation without inhibiting noradrenaline release in canine subcutaneous adipose tissue in vivo, Naunyn-Schmiedeberg's Arch. Pharmacol. 316:112-119.

Somlyo, A. P., and Somlyo, A. V., 1970, Vascular smooth muscle, II. Pharmacology of normal and hypertensive vessels, Pharmacol. Rev. 22:249-353.

Spector, S., Tarver, J., Berkowitz, B., 1972, Effects of drugs and physiological factors in the disposition of catecholamines in blood vessels, Pharmacol. Rev. 24:191-202.

Starke, K., 1983, Amezinium: a novel pattern of effects on the sympathetic nervous system, Trends in Pharmacol. Sc. 4:269-272.

Starke, K., Steppeler, A., Zumsteier, A., Henseling, M., and Trendelenburg, U., 1980, False labelling of commercially available ^{3}H-catecholamines?, Naunyn-Schmiedeberg's Arch. Pharmacol. 311:109-112.

Sweet, W. D., and Levin, J. A., 1983, Two mechanisms of ^{3}H-catecholamine accumulation in rabbit aorta media differentiated by sensitivity to temperature and corticosterone, Blood Vessels 20:265-282.

Symposium, 1983, Is dopamine a peripheral neurotransmitter?, M. Relja, and N. H. Neff, eds., Fed. Proc. 42:2998-3021.

Takimoto, G. S., Cho, A. K., and Shaeffer, J. C., 1977, Inhibition of norepinephrine accumulation by amphetamine derivatives. Studies with rat brain and rabbit aorta, J. Pharmacol. Exp. Ther. 202:267-277.

Teixeira, F., 1977, The effects of drugs and denervation on removal and accumulation of adrenaline in the perfused hind-limb of the dog, _Arch. Int. Pharmacodyn._ 225:221-231.

Teixeira, F., Almeida, L., and Figueiredo, A., 1983, "Dopaminergic effect" not due to dopamine but to one of its deaminated metabolites. New consideration, _J. Pharmacol._ (Paris) 14:507.

Teixeira, F., and Macedo, T. R. A., 1981, Influence of deaminated metabolites on the relaxing effect of dopamine dog saphenous vein, _J. Pharm. Pharmacol._ 33:529-533.

Torok, J., and Bevan, J. A., 1971, Entry of ^{3}H-norepinephrine into the arterial wall, _J. Pharmacol. Exp. Ther._ 177:613-620.

Traut, M., Brode, E., and Hoffmann, H. D., 1981, Pharmacology of amezinium a novel antihypotensive drug. IV Biochemical investigations of the mechanism of action, _Arzneim.-Forsch._ (Drug Res.) 31:1566-1574.

Trendelenburg, U., 1972, Factors influencing the concentration of catecholamines at the receptros, _in_: "Catecholamines," H. Blaschko, E. Muscholl, eds., pp 726-761, Springer, Berlin.

Trendelenburg, U., 1974, The relaxation of rabbit aortic strips after a preceding exposure to sympathomimetic amines, _Naunyn-Schmiedeberg's Arch. Pharmacol._ 281:13-46.

Trendelenburg, U., 1980, A kinetic analysis of the extraneuronal uptake and metabolism of catecholamines, _Rev. Physiol. Biochem. Pharmacol._ 87:33-115.

Trendelenburg, U., 1984, Metabolizing systems, _in_: "Neuronal and extraneuronal events in autonomic pharacology," W. W. Fleming, S. Z. Langer, K.-H. Graefe, N. Weiner, eds., pp 93-109, Academic Press, London.

Trendelenburg, U., Bönisch, H., Graefe, K.-H., and Henseling, M., 1979, The rate constants for the efflux of metabolites of catecholamines and phenethylamines, _Pharmacol. Rev._ 31:179-204.

Vanhoutte, P. M., Van Nueten, J. M., Verbeuren, T. J., and Laduron, P. M., 1977, Differential effects of the isomers of tetramisol on adrenergic neurotransmission in cutaneous vein of dog, _J. Pharmacol. Exp. Ther._ 200:127-140.

Vanhoutte, P. M., Verbeuren, T. J., and Webb, R. C., 1981, Local modulation of adrenergic neuroeffector interaction in the blood vessel wall, _Physiol. Rev._ 61:151-247.

Verbeuren, T. J., and Vanhoutte, P. M., 1982, Deamination of released ^{3}H-noradrenaline in the canine saphenous vein, _Naunyn-Schmiedeberg's Arch. Pharmacol._ 318:148-157.

Westfall, T.C., and Meldrum, M.J., 1985, Alterations in the release of norepinephrine at the vascular neuroeffector junction in hypertension. Ann. Rev. Pharmacol. 25:621-641.

Winquist, R. J., and Bevan, J. A., 1979, The effect of surgical sympathetic denervation on the development of intrinsic myogenic tone and the alpha and beta adrenergic receptor-mediated response of the rabbit facial vein, J. Pharmacol. Exp. Ther. 211:1-6.

Wyse, D. G., 1976, On the role of neuronal uptake (uptake 1) in the inactivation of noradrenaline by aortic strips, Can. J. Physiol. Pharmacol. 52:1102-1109.

Zemplenyi, T., and Fronek, K., 1981, Chemical sympathectomy by 6-hydroxydopamine and arterial enzymes and lactate in the rabbit, Exp. Mol. Path. 34:123-130.

Ca DISTRIBUTION AND ITS REGULATION IN SMOOTH MUSCLE CELLS

J. Verbist, G. Droogmans, F. Wuytack and R. Casteels

Laboratorium voor Fysiologie
Katholieke Universiteit Leuven
Campus Gasthuisberg
B-300 Leuven, Belgium

INTRODUCTION

Although there exists an important variation between different smooth muscle cells, many common mechanisms can be presumed for the regulation of their excitation-contraction coupling. There is no doubt that the final step in the activation of smooth muscle, i.e. the force development depends on an interaction of the actin and myosin filaments. The main regulatory factor of this activity is the concentration of ionized calcium in the cytoplasm. From this point of view an analysis and an understanding of the regulation of the cytoplasmic Ca concentration is essential for the elucidation of the excitation-contraction coupling in these cells. However there is ample evidence in several cell types that Ca^{2+} is not the unique regulating factor and that other intracellular mediators as cAMP, cGMP and diacylglycerol can modulate the effect of varying cytoplasmic Ca concentrations. However the importance and the exact role of these intracellular messengers has not been sufficiently determined in smooth muscle cells.

For this reason and because Ca^{2+} is the essential factor we will limit this discussion to the distribution of Ca^{2+} in smooth muscle cells and its regulation.

Ca^{2+} Exchange

The main structures regulating the cytoplasmic $[Ca^{2+}]$ are the plasma membrane and the intracellular Ca stores. These structures together with the extracellular matrix will determine the Ca exchange in smooth muscle tissues. The first studies on Ca distribution of smooth muscle by Schatzmann (1961) and by Goodford (1965) have revealed already the complexity of such analysis. Even the determination of the total Ca content of smooth muscle tissues gave rather divergent values probably because the pH of the solution and the type of buffer could affect the amount of Ca present in the extracellular space. The variability of the data on the total Ca content in smooth muscle tissues is a warning against a simplified interpretation of the flux data. Moreover we should be aware that smooth muscle cells present a small diameter and contain various not sufficiently specified intracellular Ca compartments as the endoplasmic reticulum, mitochondria, the plasma membrane, including caveolae, the cytoplasmic compartment etc. In addition these cells are surrounded by a large extracellular space containing a complex intercellular matrix of largely unknown properties.

If smooth muscle tissues are exposed to a solution containing ^{45}Ca this Ca isotope will exchange at different rates with the Ca present in extracellular and intracellular compartments. After a prolonged exposure (60–180 min) to a physiological solution containing ^{45}Ca most of the exchangeable Ca will have been replaced by ^{45}Ca. This exchange occurs as well in the cellular as in the extracellular compartment. This amount of exchangeable Ca-fraction is much better known than what is called the inexchangeable Ca. As far as the exchangeable cellular Ca is concerned, it has to be assumed that ^{45}Ca penetrates the cell membrane even in quiescent tissues across leak channels or by Ca–Ca exchange. The rather rapid exchange of ^{45}Ca with ^{40}Ca implies that the Ca-permeability of these leak channels is not negligible and that the Ca entering the cells by leakage, is continuously extruded in order to maintain a constant low value of $[Ca^{2+}]_{cyt}$. After reaching a maximal labelling of the Ca in the tissues, these can be transferred to various experimental media, each containing the same specific activity of ^{45}Ca as in the physiological solution. Any change of the tissue ^{45}Ca content should, in the framework of this experimental procedure be due to a net uptake or loss of Ca from the tissue. In view of the large amounts of Ca present in the extracellular medium, it is essential to wash off the extracellular Ca while preserving the intracellular one. Several procedures have been introduced to separate more clearly the cellular from the extracellular Ca (Aaronson et al., 1979).

It is assumed in all these procedures that the slow phase of the efflux represents the efflux of ^{45}Ca from the intracellular compartment, while the fast process is due to the washout of label from extracellular sites. This assumption might be true for some tissues under specific experimental conditions, but it should not be generalized because of the unknown compartmental structure of smooth muscle tissues.

After analogous types of loading for a prolonged period of time, it is also possible to follow the pattern of efflux and deduce some properties of cellular Ca compartments. the predominant amount of extracellular Ca makes it impossible to estimate changes of the cellular calcium during the first 20-30 min of the efflux. Aaronson et al (1979) therefore transfer the tissues after loading with ^{45}Ca to an ice-cold medium containing Ca and Ca-EGTA for 40 min. They assume that this procedure removes extracellular ^{45}Ca leaving the intracellular ^{45}Ca largely intact. The use of a solution with 6.5 mM $CaCl_2$ and 5.0 mM EGTA is meant to prevent alteration of the membrane function which might occur in the absence of free Ca^{2+}. After this procedure for elimination of the extracellular Ca, the ^{45}Ca washout was performed in the appropriate washing solutions and the numerical data were treated as usual to obtain the decrease of the radioactivity remaining in the tissue, and the fractional loss as a function of time of efflux.

Aaronson and van Breemen (1981) have also described a procedure to determine the unidirectional influx of ^{45}Ca. Tissues exposed to non-labelled experimental media were transferred for 3 min to an identical solution containing ^{45}Ca. Thereupon they were washed for 45 min in ice-cold EGTA containing solution and the amount of ^{45}Ca remaining in the tissue was determined. This remaining ^{45}Ca is assumed to be the cellular ^{45}Ca. This exposure to radioactivity for 3 min has been called "pulse labelling" and the ^{45}Ca uptake occurring during this period would be almost entirely a function of the Ca influx and should not be affected by a subsequent extrusion of Ca. A mathematical analysis of the tracer washout curves is feasible when the rate constants of the various exponential components are sufficiently different. However, the identification of these components with functional Ca compartments in the tissue is in most cases not possible (Casteels and Droogmans, 1975).

As mentioned earlier one of the aims of the ^{45}Ca efflux studies is to obtain more information on the Ca present in the cytoplasmic and other

cellular compartments and on their exchange rate. Because the exchange
rates of some extracellular Ca compartments could be similar to those of
cellular compartments several experimental procedures have been tried to
separate more clearly the extracellular Ca from the cellular Ca. The
addition of lanthanides to the washing fluid was introduced (van Breemen
and McNaughton, 1970) to reduce the amount of Ca bound to extracellular
binding sites, while reducing at the same time the passage of Ca across
the cell membrane. The La-wash procedure was further modified either by
using a much higher concentration of La^{3+} in the washing solution
(Godfraind, 1976) or by using La^{3+}-containing solutions at low
temperature. This latter procedure was based on the assumption that a
combination of La^{3+} and lowered temperature might inhibit more efficiently
the Ca efflux than a La^{3+}-containing solution at $35^{\circ}C$.

However, in our opinion, all these procedures induce some artifacts
certainly at high La^{3+}-concentration and at low temperature. The finding
that the amount of Ca calculated by the extrapolation of the slow
component to zero time, varies for the same tissue according to the
washing procedure warns us against an uncritical acceptance of the
quantitative data obtained by flux procedures.

A second procedure which has been introduced to separate
extracellular from cellular Ca, consists in washing the tissues with a
Ca-free solution containing EGTA (Droogmans and Casteels, 1979) or in
solutions containing Ca EGTA (van Breemen and Casteels, 1974). It is
assumed that in this way the extracellular bound Ca can be washed off in a
short period of time while maintaining the cellular Ca, and that low
values of $[Ca^{2+}]$ do not interfere with the selective permeability of the
plasma membrane. This procedure can again be used at lower temperature in
order to reduce the loss of Ca from the cellular compartment (Aaronson et
al., 1979).

The interpretation of the data obtained by a mathematical analysis of
the efflux curves remains controversial, not only because such procedures
are often arbitrary from point of view of the compartmental structure of
the tissue, and their serial or parallel organization in the tissue, but
also because the experimental procedures which have been introduced to
obtain a clearer separation between the extracellular and cellular calcium
can interfere with the normal function of the cells, and because the
effect of these procedures is not limited to the blockade of the
transmembrane Ca movements only.

The Ca remaining in the cells after a prolonged exposure to lanthanides or EGTA has not been precisely localized. It is probably distributed over the cellular compartments and it is slowly released across the cell membrane in all experimental conditions.

Interesting results by flux experiments have been obtained from those smooth muscle tissues which present an intracellular Ca store from which Ca can be released after a prolonged exposure to Ca-free and EGTA containing medium by addition of an agonist. Under these conditions an appreciable increase of the fractional loss of ^{45}Ca was obtained in parallel with a force development of those smooth muscle cells (Droogmans, Raeymaekers and Casteels, 1977; Casteels and Droogmans, 1981). We have proposed that the agonist elicits a release of Ca from the endoplasmic reticulum and that this released Ca activates the contractile proteins before being extruded out of the cell across the plasmalemma. The reaccumulation of Ca by the endoplasmic reticulum after such stimulation cannot be detected by the present experimental procedures. In other tissues in which the agonist-sensitive intracellular Ca store is depleted after a few minutes exposure to Ca-free solution neither Ca release, nor contraction can be obtained after 5-10 min exposure to the Ca-free solution (Casteels and Raeymaekers, 1979). The difference between the action of agonists on both types of tissues depends on the rate of depletion of this Ca store in Ca-free medium. Other intracellular Ca stores which have been proposed are the mitochondria (Batra, 1973) and the inner edge of the plasma membrane (Kolber and van Breemen, 1981). It has been assumed that the Ca present in mitochondria could be specifically released by uncoupling agents as dinitrophenol or FCCP, but also here we have to be aware that those inhibitors are not sufficiently specific. Moreover the studies of Somlyo, Somlyo and Shuman (1979) have revealed that mitochondria do not play a role in the regulation of the Ca metabolism of smooth muscle cells under normal physiological conditions and that accumulation of Ca in those organelles is most likely an expression of tissue damage.

The Plasma Membrane

Smooth muscle cells normally depend for a maintained force development on a continuous supply of calcium from the extracellular medium. On prolonged exposure to a Ca-free solution added with 2 mM EGTA or in the presence of ions as La^{3+} or Mn^{2+} which prevent the influx of Ca^{2+} across the cell membrane, a maintained force development becomes

impossible. Some smooth muscle tissues can after a limited time of
exposure to Ca-free medium still develop a transient force development
during stimulation with an agonist. This contraction is due to the
release of Ca from agonist-sensitive cellular Ca stores (Bohr, 1973;
Casteels and Raeymaekers, 1979; Droogmans, Raeymaekers and Casteels,
1977). K^+ rich solutions will depolarize the cells in the absence of Ca^{2+}
but they fail to elicit a contraction. It can be concluded that
extracellular Ca is essential to obtain a contraction by K-depolarization
and that K-depolarization is not an adequate stimulus to release Ca from
cellular stores. External Ca is also required for agonist induced
contractions but the relation between Ca entry and Ca release from
intracellular stores has to be examined further (see below).

Calcium ions enter into smooth muscle cells down their
electrochemical gradient by several types of mechanisms; either through
pores or channels as e.g. voltage dependent channels, receptor operated
channels and leak channels, or by a carrier system as Na-Ca exchange.
Spontaneous and evoked action potentials occur in most smooth muscle
cells. These action potentials are insensitive to tetrodotoxin, but they
are blocked by Mn^{2+} and organic Ca antagonists, such as D600 and verapamil
(Anderson, Ramon and Snyder, 1971; Golenhofen, Hermstein and Lammel, 1973;
Walsh and Singer, 1980; Zelcer and Sperelakis, 1981). For these reasons
it is assumed that the inward current responsible for the upstroke of the
action potential is mainly due to the influx of Ca^{2+} ions through
potential dependent channels. However in some smooth muscle tissues
action potentials have also been observed in Ca free solutions (Mangel et
al., 1982; Mironneau, Eugene and Mironneau, 1982) suggesting that these
channels are permeable both for Ca^{2+} and Na^+ ions. The relative
contribution of each ion to the inward current depends on the
extracellular concentration of these ions and on the relative selectivity
of these channels for both cations. In some tissues the action potentials
are rather abortive and in others they are completely absent. This is
probably due to an overlap of time of the inward and the outward currents,
whereby the Ca influx is maintained while the net inward current is
reduced or even completely inhibited by the simultaneously occurring
K-outward current (Casteels et al., 1977). This is consistent with the
observation that in the presence of substances which reduce the
K-permeability, such as TEA or procaine, smooth muscle preparations, which
are characterized by a stable membrane potential in normal physiological
solutions, can present action potentials either spontaneously or during
electrical or pharmacological stimulation (Droogmans et al., 1977; Harder

and Sperelakis, 1979). The difference between quiescent smooth muscle and those which present spikes is therefore less fundamental than originally thought. It is related most likely to the different kinetic behavior of the ionic channels in different tissues.

Assuming that the inward current is entirely carried by Ca^{2+} and that the net inward charge movement is a reliable measure of this inward current we can calculate from the amplitude of the action potential the amount of charge required to discharge the membrane capacity, and thereby estimate the transmembrane flux of Ca. However, such a hypothetical Ca current would only increase the intracellular $[Ca^{2+}]$ by an amount which is one order of magnitude too small to fully activate the contractile apparatus. It is thereby assumed that either the transient depolarization of the membrane or the small amount of calcium ions entering the cell during the action potential triggers a release of calcium from intracellular Ca stores. Until now there is not sufficient experimental evidence in smooth muscle cells in favour of any of these hypotheses. It needs still to be examined whether the likely presence of an early outward current Ca by K^+ ions could lead to an underestimation of the calculated calcium influx during the action potential.

A maintained depolarization of the smooth muscle membrane by increased concentrations of extracellular K also augments the Ca influx in all smooth muscle cells as indicated by the contractile response and by the increased uptake of ^{45}Ca. It is assumed that K-depolarization (Mayer, van Breemen, Casteels, 1972; van Breemen, 1977; van Breemen et al., 1973) opens the same type Ca channel, which is also responsible for the rising phase of the action potential, and this channel is therefore classified as voltage-dependent Ca channel. This hypothesis is supported by the observation that Ca antagonists inhibit both the action potentials and their accompanying contractions and the contractile response in K-depolarized tissues. The maintained force development in depolarized tissues is not incompatible with the existence of an inactivation of the Ca-channels. Such inactivation could for some smooth muscle cells remain incomplete at a lower level of depolarization as suggested by the observation that the amplitude of the tonic component of the contraction induced in the guinea-pig taenia coli by 42 mM K^+ is larger than that induced by 140 mM K^+ (Casteels and Raeymaekers, 1979).

Although the conductance of the voltage operated Ca channels might be relatively low for Na-ions, these ions might contribute significantly to

the ionic current through these channels because of their high
extracellular concentration as compared to Ca. Such a limited specificity
of the Ca channels could explain that Na-free solutions induce a
contractile response, by the augmented Ca influx at low eternal Na
concentrations through the voltage operated channels. During our study of
the rabbit ear artery we have observed that similarities exist between the
effects of K-depolarization and of Na-free solutions on the [45]Ca influx,
contractile response and on the stimulation of the [45]Ca efflux. It can
therefore be assumed that a common mechanism could play a role in this
tissue for K-rich and Na deficient solutions. The small depolarization by
10 to 20 mV, which we observed in Na free solutions could be sufficient
for a partial activation of the voltage operated Ca channels (Droogmans
and Casteels, 1979). The reduction of the competition between
extracellular Na and Ca in Na-deficient solutions could further increase
the influx of calcium up to an activation of the contractile proteins. In
other tissues however it has been reported that the Na-free induced
contractions and the concomitant increased [45]Ca uptake are not blocked by
Ca-antagonists (Kuriyama et al., 1983; Ozaki and Urakawa, 1981), and these
observations would suggest that a different mechanism could also play a
role.

An alternative pathway used by extracellular calcium to enter the
cell are the so-called receptor operated Ca-channels (Bolton, 1979).
Activation of receptors by an agonist can induce an increase of the Ca
influx and a contractile response while causing depolarization, no
depolarization at all or even hyperpolarization (Casteels et al., 1977;
Droogmans et al., 1977; Harder, 1980; Kitamura and Kuriyama, 1979; Su,
Bevan and Ursillo, 1964; Suzuki, 1981). This type of activation most
likely implies a direct effect of receptor activation on the Ca
permeability of the plasma membrane and/or on a release of cellular Ca
without acting over a change of the membrane potential. There are several
experimental arguments to differentiate in smooth muscle cells of large
arteries these receptor operated channels from the voltage-dependent ones.
Such receptor-operated channels are less sensitive to the organic Ca-
antagonists than the voltage-dependent ones (Casteels and Droogmans, 1981;
Cauvin, Loutzenhiser and van Breemen, 1983; Godfraind, 1983). Secondly
the contractile responses induced by K-depolarization and by agonists
present a different temperature sensitivity (Droogmans and Casteels, 1981;
Peiper, Griebel and Wende, 1971a,b; Schumann, Gorlitz and Wagner, 1975).
Also the finding that the stimulation of the [45]Ca-influx by
K-depolarization and by noradrenaline are additive when the two procedures

of activation are applied simultaneously support this hypothesis
(Meisheri, Hwang and van Breemen, 1981).

The small change of the membrane potential in some vascular smooth
muscle cells during exposure to exogenously applied noradrenaline is in
marked contrast with the electrical response to perivascular nerve
stimulation. The ensuing release of endogenous noradrenaline elicits
excitatory junction potentials (e.j.p.) which can sum and eventually give
rise to an action potential (Hirst, 1977; Holman and Surprenant, 1979).
These membrane phenomena are rather insensitive to α-adrenergic blocking
agents, and the existence of a special type of adrenoceptor has therefore
been postulated at the postjunctional smooth muscle membrane, the
so-called γ-receptor (Hirst and Neild, 1980). This conclusion relies on
Dale's principle that a nerve terminal can only release one type of
transmitter substance. However this principle had recently to be modified
to the co-transmitter hypothesis, because e.g. both acetylcholine and VIP
or noradrenaline and ATP can be released from the same nerve terminal.
The changes of membrane potential occurring during the stimulation of the
noradrenergic nerve terminals would be due to an interplay between the
co-transmitters. ATP could elicit the e.j.p. and the subsequent firing of
action potentials (Burnstock, 1976; Sneddon and Westfall, 1984; Suzuki,
1985). This co-transmitter hypothesis is also consistent with the
observation that nerve stimulation has a biphasic effect on the membrane
potential in some tissues, consisting of an e.j.p. and a more slowly
developing depolarization (Cheung, 1982; Suzuki, 1981). The e.j.p. is
insensitive to phentolamine, whereas the slower depolarization is blocked
by phentolamine and potentiated by cocaine. These nerve mediated
responses can be reduced by Ca-antagonists, in contrast with the effect of
these substances on the response induced by exogenous noradrenaline
(Kajiwara and Casteels, 1983; Kuriyama et al., 1983; Makita et al., 1983;
Surprenant, Neild and Holman, 1983; Suzuki, Itoh and Kuriyama, 1982).
From our observation in rabbit ear artery it was evident that the
Ca-antagonists have no effect on the release of ^{3}H-noradrenaline from the
nerve terminals (Kajiwara and Casteels, 1983). These substances would
therefore exert their effect mainly on the voltage-dependent Ca channels
of the postsynaptic smooth muscle membrane.

This indirect evidence in favor of these Ca channels has still to be
supplemented with direct biophysical evidence and with a biochemical
isolation and characterization of the structure representing the Ca
channel. For the latter procedure, the binding of Ca antagonists of the

dihydropyridine group to molecules present in the Ca channel might help to
specify those membrane structures (Glossmann et al., 1982). However
although it is presently an acceptable working hypothesis to suppose the
existence of two different entities, voltage dependent and receptor
operated channels, all present evidence is indirect and it is therefore
necessary to wait for further proof.

Intracellular Ca-stores

The essential role of extracellular calcium in the long-term
regulation of the excitation-contraction coupling in smooth muscle should
not make us forget that intracellular systems (organelles or Ca binding
proteins) might play a role in parallel or in series with the plasma
membrane.

The intracellular Ca stores play a role in excitation contraction
coupling in smooth muscle by releasing calcium. This conclusion has been
deduced from the observation that excitatory agonists are able to induce a
transient contraction in tissues exposed to Ca free solutions or in
tissues in which the entry of Ca is blocked by La^{3+}. The contractile
response elicited by agonists under these conditions is accompanied by a
transient increase of the ^{45}Ca efflux. We assume that the excitatory
agonists release Ca from some intracellular store, most likely the
endoplasmic reticulum and thereby activate the contractile apparatus (Deth
and Casteels, 1977; Deth and van Breemen, 1977; Droogmans et al., 1977).
This Ca would subsequently be extruded out of the cell probably by the
CaMg ATPase of the plasma membrane and thereby cause the increase of the
fractional loss of ^{45}Ca. It should be pointed out that La-ions do not
seem to inhibit this Ca efflux. A recycling to the sites from which this
Ca was released appears to limited because a second activation by the
agonist during the maintained exposure to Ca-free medium does not induce a
second phasic contractile response, nor a measurable stimulation of the
^{45}Ca-efflux (Casteels et al., 1981). Application of the agonist at
various times after starting the Ca free superfusion has revealed that
both the amplitude of the phasic contractile response and the amount of
released ^{45}Ca decrease in an exponential manner. Both in rabbit aorta
(Deth and Casteels, 1977) and in rabbit ear artery (Droogmans et al.,
1977) a half-time of approximately 30-40 min was found. In other smooth
muscle tissues as the guinea-pig taenia coli and the ileum, the force
development elicited by carbachol is already abolished after an exposure
for a few minutes to Ca free solution (Casteels and Raeymaekers, 1979).

This rapid loss of contractile response is due to the fast disappearance of Ca from this store in Ca-free medium and not to its limited capacity. A nearly maximal contraction can be induced after a short exposure to Ca free solutions.

The amount of Ca which can be released by the agonist and which appears in the effluent has been estimated at 21 μmoles/kg in rabbit aorta (Deth and Casteels, 1977; Deth and van Breemen, 1977) and at 60 μmoles/kg in rabbit ear artery (Droogmans et al., 1977). The value found in the aorta is too small to induce a maximal contraction, whereas the second one found in the ear artery might be sufficient to fully activate the myofilaments. In order to resolve the inconsistency between the contractile response and the amount of ^{45}Ca appearing in the effluent upon agonist stimulation van Breemen et al. (1980) have postulated that the largest part of released calcium is not immediately eliminated from the cells after activation of the contractile proteins, but that at least part of it is taken up by another Ca sequestering organelle, from which it is extruded during relaxation by an as yet unidentified pathway. This model thus implies two separate Ca fractions: one close to the inner surface of the cell membrane, from which the agonists are able to release calcium and which has to be replenished with Ca bound to the outer membrane surface, and a second fraction, possibly present in the endoplasmic reticulum.

The estimate of 60 μmoles/kg Ca in the ear artery is consistent with a simpler model, with only one type of sequestering organelle from which Ca can be released by an agonist. The observation that the released Ca does not return to its release sites during exposure to Ca-free medium is not incompatible with the possibility that Ca is accumulated in these organelles during relaxation in normal Ca containing physiological solutions. This assumption is supported by the finding that pretreatment with K-rich solutions and with β-agonists in the presence of Ca potentiate the subsequent agonist induced contractile response in Ca free solution, indicating that during such treatment more Ca is taken up in the cellular stores (Casteels and Raeymaekers, 1979; Van Eldere, Raeymaekers and Casteels, 1982).

This agonist-sensitive Ca store might correspond to the endoplasmic reticulum (ER). ER vesicles from smooth muscle can accumulate calcium in an ATP dependent way, and this has also been demonstrated for the ER in chemically skinned smooth muscle fibers (Raeymaekers and Casteels, 1981; Raeymaekers, 1982). Evidence has been provided showing that catecholamines and caffeine release Ca from the same intracellular store in intact smooth muscle (Deth and Casteels, 1977; Deth and Lynch, 1981). Caffeine was shown to induce contraction in skinned smooth muscle fibers preloaded with calcium, by releasing Ca from intracellular Ca sequestering sites, probably corresponding to the ER (Itoh et al., 1982; Itoh, Kuriyama and Suzuki, 1983; Stout and Diecke, 1983).

An interesting characteristic of the agonist sensitive Ca store in rabbit ear artery is that the refilling of this store by exposure to a Ca containing solution after its depletion with noradrenaline in Ca free solution, proceeds at a much faster rate than its depletion during exposure to Ca free solution (Casteels and Droogmans, 1981). Also the rate of filling and the final amount of calcium taken up in the store depend on $[Ca^{2+}]_o$. These observations suggest a direct pathway between the store and the extracellular space, allowing the rapid filling of that store. Consistent with such a direct pathway is the observation that the amplitude of the contraction induced by the agonist in Ca free solution is largely determined by the extracellular rather than by the intracellular Ca concentration during the preceding exposure to Ca containing solutions. The amplitude of the agonist induced contraction in Ca free solution after an exposure to a solution containing 5.9 mM K and 10 mM Ca, during which the tissues remain relaxed, is higher than that following an exposure to a solution containing 59 mM K and 0.2 mM Ca, which is accompanied by a contraction.

This direct pathway might be related to the close connection between the SR and the surface membrane observed in electron microscopic pictures (Devine, Somlyo and Somlyo, 1971). These are specialized regions where the ER membrane is separated from the surface membrane by a 10-12 nm gap traversed by periodic electron opaque processes. An attractive implication of such direct connection is that it allows an alternative picture of receptor operated channels. Activation with an agonist releases Ca^{2+} ions from its storage sites and initiates the phasic component of contraction. The ensuing depletion of this store could increase the Ca permeability of the junction between agonist sensitive store and the extracellular space and create continuous influx of calcium, which is responsible for the tonic component of the agonist-induced contraction in Ca containing solutions.

Ca-extrusion

The important contribution of extracellular calcium to the maintained
force development in smooth muscle tissue implies that in order to avoid a
cumulative deposition of Ca^{2+} to some intracellular sites, Ca^{2+} ions must
be extruded across the cell membrane against their electrochemical
gradient. Two types of Ca extrusion mechanisms have been proposed, an
ATP-dependent Ca pump and a Na-Ca exchange. Both mechanisms have been
investigated in several tissues and have also been suggested for smooth
muscle. The existence of coupled movements of Na and Ca has been
demonstrated in squid giant axons (Dipolo and Beauge, 1983), and it has
been proposed that the Na gradient provides the energy for the Ca
extrusion in this tissue. However this Na-Ca exchange mechanism seems to
function in a more complex way under physiological conditions. In smooth
muscle cells, experiments have been performed in which the electrochemical
gradient for Na ions was changed by exposure to Na-free solutions or by
inhibiting the Na-K pump by addition of ouabain or by using K-free
solutions (Ozaki and Urakawa, 1981; Sitrin and Bohr, 1971). Hereby the
intracellular Na concentration increases and on the basis of the Na-Ca
exchange hypothesis this should result in an increase of $[Ca^{2+}]_i$ and evoke
a contractile response. Restoring the Na-gradient should induce
relaxation. Although such observations would support the Na-Ca exchange
hypothesis they cannot be considered as a proof for a direct casual
relation between Na-Ca exchange and changes of $[Ca^{2+}]_i$ in smooth muscle.
Moreover contractions and relaxations have been induced by various
procedures in the complete absence of a Na-gradient. Also after complete
dissipation of the Na-gradient by increasing $[Na]_i$ following a prolonged
exposure to K-free solutions, some smooth muscles remain relaxed
(Casteels, Droogmans and Hendrickx, 1971).

Because Na-Ca exchange implies that some carrier molecule is
providing the 3 Na^+ for 1 Ca^{2+} exchange, one would expect that
Ca-antagonists do not affect this Na-Ca exchange. This principle does not
seem to apply to the ear artery of the rabbit. It was observed that in
this tissue organic Ca antagonists inhibited the contractile response
elicited by a Na-free medium (Droogmans and Casteels, 1979), while in some
other smooth muscle tissues this inhibitory effect could not be observed
(Ozaki and Urakawa, 1981). Another observation which argues against the
possible role of a Na-Ca exchange mechanism in smooth muscles is that
changing the Na-gradient does not affect the rate of extrusion of ^{45}Ca
from the rabbit ear artery during stimulation with noradrenaline

(Droogmans and Casteels, 1979). For the guinea-pig taenia coli
(Raeymaekers, Wuytack and Casteels, 1975) it was shown that the effect of
Na-free solutions on the ^{45}Ca-efflux could be due to a longer diffusional
pathway for Ca^{2+} ions leaving the cells. Such an increased pathway would
be caused by an increased capacity of Ca-binding in the extracellular
space during exposure to Na-deficient medium.

These doubts surrounding the nature of the Ca extrusion mechanism
required a more direct approach. In the above mentioned experimental
conditions using smooth muscle tissues the composition of the
intracellular compartment cannot be controlled to a sufficient extent. It
was therefore necessary to investigate the Na-Ca exchange and the
alternative mechanism i.e. an ATP-dependent Ca pump in simplified
experimental systems as microsomal preparations from smooth muscle. While
the study of Na-Ca exchange in cardiac sarcolemmal vesicles has confirmed
that such process could be quantitatively important for the Ca homeostasis
(Caroni and Carafoli, 1981), the data obtained for smooth microsomes only
indicate that the Na-Ca exchange is poorly developed in these tissues and
cannot contribute significantly to the Ca extrusion under physiological
conditions (Grover, Kwan and Daniel, 1981; Morel and Godfraind, 1984).
The rate of Ca extrusion by Na-Ca exchange in plasmalemmal vesicles of
smooth muscle is two orders of magnitude smaller than the value determined
in similar experimental conditions for cardiac muscle plasmalemmal
membranes.

Another argument against an important role of Na-Ca exchange in
smooth muscle can be deduced from a comparison of the NaK ATPase and the
CaMg ATPase activities in smooth muscle membranes as described by Casteels
et al. (1985).

In the Na-Ca exchange process at least 3 Na^+ should enter the cell
for extruding 1 Ca^{2+} and the Na^+ pump requires one 1 ATP for extruding 3
Na^+. The Na-Ca exchange, which depends finally on the transmembrane Na
gradient and thereby on the Na-K pump, could therefore maximally extrude 1
Ca^{2+} for 1 ATP hydrolyzed. In all membrane fractions isolated from smooth
muscle, the activity of the NaK-ATPase is always lower than that of CaMg
ATPase. In contrast highly purified plasma membrane fractions from
cardiac muscle contain much more NaK ATPase than the plasma membrane
fractions from smooth muscle. In addition the NaK ATPase activity in
cardiac plasmalemmal vesicles is higher than that of the CaMg ATPase.
These differences in activity of these two enzymes in plasma membrane

preparations of cardiac and smooth muscle strongly suggest that in cardiac muscle the Na-Ca exchange plays a predominant role. This hypothesis is further confirmed by the high value of V_{max} for Na-Ca exchange in plasmalemmal vesicles of cardiac muscle. From the activity in smooth muscle membranes of the NaK ATPase and of the CaMg ATPase, the activity of the pure enzymes and their molecular weight, it can be estimated that the density of the NaK- and Ca-pump sites in smooth muscle membranes amounts to 250-300 sites and to 300 sites per μm^2 respectively. After bringing forward those negative results on Na-Ca exchange in smooth muscle membrane, it is essential to investigate in more detail the alternative mechanism for Ca extrusion i.e. an ATP-dependent Ca pump.

In recent years important progress has been made in understanding this ATP-dependent Ca transport in smooth muscle membranes. A Ca^{2+} uptake with high affinity can easily be demonstrated in membrane vesicles prepared from different types of smooth muscle and this uptake requires the presence of Mg ATP (Carsten, 1969; Wuytack et al., 1978). However the demonstration of a corresponding $(Ca^{2+} + Mg^{2+})$-dependent ATPase activity proved to be more difficult (Wuytack and Casteels, 1980). In recent years we have succeeded in purifying a calmodulin stimulated Ca^{2+}-transport ATPase of 130-140 kDa which has many characteristics in common with the well known analogous enzyme from human erythrocytes (Wuytack, De Schutter and Casteels, 1981).

Recently also experimental evidence in favour of the existence in smooth muscle microsomes of a second Ca transport ATPase of 100 kDa has been brought forward. This second transport protein has many properties in common with the Ca^{2+}-transport ATPase from the sarcoplasmic reticulum of skeletal muscle (Raeymaekers et al., 1983). A limited overview of our study of these Ca-transport enzymes will be given.

Starting from 100 g of frozen porcine antral smooth muscle we can obtain approximately 110 mg of protein in a KCl-extracted microsomal fraction. This membrane fraction is a mixture of membranes from different subcellular origin. The membrane vesicles are either leaky or sealed for substrates (e.g. ATP and Ca^{2+}) and they present an orientation which is either the same or opposite to that present in the cell. The optimal procedure to estimate the total transport enzyme activities in these microsomes consists in an unmasking of CaMg ATPase- and NaK ATPase activity by addition of 0.2 mg.ml^{-1} saponin. The following values were obtained (nmoles.mg^{-1} protein.min^{-1} at 37°C, given as mean $\pm$ S.E. of mean,

the number of observations between parenthesis): Mg ATPase: 92 ± 6.7 (5) NaK
ATPase 25 ± 5.5 (2). CaMg ATPase 44 ± 5.9 (5). In the presence of saturating
concentrations of calmodulin (0.6 µM) the CaMg ATPase activity was
increased by a factor of 2.9 ± 0.25 to a value of 120 ± 2.0 nmoles.mg^{-1}
protein.min^{-1}. In human erythrocyte vesicles we observed a twofold higher
calmodulin stimulation under the same experimental conditions. This
discrepancy between the two tissues, could be due to either a partial
activation of the CaMg ATPase in the microsomes from smooth muscles by
contaminating calmodulin, to the presence of negatively charged
phospholipids (Gietzen, Sadorf and Bader 1982), or to limited proteolysis
(Sarkadi, Enyedi and Gardos, 1980) or to the fact that smooth muscle
microsomes contain different types of CaMg ATPase which cannot all be
stimulated by calmodulin.

In order to examine these different possibilities it was essential to
purify the calmodulin-dependent ATPase (Wuytack, et al., 1981). This was
done by calmodulin-affinity chromatography, a technique originally
designed for the purification of CaMg ATPase from human erythrocytes
(Gietzen, Tescka and Wolf, 1980; Niggli, Penniston and Carafoli, 1979).
Microsomal proteins were first solubilized with Triton X-100 and incubated
with the calmodulin Sepharose 4B gel in the presence of Ca^{2+}. After
extensive washing of the gel, the ($Ca^{2+} + Mg^{2+}$) ATPase and some minor
proteins that were bound to the calmodulin gel in a Ca^{2+}-dependent way
were released by eluting the calmodulin affinity gel with a Ca-free
buffer. It was observed that phospholipids had to be present continuously
during the purification in order to preserve enzyme activity. The
resulting enzyme preparation consisted largely of an ATPase with a M_R of
140-150 kDa as estimated from Laemmli-type sodium dodecyl-sulphate-
polyacrylamide gel electrophoresis. The specific activity of the purified
CaMg ATPase preparation at $37^{\circ}C$, and in the presence of $10^{-5}M$ Ca^{2+} and 0.6µ
M calmodulin was 12 µmoles ATP mg^{-1} protein.min^{-1}. This value is
comparable to that of the CaMg ATPase purified from human erythrocytes
(Gietzen et al., 1980; Niggli et al., 1979) and approximately half of that
of purified CaMg ATPase from sarcoplasmic reticulum of skeletal muscle and
of NaK ATPase from kidney (Korenbrot, 1977). Our purified CaMg ATPase can
be stimulated by calmodulin, but the level of stimulation depends on the
nature of the lipid environment of the enzyme as well as on its structural
integrity.

When the CaMg ATPase is eluted from the calmodulin affinity gel in
the presence of 0.05% Triton X-100 and 0.05% asolectin (a crude soy bean

phospholipid mixture consisting for about one fifth of negatively charged phospholipids (Niggli et al., 1981) the CaMg ATPase from smooth muscle is stimulated by 0.60 μM calmodulin (at 10^{-5}M Ca^{2+}) only by a factor of 1.42±0.03 (45). When on the other hand the elution medium contains a neutral phospholipid like phosphatidylcholine, the CaMg ATPase is stimulated by calmodulin by a factor of 2.27±0.13 (15).

Limited trypsin digestion of the CaMg ATPase purified from smooth muscle increases the CaMg ATPase activity in the absence of calmodulin While it reduces at the same time the calmodulin stimulation. This increase of the enzyme activity by partial proteolysis has first been described for erythrocyte Ca^{2+}-transport ATPase (Gietzen et al., 1982; Sarkadi et al., 1980; Schatzmann, 1982).

This purified calmodulin-dependent CaMg ATPase is a genuine Ca^{2+} transport ATPase because it mediates the accumulation of Ca^{2+} in an ATP-dependent way after reconstitution in artificial lipid vesicles. This Ca uptake occurs with a Ca^{2+}/ATP ratio of 1 (Verbist et al., 1984).

The influence of the phospholipid environment on the activation of the CaMg ATPase by calmodulin is more obvious in these vesicles than in the solubilized enzyme. Whereas the stimulation by calmodulin of the CaMg ATPase reconstituted in asolectin vesicles amounts to 1.25±0.09 (24), it reaches a value of 4.05±0.63 (12) in phosphatidylcholine vesicles.

The question now arises about the structural relationship between this CaMg ATPase and that of porcine erythrocytes or of the sarcoplasmic reticulum from porcine skeletal muscle. We have therefore examined the reaction between antibodies raised against CaMg ATPase from porcine erythrocytes or against fragmented sarcoplasmic reticulum from porcine skeletal muscle and our CaMg ATPase of smooth muscle. IgG against the erythrocyte ATPase were found to bind to the ATPase purified either from erythrocytes or from smooth muscle and these antibodies also bind to a protein of 140 kDa M_r in the smooth muscle microsomes. In contrast, IgG raised to skeletal muscle CaMg ATPase bind only to the ATPase of skeletal muscle sarcoplasmic reticulum and not to the CaMg ATPase purified from erythrocytes or smooth muscle. However, neither do these antibodies bind to a protein with M_r 100 kDa in the smooth muscle membranes.

These findings suggest that smooth muscle cells contain a Ca^{2+}-transport ATPase which is very similar to the CaMg ATPase of erythrocytes and which is probably originating from the plasma membrane. We have tested this latter hypothesis by separating membrane fractions by density gradient centrifugation and by studying the distribution of the CaMg ATPase activity and its relation to the NaK ATPase activity, an enzyme which can be considered to be the most reliable plasma membrane marker.

The analysis of the distribution of marker enzymes in the gradient fractions indicates that the activities of NaK ATPase and of CaMg ATPase measured in the presence of 0.6μM calmodulin do correlate well. Both enzymes are found over a wide density range with a maximum between densities of 25% and 30% sucrose. Assuming that NaK ATPase is a good marker for plasma membranes, this finding suggests that the largest fraction of the Ca^{2+} transport activity is located in the plasmalemma.

Ca^{2+} accumulation measured in the presence of oxalate can be ascribed to endoplasmic reticulum vesicles, due to their larger permeability for oxalate as compared to the plasma membrane. Therefore we have studied in these different fractions the rate of oxalate stimulated Ca^{2+} uptake. Because it was observed that the oxalate stimulated Ca^{2+} uptake was very sensitive to mechanical damage induced by pelleting (Raeymaekers and Casteels, 1984), we have introduced different conditions of centrifugation in which pelleting is avoided. In this way the oxalate stimulated Ca^{2+} uptake could be largely preserved. The main features of our present centrifugation method are the avoidance of pelleting by immediate application of the post-mitochondrial supernatant below a density gradient, together with the inclusion of 0.6 M KCl both in the post-mitochondrial supernatant and in the density gradient. In addition to preserving the oxalate stimulated Ca uptake, this flotation rather than sedimentation procedure combined with the inclusion of 0.6 M KCl also resulted in an effective extraction and separation of extrinsic proteins from the membranes, as was shown by sodium dodecylsulphate-polyacrylamide gel electrophoresis of the different fractions obtained from the gradient.

Under the present conditions of preparation, the rate of oxalate-stimulated Ca^{2+} uptake in the gradient fractions is several fold higher than in membranes prepared by the conventional pelleting procedure, reaching a maximum of about 80 $nmol.mg^{-1}$ $protein.min^{-1}$ at a density of about 24% sucrose. In contrast to the CaMg ATPase activity, this Ca^{2+} uptake presents a distribution which differs from that of the NaK ATPase activity. The distribution of NADH cytochrome c reductase (rotenone insensitive) a putative marker for endoplasmic reticulum, is more similar to that of the rate of oxalate-stimulated Ca^{2+} uptake. A link between both activities in microsomes of stomach smooth muscle was also suggested by earlier experiments in which the NADH cytochrome c reductase activity was found to be enriched in vesicles isolated after loading with calcium oxalate (Raeymaekers et al., 1983).

These results together with the experimental evidence indicating the presence in a mixed membrane fraction of another type of CaMg ATPase which could correspond to an endoplasmic ATPase suggest the existence in smooth muscle of two CaMg ATPases of which one is located in the plasmamembrane and the second one in the endoplasmic reticulum (Wuytack et al., 1984). Our phosphorylation experiments have shown that in smooth muscle microsomes two Ca^{2+}-dependent acyl phosphate bonds are formed one with a M_r similar to that of the erythrocyte Ca^{2+} transport ATPase (130 kDa) and another comparable to the ATPase of skeletal muscle sarcoplasmic reticulum (100 kDa) (Wuytack et al., 1982). Furthermore, both in smooth muscle membranes and in sarcoplasmic reticulum from skeletal muscle, La^{3+} decreases the phosphorylation of the M_r 100 kDa ATPase, whereas the phosphorylation levels of the 130 kDa ATPase in smooth muscle vesicles as well as in erythrocyte vesicles are increased. It is unlikely that the 100 kDa phosphoprotein in smooth muscle is a proteolytic product of the 130 kDa calmodulin binding ATPase, because the Ca^{2+}-dependent phosphorylation of the 130 kDa protein is stimulated by 100µM La^{3+} whereas this of the 100 kDa protein is inhibited. Moreover we found that by partial proteolysis with trypsin the 100 kDa ATPase phosphoprotein in smooth muscle and the corresponding protein from skeletal muscle are degraded to fragments with the same M_r. We must point out that our antibodies directed against the sarcoplasmic reticulum ATPase from skeletal muscle do not recognize a band of M_r 100 kDa in smooth muscle microsomes. This finding indicates that any CaMg ATPase in smooth muscle related to the Ca^{2+}-transport ATPase in SR of skeletal muscle is probably sufficiently different from it to prevent crossreaction with the antibodies.

REFERENCES

Aaronson, P., and van Breemen, C., 1981, Effects of sodium gradient manipulation upon cellular calcium, ^{45}Ca fluxes and cellular sodium in the guinea-pig taenia coli. J. Physiol. 319:443-461.

Aaronson, P., van Breemen, C., Loutzenhiser, R., and Kolber, M. A., 1979,

A new method for measuring the kinetics of transmembrane calcium-45 efflux in the smooth muscle of the guinea pig taenia coli. Life Science 25:1781-1790.

Anderson, N.C., Ramon, F., and Snyder, A., 1971, Studies on calcium and sodium in uterine smooth muscle excitation under current-clamp and voltage-clamp conditions. J. Gen. Physiol. 58:322-339.

Batra, S., 1973, The role of mitochondrial calcium uptake in contraction and relaxation of the human myometrium. Biochim. Biophys. Acta. 305:428-432.

Bohr, D.F., 1973, Vascular smooth muscle updated. Circulation Res. 32:665-672.

Bolton, T.B., 1979, Mechanism of action of transmitters and other substances on smooth muscle. Physiol Rev. 59:606-718.

Burnstock, G., 1976, Do some nerves release more than one transmitter. Neuroscience 1:239-248.

Caroni, P., and Carafoli, E., 1981, The Ca^{2+}-pumping ATPase of heart sarcolemma. J. Biol. Chem. 256:3263-3270.

Carsten, M.E., 1969, Role of calcium binding by sarcoplasmic reticulum in the contraction and relaxation of uterine smooth muscle. J. Gen. Physiol. 53:414-426.

Casteels, R., and Droogmans, G., 1975, Compartmental analysis of ion movements in: "Methods in Pharmacology," E.E. Daniel and D.M. Paton, ed, Vol 3, pp 663-671, Plenum Press.

Casteels, R., and Droogmans, G., 1981, Exchange characteristics of the noradrenaline-sensitive calcium store in vascular smooth muscle cells of rabbit ear artery, J. Physiol. 317:263-279.

Casteels, R., Droogmans, G., and Hendrickx, H., 1971, Membrane potential of smooth muscle cells in K-free solution, J. Physiol. 217:281-295.

Casteels, R., Kitamura, K., Kuriyama, H., and Suzuki, H., 1977, Excitation-contraction coupling in the smooth muscle cells of the rabbit main pulmonary artery, J. Physiol. 271:63-79.

Casteels, R., Kitamura, K., Kuriyama, H., and Suzuki, H., 1977, The membrane properties of the smooth muscle cells of the rabbit main pulmonary artery. J. Physiol. 271:41-61.

Casteels, R., and Raeymaekers, L., 1979, The action of acetylcholine and catecholamines on an intracellular calcium store in the smooth muscle cells of the guinea-pig taenia coli, J. Physiol. 294:51-68.

Casteels, R., Raeymaekers, L., Droogmans, G., and Wuytack, F., 1985, Na^+-K^+ ATPase, Na-Ca exchange, and excitation contraction coupling in smooth muscle, J. Cardiovasc. Pharmacol. in press.

Casteels, R., Raeymaekers, L., Suzuki, H., and Van Eldere, J., 1981,

Tension response and ^{45}Ca release in vascular smooth muscle incubated in Ca-free solution, Pflügers Arch., 392:139-145.

Cauvin, C., Loutzenhiser, R., and van Breemen, C., 1983, Mechanisms of calcium antagonist-induced vasodilatation, Ann. Rev. Pharmacol. Toxicol. 23:373-396.

Cheung, D.W., 1982, Two components in the cellular response of rat tail arteries to nerve stimulation, J. Physiol. 328:461-468.

Deth, R., and Casteels, R., 1977, A study of releasable Ca fractions in smooth muscle cells of the rabbit aorta, J. Gen. Physiol. 69:401-416.

Deth, R., and van Breemen, C., 1977, Agonist induced release of intracellular Ca in the rabbit aorta, J. Membrane Biol. 30:363-380.

Deth R., and Lynch, C. J., 1981, Mobilization of a common source of smooth muscle Ca by norepinephrine and methylxanthines, Am. J. Physiol. 240:C239-C247.

Devine, C.E., Somlyo, A.V., and Somlyo, A.P., 1971, Sarcoplasmic reticulum and excitation-contraction coupling in mammalian smooth muscle, J. Cell Biol. 52:690-715.

Dipolo, R., and Beauge, L., 1983, The calcium pump and sodium-calcium exchange in squid axons, Ann. Rev. Physiol. 45:313-324.

Droogmans, G., and Casteels, R., 1979, Sodium and calcium interactions in vascular smooth muscle cells of the rabbit ear artery, J. Gen. Physiol. 74:57-70.

Droogmans, G., and Casteels, R., 1981, Temperature-dependence of ^{45}Ca fluxes and contraction in vascular smooth muscle cells in rabbit ear artery, Pflügers Arch. 391:183-189.

Droogmans, G., Raeymaekers, L., and Casteels, R., 1977, Electro- and pharmacomechanical coupling in the smooth muscle cells of the rabbit ear artery, J. Gen. Physiol. 70:129-148.

Gietzen, K., Teschka, M., and Wolf, H.U., 1980, Calmodulin affinity chromatography yields a functional purified erythrocyte (Ca^{2+}+Mg^{2+}) dependent adenosine triphosphatase, Biochem. J. 189:81-88.

Gietzen, K., Sadorf, I., and Bader, H., 1982, A model for the regulation of the calmodulin-dependent enzymes erythrocyte Ca^{2+} transport ATPase and brain phosphodiesterase by activators and inhibitors, Biochem. J. 207:541-548.

Glossmann, H., Ferry, D.R., Lubbecke, F., Mewes, R., and Hoffman, F., 1982, Calcium channels: direct identification with radioligand binding studies, Trends Pharmacol. Sci. 3:431-437.

Godfraind, T., 1976, Calcium exchange in vascular smooth muscle, action of noradrenaline and lanthanum, J. Physiol. (London) 260:21-36.

Golenhofen, K., Hermstein, N., and Lammel, E., 1973, Membrane potential and contraction of vascular smooth muscle (portal vein) during application of noradrenaline and high potassium, and selective inhibitory effects of iproveratril (verapamil), Microvasc. Res. 5:73-80.

Goodford, P.J., 1965, The loss of radioactive ^{45}calcium from the smooth muscle of the guinea-pig taenia coli, J. Physiol. 176:180-190.

Grover, A.K., Kwan, C.Y., and Daniel, E.E., 1981, Na-Ca exchange in rat myometrium membrane vesicles highly enriched in plasma membranes, Am. J. Physiol. 240:C175-C182.

Harder, D.R., 1980, Comparison of electrical properties of middle cerebral and mesenteric artery in the cat, Am. J. Physiol. 239:C23-C26.

Harder, D.R., and Sperelakis, N., 1979, Action potentials induced in guinea-pig arterial smooth muscle by tetraethylammonium, Am. J. Physiol. 237:C75-C80.

Hirst, G.D.S., 1977, Neuromuscular transmission in arterioles of guinea-pig submucosa, J. Physiol. 273:263-275.

Hirst, G.D.S., and Neild, T.O., 1980, Evidence for two populations of excitatory receptors for noradrenaline on arteriolar smooth muscle, Nature 283:767-768.

Holman, M.E., and Surprenant, A.M., 1979, Some properties of the excitatory junction potentials recorded from spontaneous arteries of rabbits, J. Physiol. 287:337-352.

Itoh, T., Kajiwara, M., Kitamura, K., and Kuriyama, H., 1982, Roles of stored calcium on the mechanical response evoked in smooth muscle cells of the porcine coronary artery, J. Physiol. 22:107-125.

Itoh, T., Kuriyama H., and Suzuki, H., 1983, Differences and similarities in the noradrenaline- and caffeine-induced mechanical responses in the rabbit mesenteric artery, J. Physiol. 337:609-629.

Kajiwara, M., and Casteels, M., 1983, Effects of Ca-antagonists on neuromuscular transmission in the rabbit ear artery, Pflügers Arch. 396:1-7.

Kitamura, K., and Kuriyama, H., 1979, Effects of acetylcholine on the smooth muscle cells of isolated main coronary artery of the guinea-pig, J. Physiol. 293:119-133.

Kolber, M.A., and van Breemen, C., 1981, Competitive membrane adsorption of Na^+, K^+ and Ca^{2+} in smooth muscle cells, J. Membrane Biol. 58:115-121.

Korenbrot, J., 1977, Ion transport in membranes: Incorporation of biological ion-translocating proteins in model membrane systems, Ann. Rev. Physiol. 39:19-49.

Kuriyama, H., Itoh, Y., Suzuki, H., Kitamura, K., Itoh, T., Kajiwara, M., and Fukiwara, S., 1983, Actions of diltiazem on single smooth muscle cells and on neuromuscular transmission in the vascular bed, Circulation Res. supp I, 52:92-96.

Makita, Y., Kanmura, Y., Itoh, T., Suzuki, H., and Kuriyama, H., 1983, Effects of nifedipine derivates on smooth muscle cells and neuromuscular transmission in the rabbit mesenteric artery, Arch. Pharmacol. 324:302-312.

Mangel, A., Fahim, M., and van Breemen, C., 1982, Control of vascular contractility by the cardiac pacemaker, Science 215:1627-1629.

Mayer, C.J., van Breemen, C., and Casteels, R., 1972, The action of lanthanum and D600 on the calcium exchange in the smooth muscle cells of the guinea-pig coli, Pflügers Arch. 337:333-350.

Meisheri, K.D., Hwang, O., and van Breemen, C., 1981, Evidence for two separate Ca pathways in smooth muscle plasmalemma, J. Membrane Biol. 59:19-25.

Mironneau, J., Eugene, D., and Mironneau, C., 1982, Sodium action potentials induced by calcium chelation in rat uterine smooth muscle, Pflügers Arch. 395:232-238.

Morel, N., and Godfraind, T., 1984, Sodium/calcium exchange in smooth muscle microsomal fractions, Biochem. J. 218:421-427.

Niggli, V., Penniston, J.T., and Carafoli, E., 1979, Purification of the $(Ca^{2+} + Mg^{2+})$-ATPase from human erythrocyte membranes using a calmodulin affinity column, J. Biol. Chem. 254:9955-9958.

Niggli, V., Adunyah, E.S., Penniston, J.T., and Carafoli, E., 1981, Purified $(Ca^{2+} + Mg^{2+})$-ATPase of the erythrocyte membrane. Reconstitution and effect of calmodulin and phospholipids, J. Biol. Chem. 256:395-401.

Ozaki, H., and Urakawa, N., 1981, Involvement of a Na-Ca exchange mechanism in contraction induced by low-Na solution in isolated guinea-pig aorta, Pflügers Arch. 390:107-112.

Peiper, U., Griebel, L., and Wende, W., 1971a, Activation of vascular smooth muscle of rat aorta by noradrenaline and depolarization: two different mechanisms, Pflügers Arch. 330:74-89.

Peiper, U., Griebel, L., and Wende, W., 1971b, Unterschiedliche temperaturabhangigkeit der gefassmuskelkontraktion nach aktivierung durch kalium-depolarisation bzw. Noradrenalin, Pflügers Arch. 324:67-78.

Raeymaekers, L., 1982, The sarcoplasmic reticulum of smooth muscle fibers, Z. Naturforsch. 37c:481-488.

Raeymaekers, L., and Casteels, R., 1981, Measurement of Ca uptake in the endoplasmic reticulum of the smooth muscle cells of the rabbit ear artery, Arch. Int. Physiol. Biochem. 89:33-34.

Raeymaekers, L., and Casteels, R., 1984, The calcium uptake in smooth
muscle microsomal vesicles is reduced by centrifugation, Cell Calcium
5:205-210.

Raeymaekers, L., Wuytack, F., and Casteels, R., 1975, Na-Ca exchange in
taenia coli of the guinea-pig, Pflügers Arch. 347:329-340.

Raeymaekers, L., Wuytack, F., Eggermont, J., De Schutter, G., and
Casteels, R., 1983, Isolation of a highly enriched plasma membrane
fraction from gastric smooth muscle. Comparision of the Ca-uptake to
that in endoplasmic reticulum, Biochem. J. 210:315-322.

Sarkadi, B., Enyedi, A., and Gardos. G., 1980, Molecular properties of the
red cell calcium pump. I. Effects of calmodulin, proteolytic
digestion and drugs on the kinetics of active calcium uptake in
inside-out red cell membrane vesicles, Cell Calcium 1:287-297.

Schatzmanm, H.J., 1961, Calciumaufnahme und Abgabe am Darmmuskel des
Meerschweinchens, Pflügers Arch. 274:295-310.

Schatzmann, H.J., 1982, The plasma membrane calcium pump of erythrocytes
and other animal cells, in: "Membrane transport of calcium," E.
Carafoli, ed., pp. 41-108, Acad. Press, London.

Schumann, H.J., Gorlitz, B.D., and Wagner, J., 1975, Influence of
papaverine, D600, and nifedine on the effects of noradrenaline and
calcium on the isolated aorta and mesenteric artery of the rabbit,
Arch. Pharamcol. 289:409-418.

Sitrin, M.D., and Bohr, D., 1971, Ca and Na interaction in vascular smooth
muscle contraction, Am. J. Physiol. 220:1124-1128.

Sneddon, P., and Westfall, D.P., 1984, Pharmacological evidence that
adenosine triphosphate and noradrenaline are co-transmitters in the
guinea-pig vas deferens, J. Physiol. 347:561-580.

Somlyo, A.P., Somlyo, A.V., and Shuman, H., 1979, Electron probe analysis
of vascular smooth muscle. Composition of mitochondria, nuclei, and
cytoplasm, J. Cell Biol. 67:911-918.

Stout, M.A., and Diecke, F.P.J., 1983, [45]Ca distribution and transport in
saponin skinned vascular smooth muscle, J. Pharmacol. Exp. Ther.
225:102-111.

Su, C., Bevan, J.A., and Ursillo, R.C., 1964, Electrical quiescence of
pulmonary artery smooth muscle during sympathomimetic stimulation,
Circulation Res. 15:20-27.

Surprenant, A., Neild, T.O., and Holman, M.E., 1983, Effects of nifedipine
on nerve-evoked action potentials and consequent contractions in rat
tail artery, Pflügers Arch. 396:342-349.

Suzuki, H., 1981, Effects of endogenous and exogenous noradrenaline on the smooth muscle of guinea-pig mesenteric vein, J. Physiol. 321:495-512.

Suzuki, H., 1985, Electrical responses of smooth muscle cells of the rabbit ear artery to adenosine triphosphate, J. Physiol. 359:401-415.

Suzuki, H., Itoh, T., and Kuriyama, H., 1982, Effects of diltiazem on smooth muscle and neuromuscular junction in mesenteric artery, Am. J. Physiol. 242:H325-H336.

van Breemen, C., 1977, Calcium requirement for activation of aortic smooth muscle, J. Physiol. 272:317-330.

van Breemen, C., Aaronson, P., Loutzenhiser, R., and Meisheri, K., 1980, Ca movements in smooth muscle, Chest 78:157-165.

van Breemen, C., Casteels, R., 1974, The use of Ca-EGTA in measurements of ^{45}Ca efflux from smooth muscle, Pflügers Arch 348:239-245.

van Breemen, C., Farinas, B.R., Casteels, R., Gerba, P., Wuytack, F., and Deth, R., 1973, Factors controlling cytoplasmic Ca-concentration, Phil. Trans. R. Soc. Lond. B 265:57-71.

van Breemen, C., and McNaughton, E., 1970, The separation of cell membrane calcium transport from extracellular calcium exchange in vascular smooth muscle, Biochem. Biophys. Res. Commun. 39:567-574.

Van Eldere, J., Raeymaekers, L., and Casteels, R., 1982, Effect of isoprenaline on intracellular Ca uptake and on Ca influx in arterial smooth muscle, Pflügers Arch. 395:81-83.

Verbist, J., Wuytack, F., De Schutter, G., Raeymaekers, L., and Casteels, R., 1984, Reconstitution of the purified calmodulin (Ca^{2+} + Mg^{2+})-ATPase from smooth muscle, Cell Calcium 5:253-263.

Walsh, J.V., and Singer, J.J., 1980, Calcium action potentials in single freshly isolated smooth muscle cells, Am. J. Physiol. 239:C162-C174.

Wuytack, F., and Casteels, R., 1980, Demonstration of a (Ca^{2+} + Mg^{2+}) ATPase activity probably related to Ca^{2+} transport in the microsomal fraction of porcine coronary artery smooth muscle. Biochim. Biophys. Acta 595:257-263.

Wuytack, F., De Schutter, G., and Casteels, R., 1981, Purification of (Ca^{2+} + Mg^{2+})-ATPase from smooth muscle by calmodulin affinity chromatography, FEBS Letters 129:297-300.

Wuytack, F., Landon, E., Fleischer, R., and Hardman, J.G., 1978, The calcium accumulation in a microsomal fraction from porcine artery smooth muscle. A study of the heterogeneity of the fraction, Biochim. Biophys. Acta 540:253-269.

Wuytack, F., Raeymaekers, L., De Schutter, G., and Casteels, R., 1982, Demonstration of the phosphorylated intermediates of the Ca^{2+} transport ATPase in a microsomal fraction in a (Ca^{2+} + Mg^{2+})-ATPase purified from smooth muscle by means of calmodulin affinity chromatography, _Biochim. Biophys. Acta_ 693:45-52.

Wuytack, F., Raeymaekers, L., Verbist, J., Desmedt, H., and Casteels, R., 1984, Evidence for the presence in smooth muscle of two types of Ca^{2+}-transport ATPase, _Bioichem. J._ 224:445-451.

Zelcer, E., and Sperelakis, N., 1981, Ionic dependence of electrical activity in small mesenteric arteries of guinea-pig, _Pflügers Arch._ 392:72-78.

MECHANISMS INVOLVED IN THE ACTIONS OF STIMULANT DRUGS ON VASCULAR SMOOTH MUSCLE

Thomas B. Bolton

Department of Pharmacology
St. George's Hospital Medical School
London SW17 ORE
UK

INTRODUCTION

Smooth muscle tissues in the mammalian body carry out a wide variety of physiological functions. In each organ the smooth muscle is organized to provide tension development, or relaxation of existing tension, in response to varied stimuli reaching it from both local and distant parts of the body. The stimuli which reach it are of two main types - blood borne or local hormones and neurotransmitter substances. The smooth muscle in a particular organ is generally endowed with a wide variety of receptors which can recognize and respond to these hormones and neurotransmitters. The complement of receptors possessed by a particular smooth muscle to a large extent determine its responsiveness to chemical signals from other cells of the body. A second important factor in determining this responsiveness is the type of electromechanical coupling exhibited by the constituent smooth muscles of the tissue. In different tissues this takes a bewildering variety of forms. However, a simple and important division is between those smooth muscle cells which readily generate and propagate action potentials and those which do not. Where action potentials occur in smooth muscle cells they are the main determinant of tension generation by smooth muscle tissues - although not necessarily the only one.

The important factors which can be identified as contributing to the individuality of a particular smooth muscle are thus the ability to generate action potentials, the presence and influence of nerves, and the complement of receptors present on the smooth muscle cells. Subtle variations in these characteristics occur from smooth muscle to smooth muscle and account for much of the variability in the properties which is encountered. For example, the innervation may be by nerves which release more than one transmitter, perhaps two or even three, which exert opposing excitatory and inhibitory influences on smooth muscle contractility. The intimacy of relationship between smooth muscle and nerve ending can be extremely variable; in vas deferens and iris smooth muscle, close (<0.2 μ meter) apposition of nerve and muscle membranes is common whereas in some visceral muscles single nerve fibres are uncommon and bundles are more normally found of a few nerve fibres which course with blood vessels in the connective tissue trabeculae between smooth muscle bundles. Single nerve fibres penetrating between individual smooth muscle cells are uncommon (Richardson, 1958; Burnstock, 1970; Gabella, 1981).

VASCULAR MUSCLE

All blood vessels are equipped with smooth muscle cells which by their contraction regulate the calibre or resistance to distension. The tension and shortening developed by smooth muscle cells is the main regulator of blood pressure and of blood flow to particular tissues and organs. The blood vessels possess varying amounts of smooth muscle in their media and it is likely that the properties of vascular muscle depend on the size of the blood vessel (i.e. its position in the vascular tree) and on the special requirements for the control of blood flow in a particular tissue. Arteries have more smooth muscle than veins of the same size. It has been found that some blood vessels possess smooth muscle of unusual properties (Bevan & Bevan, 1984). Passing from large elastic arteries through distributing arteries to the smaller arteries and arterioles, the smooth muscle cells receive an innervation which in general increases both in amount and in intimacy of nerve–muscle relationship. In larger arteries with a thick layer of smooth muscle in their media, the nerves may be in relation to the outmost smooth muscle cells only, i.e. they occur at the medio–adventitial border and may not penetrate very far, if at all, into the media, so that the inner layers of smooth muscle are remote from the nerves. In smaller arteries and in arterioles, a plexus of nerve fibres ramifies over the one or two layers of smooth muscle cells which form the media (Keatinge, 1966; Burnstock, 1975). Transmitter release probably occurs at several points scattered over the surface of the vessel (Hirst & Neild, 1978; 1980a). Besides responding to neurotransmitter, the smooth muscle of small vessels is also very responsive to chemicals carried in the blood or released locally in the vicinity of the vessel from adjacent cells of the tissue which the blood vessels supply. There is an extremely important release of factors from the endothelial cells of blood vessels (Furchgott & Zawadzki, 1980). It also seems to be true that the smooth muscle of large arteries less readily generates and discharges action potentials than that of small vessels (but see Mekata & Keatinge, 1975). Rather few studies have been made of veins or venules with the exception of the portal vein. This readily generates action potentials in several species. Because so few electrophysiological studies have been made of veins, it is too soon to make generalizations about their properties. However, in their responses to nerve stimulation and in their membrane properties, they can show remarkable similarities to arteries, despite a generally less dense innervation (Funaki, 1960; Suzuki, 1981; Makita, 1983; Morgan 1983; Kou, Ibengwe & Suzuki, 1984; Cheung, 1985).

ACTIONS OF STIMULANT DRUGS ON VASCULAR MUSCLE

Stimulant drugs generally contract vascular muscle by combination with receptors which may normally react with hormones or neurotransmitters. The disposition of stimulant drug will depend on how it reaches the vascular muscle: blood borne agents will be present at higher concentrations in the inner layer of the media whereas neurotransmitters will be delivered at higher concentrations to the outer layer of the media. It is not my intention to deal with the problem of drug distribution in the media of a vessel in detail and much of what follows will assume that for all intents and purposes that the media is sufficiently thin for the concentration of drug to be reasonably uniform. This is more likely to be true when small vascular smooth muscle preparations are bathed in physiological solutions in which the stimulant drug or transmitter is dissolved. Much work has been done on helical strips, rings, or segments of arteries to which the drug is bath-applied in this way.

Responses to nerve-stimulation

However, disposition of the transmitter has recently been suggested to be very important in understanding the responses of arterioles to transmitter released from nerves. Hirst & Neild (1980b; 1981) suggest that the portions of vascular smooth muscle cell membrane adjacent to sites of transmitter release from autonomic nerves (junctional membrane) have a type of adrenoreceptor (gamma-adrenoreceptor, Hirst, Neild & Silverberg, 1982) which is distinct from the alpha variety present on the

much more extensive extrajunctional membrane. The mechanisms by which
these two sorts of adrenoreceptor elicited contraction were suggested to
be substantially different. The junctional gamma receptor was seen as
having a low affinity for the transmitter (noradrenaline) released from
the nerve ending in amounts sufficient to create high (millimolar)
concentrations in the junctional region. The proximity of nerve and
muscle membrane were believed to be sufficiently close to allow this
situation to exist. In contrast, the alpha extrajunctional receptors
would respond to much lower concentrations of noradrenaline (or
adrenaline) reaching them by overspill from junctional regions or blood
borne in the case of adrenaline. Because these extrajunctional receptors
are supposed to respond to low concentrations of agonist they would have a
high affinity for it, in contrast to the junctional gamma receptors (Hirst
et al., 1984).

The mechanisms linking these two sorts of adrenoreceptor, when
activated, to contraction of the vascular smooth muscle cell were
suggested to be radically different. In the case of extrajunctional alpha
receptors, the localized iontophoretic application of noradrenaline from a
micropipette caused localized contraction of the arteriolar segment
without detectable depolarization of the cells. In contrast, similar
application to junctional gamma receptors caused depolarization without
necessarily producing contraction; contraction would occur, however, if
depolarization produced was sufficiently great to trigger action potential
generation. The discharge of an action potential, triggered by a
suprathreshold depolarization to noradrenaline iontophoretically applied
to junctional membrane was found to evoke contraction of the smooth muscle
cells (Hirst & Neild, 1980a). These experiments raise the possibility
that where an authentic neurotransmitter is applied exogenously in some
way, its disposition must closely mimic that found when the transmitter is
released from nerve, for the effects of exogenous and endogenous
transmitter to be the same. In many situations, the disposition of
neurotransmitters at junctional regions may be very difficult to mimic at
a microscopic level by exogenous application.

The response of vascular smooth muscle to sympathetic nerve
stimulation seems to be more variable than was previously thought. In
some arteries such as the rat tail artery (Medgett & Langer, 1984) the
response to stimulating the sympathetic nerves is blocked by alpha-
adrenoreceptor blocking agents at low (10^{-7} M) concentrations; this is also
(Kawasaki & Takasaki, 1984). However, in other preparations the

contraction to nerve stimulation is much less affected even by
concentrations up to $5x10^{-6}M$ of prazosin (canine mesenteric artery, Kou,
Kuriyama & Suzuki, 1982; rabbit ear artery, Suzuki & Kou, 1983). This may
indicate a contribution of other transmitters or types of adrenoreceptor
other than the alpha variety, to the contractile response.

The e.j.p. has been reported to be severely reduced or abolished by
analogues of ATP which have been suggested to antagonize the actions of
this compound at its receptor (Sneddon & Burnstock, 1984; Allcorn,
Cunnane, Muir & Wardle, 1985). The implication of these results is that
ATP is released to act on junctional receptors previously believed to be
acted on by noradrenaline and designed "gamma" adrenoreceptors because of
their resistance to alpha-receptor blocking agents (Hirst & Neild, 1980b,
1981). At present whether the junctional receptors are ATP or gamma
adrenoreceptors is not clear.

The electrophysiological effects of nerve stimulation in many blood
vessels, including some veins, are a fast excitatory junction potential
(e.j.p.) resistant to blockade by alpha adrenoreceptor blockers (Neild &
Zelcer, 1982). In many vessels repetitive nerve stimulation elicits
e.j.p.s followed by a slow depolarization which is blocked by alpha-
adrenoreceptor blockers (rat tail artery, Cheung, 1984; rabbit ear artery,
Suzuki & Kou, 1983; rat saphenous vein, Cheung, 1985). The e.j.p. may
trigger the opening of potential-sensitive calcium channels if large
enough to reach threshold for the opening of these; in vascular muscles
discharging action potentials, one or more may be triggered. Sometimes
hyperpolarizing components mediated by beta-adrenoreceptors may be seen
(Morgan, 1983).

Responses to bath-applied drugs and neurotransmitters

Even thoroughly washed, vascular smooth muscle preparations contain
cell types other than smooth muscle cells in vitro. Besides nerves and
connective tissue cells in the adventia and elsewhere, arteries and veins
possess endothelial cells. These are believed to release factors which
are inhibitory to smooth muscle tension (see Vanhoutte and Houston, this
volume). Such factors are released under a variety of conditions but
particularly when certain vasoactive substances are applied e.g.
acetylcholine, substance P, thrombin. It has also been found that some
stimulants, besides contracting the vascular smooth muscle cells by a

direct action, at the same time release from the endothelium inhibitory factors which attenuate their direct spasmogenic action on the muscle cells (Egleme, Godfraind & Miller, 1984; Griffith, Henderson, Edwards & Lewis, 1984). Thus many vasorelaxants act wholly or in part by the release of endothelial-derived factors (EDRFs) and the actions of some spasmogens may be complicated by the release of these also (see Vanhoutte and Houston, this volume).

Bath application of noradrenaline to smooth muscle from various arteries produces either no detectable effect on membrane potential, or especially at higher concentrations, moderate to appreciable depolarization. Considerable dispute exists concerning the extent of noradrenaline's depolarizing action on vascular smooth muscle preparations; this may arise because of the difficulties experienced by various workers in applying microelectrode technique to these tissues (Casteels, Kitamura, Kuriyama & Suzuki, 1977a; Droogmans, Raeymaekers & Casteels, 1977; Trapani, Matsuki, Abel & Hermsmeyer, 1981; Bolton, Lang & Takewaki, 1984). It seems certain that substantial contraction can occur in some arteries with only a few millivolts depolarization or even no change in membrane potential (Fig. 1).

The contractions produced by noradrenaline are reduced by removing calcium from the bathing solution but a substantial contraction can generally be evoked by noradrenaline even if the calcium concentration in the bathing solution is reduced to very low levels by the addition of chelating agents such as EGTA. This result implies that at least a component of the noradrenaline-evoked contraction does not depend on the influx of calcium into the smooth muscle cell (Casteels, Raeymaekers, Suzuki & van Eldere, 1981). How this component of contraction is evoked is not certain: release of calcium from stores within the cell has been usually invoked to explain the persistence of noradrenaline contractions, and those to other stimulants, in calcium-free solutions (Bozler, 1969; Waugh, 1962, 1965; Hinke, 1965; Casteels & Droogmans, 1981; Manzini, Maggi & Meli, 1982; Saida & van Breemen, 1984; Cauvin, Saida & van Breemen, 1984). However, it is possible that other mechanisms not involving calcium as a second messenger are responsible for linking contraction to adrenoreceptor activation when noradrenaline evokes contractions in calcium-free conditions not associated with an increase in calcium efflux (Casteels et al., 1981). In potassium-contracted vascular muscle, alpha-receptor agonists cause contraction without a rise in net calcium levels within the cell - if we are to believe the results using the

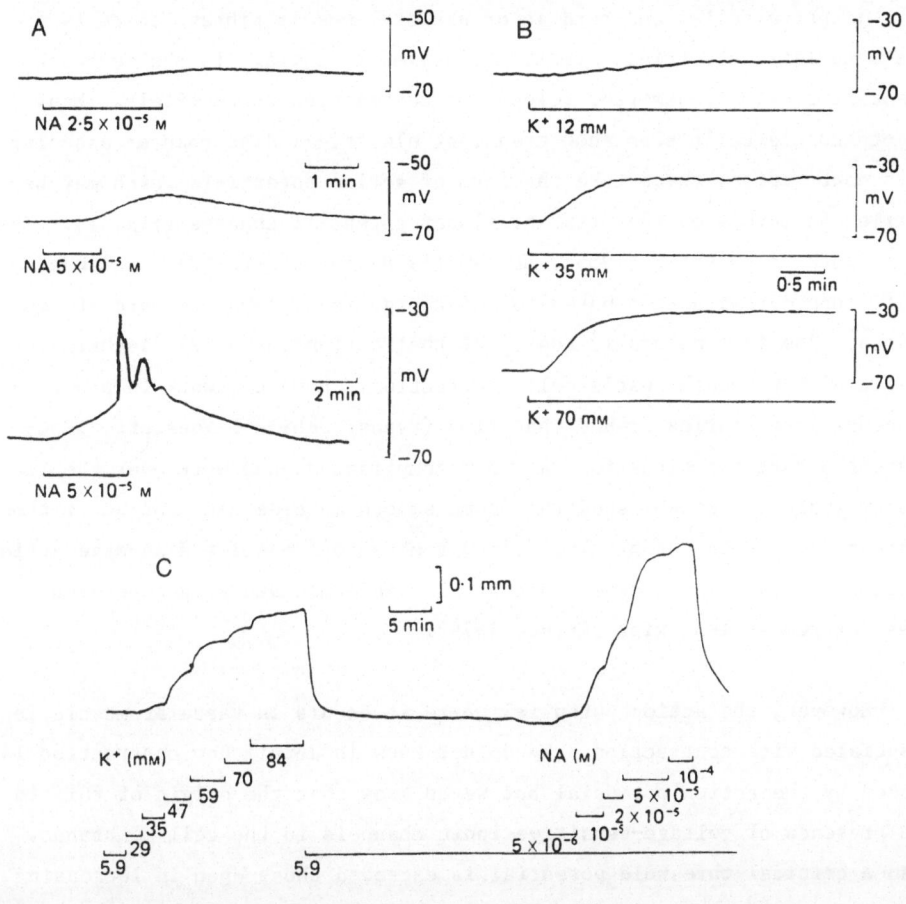

Fig. 1. Depolarizing action (A, B) and contraction of helically-cut strip (C) produced by noradrenaline (A, C) and high-K (B, C). Notice that concentrations of noradrenaline producing substantial contraction (C) produce only slight depolarization (A) and only maximally effective concentrations on tension produce an active response. The latter is not seen with high-K (B) which produced a smaller maximum contraction (C) but much greater depolarization (B). Reproduced from Bolton, Lang & Takewaki (1984) with permission of the Editors of The Journal of Physiology.

photoprotein aequorin (Fig. 3) (Morgan & Morgan, 1984a,b; Bradley & Morgan, 1985). This result might indicate compartmentalization of calcium or more likely, the existence of as yet unidentified mechanisms linking activated adrenoreceptors, and possibly other receptors, to actin–myosin interaction and tension generation.

In nerve cells, and cardiac or skeletal muscle fibres, there is a characteristic electrical event, the action potential, in the membrane associated with transmitter release or contraction respectively. Many smooth muscle cells also show transient electrical disturbances associated with contraction; these take the form of action potentials which may be carried in bursts on the crest of slower potential changes (Fig. 1). The early work of Bulbring (1955), Holman (1958) and others (West & Landa, 1956) showed that action potential discharge is a common feature of smooth muscle. The idea naturally took root that action potential discharge was essential for smooth muscle cell contraction. However, work in progress in other laboratories around that time (Evans, Schild & Thesleff, 1958) suggested that the situation was more complicated, and work over the last twenty years has established that some smooth muscles can contract without discharging action potentials. Thus, some smooth muscles discharge action potentials freely, and others not at all; intermediate stages between these extremes also exist (Creed, 1979).

However, the action potential where it occurs in vascular muscle is associated with contraction. We do not know in detail how contraction is evoked by the action potential but we do know that the origin of this is the presence of voltage-sensitive ionic channels in the cell membranes. When a critical threshold potential is exceeded these open in increasing numbers (up to a point) over a restricted range of potential beyond threshold. The mechanism is called 'regenerative' and is similar in most excitable cells that have been studied. This term arises because at threshold some channels open which allow inward currents to cross the membrane; this depolarizes the cell further and more channels open; more inward current flows depolarizing the cell more. Hence an explosive depolarization develops which is the upstroke of the action potential. These voltage-sensitive inward current channels normally seem to admit calcium, rather than sodium, in smooth muscle (Tomita, 1980).

Whether it is the calcium entry into the cell associated with the action potential which is localized to parts of the membrane in special relation to calcium stores or other mechanisms, or whether it is the depolarization, which triggers further events leading to contraction is not known. However, action potentials trigger contraction presumably by the entry of calcium into the cell and possibly also by triggering further calcium release from stores within it.

Depolarization, however produced, seems to be effective in producing smooth muscle cell contraction. If the membrane is depolarized electrically, or by reducing the potassium gradient across it by immersing the muscle in solutions with elevated potassium concentrations (high-K) then smooth muscle contracts (Casteels, Kitamura, Kuriyama & Suzuki, 1977b; Ito, Suzuki & Kuriyama, 1977). We can explain this seemingly constant effect of depolarization in eliciting tension by supposing that the voltage-sensitive calcium channels open whenever depolarization occurs; the more depolarization up to a limit, the more channels open, the more calcium enters, and the greater tension is engendered (Bolton, 1979).

However, it has long been known that smooth muscle will contract further to stimulant drugs when held depolarized by immersion in high-K solutions. Since no membrane potential change can occur under these conditions, the cells must possess some other method of producing tension. The contractile action of noradrenaline (or other stimulant) in high-K solution is attenuated if calcium is removed from the bathing solution (Bozler, 1969; Waugh, 1962; Casteels & Droogmans, 1981). This result implies that calcium enters the cell in increasing amounts when noradrenaline is applied even without a change in membrane potential; under these conditions a potential-dependent opening of voltage-sensitive calcium channels cannot be invoked. Calcium must enter in some other way i.e. through channels which open when the adrenoreceptor is stimulated but which do not open primarily in response to depolarization. We could call these channels receptor-operated channels (ROCs) to distinguish them from potential-sensitive ones admitting calcium (Bolton, 1979).

In calcium-free solution whether the potassium concentration is normal (3-6mM) or higher (up to 150mM) some contraction can still be produced by noradrenaline in many vascular muscles as alluded to above. Thus, calcium may also be released from stores within the cell as well as entering the cell either through ROCs or through potential-sensitive calcium channels when noradrenaline (and probably other stimulants) is applied.

Recent evidence concerning the sources of calcium for contraction

More recent experiments on the sources of calcium for contraction of vascular muscle depend, in my view, on the framework of hypotheses

outlined in the previous section for much of their interpretation. The
influx of calcium ions into smooth muscle, and their efflux from it, can
be traced by the use of radioactive isotopes of calcium. However, special
techniques are required and it must be said that before these were used
variable changes in the movement of calcium were obtained upon application
of noradrenaline and other stimulants. Unfortunately, the special
techniques used yield results which we cannot evaluate independently
because unknown factors are introduced; however, the results obtained on
calcium influx are internally consistent and support the hypotheses of the
sources of calcium for contraction. In the case of calcium efflux, there
is considerable uncertainty as to the mechanisms giving rise to this.
Measurement of calcium in the smooth muscle cell has been made using
either calcium sensitive dyes or photoproteins and by measuring the
amounts of calcium in various organelles of the cell in sections prepared
for electron microscopy. These techniques are extremely involved and
depend on various and different assumptions about the way calcium behaves
and can be measured using these techniques. The measurement of free
ionized calcium concentration (or at least detecting whether it increases
or decreases) in living smooth muscle cells has yielded some surprising
results.

To measure the movement of calcium across the cell membranes of
smooth muscle cells in a preparation the use of lanthanum was introduced
with a view to displacing calcium from binding sites in the extracellular
space (van Breemen, Farinas, Casteels, Gerba, Wuytack & Deth, 1973).
Later lanthanum was replaced by Ca-free EGTA solutions (Meisheri, Palmer &
van Breemen, 1980; Cauvin et al., 1984). Both those substances have
additional effects but what is uncertain is whether the known or
unsuspected actions of these substances invalidate any of the results
obtained. Using ^{45}Ca the amounts entering the muscle can be measured in a
fixed period of time and the variation in the results obtained is small if
extracellularly bound calcium is removed with lanthanum or Ca-free EGTA
solution as described. The entry of calcium can be shown to be increased
if the muscle is incubated with noradrenaline. This is also true if
incubation is in high-K solution. However, the influx in this case is
much more sensitive to a class of compounds, the calcium-entry blockers
(van Breemen, Hwang & Meisheri, 1981). The explanation for these results
is that high-K opens voltage-sensitive calcium channels which are
susceptible to blockade by calcium-entry blockers, whereas noradrenaline
opens ROCs to admit calcium and opens voltage-sensitive calcium channels
to a much smaller extent (Fig. 2). Also, as explained, at least initially

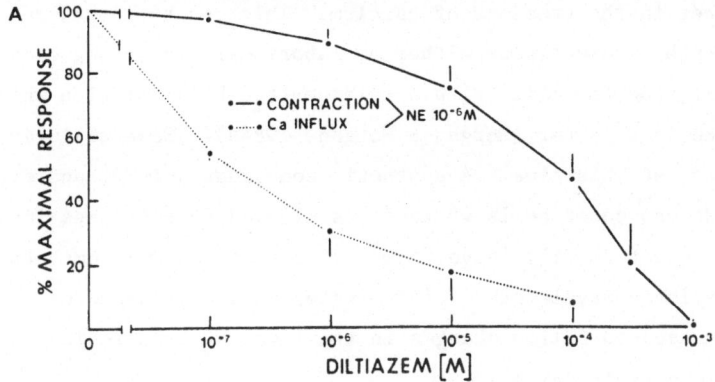

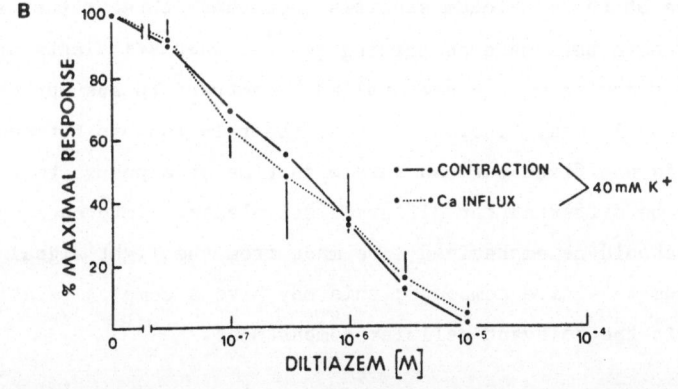

Fig. 2. Inhibitory effect of a calcium-entry blocker, diltiazem, on ^{45}Ca
influx measured by the calcium-free EGTA quenching method in
rabbit aortic strip. <u>A</u>. Diltiazem below 10^{-5}M is ineffective at
blocking contraction in response to noradrenaline (10^{-6}M) but has
a substantial and progressive effect on calcium influx. <u>B</u>. In
contrast, diltiazem has effects on tension to high-K (40mM) and
on calcium influx which increase pari passu with concentration.
Reproduced from van Breemen, Hwang, and Meisheri (1981) by
permission of the authors and publishers.

noradrenaline releases calcium from stores in the smooth muscle cell, as
well as accelerating entry (Deth & van Breemen, 1974; Cauvin,
Loutzenhiser, Hwang & van Breemen, 1982).

    Measurements of the level of free ionized calcium in the smooth
muscle cell have been made most frequently using aequorin, a protein which

fluoresces in the presence of calcium. This can be introduced into cells of a smooth muscle tissue either by laboriously injecting each one, or by permeabilizing the cells with a calcium-free EGTA solution which allows the aequorin to enter (Morgan & Morgan, 1984a). Some cellular compounds may escape at this time. A synthetic compound, quin 2, and similar compounds can enter cells where it is cleaved by esterases into a form which does not readily leave them. This compound will emit light at a characteristic wavelength if illuminated in the presence of calcium and has been used to follow changes in the level of free ionized calcium in several excitable cells.

By means of these calcium sensitive compounds three rather surprising observations have been made concerning $[Ca^{2+}]_i$ when stimulants and relaxants of vascular muscle are applied. Contrary to assumptions previously made in many tension studies, the relationship between tension and $[Ca^{2+}]_i$ is not fixed but can vary with time of exposure to a stimulant drug and may be different for different stimulants. Inferences concerning $[Ca]_i^{2+}$, it should be emphasized, are made from the light signal emitted by the calcium-sensitive compound; this may have a complex relationship to the $[Ca^{2+}]_i$ in the relevant cellular compartment.

When noradrenaline contracts vascular muscle, $[Ca^{2+}]_i$ rises rapidly initially to a peak and then declines to a much lower level which is still above basal level: tension, however, rises steadily and more slowly to a plateau level which is maintained. Clearly, tension and $[Ca^{2+}]_i$ by no means move in parallel (Fig. 3). Also, if high-K is used, $[Ca^{2+}]_i$ rises more slowly to a level which is maintained, quite unlike its behaviour when noradrenaline is applied. If alpha receptors are activated in the presence of high-K-induced tension, tension rises further but $[Ca^{2+}]_i$ apparently does not. Either this means that tension can be activated by an unknown mechanism not primarily involving a rise in $[Ca^{2+}]_i$, or $[Ca^{2+}]_i$ rises in a compartment from which the light-signal is obscured. Since in normal solution, alpha receptor activation causes a rise in $[Ca^{2+}]_i$ as tension rises, it is strange that high-K should now apparently prevent the registration of a signal from this compartment (Fig. 3). It has been also found that vasorelaxants can reduce tension created by high-K or noradrenaline without an apparent fall in $[Ca^{2+}]_i$ although they reduce the rise in $[Ca^{2+}]_i$ associated with application of noradrenaline or high-K (Morgan & Morgan, 1984a,b; Bradley & Morgan, 1985).

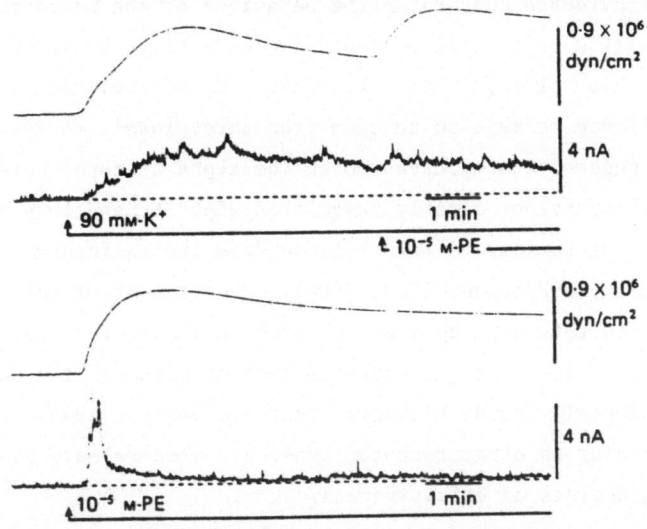

Fig. 3. Effect of activating alpha adrenoreceptors with phenylephrine on
$[Ca^{2+}]_i$, estimated from the aequorin light signal, and on tension
in normal solution (below) and in high-K (90mM) solution in
ferret portal vein. The light signal rises before tension and
then declines to a much lower level as tension rises upon
alpha-adrenoreceptor activation in normal solution. In 90mM
potassium concentration the light signal and tension rise
together but now alpha receptor activation has little effect on
the light signal implying that under these conditions $[Ca^{2+}]_i$
does not rise. Reproduced with permission of the Authors and
Editors of The Journal of Physiology from Morgan & Morgan
(1984a).

Vasoconstrictors cause an increase in the rate of loss of ^{45}Ca
previously loaded into the muscle. This increased rate of loss may
indicate that active extrusion of calcium from the cell is increased.
However, besides noradrenaline and high-K which produce this effect,
mitochondrial poisons will also do so; the interpretation of these results
is uncertain but they may indicate that whenever calcium is released or
leaks from stores within the cell due to the action of noradrenaline or
high-K or when high-energy compounds are depleted by metabolic poisons,
then cellular calcium extrusion processes are accelerated (despite the
depletion of high-energy compounds in the case of metabolic poisons) (van
Breemen, Wuytack & Casteels, 1975; Deth & van Breemen, 1977; Deth & Lynch,
1981; Leijten & van Breemen, 1984).

Further evidence concerning the behaviour of the noradrenaline-sensitive calcium store will be found in the article by Verbist, Droogmans & Casteels (this volume). This store can also be depleted by caffeine which is believed to release calcium from sarcoplasmic reticulum. Experiments suggest that activation of the alpha receptor releases calcium, perhaps rather closely associated with it, and this calcium acts as a trigger for further calcium release from the caffeine-releasable store (Saida & van Breemen, 1983; 1984). Measurement of calcium content of sarcoplasmic reticulum by electron probe analysis revealed that noradrenaline could reduce its calcium content although the changes observed were small (Bond, Kitazawa, Somlyo & Somlyo, 1984). Other stimulants acting on other receptor types can also release bound calcium with varying degrees of effectiveness.

Further evidence for the existence of ROCs

The activation of alpha receptors is supposed in the above hypotheses to open ROCs in the membrane of vascular smooth muscle cells. This can be investigated further in at least two ways.

If a constant current is injected into vascular smooth muscle cells, the size of the resulting displacement of the membrane potential can give an indication of the membrane resistance. If, upon application of noradrenaline, this falls it would be evidence supporting the opening of ROCs. Intracellular injection of current into a single cell of an electrical syncytium such as that formed by the smooth muscle cells in the media of a small artery is not a sensitive method to detect a change in resistance of that cell, because most current leaks away to other cells in the syncytium through the low-resistance pathways which must exist between smooth muscle cells (Tomita, 1980). A more satisfactory method is to apply a potential field to a strip of smooth muscle and to make use of the fact that electrically such a strip behaves as an electrical cable. The input resistance of the cable will change in a way which is relatively sensitive to changes in the resistance of the cell membrane of its constituent cells. Experiments of this type have been performed mostly on segments of arteries where the potential gradient has been applied in the longitudinal axis of the vessel. The changes in input resistance observed when noradrenaline is applied in this less than optimum arrangement have been small. Often an increase in resistance has been detected, sometimes a decrease (Kajiwara, Kitamura & Kuriyama, 1981; Bolton et al., 1984). The latter is consistent with an opening of additional channels in the

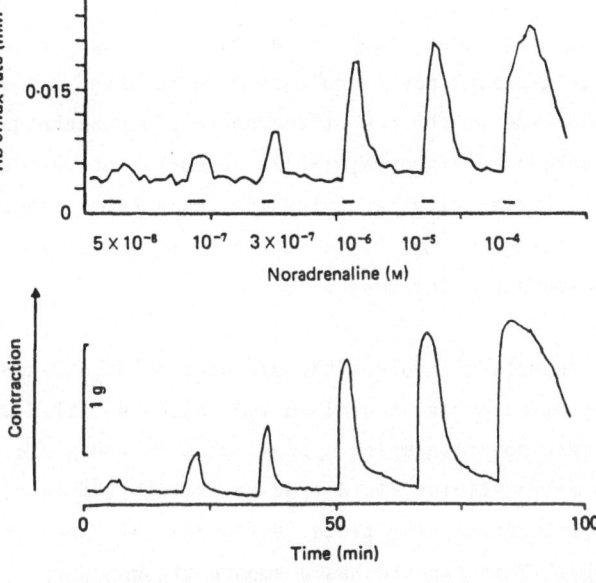

Fig. 4.   Effect of noradrenaline on transverse strip of rabbit aorta. The
strip was first loaded with ^{86}Rb and the rate of loss of this
tracer then followed into rubidium-free solution containing 5.9mM
potassium concentration. The rate of efflux of rubidium is
increased by noradrenaline over a wide concentration range in a
way that parallels the effects on tension closely. Reproduced
from Bolton & Clapp (1984) with permission of the Editors of the
Journal of Physiology.

vascular smooth muscle cell membrane. These could be ROCs.

If vascular smooth muscle strips cut from arteries are loaded with
[42]K or ^{86}Rb, the rate of efflux of these ions can trace changes in the
rate of loss of potassium ions (Fig. 4). From this, inferences concerning
the effects on potassium permeability can be made. Noradrenaline
increases the rate of loss of ^{86}Rb or [42]K from some arteries e.g. rabbit
ear artery at concentrations which microelectrode recordings show do not
depolarize the cells (Bolton & Clapp, 1984). This suggests that alpha
receptor activation increases the potassium permeability of vascular
smooth muscle cells. This would be expected to hyperpolarize them but at
low concentrations membrane potential changes are not generally seen. It
may be that the permeability to sodium or calcium also increases so that
no net change in membrane potential is observed. The electrophysiological
methods of measuring the cell membrane resistance are relatively
insensitive so the increase in potassium permeability may not be
detectable.

Patch clamp studies on single smooth muscle cells
==========

In order to learn more about the actions of drugs on smooth muscle cells at a cellular level new techniques were required. A significant advance has been made by the use of the patch clamp technique (Hamill, Marty, Neher, Sakmann & Sigworth, 1981). A number of variations of this allow study of the electrical activity of single cells, their membrane currents under voltage clamp, or the activity of single ionic channels in the membrane responsible for these.

The basic technique involves the preparation of a patch clamp pipette which has a tip diameter of about 1 µm and which is filled with a physiological salt solution which approximates in inorganic ion composition to extracellular fluid (for recording from inside-out patches, see later) or to intracellular fluid (for whole-cell and for outside-out patch recording). This pipette has a smooth tip which when brought in contact with the smooth muscle cell surface, cleaned by treatment with collagenase (and in the case of arterial cells, elastase) can be induced to form a very high resistance seal to the cell membrane by the application of some negative pressure to the interior of the pipette. In this configuration recording can be made of the activity of ionic channels in the patch of membrane still attached to the cell ("cell-attached" patch) which is isolated by the high resistance seal formed as described (Benham & Bolton, 1983).

If the patch is ruptured by applying sharp pulses of negative pressure to the interior of the pipette, access can be gained to the interior of the cell and recording can then be made in the "cell-attached mode". Either potential can be recorded or the voltage can be constrained ("voltage-clamped") to adopt various waveforms, a rectangular depolarizing, or hyperpolarizing, command being most useful (Benham, Bolton & Lang, 1984). Voltage-clamp analysis is now a standard technique that can be applied to excitable cells.

As far as whole-cell recording is concerned, we have so far not studied vascular smooth muscle cells very extensively. We have done experiments on the action-potential discharging longitudinal muscle of rabbit jejunum. In single cells obtained from this tissue by collagenase digestion, membrane potentials up to -60mV can be obtained. The cells can discharge normal action potentials very similar to those recorded from the whole tissue by microelectrode. Single vascular smooth muscle cells from

rabbit ear artery or guinea-pig small mesenteric artery do not generate
action potentials either in the whole tissue or when single cells are
isolated by collagenase/elastase digestion. The action of acetylcholine
on jejunal cells has also been studied in the whole-cell mode (Benham,
Bolton and Lang, 1985) and this technique should provide a method of
investigating the actions of drugs on single vascular smooth muscle cells.

In both jejunal, and in mesenteric artery smooth muscle cells,
recordings from isolated patches revealed a population of potassium
channels which were activated by the presence of calcium (10^{-9}-10^{-7}M) on
the inner surface of the membrane (Benham, Bolton, Lang & Takewaki, 1984).
These calcium-activated potassium channels also opened more frequently
when the patch was depolarized. These channels are likely to contribute
to the outward potassium current which is responsible for the repolarizing
phase of the action potential. The calcium entering during the upstroke
of the action potential, due to the opening of the voltage-sensitive
calcium channels, is likely to increase the concentration of calcium close
to the internal surface of the membrane. This rise, in combination with
the depolarization which is occurring, is likely to cause the
calcium-activated potassium channels to open. These will exert a strong
repolarizing effect on the membrane. We have also observed a population
of potassium channels of much slower kinetics which open upon
depolarization but which are not calcium-sensitive.

These patch clamp studies are only in their infancy. We do not yet
fully understand the varieties and properties of the ionic channels
present in the membrane of the different smooth muscles. It is therefore
not possible to say anything, as yet, about whether separate populations
of channels exist (ROCs) which are 'gated' by receptor activation but not
by changes in potential as predicted on the basis of experiments on the
mechanical and electrical properties of whole smooth muscle tissues. It
will be of interest to discover whether receptor activation also affects
the behaviour of channels normally 'gated' by potential.

CONCLUSIONS

Most studies on the mechanism of action of stimulant drugs on smooth
muscle have used noradrenaline. We know that this transmitter may
contract smooth muscle with little change in membrane potential when
bath-applied, or with a transient depolarization of the membrane if the
sympathetic nerves are stimulated. The latter observation assumes that

only noradrenaline, and not some other compound such as ATP, is released when the nerves are stimulated in this way. When bath applied, noradrenaline activates receptors which apparently allow calcium to enter the cell through ROCs, although direct evidence for the existence of channels gated only by receptor activation is lacking as yet. Sometimes, noradrenaline depolarizes sufficiently to bring into play the opening of voltage-sensitive calcium channels. These are involved in the action potential in vascular smooth muscles displaying this but are also present in non-excitable vascular muscles. However, in calcium-free solutions substantial noradrenaline-evoked contraction can occur. Release of calcium from storage sites in the muscle is believed to contribute to the contractions under these conditions and, by implication, also in normal solution. This store is believed to be located in the sarcoplasmic reticulum and activation of alpha receptors has been suggested to first release a small amount of trigger calcium associated with them. This in turn acts on the caffeine-releasable calcium stores in the sarcoplasmic reticulum.

However, the above hypotheses revolve around calcium as the sole second messenger. Recent work suggests that other pathways, or second messengers, may exist which we as yet cannot identify. If these exist, they may shift calcium from its pivoted role in our hypotheses concerning the stimulant action of drugs on vascular muscle.

REFERENCES

Allcorn, R. J., Cunnane, T. C., Muir, T. C., and Wardle, K. A., 1985, Does contraction in the rabbit ear artery require excitatory junction potentials (e.j.p.s.) and 'spikes'?, J. Physiol., In Press.

Benham, C. D., and Bolton, T. B., 1983, Patch-clamp studies of slow potential-sensitive potassium channels in longitudinal smooth muscle cells of rabbit jejunum, J. Physiol., 340:469-486.

Benham, C. D., Bolton, T. B., and Lang, R. J., 1984, Membrane potential and voltage-clamp recording from single smooth muscle cells of rabbit jejunum, J. Physiol., 353:67P.

Benham, C. D., Bolton, T. B., and Lang, R. J., 1985, The action of acetylcholine on single cells of longitudinal smooth muscle of rabbit jejunum studied by whole-cell voltage-clamp, J. Physiol., 358:84P.

Benham, C. D., Bolton, T. B., Lang, R. J., and Takewaki, T., 1984, Calcium-dependent K^+ channels in dispersed intestinal and arterial smooth muscle cells of guinea-pig and rabbits studied by the patch-clamp technqiue, J. Physiol., 350:51P.

Bevan, J. A., and Bevan, R. D., 1984, Patterns of α-adrenoceptor regulation of the vasculature, Blood Vessels, 21:110-116.

Bolton, T. B., 1979, Mechanisms of action of transmitters and other substances on smooth muscle, Physiol. Rev., 59:606-718.

Bolton, T. B., Lang, R. J., and Takewaki, T., 1984, Mechanisms of action of noradrenaline and carbachol on smooth muscle of guinea-pig anterior mesenteric artery, J. Physiol. 351:549-572.

Bolton, T. B., and Clapp, L. H., 1984, The diverse effects of noradrenaline and other stimulants on ^{86}Rb and ^{42}K efflux in rabbit and guinea-pig arterial muscle, J. Physiol. 355:43-63.

Bond, M., Kitazawa, T., Somlyo, A. P., and Somlyo, A. V., 1984, Release and recycling of calcium by the sarcoplasmic reticulum in guinea-pig portal vein smooth muscle, J. Physiol. 355:677-695.

Bozler, E., 1969, Role of calcium in initiation of activity of smooth muscle, Am. J. Physiol. 216:671-674.

Bradley, A. B., and Morgan, K. G., 1985, Cellular Ca^{2+} monitored by aequorin in adenosine-mediated smooth muscle relaxation, Am. J. Physiol. 248:H109-H117.

Bulbring, E., 1955, Correlation between membrane potential, spike discharge and tension in smooth muscle, J. Physiol. 128:200-221.

Burnstock, G., 1970, Structure of smooth muscle and its innervation, in: "Smooth Muscle," E. Bulbring, A. F. Brading, A. W. Jones, and T. Tomita, eds., pp. 1-69, Edward Arnold, London.

Burnstock, G., 1975, Innervation of vascular smooth muscle: histochemistry and electron microscopy, Clin. Exp. Pharmacol. Physiol. Supp. 2:7-20.

Casteels, R., Kitamura, K., Kuriyama, H., and Suzuki, H., 1977a, The membrane properties of the smooth muscle cells of the rabbit main pulmonary artery, J. Physiol. 271:41-61.

Casteels, R., Kitamura, K., Kuriiyama H., and Suzuki, H. 1977b, Excitation-contraction coupling in the smooth muscle cells of the rabbit main pulmonary artery, J. Physiol. 271:63-79.

Casteels, R., and Droogmans, G., 1981, Exchange characteristics of the noradrenaline-sensitive calcium store in vascular smooth muscle cells of rabbit ear artery, J. Physiol. 317:263-279.

Casteels, R., Raeymaekers, L., Suzuki, H., and van Eldere, J., 1981, Tension response and ^{45}Ca release in vascular smooth muscle incubated in Ca-free solution, Pflugers Arch. 392:139-145.

Cauvin, C., Loutzenhiser, R., Hwang, O., and van Breemen, C., 1982,

α_1-adrenoceptors induce Ca influx and intracellular Ca release in isolated rabbit aorta, Eur. J. Pharmacol. 84:233-235.

Cauvin, C., Saida, K., and van Breemen, C., 1984, Extracellular Ca^{2+} dependence and diltiazem inhibition of contraction in rabbit conduit arteries and mesenteric resistance vessels, Blood Vessels, 21:23-31.

Cheung, D. W., 1984, Neural regulation of electrical and mechanical activities in the rat tail artery, Pflugers Arch. 400:335-337.

Cheung, D. W., 1985, An electrophysiological study of α-adrenoceptor mediated excitation-contraction coupling in the smooth muscle cell of the rat saphenous vein, Br. J. Pharmac. 84:265-271.

Creed, K. E., 1979, Functional diversity of smooth muscle, Br. Med. Bull. 35:243-247.

Deth, R., and van Breemen, C., 1974, Relative contributions of Ca^{2+} influx and cellular Ca^{2+} release during drug induced activation of the rabbit aorta, Pflugers Arch. 348:13-22.

Deth, R., and van Breemen, C., 1977, Agonist induced release of intracellular Ca^{2+} in the rabbit aorta, J. Membr. Biol. 30:363-380.

Deth, R. C., and Lynch, C. J., 1981, Mobilization of a common source of smooth muscle Ca^{2+} by norepinephrine and methylxanthines, Am. J. Physiol. 240:C239-C247.

Droogmans, G., Raeymaekers, L., and Casteels, R., 1977, Electro- and pharmacomechanical coupling in the smooth muscle cells of the rabbit ear artery, J. Gen. Physiol. 70:129-148.

Egleme, C., Godfraind, T., and Miller, R. C., 1984, Enhanced responsiveness of rat isolated aorta to clonidine after removal of the endothelium cells, Br. J. Pharmac. 81:016-018.

Evans, D. H. L., Shild, H. O., and Thesleff, S., 1958, Effects of drugs on depolarized plain muscle, J. Physiol. 143:474-485.

Funaki, S., 1960, Electrical activity of single vascular smooth muscle fibers, in: "Electrical Activity of Single Cells," Igakushoin, Tokyo.

Furchgott, R. F., and Zawadzki, J. V., 1980, The obligatory of endothelian cells in the relaxation of arterial smooth muscle of acetylcholine, Nature 288:373-376.

Gabella, G., 1981, Structure of smooth muscle, in: "Smooth Muscle", E. Bulbring, A. F. Brading, A. W. Jones, and T. Tomita, eds, pp. 1-46, Edward Arnold, London.

Griffith, T. M., Henderson A. H., Edwards, D., and Lewis, M. J., 1984, Isolated perfused rabbit coronary artery and aortic strip preparations: the role of endothelium-derived relaxant factor, J. Physiol. 351:13-24.

Hamill, O. P., Marty, A., Neher, E., Sakmann, B., and Sigworth, F. J., 1981, Improved patch-clamp techniques for high-resolution current recording from cells and cell-free membrane patches, Pflugers Arch. 391:85-100.

Hinke, J. A. M., 1965, Calcium requirements for noradrenaline and high potassium ion contraction in arterial smooth muscle, in: "Muscle," W. M. Paul, E. E. Daniel, C. M, Kay, and G. Monckton, eds., pp. 269-284, Pergamon Press.

Hirst, G. D. S., and Neild, T. O., 1978, An analysis of excitatory junctional potentials recorded from arterioles, J. Physiol. 280:87-104.

Hirst, G. D. S., and Neild, T. O., 1980a, Some properties of spontaneous excitatory junction potentials from arterioles of guinea pigs, J. Physiol. 303:43-60.

Hirst, G. D. S., and Neild, T: O., 1980b, Evidence for two populations of excitatory receptors for noradrenaline on arteriolar smooth muscle, Nature 283:5749-5750.

Hirst, G. D. S., and Neild, T. O., 1981, Localization of specialized noradrenaline receptors at neuromuscular junctions on arterioles of the guinea-pig, J. Physiol. 313:343-350.

Hirst, G. D. S., Neild, T. O., and Silveberg, G. D., 1982, Noradrenaline receptors on the rat basilar artery, J. Physiol. 328:351-360.

Holman, M. E., 1958, Membrane potentials recorded with high-resistance micro-electrodes; and the effects of changes in ionic environment on the electrical and mechanical activity of the smooth muscle of the taenia coli of the guinea-pig, J. Physiol. 141:464-488.

Ito, Y., Suzuki, H., and Kuriyama, H., 1977, On the roles of calcium ion during potassium induced contracture in the smooth muscle cells of the rabbit main pulmonary artery, Jap. J. Physiol. 27:755-770.

Kajiwara, M., Kitamura, K. and Kuriyama, H., 1981, Neuromuscular transmission and smooth muscle membrane properties in the guinea-pig ear artery, J. Phsiol. 315:283-302.

Kawasaki, H., and Takasaki, K., 1984, Vasoconstrictor response induced by 5-hydroxytryptamine released from vascular adrenergic nerves by periarterial nerve stimulation, J. Pharmacol. Exp. Ther. 229:816-822.

Keatinge, W. R., 1966, Electrical and mechanical response of arteries to stimulation of sympathetic nerves, J. Physiol. 185:701-715.

Kou, K., Kuriyama, H., and Suzuki, H., 1982, Effects of 3,4-dihydro-8-(2-hydroxy-3-isopropylaminopropoxy)-3-nitroxy-2H-1-benzopyran (K-351) on smooth muscle cells and neuromuscular transmission in the canine mesenteric artery, Br. J. Pharmac. 77:679-689.

Kou, K., Ibengwe, J., and Suzuki, H., 1984, Effects of alpha-adrenoceptor antagonists on electrical and mechanical responses of the isolated dog mesenteric vein to perivascular nerve stimulation and exogenous noradrenaline, Naunyn-Schmiedeberg's Arch. Pharmacol. 326:7-13.

Leijten, P. A. A., and van Breemen, C., 1984, The effects of caffeine on the noradrenaline-sensitive calcium store in rabbit aorta, J. Physiol. 357:327-339.

Makita, Y., 1983, Effects of adrenoceptor agonists and antagonists on smooth muscle cells and neuromuscular transmission in the guinea-pig renal artery and vein, Br. J. Pharmac. 80:671-679.

Manzini, S., Maggi, C. A., and Meli, A., 1982, A simple procedure for assessing norepinephrine-induced cellular and extracellular Ca^{++} mobilization in rabbit ear artery, J. Pharmacol. Methods 8:047-057.

Medgett, I. C., and Langer, S. Z., 1984, Heterogeneity of smooth muscle alpha adrenoceptors in rat tail artery in vitro, J. Pharmacol. Exp. Ther. 229:823-830.

Mekata, F., and Keatinge, W. R., 1975, Electrical behaviour of inner and outer muscle of sheep carotid artery, Nature 258:534-535.

Meisheri, K. D., Palmer, R. F., and van Breemen, C., 1980, The effects of amrinone on contractility, Ca^{2+} uptake, and cAMP in smooth muscle, Europ. J. Pharmacol. 61:159.

Morgan, K. G., 1983, Comparsion of membrane electrical activity of cat gastric submucosal arterioles and venules, J. Physiol. 345:135-147.

Morgan, J. P., and Morgan, K. G., 1984a, Stimulus-specific patterns of intracellular calcium levels in smooth muscle of ferret portal vein, J. Physiol. 351:155-167.

Morgan, J. P., and Morgan, K. G., 1984b, Alteration of cytoplasmic ionized calcium levels in smooth muscle by vasodilators in the ferret, J. Physiol. 357:539-551.

Neild, T. O., and Zelcer, E., 1982, Noradrenergic neuromuscular transmission with special reference to arterial smooth muscle, Prog. Neurobiol. 19:141-158.

Richardson, K. C., 1958, Electronmicroscopic observations on auerbach's plexus in the rabbit, with special reference to the problem of smooth muscle innervation, Am. J. Anat. 103:99-136.

Saida, K., and van Breemen, C., 1983, A possible Ca^{2+}-induced Ca^{2+} release mechanism mediated by norepinephrine in vascular smooth muscle, Pflugers Arch. 397:166-167.

Saida, K., and van Breemen, C., 1984, Characteristics of the norepinephrine-sensitive Ca^{2+} store in vascular smooth muscle, Blood

Vessels 21:43–52.

Sneedon, P., and Burnstock, G., 1984, ATP as a co-transmitter in rat tail artery, Europ. J. Pharmacol. 106, 149–152.

Suzuki, H., 1981, Effects of endogenous and exogenous noradrenaline on the smooth muscle of guinea-pig mesenteric vein, J. Physiol. 321:495–512.

Suzuki, H., and Kou, K., 1983, Electrical components contributing to the nerve-mediated contractions in the smooth muscles of the rabit ear artery, Jap. J. Physiol. 33:743–756.

Tomita, T., 1980, Electrical properties of mammalian smooth muscle, in: "Smooth Muscle," E. Bulbring, A. F. Brading, A. W. Jones, and T. Tomita, eds., pp. 197–243, Edward Arnold, London.

Trapani, A., Matsuki, N., Abel, P. W., and Hermsmeyer, K., 1981, Norepinephrine produces tension through electromechanical coupling in rabbit ear artery, Europ. J. Pharmacol. 72:87–91.

van Breemen, C., Farinas, B. R., Casteels, R., Gerba, P., Wuytack, F., and Deth, R., 1973, Factors controlling cytoplasmic Ca^{2+} concentration, Phil. Trans. R Soc. Lond. B265:57–71.

van Breemen, C., Wuytack, F., and Casteels, R., 1975, Stimulation of ^{45}Ca efflux from smooth muscle cells by metabolic inhibition and high K depolarization, Pflugers Arch. 359:183–196.

van Breemen, C., Hwang, O., and Meisheri, K. D., 1981, The mechanisms of inhibitory action of diltiazem on vascular smooth muscle contractility, J. Pharmacol. Exp. Ther. 218:459–463.

Waugh, W. H., 1962, Role of calcium in contractile excitation of vascular smooth muscle by epinephrine and potassium, Circ. Res. XI:927–940.

Waugh, W. H., 1965, Calcium and contraction of arterial smooth muscle, in: "Muscle," Proc. of Symp. at Fac. of Med. Univ. of Alberta, June 1–4, 1964, W. M. Paul, E. E. Daniel, C. M. Kay, and G. Monkton, Pergamon Press.

West, T. C., and Landa, J., 1956, Transmembrane potentials and contractility in the pregnant rat uterus, Am. J. Physiol. 187, 333–337.

VASCULAR SMOOTH MUSCLE: STRUCTURE AND FUNCTION

Michael J. Mulvany

Biophysics Institute
Aarhus University
8000 Aarhus C.
Denmark

INTRODUCTION

The hemodynamic characteristics of the vasculature are determined
primarily by the vascular characteristic impedance, flow resistance, and
wall stiffness. The first is of most importance in the larger Windkessel
vessels and determines the pulse pressure for a given oscillatory flow.
The second concerns primarily the vessels responsible for flow
distribution (the small arteries and arterioles) and gives the steady flow
maintained by a given perfusion pressure. The third is of importance as
regards the capacitance vessels (the small veins) as well as the vessels
which determine the venous return pressure (the large veins) and gives the
change in transmural pressure for a given change in vascular volume.
These characteristics are functions of the vessel caliber and the elastic
modulus of the vascular wall (McDonald, 1974; Dobrin, 1978; Caro et al.,
1978; Gow, 1980; Mulvany, 1984b).

The components of the vascular wall which determine vascular caliber
and elastic modulus are the connective tissue and, if activated, the
vascular smooth muscle cells. It is the purpose of this review, first, to
discuss how these components exert their effects, and then to discuss some
of the mechanisms by which the activation level of the vascular smooth
muscle cells is controlled.

VASCULAR STRUCTURE

Despite the large variability in the size and function of blood
vessels, there is a remarkable similarity in their basic construction, the

main difference lying in the proportion of the different components. Most vessels consist of 4 main layers: the endothelium, the internal elastic lamina, the tunica media and the tunica adventitia (Fig. 1).

The endothelium consists of a monolayer of squamous cells in close apposition and joined by junctional complexes, but separated from the internal elastic lamina by the subendothelial space (Gabbiani and Majno, 1977). Previously it had been thought that the endothelium was primarily responsible for the regulation of the passage of plasma components into the vascular wall (Rhodin, 1980). However, it now appears that the endothelium can play an important role in the regulation of vascular tone through the release of an "endothelium dependent relaxing factor" (Furchgott and Zawadzki, 1980; Cocks and Angus, 1983). Disturbance of the endothelium also plays an important role in pathological states, particularly as regards its effect on blood platelets, but also in the development of atherosclerosis, where smooth muscle cells can migrate into the subendothelial space and proliferate (Schwartz and Ross, 1984). In dog saphenous vein, the internal elastic lamina consists of a single layer of elastic fibres, and are reported to have a "fishnet-like" arrangement around the periphery of the luminal surface (Crissman, 1984).

The tunica media contains the vascular smooth muscle cells as well as a large proportion of the connective tissue. The smooth muscle cells are in general spindle-shaped, although recent investigations show them to have many invaginations and processes (Todd et al.,1983), as in visceral smooth muscle (Gabella, 1976). The reported lengths vary from ca. 40 µm (Baandrup et al., 1985) in rat resistance vessels to over 250 µm in rabbit portal mesenteric vein (Berner et al., 1981). The diameter at the widest point, in the region of the nucleus, is 3–5 µm. The cells are in general circumferentially arranged (Canham and Mullin, 1978; Walmsley et al., 1982) but there is considerable variation in their orientation to either side of the circumferential direction (Todd et al., 1983). In some vessels (e.g. portal vein), there are layers of cells running parallel to the long axis of the vessel (Mulvany et al., 1980), while in other vessels (e.g. coronary resistance vessels, (N. Nyborg, personal communication) the pattern is quite irregular. Force is transmitted between the smooth muscle cells via intermediate junctions, having intercellular gaps of 40-60 nm and containing a single ill-defined layer of electron dense material (Gabella, 1984a). The smooth muscle cells are also electrically connected to form a syncytium through nexuses or gap junctions, where the

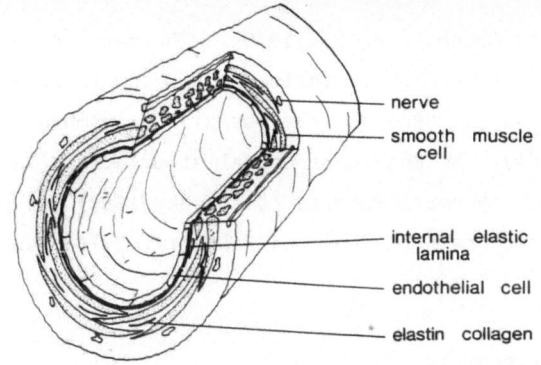

Fig. 1.   Sketch of a small artery (e.g. with 100 μm lumen diameter),
          showing principle components.

gaps are only 2-3 nm and contain small intercellular particles which
presumably provide the low resistance pathways (Litwin, 1980). The smooth
muscle cells are embedded in a network of connective tissue, consisting
mainly of elastin and collagen fibres (Cox, 1978; Greenwald and Berry,
1978). From work with visceral smooth muscle it seems that there may be
some connections between the smooth muscle cells and the elastin and
collagen fibres (Gabella, 1984b), but their function is not understood.

     In most vascular smooth muscle, the perivascular nerves are contained
primarily within the tunica adventitia, with the nerve endings from which
transmitter is released lying in the media-adventitia border (Burnstock et
al. 1984). Thus, in general, it is only the outer layer of smooth muscle
cells which is directly innervated. The cleft gap in the synapse varies
with the type of blood vessel, being several microns in the larger
arteries and under 100 nm in the resistance vasculature (Rowan and Bevan,
1983). Potential changes induced by nerve stimulation in the innervated
cells are transmitted to the non-innervated cells via the gap-junctions:
the space constant for vascular smooth muscle is typically of the order of
1.5 mm in e.g. arterioles (Hirst and Neild, 1978). The adventitia also
contains, besides connective tissue, many other cell types such as
fibroblasts and mast cells.

     All the major structural components of the vascular wall (i.e.
collagen, elastin and smooth muscle cells) are synthesized by smooth

muscle cells (Chamley-Campbell et al.,1979). The rate at which this synthesis takes place is under the influence of many factors of which the most important (Mulvany, 1984c) are the vascular transmural pressure (Uvelius et al., 1981), the degree of neural stimulation (Bevan and Tsuru, 1981), the presence of humoral factors (Overbeck, 1980) and probably intrinsic cellular factors (Friedman and Friedman, 1976; Kanbe et al., 1983).

SMOOTH MUSCLE CELL STRUCTURE

The force development of the vascular smooth muscle cells is a function of the interaction of the contractile apparatus and of the intracellular concentrations of the messengers which regulate this interaction. The most important of these messengers is calcium, although other factors are probably also involved (Morgan and Morgan, 1984). This section is concerned with a description of some of the main structures involved in the determination of intracellular calcium and force of contraction.

Plasma Membrane

As in other excitable cells, the potential across the plasma membrane plays an important role in the control of cytoplasmic calcium and hence vascular tone. The resting membrane potential of cells is a consequence of the activity of the Na, K-pump, and the presence of this mechanism in vascular smooth muscle is now well established (Wei et al., 1976; Bukoski et al., 1983; Adams et al., 1983; Aalkjaer and Mulvany, 1985). Modulation of the membrane potential is mediated primarily through a variety of specific channels in the plasma membrane, in particular potassium channels (Benham et al., 1985; Singer and Walsh, 1984) and apparently calcium channels (Harder and Sperelakis, 1979; Walsh and Singer, 1980; Su et al., 1984). The receptor control of these channels is considered in detail in the chapter by T. B. Bolton, elsewhere in this volume. Modulation of the membrane potential, at least acutely, can also be affected by alteration of the activity of the Na, K-pump (Hirst and van Helden, 1982; Aalkjaer and Mulvany, 1985).

Caveoli

The plasma membrane contains numerous invaginations, known as

86

caveoli, up to 50% of the cell surface being covered by these structures (Bevan and Verity, 1967). The caveoli are arranged in groups and in the smooth muscle of the mouse coronary artery they are reported to be some 70 nm across and 120 nm long (Forbes et al., 1979). Often, the caveoli appear connected to other vesicles deeper into the smooth muscle cells (Devine et al., 1972). The caveoli membrane forms an integral part of the plasma membrane, such that up to 40% of the smooth muscle membrane surface lies in the caveoli (Gabella, 1981). The role of caveoli is not clear. They do not seem to be affected by stretch (Gabella and Blundell, 1978), but it has been suggested that they might be the sites for stretch receptors (Prescott and Brightman, 1976; see below). Another suggestion is that they perform a function analogous with the T-tubule system of skeletal and cardiac muscle (Devine et al., 1972). This attractive possibility is supported by the fact that caveoli are much more numerous in smooth muscle than in other muscle types, and because they appear to be associated with the sarcoplasmic reticulum within the cells. Thus, as in other muscle types, the changes in membrane potential associated with activation could cause release of calcium from the sarcoplasmic reticulum through a mechanism depending on the close juxtaposition of an internalized plasma membrane, but further work is needed to test this possibility.

Sarcoplasmic Reticulum

The sarcoplasmic reticulum of smooth muscle occupies up to 7.5% of the cellular volume in large elastic arteries, but only in 2% in small arteries (Somlyo, 1978). As in other muscle types, the sarcoplasmic reticulum of smooth muscle is a store of calcium, the concentration in the sarcoplasmic tubules being at least 28 mmol/(kg dry weight) (Bond et al., 1984). Furthermore, although its capacity is quantitatively less than that of other intracellular organelles, such as the mitochondria, it appears that during activation, it is primarily calcium from the sarcoplasmic reticulum which is involved in raising the cytoplasmic calcium concentration (Somlyo et al., 1979).

Contractile Apparatus

In the past decade it has become increasingly clear that, even though smooth muscle lacks striations, the contractile apparatus has many similarities with that of other muscle types. Like skeletal muscle, for example, smooth muscle contains both actin and myosin, and both X-ray diffraction and electron microcope studies show that these proteins are arranged in filamentous form (Lowy et al., 1970; Shoenberg and Haselgrove, 1974; Somlyo et al., 1973). In vascular smooth muscle, the length of the myosin filaments is reported to be 2.2 μm (Ashton et al., 1975). These authors also provided evidence that the filaments are arranged in quasi-sarcomeres, with dense bodies performing the function of the Z-discs of skeletal muscle. Like Z-discs, these dense bodies consist of alpha-actinin, and using alpha-actinin antibodies, the length of the quasi-sarcomeres in visceral smooth muscle is estimated to be about 2 μm (Fay et al., 1984). The available evidence, therefore, suggests that filament and sarcomere lengths in smooth muscle are generally similar to that found in skeletal muscle. The quasi-sarcomeres are thought to be arranged helically, in chains, around the nucleus (Fig. 2). The ends of the chains are thought to be attached to the plasma membrane at so-called dense bands, which extend up to 1 μm into the cytoplasm (Gabella, 1981). These dense bands are associated with the intermediate junctions, to provide the mechanical connection between cells. That smooth muscle does not have striations is thus probably not due to lack of sarcomeres, but because the sarcomeres either are heterogeneous in length or because there is a lack of lateral register between sarcomeres in adjacent contractile units.

The fine structure of the actin filaments seems to be quite similar to that of the actin filaments of skeletal muscle, with the globular actin molecules arranged in a double helix (Vibert et al., 1972). The fine structure of the myosin filaments is still not however clear. Earlier evidence had suggested that at least under certain conditions (e.g. high osmolarity, Small and Squire, 1972), smooth muscle myosin can aggregate into "face-polarized" filaments, with the myosin molecules oriented in opposite directions on each side of the filament (Craig and Mergerman, 1977; Fig. 2). This matter is, however, still the subject of debate and it cannot be excluded that the myosin filaments are end-polarized as in skeletal muscle, the myosin molecules all pointing away from the mid-point of the filament. In any event, by analogy, it seems likely that the basis of contraction in vascular smooth muscle is similar to that found in skeletal muscle. Thus, to the extent that the crossbridge theory holds for skeletal muscle (Huxley, 1974), smooth muscle contraction is due to a cyclic interaction of the globular heads of the myosin molecules (the crossbridges) with the actin filaments.

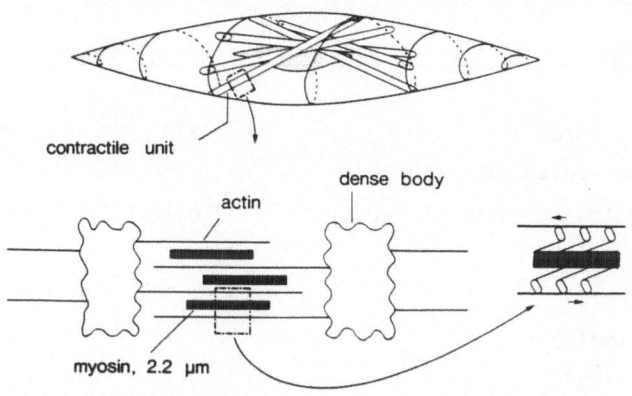

Fig. 2. Sketch showing possible arrangement of contractile apparatus in
vascular smooth muscle. From Mulvany (1984b).

In addition to the actin and myosin filaments, a third set of
filaments - the intermediate filaments - is seen. These have a diameter of
ca. 10 nm, and form the cytoskeleton. They probably serve to keep the
dense bodies in register (Ashton et al., 1975).

MECHANICAL PROPERTIES

Qualitatively the mechanical properties of vascular smooth muscle are
similar to those of other muscle types (Table 1). The force development
per unit cross-section of muscle is similar to that of skeletal muscle,
the main difference concerning the velocity of shortening which is much
slower. This decreased rate of shortening is matched by a much slower
metabolic rate.

Considering, as indicated above, the apparent structural similarity
of the contractile apparatus of smooth and skeletal muscle, and the
similarity of the ability to develop force, it is surprising that the
quantity of myosin in smooth muscle is only about 20% of that in skeletal
muscle (Table 1). The actin concentration is about twice that of skeletal
muscle, so that the actin:myosin ratio is about 10 times greater in smooth
muscle than in skeletal muscle, and this may be one of the reasons for the
greater force production per myosin molecule. The biochemical reactions
associated with the contractile cycle in smooth muscle are similar to that
of skeletal muscle (Marston and Taylor, 1980), the only substantial

Table 1

Characteristics of Smooth and Skeletal Muscle

	Smooth	Skeletal
Isometric force (mN/mm^2)	370 [1]	100–300 [4]
Corresponding metabolic rate (mW/g)	1.5 [2]	>180 [5]
Maximum contraction velocity (length/s)	0.12 [3]	>10 [4]
Actin content (mg/g)	42 [1]	22 [1]
Myosin content (mg/g)	13 [1]	62 [1]
Actin/myosin (mg/mg)	3.2 [1]	0.35 [1]

References

(1), Murphy et al., 1974;   (2), Herlihy and Murphy, 1974;   (3), Paul et
    al., 1976; all hog carotid artery, 37°C.

(4), Close, 1972   )   fast skeletal muscle

(5), Canfield, 1971)   corrected to 37°C

difference being that one of the reactions is much slower: that in which
products of ATP hydrolysis are released from the actomyosin complex.
Under these conditions, the fraction of myosin molecules which is attached
to the actin filaments can be shown to be alone due to the relative
concentration of actin to myosin. Therefore, in a muscle in which the
actin:myosin ratio is high, as in smooth muscle, a high proportion of the
myosin molecules will be attached and producing force. Thus, the
proportion of time for which smooth muscle myosin produces force may be
higher than in skeletal muscle, and this may account for the higher force
production per myosin molecule.

The relation between the force which a vascular smooth muscle can
develop and the degree of extension (the force–length characteristic) is
also similar to that of skeletal muscle (Fig. 3a). The shape of the
force–length characteristic in the unstimulated muscle is approximately
exponential. The active force–length curve is bell-shaped, having a length
(length L_o) at which maximum force is developed and falling to zero
tension at about $0.4 L_o$ and $1.8 L_o$, values remarkably similar to those
seen in skeletal muscle (e.g. Deleze, 1961; Gordon et al., 1966). The
relation between the rate at which a vascular smooth muscle can shorten
and the load to which it is subjected (the force–velocity characteristic)

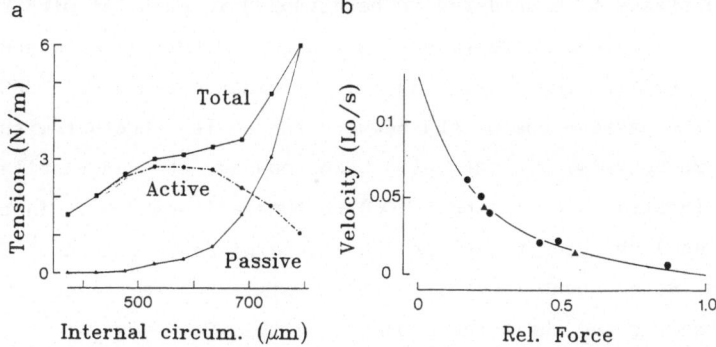

Fig. 3.   (a), force-length and (b), velocity of shortening-force
relations of rat mesenteric small arteries.   From Mulvany and
Warshaw (1979) and Mulvany (1979).

is hyperbolic (Fig. 3b), as for the corresponding characteristics of
skeletal muscle.   However, it should be noted that the rate of shortening
is very slow: only about 0.1 length/s for shortening under no-load
conditions and about 0.01 length/s when shortening against half-maximal
load (probably the least load to be expected under in vivo conditions).   A
discussion of the molecular basis of these mechanical properties is most
easily made with reference to a Maxwell model (Fig. 4), which as a number
of authors have shown (Dobrin, 1978; Mulvany and Warshaw, 1979) provides
an adequate description of vascular smooth muscle.   In this model, a
component (the parallel elastic component, PEC) representing the

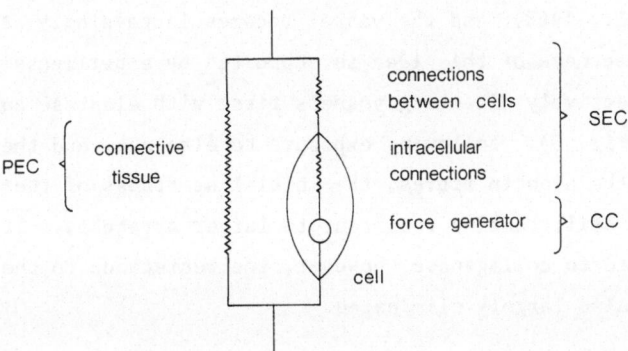

Fig. 4.   Maxwell model of vascular smooth muscle.

connective tissue is considered to be arranged in parallel with the series
connection of components representing the smooth muscle cells and the
connections between the cells. The series elastic component (SEC)
comprises the passive components between the cells, together with the
passive components within the cells which connect the contractile
apparatus (contractile component, CC) to the cell membrane. In the
relaxed vessel the properties are alone determined by the PEC. Upon
activation, at constant length, the force will thus be the sum of the
force in the PEC and the force produced by the CC.

One feature of this model is that the smooth muscle cell should
present no resistance to stretch under relaxed conditions. That this was
the case in visceral smooth muscle was demonstrated some years ago by Fay
(1975) using isolated cells. Recently the same has been demonstrated for
vascular smooth muscle cells by Van Dijk and colleagues (1984). In these
extremely fine experiments, isolated vascular smooth muscle cells – length
ca. 74 μm – were mounted on a myograph which would measure the force in
them when they were stretched. It was found that although they showed a
small resistance to fast stretch, they showed no measurable resistance to
slow stretch. Thus it would seem that the resting characteristic of whole
smooth muscle must be due entirely to extracellular structures.

The shape of the resting characteristic has been explained in terms
of the passive wall components being largely elastin and collagen. At
short lengths, the collagen fibres are unstretched and present no
resistance to stretch: the resting characteristic is thus thought to be
entirely that of the elastin fibres. Elastin fibres have a relatively low
elastic modulus (Young's modulus = 0.1 - 1.0 MPa, Carton et al., 1962) and
the vessel has a corresponding low stiffness. As the vessel is stretched
out, so are successively more collagen fibres recruited. The collagen
fibres have a much higher elastic modulus (Young's modulus = 100 MPa,
Benedict et al., 1968), and the vessel becomes increasingly stiffer. The
probable correctness of this idea is supported by experiments where the
effect of selectively digesting vessels first with elastase and then with
collagenase (Fig. 5). Following exposure to elastase, and the consequent
digestion of the elastin fibres, the initial stiffness of the vessel falls
markedly, but still remains resistant to larger stretches. If the vessel
is then exposed to collagenase, however, the resistance to the larger
stretches is also largely eliminated.

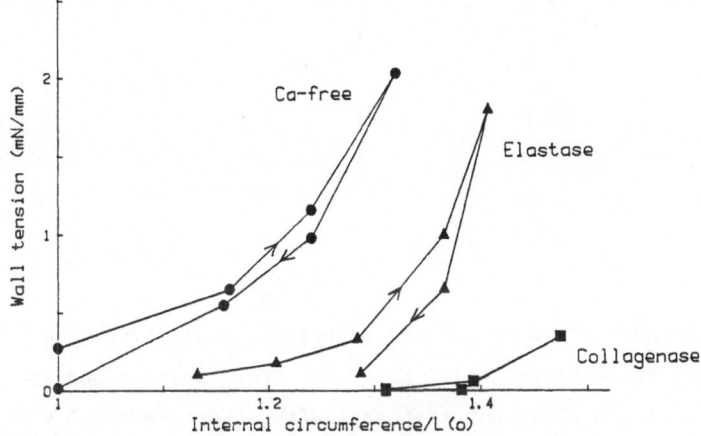

Fig. 5. Effect of enzymatic digestion on force-length characteristic of
an unstimulated mesenteric resistance vessel from the rat.
Left-hand curve shows stretch-release cycle in calcium-free
solution. Vessel then exposed to elastase (8 U/ml) for 10 min,
and stretch-release cycle repeated. Vessel then exposed to
collagenase (20 U/ml) for 10 min, and stretch-release cycle again
repeated. Mulvany, unpublished experiments.

        The characteristics of the SEC can be determined in experiments where
a vessel is first activated and then subjected to a rapid shortening, at a
rate fast enough to prevent substantial shortening of the contractile
apparatus during the release (Johansson et al., 1978; Mulvany, 1979; see
Fig. 6). The relation between the force at the end of the release and the
extent of the release then provides the undamped characteristic of the
SEC. From such experiments it is found that the force-length relation of
the SEC is approximtely exponential (Mulvany, 1984a) and rather larger
than that of skeletal muscle. Using a photoflash technique we established
that in resistance vessels this undamped characteristic of the SEC was
approximately equally divided between extracellular and intracellular
structures (Mulvany and Warshaw, 1981). The SEC is, however,
visco-elastic, such that sudden changes in load cause not only acute
changes in the length of the SEC, but also a more gradual creep. In
resistance vessel we have found that up to 0.3 s are required for this
creep to be completed (Fig. 6). It is not clear to what extent this creep
can be related to transients in the crossbridge cycle or to the passive
visco-elastic properties of the structures connecting the contractile

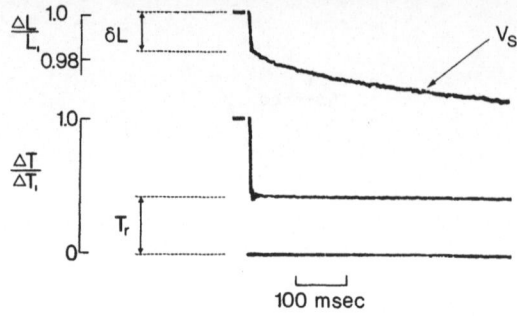

Fig. 6. Response of rat activated mesenteric small artery to a rapid (5
ms) decrease in load. Note that there is an immediate decrease
in length (dL), but that thereafter there is a 0.3 s period
before a steady shortening velocity (V_s) is seen. From Mulvany
(1979).

elements to the plasma membrane (Hellstrand and Arner, 1980; Warshaw and
Fay, 1983; Mulvany, 1984a).

HEMODYNAMIC EFFECTS OF ACTIVATION

Under isometric conditions, the effect of activation is to cause an
increase in the vascular wall force. Moreover, since – at least for rapid
length or load changes – the elastic modulus of active vascular smooth
muscle cells (Young's modulus up to 10 MPa, Warshaw et al., 1979) is much
greater than that of the vascular wall up to moderate extensions,
activation will also tend to increase the elastic modulus under isometric
conditions. However, for the in vivo situations, most blood vessels are
under approximately isobaric conditions. Here, therefore, activation will
normally cause a reduction in lumen diameter. Furthermore, since this
diameter reduction reduces the load supported by the parallel connective
tissue, the effect of activation is also to cause a reduction in the
elastic modulus (Gow, 1980; Mulvany, 1984b).

It was pointed out in the Introduction that the main hemodynamic
characteristics of interest are the characteristic impedance, the flow
resistance and the wall stiffness. As will be seen from Fig. 7, all these
characteristics are strongly dependent on the inverse of the lumen radius.
Indeed, this dependence is so strong that it exceeds the direct dependence
of the characteristic impedance and wall stiffness on the elastic modulus.

$$\text{Characteristic impedance} \quad Z = \frac{\tilde{P}}{\tilde{Q}} \propto \sqrt{\frac{E}{r^5}}$$

$$\text{Flow resistance} \quad R = \frac{\Delta p_l}{\dot{Q}} \propto \frac{1}{r^4}$$

$$\text{Wall stiffness} \quad k = \frac{d(\Delta p_t)}{dr_i/r_i} \propto \frac{E}{r^2}$$

Fig. 7. Hemodynamic characteristics of blood vessels. $\tilde{P}$, pulse pressure; $\tilde{Q}$, pulsatile flow; Δp_l, longitudinal pressure drop; $\dot{Q}$, steady flow; Δp_t, transmural pressure; $\underline{E}$, elastic modulus, $\underline{r}$, internal radius. For further details see Mulvany (1984b).

Thus even though under in vivo conditions activation causes a reduction in elastic modulus, the effect of activation is to cause an increase in all three hemodynamic parameters. That is, the effects of ativation will, for a given flow rate, increase the pulse pressure, increase the mean blood pressure and increase the venous return pressure. It is, therefore, clear that although vascular smooth muscle plays an important role in exercising control of the circulation, overactivity – be it due either to abnormally high activation levels or to abnormally high quantities of vascular smooth muscle – will have a detrimental effect on circulatory hemodynamics (Mulvany, 1984c).

REGULATION

Although smooth muscle thin filaments do, as in skeletal muscle contain tropomyosin, it is not generally thought that this exerts any regulatory function, as it does in skeletal muscle (Sobieszek, 1977; Hartshorne and Mrwa, 1982; Ruegg et al., 1984). Regulation of the interaction of myosin with actin in smooth muscle is instead mediated through the light chains which are positioned on the globular heads of the myosin molecule. In skeletal muscle, these light chains are thought to regulate the rate of interaction of myosin with actin, but in smooth muscle they appear to play the even more important role of determining whether there is any interaction at all. The light chain regulation is determined by the degree to which the light chains are phosphorylated, and light chain phosphorylation is effected by another enzyme, myosin light chain kinase (MLCK, see Fig. 8). The activity of this kinase is in turn under the influence of a number of factors, in particular, calmodulin and cyclic AMP-dependent protein kinase.

As in other cells, the cytoplasmic calcium concentration plays an important role in the determination of the level of activation. In smooth muscle it appears that this calcium sensitivity is mediated through the effect of calmodulin on MLCK being greatly enhanced if calcium is bound to the calmodulin (4 Ca^{++} per calmodulin). The cytoplasmic calcium concentration is, however, not the only modulator of myosin light chain phosphorylation. This is because the effectiveness of MLCK is in turn dependent on the degree to which it is itself phosphorylated, in that the phosphorylated form of MLCK is less active. Phosphorylation of MLCK can be affected by cyclic AMP-dependent protein kinase, and this provides a mechanism whereby the intracellular level of cyclic AMP can decrease the interaction of the contractile filaments.

These effects of intracellular calcium, calmodulin, cyclic AMP and cyclic AMP-dependent protein kinase have been demonstrated in "skinned" vascular smooth muscle preparations, where the plasma membrane has been made permeable by exposure to a detergent (Ruegg and Paul, 1982). In such experiments, calmodulin can be shown to enhance the calcium sensitivity such that half-maximal activation is obtained with about 0.1 μM. Furthermore, addition of cyclic AMP-dependent protein kinase in the presence of cyclic AMP causes a small reduction of the calcium sensitivity.

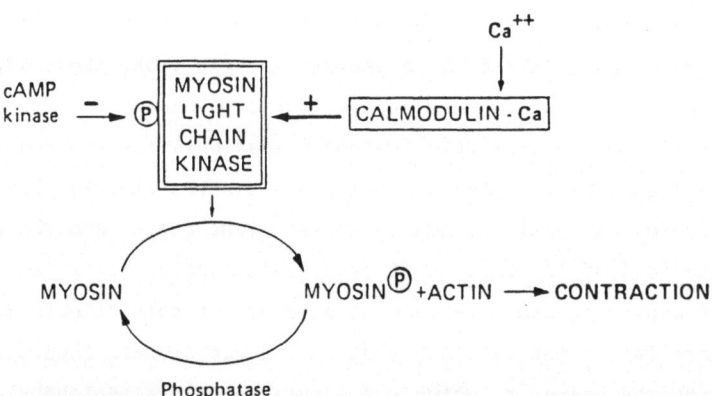

Fig. 8. Calcium regulation of smooth muscle contraction. From Ruegg et al. (1984).

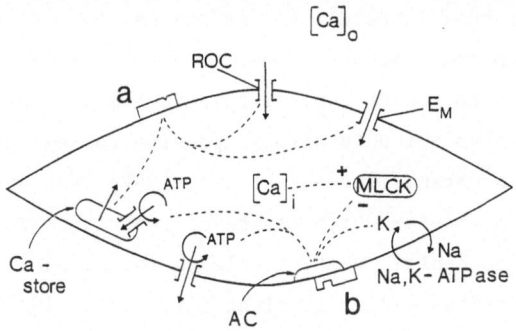

Fig. 9. Some mechanisms for hormonal control of vascular tone. Receptor
"a" at top could mediate constrictor responses; receptor "b" at
bottom could mediate dilator responses. Straight arrows indicate
calcium movement. AC, adenyl cyclase; E_m, potential dependent
channel; MLCK, myosin light chain kinase; ROC, receptor operated
channel.

CONTROL OF VASCULAR TONE

As described in the chapters by Drs. T.B. Bolton and R. Casteels in
this volume, the mechanisms which cause changes in vascular tone are
extremely complex, both as regards the cellular mechanisms and the
responses to extracellular influences. In this section, some of the basic
cellular mechanisms will first be discussed. Thereafter, as examples, the
effects of some extracellular influences on resistance vessels will be
described.

Cellular Mechanisms

As we have seen, the interaction of the contractile proteins is
dependent in part on the cytoplasmic calcium concentration and in part on
the level of other second messengers, such as cyclic AMP (Fig. 9).

The level of the cytoplasmic calcium is the result of the balance
between the influx of extracellular calcium, the release or uptake of
calcium into the intracellular stores (mainly the sarcoplasmic reticulum)
and the extrusion of calcium across the plasma membrane. The influx of
calcium is mediated by potential dependent channels (Bolton, 1979) and,
possibly, by receptor operated channels (ROCs). As indicated above,

release of calcium from intracellular stores may be mediated through some interaction between the membrane caveoli and the sarcoplasmic reticulum. Uptake of calcium into the sarcoplasmic reticulum is mediated by a Ca-ATPase, a mechanism which is also responsibe for extrusion of calcium across the plasma membrane (Wuytack et al., 1982; Kwan et al., 1983). Another extrusion mechanism which may be of importance in some vessels is a Na, Ca-exchange mechanism, whereby calcium is extruded using the energy of the inwardly directed sodium gradient (Blaustein, 1977). The importance of this mechanism, in peripheral vessels has, however, been questioned (Droogmans and Casteels, 1979; Hermsmeyer, 1983; Mulvany et al., 1984). In any event, increases in cytoplasmic calcium may be expected to increase the interaction of the contractile proteins through binding to calmodulin and hence affecting the phosphorylation level of myosin light chain kinase (MLCK, see above).

The control of the level of other second messengers is less understood, but it is for example known that stimulation of some receptors (e.g., beta-adrenergic receptors) can stimulate membrane bound adenyl cyclase and thus increase the synthesis of cyclic AMP (Kramer and Hardman, 1980). This increase in cyclic AMP can then cause a decrease in the interaction of the contractile proteins through its effect on MLCK (see above). There is also evidence that cyclic AMP can reduce cytoplasmic calcium either by the stimulation of calcium extrusion, or by stimulation of the Na, K-pump and hence hyperpolarization due, at least in part, to the electrogenicity of this pump (Hirst and van Helden, 1982; Aalkjaer and Mulvany, 1985).

We may now consider some of the ways in which extracellular influences can affect vascular tone. The discussion is confined to considering some effects of receptor activation, of alteration in the physical environment and of the myogenic response.

Receptor Activation

Activation of receptors through neural stimulation causes in most blood vessels a release of calcium from intracellular stores and a depolarization (either an action potential or a graded decrease in potential), and this is accompanied by a vasoconstriction. The action potential is probably due to the influx of calcium ions, not sodium as in other muscle types (Harder and Sperelakis, 1979). The effect of neural stimulation on rat mesenteric resistance vessels has been recently investigated (Nilsson et al., 1985). These authors found that with a constant rate of stimulation, frequencies of up to 10-15 Hz were required to obtain a maximal response. These frequencies are surprisingly high considering that the mean firing rate in autonomic nerves is probably much lower (Thoren and Ricksten, 1979). However, Nilsson and colleagues also found (Fig. 10) that the pattern of stimuli played an important role in the response of these small vessels. Thus, if vessels were stimulated at the normal (slow) firing rate, but with the usual irregular pattern of neural firing the vessels did respond, even though there was no response if the same rate of stimulation was given at a constant rate. This difference is presumably related to the mechanisms which reduce cytoplasmic calcium being slower than those which increase it. In any case, these findings clearly have important implications as regards the relation between the autonomic nervous system and the vascular response.

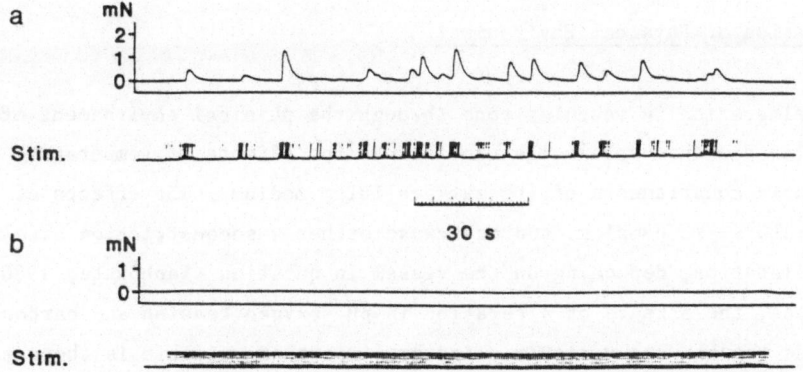

Fig. 10. Neural stimulation of a rat mesenteric small artery in vitro.
(a), effect of stimulation with pattern typically seen in autonomic nerves in vivo. (b), effect of constant frequency stimulation at a rate equal to the mean frequency used (a). From Nilsson et al. (1985).

Receptor activation with exogenous stimuli may or may not be accompanied by alterations in membrane potential. In some vessels, such as for example rabbit ear artery, the vasoconstrictor response to

exogenous noradrenaline produces no change in membrane potential
(Droogmans et al., 1977). In others, for example rat mesenteric
resistance vessels (see Fig. 11a), noradrenaline activation does produce
depolarization. Moreover, in these vessels, at least, it appears that the
depolarization is a potentiating factor as regards the vasoconstrictor
response, for the oscillatory mechanical response caused by noradrenaline
stimulation is accompanied by similar changes in the membrane potential
which precede the mechanical changes by about 1 s. This, therefore,
suggests that the electrical changes are responsible for the mechanical
changes. As another example of how mebrane potential can affect vascular
tone we can consider the effect of ouabain. Exposure of resistance
vessels to this cardiac glycoside causes potentiation of tone, and at the
same time a depolarization. Since there is an excellent temporal
correspondence between the depolarization and the vascular response (Fig.
11b), it has been concluded (Mulvany et al.,1984) that the vasoconstrictor
effect is due to the depolarization. Other hormones, however, cause
changes in the tone of small vessels without changes in membrane
potential. For example, atrial natriuretic factor causes a pronounced
relaxation of renal resistance vessels, but there is no change in their
membrane potential (Fig. 11c).

Alteration of Physical Environment

Alteration in vascular tone through the physical environment of the
vascular smooth muscle cells can include the effects of temperature and
the ionic constituents of the extracellular medium. The effects of
temperature are complex, and may cause either vasoconstriction or
vasodilatation, depending on the vessel in question (Vanhoutte, 1980).
Likewise, the effects of alteration in pH, oxygen tension and carbon
dioxide tension are variable. Another important variable is the potassium
concentration of the extracellular medium. For large increases in
extracellular potassium, such as can be applied in in vitro experiments
using potassium saline, the membrane will be depolarized according to the
diffusion potential equation, and vasoconstriction will ensue. For more
modest increases in the potassium concentration though, e.g. up to 6-8 mM
- such as for example can occur in the plasma perfusing muscular tissue
during activation - then vasodilatation can result. This vasodilatation
is thought to be primarily due to the raised potassium stimulating the Na,
K-pump, which due to its electrogenicity will then produce a
hyperpolarization (Webb and Bohr, 1978).

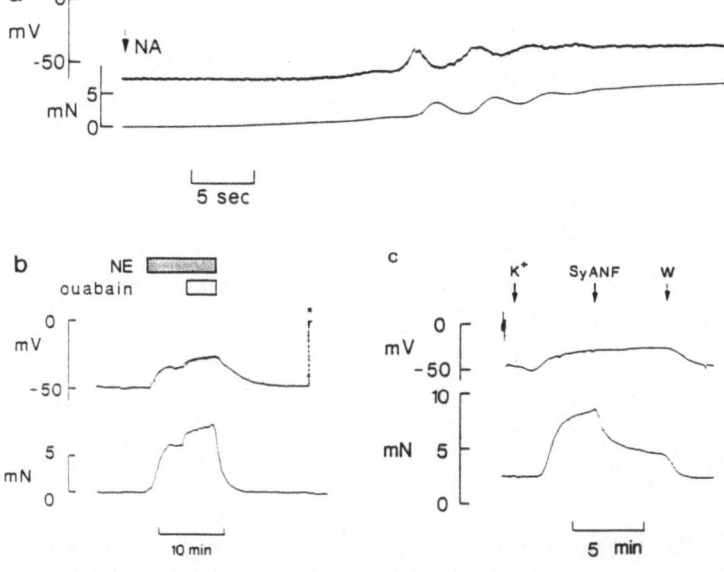

Fig. 11. Simultaneous measurements of membrane potential and wall tension. (a), rat mesenteric resistance vessel in response to noradrenaline (NA); Mulvany et al. (1982a). (b), rat mesenteric resistance vessel in response to ouabain (1 mM) in addition to noradrenaline (NE); Mulvany et al. (1982b). (c), rat renal resistance vessel in response to synthetic atrial natriuretic factor (0.2 μM, SyANF); Aalkjaer et al. (1985).

Myogenic Response

The myogenic response, first described by Bayliss (1902), concerns the vasoconstrictor response which is seen if the load on a vessel is increased. The autoregulatory response of the cerebral circulation is thought in large part to be due to myogenicity, and the phenomenon can be demonstrated in isolated cerebral resistance vessels (Halpern et al., 1984) as well as in other small vessels (Duling et al., 1981). If these vessels are cannulated and the intravascular pressure is raised, then the vessels respond by contracting (Fig. 12). In cannulated cat middle cerebral artery, increases in intravascular pressure are accompanied by a depolarization (Harder, 1984). It, therefore, seems likely that the myogenic response is associated with the opening of potential dependent

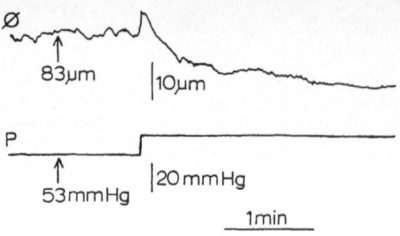

Fig. 12. Myogenic diameter change of a cannulated posterior cerebral
          artery in response to a pressure step from 53 to 70 mm Hg. Note
          vessel contracts in response to increased pressure. Halpern et
          al. (1984).

channels. The stimulus for the depolarization is still unknown, but it
has been suggested that the parameter of importance is the _force_ in the
muscle, not the length (Johnson, 1980): it is otherwise hard to explain
how the myogenic response can result in a vasoconstriction. One
possibility is that the sites of attachment of the contractile filaments
to the plasma membrane (the dense bands, see above) are associated with
plasma membrane channels which can open to allow the influx of some
cation, a suggestion in line with that of Prescott and Brightman (1976) as
mentioned above, concerning the possible presence of stretch receptors.

SUMMARY

     This review has been concerned with how the components of the
vascular wall affect the hemodynamic characteristics of the circulation.
The vascular components which affect these characteristics are primarily
the passive connective tissue (consisting of elastin and collagen) and the
active vascular smooth muscle cells. The activation level of the vascular
smooth muscle cells is dependent on the state of the contractile
apparatus. The contractile apparatus consists of actin and myosin
filaments, as in other muscle types, but it seems that there are
differences in the biochemistry of the contractile cycle, for the force
developed per myosin molecule is much greater than in other muscle types.
The regulatory system is also different. In vascular smooth muscle, the
calcium dependence of the contractile apparatus is mediated through myosin
light chain kinase which controls the state of phosphorylation of the
contractile apparatus. The state of phosphorylation can also be affected

by cyclic nucleotides. Receptor control of vascular tone is complex, and can be mediated, through alterations in the membrane potential, through release of calcium from intracellular stores and probably also through calcium-independent processes.

REFERENCES

Aalkjaer, C. and Mulvany, M.J., 1985, Effect of ouabain on tone, membrane potential and sodium efflux compared with ^{3}H-ouabain binding in rat resistance vessels, J. Physiol., 362:215-231.

Aalkjaer C., Mulvany, M.J. and Nyborg, N., 1985, Atrial natriuretic factor causes specific relaxation of rat renal arcuate arteries, Br. J. Pharmacol., In Press.

Adams, R.J., Wallick, E.T., Asno, G., Disalvo, J., Fondacaro,J.D. and Jacobsen, E.D., 1983, Canine mesentery artery Na$^+$, K$^+$-ATPase: Vasopressor receptor for digitalis?, J. Cardiovasc. Pharmacol., 5:468-482.

Ashton, F.T., Somlyo, A.V. and Somlyo, A.P., 1975, The contractile apparatus of vascular smooth muscle: Intermediate high voltage stereo electron microscopy, J. Molec. Biol. 98:17-29.

Baandrup, U., Gundersen, H.J.G. and Mulvany, M.J., 1985, Is it possible to solve the problem: Hypertrophy/hyperplasia of smooth muscle in the vessel wall of hypertensive subjects? Adv. Appl. Microcirc., 8:122-128.

Bayliss, W.M., 1902, On the local reactions of the arterial wall to changes of internal pressure, J. Physiol., 28:220-231.

Benedict, J.V., Walker, L.B. and Harris, E.H., 1968, Stress-strain characteristics and tensile strength of unembalmed human tendon, J. Biochem., 1:53-63.

Benham, C.D., Bolton, T.B., Lang, R.J. and Takewaki, T., 1985, The mechanism of action of barium and TEA on single calcium-activated potassium channels in arterial and intestinal smooth muscle cell membranes, Pflugers Archiv., 403:120-127.

Berner, P.F., Somlyo, A.V., Somlyo, A.P., 1981, Hypertrophy-induced increase of intermediate filaments in vascular smooth muscle, J. Cell Biol., 88:96-101.

Bevan, J.A. and Verity, M.A., 1967, Sympathetic nerve-free vascular muscle. J. Pharmacol. Exp. Therap., 157:117-124.

Bevan, R.D. and Tsuru, H., 1981, Functional and structural changes in the rabbit ear artery after sympathetic denervation. Circ. Res., 49:478-485.

Blaustein, M.P., 1977, Sodium ions, calcium ions, blood pressure regulation and hypertension: A reassessment and a hypothesis, Am. J. Physiol., 232:C165-C173.

Bolton, T.B., 1979, Mechanisms of action of transmitters and other substances on smooth muscle, Physiol. Rev., 59:606-718.

Bond, M., Kitazawa, T., Somlyo, A.P. and Somlyo, A.V., 1984, Release and recycling of calcium by the sarcoplasmic reticulum in guinea-pig portal vein smooth muscle, J. Physiol., 355:677-695.

Bukoski, R.D., Seidel, C.L. and Allen, J.C., 1983, Ouabain binding, Na^+-K^+-ATPase activity and ^{86}Rb uptake of canine arteries, Am. J. Physiol., 245:H604-H609.

Burnstock, G., Griffith, S.G. and Sneddon, P., 1984, Autonomic nerves in the precapillary vessel wall, J. Cardiovasc. Pharmacol., 6 (Suppl. 2):S344-S353.

Canfield, S.P., 1971, The mechanical properties and heat production of chicken latissimus dorsi muscles during tetanic contractions, J. Physiol., 219:281-302.

Canham, P.B. and Mullin, K., 1978, Orientation of medial smooth muscle in the wall of systemic muscular arteries, J. Microscopy, 114:307-318.

Caro, C.G., Pedley, T.J., Schroter, R.C. and Seed, W.A., 1978, The mechanics of the circulation, Oxford University Press, Oxford: pp. 1-527.

Carton, R.W., Dainauskas, J. and Clark, J.W., 1962, Elastic properties of single fibres, J. Appl. Physiol., 17:547-551.

Chamley-Campbell, J., Campbell, G.R. and Ross, R., 1979, Smooth muscle cell in culture, Physiol. Rev., 59:1-61.

Close, R.I., 1972, Dynamic properties of mammalian skeletal muscle, Physiol. Rev., 52:129-197.

Cocks, T.M. and Angus, J.A., 1983, Endothelium-dependent relaxation of coronary arteries by noradrenaline and serotinin, Nature, 305:627-629.

Cox, R.H., 1978, Passive mechanics and connective tissue composition of canine arteries, Am. J. Physiol., 234:H533-H541.

Craig, R. and Mergerman, J., 1977, Assembly of smooth muscle myosin into side-polar filaments. J. Cell. Biol., 75:990-996.

Crissman, R.S., 1984, The three-dimensional configuration of the elastic fiber network in canine saphenous vein, Blood Vessels, 21:156-170.

Deleze, J.B., 1961, The mechanical properties of the semitendinosus muscle at lengths greater than its length in the body, J. Physiol., 158:154-164.

Devine, C.E., Somlyo, A.V. and Somlyo, A.P., 1972, Sarcoplasmic reticulum and excitation-contraction coupling in mammalian smooth muscle, J. Cell. Biol., 52:690-718.

Dobrin, P.B., 1978, Mechanical properties of arteries, Physiol. Rev., 58:397-460.

Droogmans, G. and Casteels, R., 1979, Sodium and calcium interactions in vascular smooth muscle cells of the rabbit ear artery, J. Gen. Physiol., 74:57-70.

Droogmans, G., Raeymakers, L. and Casteels, R., 1977, electro- and pharmacomechanical coupling in the smooth muscle cells of the rabbit ear artery, J. Gen. Physiol., 70:129-148.

Duling, B.R., Gore, R.W., Dacey, R.G. and Damon, D.N., 1981, Methods for isolation, cannulation and in vitro study of single microvessels, Am. J. Physiol., 241:H108-H116.

Fay, F.S., 1975, Mechanical properties of single isolated smooth muscle cells, INSERM, 50:327-342.

Fay, F.S., Fogarty, K. and Fujiwara, K., 1984, The organization of the contractile apparatus in single isolated smooth muscle cells, in: "Smooth Muscle Contraction," N.L. Stephens (ed.), Marcel Dekker, pp. 75-90, New York.

Forbes, M.S., Rennels, M.L., and Nelson, E., 1979, Caveolar system and sarcoplasmic reticulum in coronary smooth muscle cells, J. Ultrastruct. Res., 67:325-339.

Friedman, S.M. and Friedman, C.L., 1976, Cell permeability sodium transport and the hypertensive process in the rat, Circ. Res., 39:433-440.

Furchgott, R.F. and Zawadzki, J.V., 1980, The obligatory role of endothelial cells in the relaxation of arterial smooth muscle by acetycholine, Nature, 288:373-376.

Gabbiani, G. and Majno, G., 1977, Fine structure of the endothelium. In: "Microcirculation", G. Kaley and B.M. Altura (eds.), pp. 133-144, Univ. Park, Baltimore.

Gabella, G., 1976, Quantitative morphological study of smooth muscle cells of the guinea-pig taenia coli, Cell Tissue Res., 170:161-186.

Gabella, G., 1981, Structure of smooth muscles. in: "Smooth Muscle; An Assessment of Current Knowledge," Bulbring, E., et al. (eds.), pp. 1-46, Arnold, London.

Gabella, G., 1984a, Structural apparatus for force transmission in smooth muscles, Physiol. Rev., 64:455-477.

Gabella, G., 1984b, Smooth muscle cell membrane and allied structures. in: "Smooth Muscle Contraction," N.L. Stephens (ed.), pp. 21-46, Marcel Dekker, New York.

Gabella, G. and Blundell, D., 1978, Effect of stretch and contraction on caveoli of smooth muscle cells, Cell Tissue Res., 190:255-271.

Gordon, A.M., Huxley, A.F. and Julian, F.J., 1966, The variation in isometric tension with sarcomere length in vertebrate muscle fibres, J. Physiol., 184:170-192.

Gow, B.S., 1980, Circulatory correlates: vascular impedance, resistance and capacity, in: Handbook of Physiology, Section 2, Volume 2, Vascular Smooth Muscle, D.F. Bohr, A.P. Somlyo, H.V. Sparks (eds.), pp. 353-408, American Physiological Society, Bethesda.

Greenwald, S.E. and Berry, C.L., 1978, Static mechanical properties and chemial composition of the aorta of spontaneously hypertensive rats: A comparison with the effects of induced hypertension, Cardiovasc. Res., 12:364-372.

Halpern, W., Osol, G. and Coy, G.S., 1984, Mechanical behaviour of pressurized in vitro prearteriolar vessels determined with a video-system, Annals Biomed. Eng., In Press.

Harder, D.R., 1984, Pressure-dependent membrane depolarization in cat middle cerebral artery, Circ. Res., 55:197-202.

Harder, D.R. and Sperelakis, N., 1978, Action potentials induced in guinea pig arterial smooth muscle by tetraethylammonium, Am. J. Physiol., 237:C75-C80.

Hartshorne, D.J. and Mrwa, U., 1982, Regulation of smooth muscle actomyosin, Blood Vessels, 19:1-18.

Hellstrand, P. and Arner, A., 1980, Contraction of the rat portal vein in hypertonic and isotonic medium: Mechanical properties and effects of magnesium, Acta Physiol. Scand., 110:59-69.

Herlihy, J.T. and Murphy, R.A., 1974, Force-velocity and series elastic characteristics of smooth muscle from hog carotid artery, Circ. Res., 34:461-466.

Hermsmeyer, K., 1983, Sodium pump hyperpolarization-relaxation in rat caudal artery, Fedn. Proc., 42:246-252.

Hirst, G.D.S. and Neild, T.O., 1978, An analysis of excitatory junction potentials recorded from arterioles, J. Physiol., 280:87-104.

Hirst, G.D.S. and van Helden, D.F., 1982, Ionic basis of the resting potential of submucosal arterioles in the ileum of the guinea-pig, J. Physiol., 333:53-67.

Huxley, A.F., 1974, Muscular contraction, J. Physiol., 243:1-44.

Johansson, B., Hellstrand, P. and Uvelius, B., 1978, Responses of smooth muscle to quick load change at high time resolution, Blood Vessels, 15:65-82.

Johnson, P.C., 1980, The myogenic response in: Handbook of Physiology,

Section 2, Volume 2, Vascular Smooth Muscle, D.F. Bohr, A.P. Somlyo, H.V. Sparks (eds.), pp. 475-514, American Physiological Society, Bethesda.

Kanbe, T., Nara, Y., Tagami, M. and Yamori, Y., 1983, Studies on hypertension-induced vascular hypertrophy in cultured smooth muscle cells from spontaneously hypertensive rats, Hypertension, 5:887-892.

Kramer, G.L. and Hardman, J.G., 1980, Cyclic nucleotides and blood vessel contraction in: Handbook of Physiology, Section 2, Volume 2, Vascular Smooth Muscle, D.G.Bohr, A.P. Somlyo, H.V. Sparks (eds.), pp. 179-200, American Physiological Society, Bethesda.

Kwan, C.Y., Triggle, C.R., Grover, A.K., Lee, R.M.K.W., Daniel, E.E., 1983, An analytical approach to the preparation and characterization of subcellular membranes from canine mesenteric arteries, Preparative Biochem., 13:275-314.

Litwin, W.J., 1980, Cell membrane features of rabbit arterial smooth muscle, Cell Tissue Res., 212:341-350.

Lowy, J., Poulsen, F.R. and Viebert, P.J., 1970, Myosin filaments in vertebrate smooth muscle, Nature, 225:1053-1054.

McDonald, D.A., 1974, Blood Flow in Arteries, London; Arnold, pp. 1-496.

Marston, S.B. and Taylor, E.W., 1980, Comparison of the myosin and actomysin ATPase mechanisms of the four types of vertebrate muscles, J. Molec. Biol., 139:573-600.

Morgan, J.P. and Morgan, K.G., 1984, Stimulus specific patterns of intracellular calcium levels in smooth muscle of ferret portal vein, J. Physiol., 351:155-167.

Mulvany, M.J., 1979, The undamped and damped series elastic components of a vascular smooth muscle, Biophys. J., 26:401-413.

Mulvany, M.J., 1984a, Crossbridges in smooth muscle, in: "Smooth Muscle Contraction," N.L. Stephens (ed.), Marcel Dekker, pp. 145-150, New York.

Mulvany, M.J., 1984b, Determinants of vascular hemodynamic characteristics, Hypertension 6 (Suppl. 3):III-13 - III-18.

Mulvany, M.J., 1984c, Pathophysiology of vascular smooth muscle in hypertension, J. Hypertension 2 (Suppl. 3):413-420.

Mulvany, M.J., Aalkjaer, C. and Petersen, T.T., 1984, Intracellular sodium, membrane potential and contractility in rat mesenteric small arteries, Circ. Res., 54:740-749.

Mulvany, M.J., Ljung, B., Stoltze, M. and Kjellstedt, A., 1980, Contractile and morphological properties of the portal vein in spontaneously hypertensive and Wistar-Kyoto rats, Blood Vessels, 17:202-215.

Mulvany, M.J., Nilsson, H. and Flatman, J.A., 1982a, Role of membrane potential in the response of rat small mesenteric arteries to exogenous noradrenaline stimulation, J. Physiol., 332:363-373.

Mulvany, M.J., Nilsson, H., Flatman, J.A., Korsgaard, N., 1982b, Potentiating and depressive effects of ouabain and potassium-free solutions on rat mesenteric resistance vessels, Circ. Res., 51:514-524.

Mulvany, M.J. and Warshaw, D.M., 1979, The active tension-length curve of vascular smooth muscle related to its cellular components. J. Gen. Physiol., 74:85-104.

Mulvany, M.J. and Warshaw, D.M., 1981, The anatomic location of the series elastic component in rat vascular smooth muscle, J. Physiol., 314:321-330.

Murphy, R.A., Herlihy, J.T. and Mergerman, J., 1974, Force-generating capacity and contractile protein content of arterial smooth muscle, J. Gen. Physiol., 64:691-705.

Nilsson, H., Ljung, B., Sjoblom, N. and Wallin, B.G., 1985, The influence of the sympathetic impulse pattern on contractile responses of rat mesenteric arteries and veins, Acta Physiol. Scand., 123:303-309.

Overbeck, H.W., 1980, Pressure-independent increases in vascular resistance in hypertension: Role of sympathoadrenergic influences, Hypertension 2:780-786.

Paul, R.J., Gluck, E. and Ruegg, J.C., 1976, Crossbridge ATP utilization in arterial smooth muscle, Pflugers Archiv., 361:297-300.

Prescott, L. and Brightman, M.W., 1976, The sarcolemma of Aplaysia smooth muscle in freeze-fracture preparations, Tissue and Cell, 8:241-258.

Rhodin, J.A.G., 1980, Architecture of the vessel wall, in: Handbook of Physiology, Section 2, Volume 2, Vascular Smooth Muscle, D.F. Bohr, A.P. Somlyo, H.V. Sparks (eds.), pp. 1-31, American Physioloical Society, Bethesda.

Rowan, R.A. and Bevan, J.A., 1983, Distribution of adrenergic synaptic cleft width in vascular and non-vascular smooth muscle, in: "Vascular Neuroeffector Mechanisms," J.A. Bevan, et al. (eds.), pp.75-83, Raven Press, New York.

Ruegg, J.C. and Paul, R.J., 1982, Vascular smooth muscle, calmodulin and cyclic AMP-dependent protein kinase alter calcium sensitivity in porcine carotid skinned fibres, Circ. Res., 50:394-399.

Ruegg, J.C., Sparrow, M.P., Mrwa, U., Schneider, M. and Pfitzer, G., 1984, Calcium and calmodulin dependent regulatory mechanisms in chemically skinned smooth muscle, in: "Smooth Muscle Contraction," N.L. Stephens (ed.), Marcel Dekker, pp. 361-372, New York.

Schwartz, S.M. and Ross, R., 1984, Cellular proliferation in atherosclerosis and hypertension, Prog. Cardiovasc. Diseases, 26:355-372.

Shoenberg, C.F. and Haselgrove, J.C., 1974, Filaments and ribbons in
   vertebrate smooth muscle, Nature, 249:152-154.

Singer, J.J. and Walsh, J.V., 1984, Large conductance calcium-activated
   potassium channels in smooth muscle membrane. Biophys. J.,
   45:68-70.

Small, J.V. and Squire, J.M., 1972, Structural basis of contraction in
   vertebrate smooth muscle, J. Molec. Biol., 67: 117-149.

Sobieszek, A., 1977, Vertebrate smooth muscle myosin: Enzymatic and
   structural properties, in: The Biochemistry of Smooth Muscle," N.L.
   Stephens (ed.), pp. 413-443, Univ. Park, Baltimore.

Somlyo, A.P., 1978, The role of organelles in regulating cytoplasmic
   calcium in vascular smooth muscle, in: "Mechanisms of
   Vasodilatation," P.M. Vanhoutte, and I. Leusen (eds.), pp. 21-29,
   Karger, Basel.

Somlyo, A.P., Devine, C.E., Somlyo, A.V. and Rice, R.V., 1973, Filament
   organization in vertebrate smooth muscle, Phil. Trans. Roy. Soc.
   Lond. B, 265:223-229.

Somlyo, A.P., Somlyo, A.V. and Shuman, H., 1979, Electron probe analysis
   of vascular smooth muscle. Composition of mitochondria, nuclei and
   cytoplasm, J. Cell. Biol., 81:316-335.

Su, C.M., Swamy, V.C. and Triggle, D.J., 1984, Calcium channel activation
   in vascular smooth muscle by Bay K 8644, Can. J. Physiol.
   Pharmacol., 62, 1401-1410.

Thoren, P. and Ricksten, S.E., 1979, Recordings of renal and splanchnic
   sympathetic nervous activity in normotensive and spontaneously
   hypertensive rats, Clin. Sci., 57:197s-199s.

Todd, M.E., Laye, C.G. and Osborne, D.N., 1983, Dimensional characteristics of smooth muscle in rat blood vessels: A computer-assisted
   analysis, Circ. Res., 53:319-331.

Uvelius, B., Arner, A. and Johansson, B., 1981, Structural and mechanical
   alteration in hypertrophic venous smooth muscle, Acta Physiol.
   Scand., 112:463-471.

Van Dijk, A.M., Wieringa, P.A., Van Meer, M. and Laird, J.D., 1984,
   Mechanics of resting isolated single vascular smooth muscle cells
   from bovine coronary artery, Am. J. Physiol., 246:C277-C287.

Vanhoutte, P.M., 1980, Physical factors of regulation in: Handbook of
   Physiology, Section 2, Volume 2, Vascular Smooth Muscle, D.F. Bohr,
   A.P. Somlyo, H.V. Sparks (eds.), pp. 443-4745, Amerian Physiological
   Society, Bethesda.

Vibert, P.J., Haselgrove, J.C., Lowy, J. and Poulsen, F.R., 1972, Structural changes in actin-containing filaments of muscle, _J. Molec. Biol._, 71:757-767.

Walmsley, J.G., Gore, R.W., Dacey, R.G., Damon, D.N. and Duling, B.R., 1982, Quantitative morphology of arterioles from the hamster cheek pouch related to mechanical analysis, _Microvasc. Res._, 24:249-71.

Walsh, J.V. and Singer, J.J., 1980, Calcium action potentials in single freshly isolated smooth muscle cells, _Am. J. Physiol._, 239:C162-C174.

Warshaw, D.M. and Fay, F.S., 1983, Crossbridge elasticity in single smooth muscle cells, _J. Gen. Physiol._, 82:157-199.

Warshaw, D.M., Mulvany, M.J. and Halpern, W., 1979, Mechanical and morphological properties of arterial resistance vessels in young and old spontaneously hypertensive rats, _Circ. Res._, 45:250-259.

Webb, R.C. and Bohr, D.F., 1978, Potassium-induced relaxation as an indicator of Na^+-K^+-ATPase activity in vascular smooth muscle, _Blood Vessels_, 15:198-207.

Wei, J.W., Janis, R.A. and Daniel, E.E., 1976, Studies on subcellular fractions from mesenteric arteries of spontaneously hypertensive rats: Alterations in both calcium uptake and enzyme activities, _Blood Vessels_, 13:293-308.

Wuytack, F., Raeymakers, L., De Schutter, G. and Casteels, R., 1982, Demonstration of the phosphorylated intermediates of the Ca^{2+}-transport ATPase in a microsomal fraction and in a (Ca^{2+} + Mg^{2+})-ATPase purified from smooth muscle by means of a calmodulin affinity chromatography, _Biochim. Biophys. Acta_, 693:45-52.

LOCAL MODULATION OF NORADRENERGIC NEUROTRANSMISSION IN BLOOD VESSELS OF

NORMOTENSIVE AND HYPERTENSIVE ANIMALS

Thomas C. Westfall, Shi-Qing Zhang, Suzanne Carpentier,
Linda Naes and Michael J. Meldrum

Department of Pharmacology
St. Louis University School of Medicine
1402 South Grand Boulevard
St. Louis, MO 63104

INTRODUCTION

Our concepts of adrenoceptors and their anatomical location has
progressively changed since the initial classification by Ahlquist (1948)
(see Table 1). Since that time evidence has been provided for the concept
of subtypes of β_1- and β_2-adrenoceptors, Lands et al. (1967a,b); the
discovery of presynaptic α - and β-adrenoceptors, Farnebo and Hamberger
(1971); Kirpekar and Puig (1971); Langer et al. (1971); Starke (1971);
Adler-Graschinsky and Langer (1975); Westfall et al. (1979), with the
presynaptic α-adrenoceptor given the designation of α_2 and the postsynaptic adrenoceptor α_1, Langer (1974), the discovery of postsynaptic
α_2-adrenoceptors, Drew and Whiting (1979); Docherty et al. (1979) and
Langer and Hicks (1984) and finally evidence for presynaptic α_1-adreno-
ceptors, Docherty (1984). Therefore, at the present state of our
knowledge, the situation appears to be quite complex with a lack of

TABLE 1

Evolution of Adrenoceptors

- Classification into α and β (1948)
- Subclassification of β-adrenoceptor into β_1 and β_2 (1967)
- Subclassification of α-adrenoceptors into α_1 (post) and α_2 (pre) (1974)
- Evidence for prejunctional α-adrenoceptors (70's)
- Evidence for postjunctional α_2-adrenoceptors (late 70's)
- Evidence for prejunctional α_1-adrenoceptors (early to mid-80's)

anatomical specificity. The precise physiological, pathophysiological and pharmacological importance of the presynaptic and postsynaptic location of the various adrenoceptors is unclear and varies largely depending upon the neuroeffector junction in question. A further complication is the presence of various peptides which may be co-localized with norepinephrine in peripheral noradrenergic neurons and modulate presynaptic or postsynaptic function, Hokfelt et al. (1980).

In addition to our understanding of the anatomical localization and function of adrenoceptors, there have also been tremendous advances in recent years in our knowledge of the events that influence the release of norepinephrine from noradrenergic neurons. As just mentioned, there is convincing evidence for the existence of receptors on terminal varicosities which when activated lead to an increase or a decrease in the evoked release of norepinephrine. In the present article, a brief discussion of the evidence for the existence of presynaptic receptors, especially α_2-adrenoceptors will be presented. Attention will also be given to the role of these receptors pharmacologically and physiologically as well as a discussion of possible mechanisms by which activation of α_2-adrenoceptors leads to a decrease in norepinephrine release. Secondly, a discussion of possible alterations in certain presynaptic receptors or receptor function in hypertension will be discussed.

Presynaptic Modulation of Noradrenergic Neurotransmission

It is now well-established that neural and hormonal substances can influence the quantitative release of norepinephrine during nerve stimulation by an action on receptors located on terminal varicosities. Substances reported to decrease adrenergic neurotransmission include α_2-adrenoceptor agonists (including norepinephrine itself), purines, such as ATP and adenosine, prostaglandins of the E series, acetylcholine via muscarinic receptors, dopamine, serotonin and opioids. The newest type of substance that can modulate norepinephrine release are peptides such as neuropeptide Y. Fig. 1 shows the results of experiments demonstrating that neuropeptide Y possibly co-localized with norepinephrine in noradrenergic neurons can decrease the release of endogenous norepinephrine from the perfused mesenteric artery of the rat. Substances facilitating adrenergic neurotransmission include β-adrenoceptor agonists, acetylcholine via nicotinic receptors, angiotensin prostaglandins of the F series and thromboxane. The presynaptic receptor system that has received the greatest attention and the one that may be the most important physio-

112

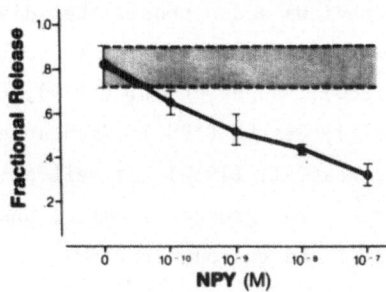

Fig. 1:   The effects of neuropeptide Y (NPY) on the periarterial nerve
stimulation induced release of norepinephrine from the perfused
mesenteric arterial vascular bed. Data are plotted as the
fractional release of endogenous norepinephrine versus
concentration of NPY in M. Periarterial nerves were stimulated
for 30 sec at a frequency of 16 Hz (supramaximal voltage, 2 msec
duration) at 15 min intervals. Each point represents the mean f
S.E.M. of at least 5 animals. Fractional release equals the
amount of norepinephrine (in ng) released in the medium, divided
by the amount of norepinephrine present in the tissue prior to
nerve stimulation. The amount of norepinephrine in the tissue at
the start of any time interval was calculated by adding cumulatively the amount of norepinephrine released into the medium to
the amount of norepinephrine in the mesenteric arterial bed at
the end of the experiment.

logically is the α_2-adrenoceptor. There is some controversy as to the
physiological role of the α_2-adrenoceptor as well as to some of the
details of the operation of this system, but it has been hypothesized that
presynaptic α_2-adrenoceptors act to operate a negative feedback system,
whereby activation of presynaptic α_2-adrenoceptors inhibits norepinephrine
release and thus tends to regulate the secretion of the transmitter. This
has lead to open discussions in the literature of the arguments for and
against the hypothesis, Angus and Korner (1981); Kalsner (1982); Langer
(1981); Rand et al. (1982); Westfall (1984); Kalsner (1984); Majewski et
al. (1983); Limberger and Starke (1984), and has led to several debates
(Fed. Proc. 1984; International Pharmacol. Congress, 1984).

There are numerous reviews and symposia that discuss the evidence for presynaptic modulation of noradrenergic transmission, Gillispie (1980); Langer (1977); Langer (1980); Langer et al. (1979); Patel et al. (1981); Paton (1979); Starke (1977); Starke (1981); Vanhoutte et al. (1981); Vizi (1979); Westfall (1977); Westfall (1980), as well as the controversies concerning their existence. For greater details, the reader should consult one or another of these thorough reviews.

Evidence for Presynaptic α_2-Adrenoceptors Modulation of Noradrenergic Neurotransmission

The evidence for the existence of presynaptic α_2-adrenoceptors comes largely from experiments utilizing α_2-adrenoceptor agonists and antagonists. This evidence has been recently reviewed by Westfall (1977); Westfall (1980); Starke (1977); Starke (1981); Langer (1980). The use of α-adrenoceptor agonists can be summarized as follows: 1) a wide range of α-adrenoceptor agonists decrease the evoked release of norepinephrine in the vast majority of noradrenergically innervated tissues; 2) inhibition of the evoked release of norepinephrine by α-adrenoceptor agonists is seen irrespective of the postsynaptic nature of the receptor in the effector cell; 3) α-adrenoceptor antagonists block or reduce the inhibitory effects on norepinephrine release produced by α-adrenoceptor agonists and 4) inactive enantiomers such as (+) norepinephrine fail to depress the evoked release of the neurotransmitter.

Evidence obtained by the use of α-adrenoceptor antagonists can be summarized as follows: 1) in contrast to what is seen with α-adrenoceptor agonists, α-adrenoceptor antagonists enhance the stimulation evoked release of norepinephrine in the majority of noradrenergically innervated tissues; 2) similar to what is seen with agonists, enhancement of the evoked release of norepinephrine by α-adrenoceptor antagonists is seen irrespective of the nature of the postsynaptic receptor of the effector cells; 3) enhancement of the evoked release by α-adrenoceptor antagonists is seen in the presence of norepinephrine uptake blockers and appears unrelated to uptake blockade; 4) α-adrenoceptor agonists abolish the facilitation of the evoked release of norepinephrine seen with α-adrenoceptor antagonists and 5) receptor protection experiments (with antagonists) are consistent with the general hypothesis.

In addition to the evidence provided by the use of α-agonists and antagonists, there is considerable evidence that the α-adrenoceptors are

located on noradrenergic nerve varicosities. For instance: 1)
enhancement or inhibition of noradrenergic transmission is independent of
the α or β type of the postsynaptic adrenoceptors that mediates the response
of the effector cell; 2) enhancement of the stimulation-evoked release of
norepinephrine by α-adrenoceptor antagonists is not affected by atrophy of
the postsynaptic cell (i.e., duct ligation of the rat submaxillary gland,
Filanger et al. (1978); 3) enhancement of the stimulation-evoked release
of norepinephrine by α-adrenoceptor antagonists is seen in recently formed
nerve endings in cultured rat superior cervical ganglia, Vogel et al.
(1972); 4) α-adrenoceptor agonists and antagonists influence the evoked
release from norepinephrine from synaptosomes, Starke (1979) and 5) there
is a reduction in the binding of ^{3}H- αligands to various membranes after
degeneration of noradrenergic terminals, Story et al. (1979).

Pharmacological and Physiological Role of Presynaptic Alpha Adrenoceptors in Influencing Noradrenergic Transmission

The pharmacological importance of presynaptic α_2-adrenoceptors
appears well accepted, Langer (1980); Starke (1981); Westfall (1984).
Similar statements could also be made for agonists and antagonists of the
other types of receptors that have been identified in noradrenergic
varicosities. Activation or antagonism of these receptors appears to
underlie the mechanism of action of numerous drugs. The physiological
role of presynaptic α_2-adrenoceptors on the other hand is less clearly
defined and obviously more difficult to establish. Despite this
uncertainty, there is considerable evidence that presynaptic
α_2-adrenoceptors exert an important physiological role at the
noradrenergic neuroeffector function. Several lines of evidence suggest
that activation of presynaptic α_2-adrenoceptors play a role in the
regulation or modulation of norepinephrine release at rest or during the
period of increased nerve activity. For instance: 1) α -adrenoceptor
antagonists enhance both transmitter release and effector organ response
in some cases, Starke et al. (1975); Langer et al. (1971); 2) similarly,
α-adrenoceptor agonists decrease both transmitter release and effector
organ response, Sax and Westfall (1981); 3) there are numerous in vivo
studies that are consistent with presynaptic α -adrenoceptor modulation of
noradrenergic neurotransmission. Armstrong and Boura (1973); Robson and
Antonaccio (1974); Scriabine and Stavorski (1973); Steppelar et al.
(1978); Madjar (1980); Cavero et al. (1979); Graham et al. (1980);
Yamaguchi et al. (1977); 4) a threshold concentration of norepinephrine or
other α-adrenoceptor agonists for activation of presynaptic

α-adrenoceptors appears to exist and is consistent with the presynaptic receptor hypothesis, Cubeddu and Weiner (1975); Dixon et al. (1979); Enero and Langer (1973); 5) presynaptic α_2-adrenoceptors have been demonstrated in human blood vessels, Stjarne and Brundin (1977); 6) modulation of norepinephrine release by presynaptic α_2-agonists is seen only for calcium dependent processes of transmitter release, Langer (1977); Langer (1981); Starke (1977); Starke (1981); Westfall (1977); Westfall (1980).

Although the exact physiological role of presynaptic α_2-adrenoceptors in unclear, several possibilities have recently been proposed, Westfall (1984). The system may operate the pulse-to-pulse release of norepinephrine during nerve stimulations. Such regulation would be similar to the original hypothesis of the role of presynaptic α-adrenoceptors. Second, presynaptic α-adrenoceptors may act to modulate the amount of transmitter released, perhaps in conjunction with neuropeptide Y (see Fig. 1). The system may operate only at periods of rest or during periods in which the frequency of nerve stimulation is very low. In this case, the system would operate to prevent undue stimulation of the neuroeffector cell from excessive stimulation by norepinephrine. Finally, activation of presynaptic α_2-adrenoceptors might act as a physiological antagonist to the facilitation of norepinephrine release reported to occur upon intensive nerve stimulation.

As mentioned above, in addition to presynaptic α_2-adrenoceptors, there are numerous other release modulatory substances that can increase or decrease the evoked release of norepinephrine from noradrenergic nerve varicosities. These substances are summarized in Table 2. Although the pharmacological role of these various presynaptic receptors can be appreciated, the physiological role for most is still unclear.

Mechanism of α_2-Adrenoceptor Inhibition of Noradrenergic Transmission

The mechanisms(s) by which activation of presynaptic α_2-adrenoceptors leads to inhibition of norepinephrine release is also unclear. There are several possible mechanisms that have received experimental support, and these are summarized in Table 3. The possibility exists that α-adrenoceptor agonists produce a blockade of action potential propagation by increasing potassium permeability, Morita and North (1981); Blakely et al. (1982); Cunnane and Stjarne (1982). In addition to a propagation block or an indirect modulation of axoplasmic Ca^{2+} levels, activation of presynaptic α_2-adrenoceptors may also produce a decrease in the

TABLE 2

Summary of Presynaptic Receptors on Noradrenergic Receptors

Inhibitory Receptors:	Facilitatory Receptors:
α-Adrenoceptor	β-Adrenoceptor
Dopamine	Serotonin
Serotonin	Nicotine-cholinergic
Histamine	GABA
Muscarinic-cholinergic	Angiotensin II
GABA	
Opiates	
Neuropeptide Y	
Prostaglandins	
Adenosine, ATP	

availability of extracellular calcium for neurosecretion coupling. There could be a decrease in Ca^{2+} entry through voltage sensitive channels or a decrease in calcium entry through an effect on the calcium channel. Evidence exists for both possibilities, Vizi (1985); Illes and Dorge (1985).

At least two-second messenger systems have been implicated in the mechanism by which activation of presynaptic receptors results in a decrease in the release of norepinephrine. There is evidence that activation of α_2-adrenoceptors decreases the activity of adenylate cyclase resulting in a reduction of intraneuronal cAMP concentration, Mulder et al. (1984). An alternate mechanism is that activation of presynaptic

TABLE 3

Possible Mechanism of α_2-Adrenergic Inhibition of
Neuroeffector Transmission

- Blockade of action potential propagation by increasing potassium permeability
- Decrease in Ca^{2+} entry through voltage sensitive channels
- Decrease in Ca^{2+} entry through an effect in the calcium channel
- Stimulation of Na-K-ATPase
- Inhibition of cAMP

α_2-adrenoceptors results in a stimulation of Na-K-ATPase activity in the adrenergic neuron resulting in a decrease in transmitter release, Schaefer et al. (1972); Vizi (1978); Powis (1981).

Despite these various attractive mechanisms, it is safe to say that the precise mechanism by which activation of presynaptic α_2-adrenoceptors results in a decrease in norepinephrine release is still unclear.

Pathophysiological Role of Presynaptic Receptors

In addition to the pharmacological and physiological role of presynaptic receptors, a question of great importance is whether or not they play a pathophysiological role in contributing to the mechanism of various disease states. There is evidence that there may be alterations in various presynaptic receptor systems contributing to enhanced release of norepinephrine in experimental hypertension.

Evidence for Increased Release of Norepinephrine from Blood Vessels of Hypertensive Animals

There is now considerable evidence for a greater evoked release of norepinephrine from blood vessels of genetically hypertensive rats compared to normotensive controls, such as the spontaneously hypertensive rats (SHR) thought by many to be a good model of essential hypertension in humans. This has been seen both in vivo and in vitro. It has been observed that sympathetic nerve stimulation results in an enhancement of vasoconstriction with or without a parallel increase in the response to exogenously administered norepinephrine. This has been demonstrated in young SHR, Bunag and Takeda (1979); Collis and Vanhoutte (1978); Collis et al. (1979,1980) rabbits made hypertensive by partial constriction of the abdominal aorta above the kidneys, Bevan et al. (1975); the DAHL-salt sensitive genetic model of hypertension Takeshita and Mark (1978); one clip-one kidney or two-kidney renal hypertension in the dog, Brody et al. (1970); Zimmerman et al. (1969); and one clip-one kidney hypertension in the rat, Dargie et al. (1976). Such a situation is consistent with the increased release of norepinephrine from adrenergic nerve endings. There is also evidence of an increase in plasma catecholamines and dopamine-β-hydroxylase in young SHR, Palermo et al. (1981); Affara et al. (1980); Nagatsu et al. (1974) as well as in other models of hypertension, Zimmerman (1983); Zimmerman et al. (1969).

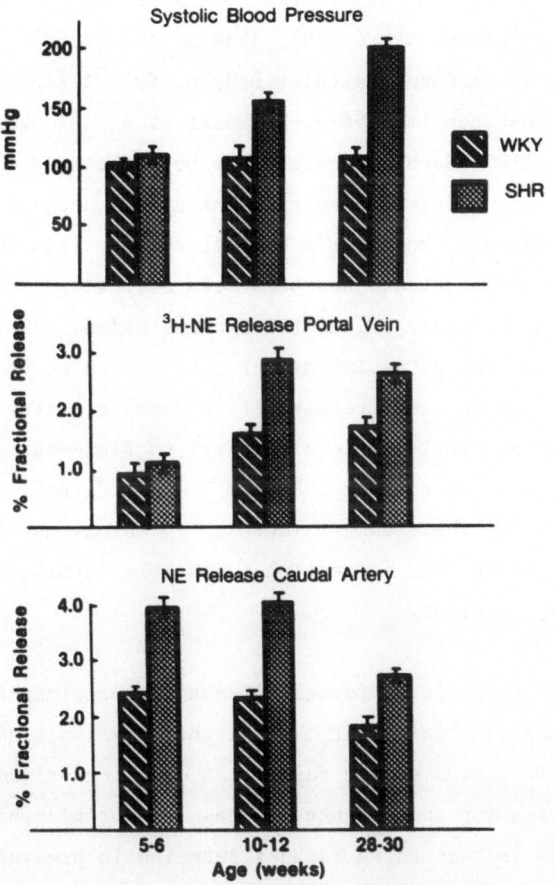

Fig. 2: The field stimulation induced release of ^{3}H-norepinephrine (portal vein) or endogenous norepinephrine (caudal artery) from these blood vessels obtained from SHR or WKY at 5-6, 10-12 or 28-30 weeks of age. Systolic blood pressure (measured by the tail cuff method) of animals of the three ages is also included. Data are plotted as % fractional release of ^{3}H-norepinephrine, endogenous norepinephrine a blood pressure (mmHg) versus ages of the animals in weeks. There was a significantly greater release of norepinephrine to field stimulation (10 Hz) of the caudal artery of the SHR compared to WKY at all three ages while there was a greater release of ^{3}H-norepinephrine from the portal vein of SHR compared to WKY at 10-12 and 28-30 weeks of age. Blood pressure was significantly elevated in the SHR at 10-12 and 28-30 weeks of age but not 5-6 weeks of age.

A direct demonstration of enhanced release of norepinephrine has been observed in studies utilizing isolated blood vessels, Galloway and Westfall (1982); Westfall et al. (1984a,b); Zsoter et al. (1982), isolated perfused organs, Collis et al. (1979); Ekas et al. (1982); Ekas et al. (1983a) and isolated perfused vascular beds of SHR, Eikenberg et al. (1981); Ekas and Lokhandwala (1981); Kawasaki et al. (1982a) compared to normotensive controls. Such increases have been observed where both ^{3}H-norepinephrine and endogenous norepinephrine are used as the markers for transmitter release. Moreover a greater release from the blood vessels of hypertensive animals has been seen following various types of stimulation, including field stimulation, nerve stimulation, depolarization with high potassium or the application of veratradine. Administration of the enhanced release of H-norepinephrine (portal vein) and endogenous norepinephrine (caudal artery) to field-stimulation of these blood vessels obtained from SHR or WKY of different ages is summarized in Fig. 2. The greatest increase in norepinephrine release occurs in pre- or young SHR, suggesting a possible involvement in the development of hypertension.

The fact that there is increased release of norepinephrine from isolated blood vessels obtained from hypertensive animals suggests that there may be an alteration at the vascular neuroeffector junction. One possible explanation for the enhanced release of norepinephrine from blood vessels of the SHR is that there is an alteration in presynaptic receptor function. An enhanced release of norepinephrine could be produced by either an enhancement of presynaptic facilitory function or a decreased inhibitory function.

Presynaptic Modulation of Norepinephrine Release from Blood Vessels of Hypertensive Animals

An examination of whether there are alterations in the activity of various release modulatory substances in hypertension has been undertaken by various laboratories as a possible explanation for the increased release of norepinephrine from the blood vessels of hypertensive animals. Results of many of these studies is summarized in Table 4, which has been adapted from Westfall and Meldrum (1985).

Angiotensin is known to enhance the evoked release of norepinephrine from noradrenergic nerve varicosities, Zimmerman (1978); Zimmerman and

TABLE 4

The Role of Presynaptic Receptors on Noradrenergic Neurotransmission in Blood Vessels of Hypertensive Animals

Receptor Type	Type of Hypertension	Blood Vessel or Vascular Bed	Observations	References
Angiotensin	Genetic (SHR)	Caudal Artery	5-6-week, SHR = WKY; 8-10-week, SHR > WKY; 28-30-week, SHR > WKY	Westfall et al., 1984b; Westfall et al., 1985
Angiotensin	Genetic (SHR)	Portal Vein	5-6 week, SHR = WKY; 8-10-week, SHR > WKY; 28-30-week, SHR > WKY	Westfall et al., 1984a;1984b
Angiotensin	Genetic (SHR)	Perfused Mesentery	18-week, SHR > WKY	Eikenberg et al., 1981
Angiotensin	Genetic (SHR)	Perfused Mesentery	15-16-week, SHR > WKY	Kawasaki et al,, 1982b
β-Adrenoceptor	Genetic (SHR)	Portal Vein	10-12-week, SHR > WKY; 28-30-week, SHR = WKY	Westfall et al., 1985b
β-Adrenoceptor	Genetic (SHR)	Perfused Kidney	14-week, SHR = WKY	Ekas et al., 1983b
β-Adrenoceptor	Genetic (SHR)	Perfused Mesentery	14-16-week, SHR > WKY	Kawasaki, 1982b
α_2-Adrenoceptor	Genetic (SHR)	Caudal Artery	5-6-week, SHR = WKY; 10-12-week, SHR = WKY; 28-30-week, SHR > WKY	Galloway and Westfall, 1982; Westfall et al., 1984a; 1984b
α_2-Adrenoceptor	Genetic (SHR)	Portal Vein	5-6-week, SHR = WKY; 10-12-week, SHR = WKY; 28-30-week, SHR > WKY	Westfall et al., 1984a; 1984b
α_2-Adrenoceptor	One Clip-One Kidney; DOCA-salt	Caudal Artery; Portal Vein	Hypertensive = Sham control	Westfall et al., 1984b, 1985b
α_2-Adrenoceptor	Neurogenic Baroceptor	Mesenteric Artery	Hypertensive = Sham control	Granata et al., 1983
α_2-Adrenoceptor	Genetic (SHR)	Perfused Mesentery	18-week, SHR = WKY	Ekas et al., 1983b; Su et al., 1982
α_2-Adrenoceptor	Genetic (SHR)	Perfused kidney	18 week, SHR < WKY	Ekas et al., 1982
Purine (Adenosine, ATP)	Genetic (SHR)	Perfused Mesentery	5-week, SHR < WKY; 15-18-week, SHR < WKY	Kamikawa et al., 1980; Kubo and Su, 1983a
Purine (Adenosine)	Renal Artery	Perfused Mesentery	15-18 week, SHR = WKY	Ekas et al., 1983a
Serotonin	Genetic (SHR)	Portal Vein	8-week, SHR = WKY	This paper
Serotonin	Genetic (SHR)	Perfused Mesentery	15-18-week, SHR < WKY	Kubo and Su, 1983b
Muscarinic	Genetic (SHR)	Portal Vein	8-10-week, SHR = WKY; 28-30-week, SHR = WKY	
PGE_2	Genetic (SHR)	Portal Vein	8-10-week, SHR = WKY	This paper

Whitmore (1967). Evidence for an increased activity of presynaptic
angiotensin receptor function in blood vessels of SHR has been obtained
from several laboratories as summarized in Table 4, Westfall et al.
(1984a,b); Westfall et al. (1985); Eikenberg et al. (1981); Kawasaki et
al. (1982).

It has been observed that angiotensin enhances the pressor response
of the mesenteric vascular bed to periarterial nerve stimulation to a
greater extent in SHR than it does in WKY, Kawasaki et al. (1982a). In
addition, the angiotensin response to periarterial nerve stimulation is
potentiated in the presence of cocaine. These results suggest that the
presynaptic facilitatory modulation of adrenergic vascular
neurotransmission mediated by angiotensin receptors is enhanced in the
perfused mesenteric vascular bed of SHR.

Other studies have shown that subpressor concentrations of
angiotensin potentiate the responses to nerve stimulation in SHR to a
greater extent than they do in WKY, Eikenberg et al. (1981). A direct
demonstration that angiotensin causes a greater enhancement of the field
stimulation-induced release of norepinephrine has been obtained in the
isolated superfused portal vein, Westfall et al. (1984a,b); Westfall et
al. (1985). In these studies, angiotensin was observed to enhance the
field stimulation-induced release of ^{3}H-norepinephrine from vessels
obtained from SHR to a greater extent than those of normotensive controls.
Enhancement of the facilitatory effect of angiotensin was apparent in SHR
at 10 weeks of age as it was in older animals (28-30 weeks of age).

Studies to determine whether or not there is an alteration in the
β-mediated enhancement of norepinephrine release from blood vessels of
hypertensive animals is unclear. There are studies that report no
difference in the β-mediated release of norepinephrine, Westfall et al.
(1984b; Ekas et al. (1983b), while others suggest there is an enhancement
of this response, Kawasaki et al. (1982b). Further studies are needed to
fully answer this question.

Several studies have been carried out to examine whether or not there
are changes in presynaptic inhibitory receptor function in blood vessels
of hypertensive animals. The selective prejunctional α_2-adrenoceptors
antagonist yohimbine has been used to evaluate whether or not changes in
the presynaptic α_2-adrenoceptors could help explain the increased field
stimulation-induced release of norepinephrine from the isolated portal

vein or caudal artery of SHR, Galloway and Westfall (1982); Westfall et al. (1984a,b). Yohimbine was seen to produce the same degree of enhancement in the evoked release of ^{3}H-norepinephrine and endogenous norepinephrine from SHR (at six and ten weeks of age), one clip-one kidney renal hypertensive rats or DOCA-salt hypertensive rats compared to their respective sham controls. However, the effect of yohimbine to enhance the overflow of norepinephrine was greatly attenuated when examined in 28-week-old SHR. The attenuation of the yohimbine effect was observed regardless of the way transmitter release was produced (field stimulation, potassium depolarization) and in both the caudal artery and the portal vein. These results are consistent with a decreased function activity of presynaptic α_2-adrenoceptors in mature SHR.

The results obtained in the portal vein and caudal artery of six- and ten-week-old SHR are consistent with the studies carried out in the mesenteric vascular bed in young SHR, which suggests that presynaptic α_2-adrenoceptor mediated inhibition is similar in SHR and age-matched WKY at these ages, Eikenberg et al. (1981); Su et al. (1982). These investigators did not examine presynaptic α_2-adrenoceptor function in older animals, so it is not known whether the results obtained in the caudal artery and portal vein in 28-week-old SHR represent only an age-related change or a difference in the response of different blood vessels.

Adrenergic neurotransmission has been shown to be decreased by purine compounds, such as adenosine and ATP, Su (1978); Moylan and Westfall (1979) which are presumably acting on purinergic receptors located in noradrenergic nerve terminals. It has been reported that both adenosine and ATP are less effective in inhibiting norepinephrine release from blood vessels of the SHR compared to WKY, Kamikawa et al. (1980); Kubo and Su (1983a). In contrast, a decrease in the adenosine effect was not seen in Wistar rats made hypertensive by left renal artery occlusion. Studies carried out in the perfused kidney are at variance with the observations in the mesenteric bed, Ekas et al. (1983a). In the latter studies, it was reported that adenosine was equally effective in causing an inhibition of the stimulation induced release of norepinephrine in both WKY and SHR.

Conflicting results have also been obtained for serotonin. It has been reported that serotonin was less effective in inhibiting the release of norepinephrine from the perfused mesenteric artery of 15-18-week-old SHR compared to WKY, Kubo and Su (1983b), while a similar inhibitory

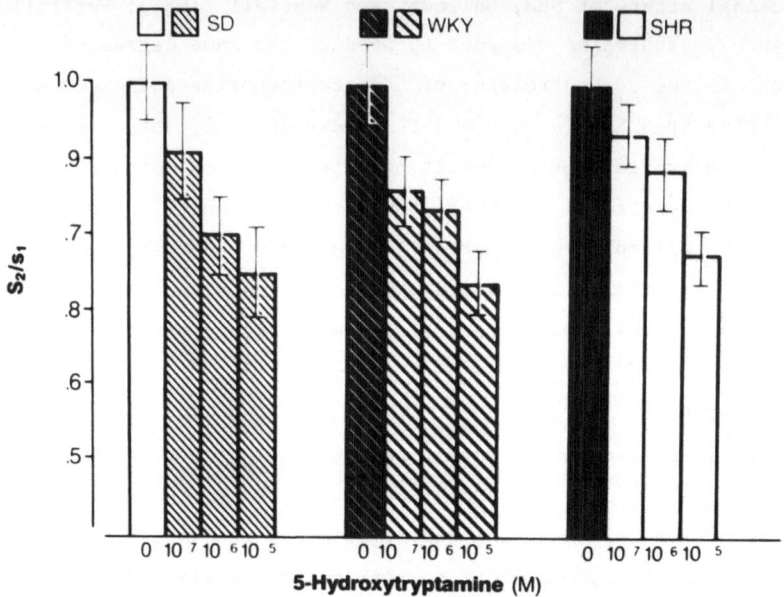

Fig. 3: Effect of 5-hydroxytryptamine (serotonin) on the field stimulation induced release of ^{3}H-norepinephrine from the portal vein obtained from Sprague-Dawley, WKY and SHR at 10-12 weeks of age. Data are plotted as a S_2/S_1 ratio depicting release in the absence (S_1) or presence (S_2) of increasing concentrations of 5-hydroxytryptamine in M. 5-hydroxytryptamine produced a similar concentration dependent inhibition of ^{3}H-norepinephrine release from the portal vein obtained from all three strains of rats.

effect has been observed from the portal vein of 8-10-week-old SHR compared to WKY (Fig. 3).

The ability of muscarinic-cholinergic agonists as well as prostaglandin E_2 to inhibit the field stimulation-induced release of norepinephrine from the portal vein of 8-10-week-old or 28-30-week-old SHR appears similar to that seen from normotensive controls (Fig. 4).

SUMMARY

This paper has provided evidence in support of the existence of prejunctional α-adrenoceptors on noradrenergic nerve terminals as well as the evidence for their physiological importance. The use of α-adrenoceptor agonists and antagonists has provided convincing data in

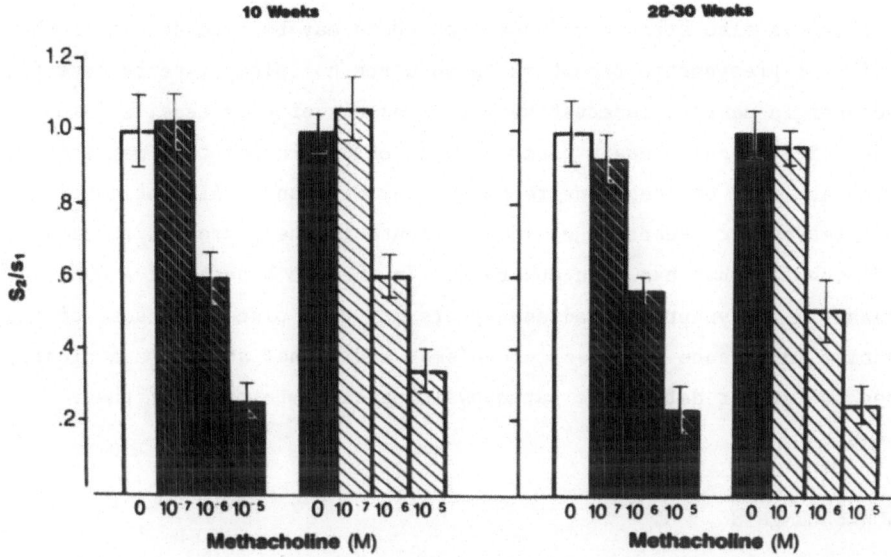

Fig. 4: The effect of the muscarinic agonists, methacholine on the field stimulation induced release of ^{3}H-norepinephrine from the portal vein obtained from 10-week-old or 28-30-week-old SHR or WKY. Data are plotted as a S_2/S_1 ratio depicting the evoked release in the absence (S_1) or presence (S_2) of increasing concentrations of methacholine. Methacholine produced a similar concentration dependent inhibition of the evoked release of ^{3}H-norepinephrine release from the portal vein obtained from both strains of rats.

support of the presynaptic receptor hypothesis. Moreover, there is ample evidence for the location of α-adrenoceptors on nerve terminals. This evidence has often been forgotten in arguments opposing the presynaptic α-adrenoceptor hypothesis. The precise physiological role of presynaptic α-adrenoceptors is still an open question, but there is support from a wide range of experiments in favor of a physiological role. Although it is not known which of these functions is most important, presynaptic α-adrenoceptors may: regulate the pulse-to-pulse regulation of norepinephrine release during nerve stimulation, prevent noise, and protect the neuroeffector cell from excessive activation by transmitter during periods of rest or as physiological antagonists to the facilitation of transmitter release. In summary, evidence reviewed here strongly supports the existence of pre-synaptic α-adrenoceptors. These receptors are clearly important pharmacologically and may play a physiological role in noradrenergic transmission. The exact physiological function must await further experimentation.

There is also strong evidence that there may be alterations in the activity of presynaptic receptors to modulate norepinephrine release from noradrenergic neurons innervating blood vessels of hypertensive animals. The activity of angiotensin receptors and of β-adrenoceptors may increase and the activity of adenosine receptors may decrease. Alterations in these presynaptic receptors could contribute to the increased release of norepinephrine that has been observed. In chronic hypertensive SHR, a decrease in presynaptic α_2-adrenoceptors may take place. Because of the potential importance of these mechanisms, additional studies are clearly needed to further define the pathophysiological importance of these systems.

ACKNOWLEDGEMENTS

This work was supported in part by USPHS grants NHLBI 26319 and NHLBI 35202.

REFERENCES

Adler-Graschinsky, E. and Langer, S.G. (1975): Possible role of a β-adrenoceptor in the regulation of noradrenaline release by nerve stimulation through a positive feedback mechanism, Brit. J. Pharmacol. 53:43-50.

Affara, S., Denoroy, L., Renaud, B., Vincent M. and Sassard, J. (1980): Serum dopamine β-hydroxylase activity in a new strain of spontaneously hypertensive rat, Experentia 36:1207-1208.

Ahlquist, R.P. (1948): A study of adrenotropic receptors, Am. J. Physiol. 153:586-600.

Angus, J.A. and Korner, P.I. (1981): Presence and physiological role of presynaptic inhibitory α-adrenoceptors in guinea pig atria, Nature (Lond) 294:671-672.

Armstrong, J.M. and Boura, A.L.A. (1973): Effect of clonidine and guanethidine on peripheral sympathetic nerve function in the pithed rat. Brit. J. Pharmacol. 47:850-852.

Bevan, R.D., Purdy, R.E., SU, C. and Bevan, J.A. (1975): Evidence for an increase in adrenergic nerve function in blood vessels from hypertensive rabbits, Circ. Res. 37:503-508.

Blakely, A.G.H., Cunnane, T.C. and Petersen, S.A. (1982): Local regulation of transmitter release from rodent sympathetic nerve terminals, J. Physiol. 325:93-109.

Brody, M.J., Dorr, L.D. and Shaffer, R.A. (1970): Reflex vasodilation
and sympathetic transmission in the renal hypertensive dog, Am. J.
Physiol. 219:1746-1750.

Bunag, R.D. and Takeda, K. (1979): Sympathetic hyperresponsiveness to
hypothalamic stimulation in young hypertensive rats, Am. J.
Physiol. 237R:39-44.

Cavero, I., Dennis, T., LeFivre-Borg, F., Perrot, P., Roach, A.G. and
Scatton, B. (1979): Effects of clonidine, prazosin and
phentolamine on heart rate and coronary sinus catecholamine
concentration during cardioaccelerator nerve stimulation in spinal
dogs, Br. J. Pharmacol. 67:283-292.

Collis, M.G. and Vanhoutte, P.M. (1978): Neuronal and vascular reactivity
in isolated perfused kidney during the development of spontaneous
hypertension, Clin. Sci. 55:233S-235S.

Collis, M.G., DeMey, C. and Vanhoutte, P.M. (1979): Enhanced release of
noradrenaline in the kidney of the young spontaneously hypertensive
rat, Clin. Sci. 57:233S-234S.

Collis, M.G., DeMey, C. and Vanhoutte, P.M. (1980): Renal vascular
reactivity in the young spontaneously hypertensive rat,
Hypertension 2:45-52.

Cubeddu, L.X. and Weiner, N. (1975): Nerve stimulation mediated overflow
of norepinephrine and dopamine β hydroxylase III. Effects of
norepinephrine depletion on the α-presynaptic regulation of
release. J. Pharmacol. Exp. Ther. 172:1-14.

Cunnane, T.C. and Stjarne, L. (1982): Secretion of transmitter from
individual varicosities of guinea pig and mouse vas deferens: all
or none and extremely intermittant, Neurosci. 7:2565-2576.

Dargie, H.J., Franklin, S.S. and Reid, J.L. (1976): The sympathetic
nervous system and neurovascular hypertension in the rat, Br. J.
Pharmacol. 56:365.

Dixon, W.F., Mosimann, W.F. and Weiner, N. (1979): The role of
presynaptic feedback mechanisms in regulation of norepinephrine
release by nerve stimulation, J. Pharmacol. Exp. Ther. 209:196-204.

Docherty, J.R., MacDonald, A. and McGrath, J.C. (1979): Further
classification of α-adrenoceptor in the cardiovascular system, vas
deferens and anococcygeus of the rat, Br. J. Pharmacol. 67:421-422.

Docherty, J.R. (1984): An investigation of presynaptic α-adrenoceptor
subtypes in the pithed heart and in the rat isolated vas deferens,
Br. J. Pharmacol. 82:015-023.

Drew, G.M. and Whiting, S.B. (1979): Evidence for two distinct types of postsynaptic α-adrenoceptor in vascular smooth muscle in vivo, Br. J. Pharmacol. 67:207-215.

Eikenberg, D.C., Ekas, R.D. and Lokhandwala, M.F. (1981): Presynaptic modulation of norepinephrine release by α-adrenoceptors and angiotensin II receptors in spontaneous rats. in: "Central Nervous System Mechanism in Hypertension," J.P. Buckley and C.M. Ferrario, eds., pp. 215-228, Raven, New York, NY.

Ekas, R.D. and Lokhandwala, M.F. (1981): Sympathetic nerve function and vascular reactivity in spontaneously hypertensvie rats, Am. J. Physiol. 241:R379-R384.

Ekas, R.D., Steenberg, M.L. and Lokhnadwala, M.F. (1982): Increased presynaptic α-adrenoceptor mediated regulation of noradrenaline release in the isolated perfused kidney of the spontaneously hypertensive rats, Clin. Sci. 63:309-311 (Suppl. 8).

Ekas, R.D., Steenberg, M.L. and Lokhandwala, M.F. (1983a): Increased norepinephrine release during sympathetic nerve stimulation and its inhibition by adenosine in the isolated perfused kidney of spontaneously hypertensive rats, Clin. Exp. Hyperten. A5:41-48.

Ekas, R.D., Steenberg, M.L., Woods, M.D. and Lokhandwala, M.F. (1983b): Presynaptic α and β-adrenoceptor stimulation and norepinephrine release in the spontaneously hypertensive rat, Hypertension 5:198-204.

Enero, M.A. and Langer, S.Z. (1973): Influence of reserpine induced depletion of noradrenaline on the negative feedback mechanism for transmitter release during nerve stimulation, Br. J. Pharmacol. 49:214-225.

Farnebo, L.O. and Hamberger, B. (1971): Drug induced changes in the release of ^{3}H-monoamines from field stimulated rat brain slices, Acta Physiol. Scand. Suppl. 371:35-44.

Filanger, E.J., Langer, S.Z. and Stefano, F.J.E. (1978): Evidence of the presynaptic location of the α-adrenoceptors which regulate norepinephrine release in the rat submaxillary gland, Naunyn-Schmiedeberg's Arch. Pharmacol. 304-21:26.

Galloway, M.P. and Westfall, T.C. (1982): The release of endogenous norepinephrine from the coccygeal artery of spontaneously hypertensive and Wistar-Kyoto rats, Circ. Res. 51:225-232.

Gillispie, J. (1980): Presynaptic receptors in the autonomic nervous system, Handb. Exp. Pharmacol. 54:169-205.

Graham, R.M., Stephenson, W.H. and Pettinger, N. (1980): Pharmacological evidence for a functional role of the prejunctional α-adrenoceptor in noradrenergic neurotransmission in the conscious rat, Arch. Pharmacol. 311:129-138.

Granata, A.R., Energ, M.A., Krieger, E.M. and Langer, S.Z. (1983):
Norepinephrine release and vascular response elicited by nerve
stimulation in rats with chronic neurogenic hypertension, J.
Pharmacol. Exp. Ther. 227:187-193.

Hokfelt, T., Johansson, D., Ljungdahl, A. Landberg, J.M. and Schaltzberg,
M. (1980): Peptidergic neurones, Nature 284:515-521.

Illes, P. and Dorge, L. (1985): Mechanism of α_2-adrenergic inhibition of
neuroeffector transmission in the mouse vas deferens,
Naunyn-Schmiedeberg's Arch. Pharmacol. 328:241-247.

Kalsner, S. (1982): The presynaptic receptor controversy, Trends in
Pharmacol. Sci. 3:11-16.

Kalsner, S. (1984): Limitations of presynaptic theory: No support for
feedback control of autonomic effectors, Fed. Proc. 43:1358-1364.

Kamikawa, Y., Cline, W.H., JR. and Su, C. (1980): Diminished purinergic
modulation of the vascular adrenergic neurotransmission in the
spontaneously hypeertensive rat, Eur. J. Pharmacol. 66:347-353.

Kawasaki, H., Cline, W.H., Jr. and Su, C. (1982a): Enhanced angiotensin
mediated facilitation of adrenergic neurotransmission in
spontaneously hypertensive rats, J. Pharmacol. Exp. Ther.
221:112-116.

Kawasaki, I.H., Cline, W.H., Jr. and Su, C. (1982b): Enhanced presynaptic
beta-adrenoceptor mediated modulation of vascular adrenergic
neurotransmission in spontaneously hypertensive rats, J. Pharmacol.
Exp. Ther. 223:721-728.

Kirpekar, S.M. and Puig, M. (1971): Effect of flow stop on noradrenaline
release from normal spleens and spleens treated with cocaine or
phenoxybenzamine, Br. J. Pharmacol. 43:359-363.

Kubo, T. and Su, C. (1983a): Effects of adenosine on ^{3}H-norepinephrine
release from the perfused mesentric arteries of SHR and renal
hypertensive rats, Eur. J. Pharmacol. 87:349-352.

Kubo, T. and Su, C. (1983b): Effects of serotonin and some other
neurohumoral agents on adrenergic neurotransmission in
spontaneously hypertensive rat vasculature, Clin. Exp. Hyper.
A5:1501-1510.

Lands, A.M., Arnold, A., McAuliff, J.P., Luduena, F.P. and Brown, T.G.,
Jr., (1967a): Differentiation of receptor systems activated by
sympathomimetic amines, Nature 214:597-598.

Lands, A.M., Luduena, F.P. and Buzzo, H.J. (1967b): Differentiation of
receptors responsive to isoproterenol, Life Sci. 6:2241-2249.

Langer, S.Z., Adler, E., Enero, A. and Stefano, F.J.E. (1971): The role
of α-receptors in regulating noradrenaline overflow by nerve

stimulation, Proc. 25th Int. Cong. Physiol. Sci. 9:335 (Abstract).

Langer, S.Z., Adler-Graschinski, E. and Giorgi, O. (1971): Physiological significance of α-adrenoceptor mediated negative feedback mechanism regulating noradrenaline release during nerve stimulation, Nature (Lond) 265:648-650.

Langer, S.Z. (1974): Presynaptic regulation of catecholamine release, Biochem. Pharmacol. 23:1793-1800.

Langer, S.Z. (1977): Presynaptic receptors and their role in the regulation of transmitter release, Br. J. Pharm. 60:481-497, 1977.

Langer, S.Z., Starke, K. and Dubocovich, M. (1979) in: "Presynaptic Receptors," Proceedings of the Satellite Syposium, Paris, July 22-23, 1978, 7th International Congress of Pharmacology, Pergamon, Elmsford, NY.

Langer, S.Z. (1980): Presynaptic regulation of the release of catecholamines, Pharmacol. Res. 32:337-362.

Langer, S.Z. (1981): Presence and physiological role of presynaptic inhibitory α_2-adrenoceptors in guinea pig atria, Nature (Lond) 194:671-672.

Langer, S.Z. and Hicks, P.E. (1984): Alpha-adrenoceptor subtypes in blood vessels: physiology and pharmacology, J. Cardiovas. Pharmacol. 6:S547-558.

Limberger, N. and Starke, K. (1984): Further study of prerequisites for the enhancement by α-adrenoceptor antagonists of the release of norepinephrine, Naunyn-Schmiedeberg's Arch. Pharmacol. 325:240-246.

Madjar, H. (1980): An examination of the pre and postsynaptic α-adrenoceptors in the autoperfused rabbit hindlimb, J. Cardiov. Pharmacol. 2:619-627.

Majewski, H., Hedler, L. and Starke, K. (1983): Evidence for a physiological role of presynaptic α-adrenoceptors: modulation of noradrenaline release in the pithed rabbit, Naunyn-Schmiedeberg's Arch. Pharmacol. 324:256-263.

Morita, K. and North, R.A. (1981): Clonidine activates membrane potassium conductances in myenteric neurons, Br. J. Pharmacol. 14:419-428.

Moylan, R.D. and Westfall, T.C. (1979): Effect of adenosine on adrenergic neurotransmission in the superfused rat portal vein, Blood Vessels 16:302-310.

Mulder, A.H., Frankhuyzen, A.L., Stoof, J.C., Weiner, J. and Schoffelmeer, A.N.M. (1984): Catecholamine receptors, opiate receptors and presynaptic modulation of neurotransmitter release in the brain, in: "Catecholamine: Neuropharmacology and Central Nervous System," E. Usdin, ed., pp. 47-58.

Nagatsu, T., Kato, T., Numata, Y., Ikuta, K. and Umezawa, H. et al. (1974): Serum dopamine β hydroxylase activity in developing hypertensive rats, Nature 251:630-631.

Palermo, A., Constantini, C., Mara, G. and Libretti, A. (1981): Role of the sympathetic nervous system in spontaneous hypertension: changes in central adrenoceptors and plasma catecholamine levels, Clin. Sci. 6:195-198.

Patel, S., Patel, U., Vithalini, D. and Verma, S.C. (1981): Regulation of catecholamine release by presynaptic receptor system, Gen. Pharmacol. 12:405-422.

Paton, D.M. (1979): "The Release of Catecholamine from Adrenergic Neurons," pp. 1-393, Pergamon, Elmsford, NY.

Powis, D.A. (1981): Does Na, K-ATPase play a role in the regulation of neurotransmitter release by prejunctional α-adrenoceptors? Biochem. Pharmacol. 30:2389-2397.

Rand, M.J., McCulloch, M.W. and Story, D.F. (1982): Comments on the presynaptic receptor controversy, Trends in Pharmacol Sci. 3:16-18.

Robson, R.D. and Antonaccio, M.M. (1974): Effects of clonidine on responses to cardiac nerve stimulation as a function of impulse frequency and stimulation duration in vagotomized dogs, Eur. J. Pharmacol. 29:182-186.

Sax, R.D. and Westfall, T.C. (1981): Alterations in prejunctional α-adrenoceptor function in guinea pig and rat vas deferens after chronic ganglionic blockade, J. Pharmacol. Exp. Ther. 219:21-26.

Schaefer, A., Unyi, G., and Pfeifer, A.K. (1972): The effects of a soluble factor and of catecholamine on the activity of adenosine triphosphate in subcellular fractions of rat brain, Biochem. Pharmacol. 21:2289-2294.

Scriabine, A. and Stavorski, J.M. (1973): Effect of clonidine on cardiac acceleration in vagotomized dogs, Eur. J. Pharmacol. 24:101-104.

Starke, K. (1971): Influence of α-receptor stimulants on noradrenaline release, Naturwissenschaften 58:420.

Starke, K., Borowski, E. and Endo, T. (1975): Preferential blockage of presynaptic α-adrenoceptors by yohimbine, Eur. J. Pharmacol. 34:385-388.

Starke, K. (1977): Regulation of noradrenaline release by presynaptic receptor systems, Rev. Physiol. Biochem. Pharmacol., 77:1-124.

Starke, K. (1979): Presynaptic regulation of release in the central nervous system, in: "The Release of Catecholamines from Adrenergic Nerves," D.M. Paton, ed., pp. 143-183, Pergamon.

Starke, K. (1981): Presynaptic receptors, Ann. Rev. Pharmacol. Tox. 21:7-30.

Steppeler, A., Tanaka, T. and Starke, K. (1978): A comparison of pre and postsynaptic α-adrenergic effects of phenylephrine and tramazoline on blood vessels of the rabbit in vivo, Naunyn-Schmiedeberg's Arch. Pharmacol. 304:223-230.

Stjarne, L. and Brundin, J. (1977): Frequency dependence of ^{3}H-noradrenaline secretion from human vasoconstrictor nerves: Modification by factors interfering with α and β-adrenoceptor or prostaglandin E_2 mediated control. Acta Physiol. Scand. 101:199-210.

Story, P.F., Briley, M.S. and Langer, S.Z. (1979) The effects of chemical sympathectomy with 6-hydroxydopamine on α-adrenoceptor and muscarinic cholinoceptor binding in rat heart ventricle, Eur. J. Pharmacol. 57:423-426.

Su, C. (1978): Purinergic inhibition of adrenergic transmission in rabbit blood vessels, J. Pharmacol. Exp. Ther. 204:351-361.

Su, C., Kamikawa, Y., Kubo, T. and Cline, W.H. Jr. (1982): Inhibitory modulation of vascular adrenergic transmission in the SHR, Proc. Springfield Blood Vessel Sym., C. Su, ed., pp. 33-37, Springfield, Ill., Univ. Press

Takeshita, A. and Mark, A.L. (1978): Neurogenic contributions to hindquarters vasoconstriction during high sodium intake in Dahl strain of genetically hypertensive rat, Circ. Res. 43:186-191.

Vanhoutte, P.M., Verbenen, T.J. and Webb, R.C. (1981): Local modulation of adrenergic neuroeffector interaction in the blood vessel wall, Physiol. Res. 61:151-247.

Vizi, E.S. (1978): Na^+-K^+-activated adenosine-triphosphate as a trigger in transmitter release, Neurosci. 3:367-384.

Vizi, E.S. (1979): Presynaptic modulation of neurochemical transmission, Prog. Neurobiol. 12:181-290.

Vizi, E.S. (1985): "Non-Synaptic Interactions Between Neurons: Modulation of Neurochemical Transmission," pp. 1-260, Wiley and Sons.

Vogel, S.A., Silberstein, S.D., Berr, V. and Kopin, I.J. (1972): Stimulation induced release of norepinephrine from rat superior cervical ganglia in vitro, Eur. J. Pharmacol. 20:308-311.

Westfall, T.C. (1977): Local regulation of adrenergic neurotransmission, Physiol. Rev. 57:659-728.

Westfall, T.C. (1980): Neuroeffector mechanisms, Ann. Rev. Physiol. 42:383-387.

Westfall, J.C. (1984): Evidence that noradrenergic transmitter release is regulated by presynaptic receptors, Fed. Proc. 43:1352-1357.

Westfall, T.C., Peach, M.J. and Tittermary, V. (1979): Enhancement of the electrically induced release of norepinephrine from the rat protal vein: Mediation by beta$_2$ adrenoceptors, Eur. J. Pharmacol. 58:67-74.

Westfall, T.C., Meldrum, M.J., Badino, L. and Earnhardt, J.T. (1984a): Noradrenergic transmission in the isolated portal vein of the spontaneously hypertensive rat, Hypertension 6:267-274.

Westfall, T.C., Meldrum, M.J., Badino, L. and Xue, C.S. (1984b): Noradrenergic transmission in blood vessels: influence of hypertension and dietary sodium, in: "Catecholamines: Basic and Peripheral Mechanisms," C. Usdin, ed., pp. 319-326, Liss.

Westfall, T.C., Xue, C.S., Meldrum, M.J. and Badino, L. (1985): The effect of sodium depletion on the facilitory effect of angiotensin on adrenergic neurotransmission in the rat portal vein, Blood Vessels 22:15-24.

Westfall, T.C. and Meldrum, M.J. (1985): Alterations in the release of norepinephrine at the vascular neuroeffector junction in hypertension, Ann. Rev. Pharmacol. 25:621-641.

Yamaguchi, N., DeChamplain, J. and Nadeau, R.A. (1977): Regulation of norepinephrine release from cardiac sympathetic fibers in the dog by presynaptic α and β receptors, Circ. Res. 41:108-117.

Zimmerman, B.G. and Whitmore, K. (1967): Effect of angiotensin and phenoxy benzamine on the release of norepinephrine in vessels during sympathetic nerve stimulation, Int. J. Neuropharmacol. 6:27-38.

Zimmerman, B.G., Rolewicz, Dunham, E.W. and Gisslen, J.L. (1969): Transmitter release and vascular responses in skin and muscle in hypertensive dogs, Am. J. Physiol. 217:798-804.

Zimmerman, B.G. (1978): Action of angiotensin in adrenergic nerve endings, Fed. Proc. 37:189-202.

Zimmerman, B.G. (1983): Peripheral neurogenic factors in acute and chronic alterations in arterial pressure, Circ. Res. 53:121-130.

Zsoter, T.T., Wolchinsky, C., Lawrin, M. and Sirko, S. (1982): Norepinephrine release in arteries of spontaneously hypertensive rats, Clin. Exp. Hyperten. A4:431-444.

SEROTONIN, PLATELETS, AND VASCULAR FUNCTION

P.M. Vanhoutte and D.S. Houston

Department of Physiology and Biophysics
Mayo Clinic and Mayo Foundation
Rochester, Minnesota 55905

ABSTRACT

Serotonin (5-hydroxytryptamine) causes contraction of most large
blood vessels and of venules. The vasoconstrictor effects of serotonin
can be due to direct activation of the smooth muscle, to amplification of
the response to other neurohumoral mediators, or to the liberation of
other endogenous vasoconstrictors (e.g. norepinephrine from adrenergic
nerves). Vasodilator responses to serotonin are seen mainly at the
arteriolar level, but can also be observed in larger blood vessels. They
can be due to the release of other endogenous vasodilators (e.g vasoactive
intestinal polypeptide from peptidergic nerves), direct relaxation of
vascular smooth muscle, inhibition of adrenergic neurotransmission, or
production of inhibitory signals by the endothelium. Aggregating
platelets release a number of vasoactive materials, including serotonin,
and can evoke both vasoconstrictor and vasodilator responses. Hence the
absence of endothelial cells or of adrenergic nerve activity may change
the primary response to aggregating platelets from dilatation to
constriction. Vasconstrictor responses to serotonin released from
aggregating platelets may play a role in the etiology of spasm in
cerebral, digital and coronary vessels, and in the maintenance of the
elevated peripheral resistance in arterial hypertension. A possible
involvement of serotonin in the increase in peripheral resistance
characteristic of chronic hypertension is suggested by the following
observations in hypertensive man or animals: (a) the turnover rate of
platelets, the main peripheral source of serotonin, is accelerated; (b)
the uptake of serotonin by platelets is reduced; (c) the metabolism of

serotonin by the endothelial cells is decreased; (d) the vascular smooth muscles are hyperresponsive to the constrictor effects of serotonin and other serotonergic agonists; (e) the S_2-serotonergic antagonist ketanserin lowers arterial blood pressure in hypertensive humans.

1. INTRODUCTION

In this paper we will consider the effects of serotonin in the peripheral vasculature: where does it originate, where and how does it act, and what may its function be? In particular we shall focus on the role blood platelets probably play in serotonin's physiological and pathological actions, and the experimental evidence for the activity of platelet-released substances, in particular serotonin, on blood vessels.

2. SOURCES OF SEROTONIN

Serotonin was the term aptly coined for a vasoconstrictor factor isolated from clotted blood, which was identified in 1949 as 5-hydroxytryptamine (Rapport, 1949; Page, 1954). In 1952, "enteramine," the gut-stimulating autocoid found in enterochromaffin cells of the intestinal tract, was recognized as serotonin (Erspamer and Asero, 1952). Almost all of the serotonin in the blood is contained by the blood platelets and originates from the enterochromaffin tissues (Thompson, 1971).

Serotonin is released from enterochromaffin cells in the gut in response to vagal or splanchnic nerve stimulation, peristalsis, or intestinal filling (Burks and Long, 1966; Ahlman, et al., 1978; Thompson, 1971; Larsson, et al., 1979). It may contribute locally to regulation of blood flow in the gut (Thompson, 1977), but some escapes to the plasma. It does not remain free in plasma long. It is avidly taken up and destroyed by the liver and the pulmonary endothelial monoamine oxidase (Thompson, 1971; Strum and Junod, 1972); it is also actively removed by platelets (Lingjaerde, 1977). It is unlikely that appreciable free serotonin is normally present in peripheral plasma given these disposition mechanisms: estimates of free plasma concentrations range from 20 to as low as 3 ng/ml (Somerville, 1976; Crawford, 1965; Genefke, et al., 1968; Engbaek and Voldby, 1982), but even these levels are in question because of the difficulty in preventing contamination by platelets during the preparation of the specimen (Sjaastad, 1975).

In the platelets, serotonin is taken up by an active transport mechanism and stored in the dense granules. It is released, along with other vasoactive substances such as thromboxane A_2, lipids, and adenosine di- and triphosphate (ADP and ATP), upon platelet activation (Holmsen, 1975). In view of serotonin's rapid disposition and the localized nature of platelet activation, one is led to suspect that serotonin would exert its effects predominantly at discrete sites as opposed to playing a hormonal role in the general homeostasis of the cardiovascular system. Only if disposition mechanisms were impaired and responsiveness of the blood vessel wall heightened might serotonin contribute to a systemic pathological process.

Serotonergic neurones may innervate blood vessels, particularly in the brain (Reinhard, et al., 1979; Edvinsson, et al., 1983). These nerves likely originate in the brainstem, but their function is uncertain; serotonin released from such nerves may cause contraction at least in some instances by acting on S_2-serotonergic receptors on the smooth muscle cells (Griffith, et al., 1982).

3. PHARMACOLOGY OF SEROTONIN

3.1. "Amphibaric Hormone"

Many different- -and often opposite- -effects of serotonin can be observed in the vascular system, depending on the dose, species, route of administration, and experimental conditions; the designation "amphibaric hormone" was entertained for the substance in recognition of its ability to either raise or lower blood pressure in differential experimental circumstances (Page and McCubbin, 1953). This complexity is due to the fact that serotonin can act on many tissues and on a variety of different cell types within one tissue such as the blood vessel wall. It can even apparently produce opposing effects on a given cell, mediated by different subtypes of serotonergic receptor. The complexity of the issue has been compounded by the lack of selective agonists and antagonists against the various receptors, making their classification difficult and the unravelling of their functions even more so. The availability of ketanserin, a selective antagonist of receptors that have been designated as S_2-serotonergic (corresponding to $5\text{-}HT_2$ binding sites in the brain; Leysen, et al., 1981) has improved matters.

3.2. Other Actions

Before concentrating on the actions of serotonin on the various cells
in the blood vessel wall, it should be noted that serotonin has a number
of actions on the intact cardiovascular system. For one, it can stimulate
a vagally-mediated van Bezold-Jarisch chemoreflex when injected into the
coronary artery (Page, 1954). It can also have a positive inotropic
effect, by increasing norepinephrine release from cardiac adrenergic
nerves (Buccino, et al., 1967; Fillion, et al., 1971). Serotonergic
pathways in the brainstem play an important but complex role in
cardiovascular regulation (see reviews by Kuhn, et al., 1980; Wolf, et
al., 1985).

3.3. Smooth Muscle

In most arterial smooth muscle serotonin induces contractions. In
many (such as the rat tail artery) but not all, ketanserin acts as a
competitive antagonist, indicating that the contraction is mediated by
S_2-serotonergic receptors (Van Nueten, et al., 1981). In other vessels,
including the coronary and cerebral arteries of the dog, ketanserin does
not display competitive antagonism, suggesting mediation of the
contraction by another serotonergic receptor subtype, or at least the
presence of a mixed population of receptors. Methysergide, a less
selective S_2-serotonergic antagonist, can antagonize serotonin
competitively in some vessels, yet act as a partial agonist and cause
contraction in others (Apperley, et al., 1980; Webb and Vanhoutte, 1985),
indicating variability in serotonergic receptor subtypes or possibly
differences in serotonergic receptor reserve between tissues.

Perhaps of equal importance, serotonin appears to be able to amplify
the effects of other vasoconstrictor agents such as histamine, angiotensin
II and norepinephrine (Fig. 1). Ketanserin has been shown in some tissues
to antagonize this amplifying action, indicating that the amplification is
mediated by S_2-serotonergic receptors (de la Lande, et al., 1966; Van
Nueten, et al., 1981, 1982).

Serotonergic receptors on vascular smooth muscle can also initiate
relaxation, as in the feline saphenous vein (Feniuk, et al., 1983) and the
canine femoral vein (Fig. 2). This effect seems not to be mediated
through S_2-serotonergic receptors since it is not antagonized by
ketanserin.

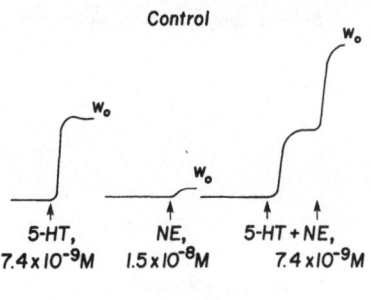

Control

Ketanserin

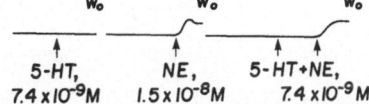

Fig. 1. In these isotonic recordings of strips of rat caudal artery,
serotonin (5-HT) induces a contraction which is abolished by
ketanserin (1.6×10^{-9}M; below). A near-threshold concentration
of norepinephrine (NE) produces a greatly augmented contraction
when added on top of a serotonin-induced contraction. Ketanserin
blocks this augmenting effect, though at this dose has almost no
effect on the norepinephrine-induced contraction. These findings
suggest that serotonin's amplifying effect on adrenergic
contraction of vascular smooth muscle is mediated by the
S_2-serotonergic receptor. (Modified from Van Nueten, et al.,
1981, by permission.)

3.4. Endothelium

In the coronary arteries of the dog and pig and jugular vein of the
chick, serotonin induces endothelium-dependent relaxations (Fig. 3; Cohen,
et al., 1983b; Cocks and Angus, 1983; Imaizumi, et al., 1984). This is
presumably mediated by release from the endothelial cells of an as-yet
uncharacterized "relaxing factor" as has been demonstrated with
acetylcholine and a number of other substances (Furchgott, 1983, 1984;
Vanhoutte and Rimele, 1983). Other, more potent (at least in the canine
coronary artery) mediators of endothelium-dependent relaxation, such as
adenosine di- and triphosphate, are also released during platelet
aggregation (Fig. 3; Houston, et al., 1985); serotonin though has the
peculiar characteristic that it would exert an opposite effect acting on
the endothelial surface of an intact blood vessel as opposed to the

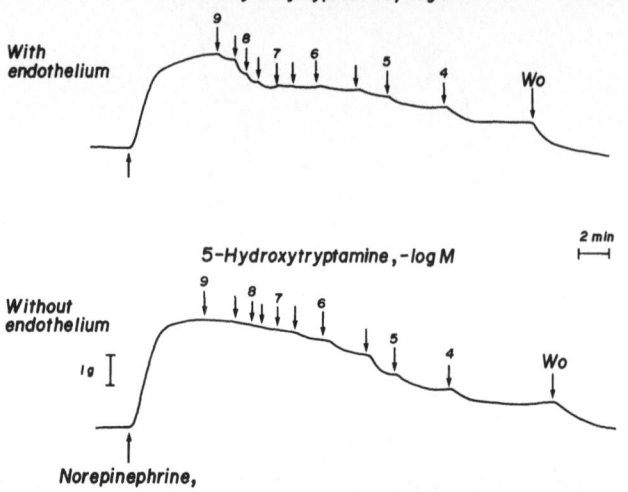

Fig. 2.   In rings of canine femoral vein contracted with norepinephrine
(10^{-6}M), serotonin induces relaxation by two disinct mechanisms.
In rings with endothelium (above), cumulative addition of
serotonin causes relaxation beginning at very low concentrations
(10^{-9}M) and reaching maximum at about 10^{-7}M.   In both rings with
and without endothelium, concentrations above 3×10^{-7}M produce a
further relaxation.   It has not been determined if these
different phases (indirect, mediated by the endothelium, and
direct on the smooth muscle) are mediated by the same receptor;
neither is blocked by the S_2-serotonergic antagonist, ketanserin
(not shown).

contractile action it would cause on direct exposure to smooth muscle.
The endothelium-dependent relaxation evoked by serotonin is inhibited by
the S_1- and S_2-serotonergic antagonist, methiothepin, but not by the
selective S_2 antagonist, ketanserin, indicating that it is mediated by a
receptor which may be of the S_1-subtype (Houston, et al., 1985).

3.5.   Adrenergic Nerves

The release of norepinephrine from vascular adrenergic nerve
varicosities can be inhibited by serotonin (McGrath, 1977; Feniuk, et al.,
1979).   This inhibition has pharmacological characteristics that indicate
mediation by an S_1-serotonergic receptor (i.e. with similarities to the

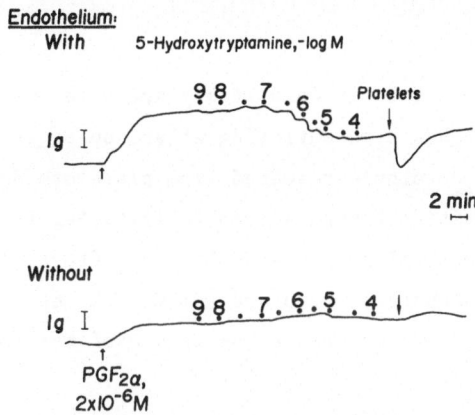

Fig 3.   Rings of canine coronary artery contracted with prostaglandin $F_{2\alpha}$
          ($PGF_{2\alpha}$) display an endothelium–dependent relaxation to cumulative
          additions of serotonin; rings without endothelium (below) respond
          with a slight contraction followed at high concentrations by a
          slight endothelium–independent relaxation. In the coronary
          artery, the endothelium–dependent relaxation is blocked by the
          S_1–serotonergic antagonist, methiothepin, but not by the S_2
          antagonist, ketanserin (not shown) suggesting that it may be
          mediated by an S_1 receptor. Addition of aggregating platelets,
          in the face of a concentration of serotonin that apparently
          saturates the endothelial serotonergic receptors, results in a
          further relaxation only in rings with endothelium. This
          indicates that much of the endothelium–dependent relaxation
          produced by aggregating platelets is due to substances other than
          serotonin (such as ADP and ATP). (From Houston, et al., 1985, by
          permission.)

$5-HT_1$ binding site that has been identified in the brain) (Engel, et al.,
1983). At higher concentrations, however, serotonin appears to act as an
indirect sympathomimetic, being taken up into the nerve ending and
displacing norepinephrine (McGrath, 1977). Serotonin taken up by
adrenergic nerve endings can be released upon stimulation of the nerves
(Verbeuren, et al., 1983; Kawasaki and Takasaki, 1984).

     In certain other nerves, including the adrenergic nerves to the heart
of the rabbit, serotonin is excitatory rather than inhibitory; the
receptor mediating this effect is of yet a different subtype and is
specifically blocked by MDL 72222 (Fozard, 1984a).

3.6. Platelets

Although in man serotonin only weakly activates platelets (Glusa and
Markwardt, 1984), it has a potentiating effect on aggregation induced by
other agents. Thus serotonin released from platelets during aggregation
contributes to a positive feedback loop facilitating further platelet
activation. Ketanserin blocks this amplifying effect, indicating its
mediation by S_2-serotonergic receptors (De Clerck, et al., 1982).
Ketanserin does not, however, block the uptake of serotonin into
platelets.

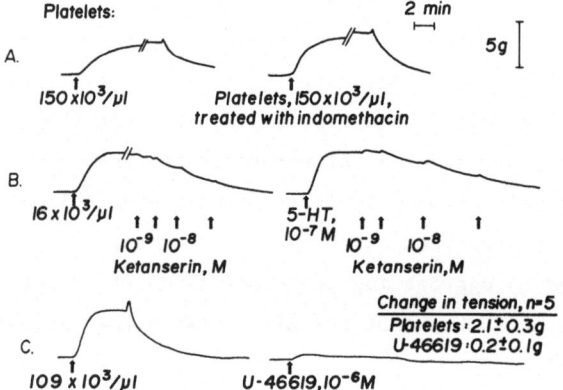

Fig. 4. Contraction of isolated canine pulmonary artery rings produced by
aggregating platelets and serotonin. Addition of platelets to
the organ bath produces a contraction (A) which is not inhibited
by pretreatment of the platelets with indomethacin to inhibit
production of thromboxane A_2 (A, right). Both the contractions
produced by aggregating platelets and by serotonin are
dose-dependently antagonized by the S_2-serotonergic antagonist,
ketanserin (B), indicating that in this blood vessel, the
contractile effect exerted by platelets is mediated principally
be serotonin. Further confirmation that thromboxane A_2 does not
play an important part in the response to platelets in this
vessel is the observation that the thromboxane mimetic, U-46619,
produces almost no contraction (C, right). (Modified from McGoon
and Vanhoutte, 1984, with permission.)

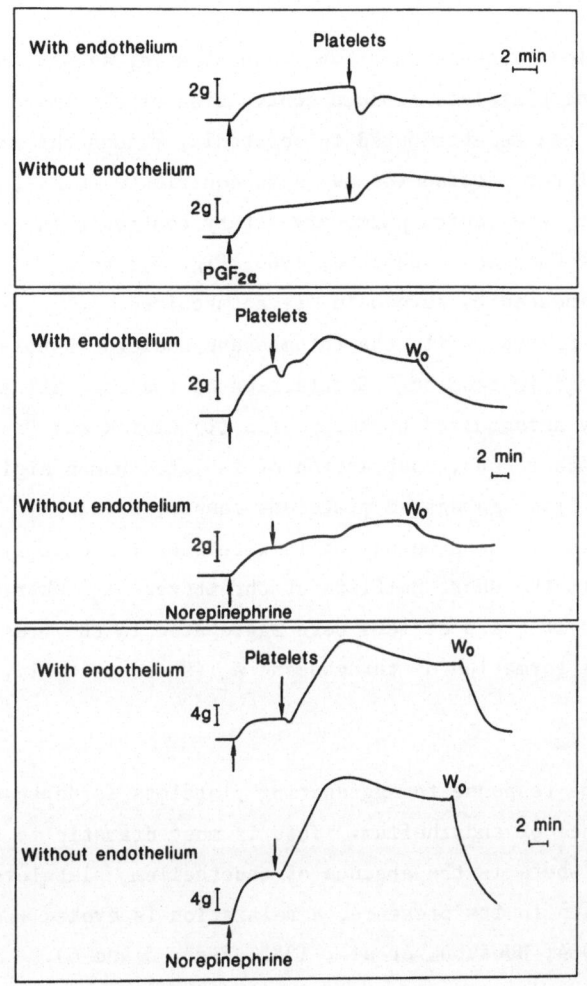

Fig. 5. Varying
endothelial
dependence of the
responses of
different canine
blood vesels to
aggregating
platelets. In the
coronary artery
contracted with
prostaglandin $F_{2\alpha}$

($PGF_{2\alpha}$, $2 \times 10^{-6}M$) platelets produce a marked relaxation when endothelium
is present, but a contraction in rings from which the endothelium has been
removed (top). (A similar response is seen if the coronary artery is
contracted with norepinephrine in the presence of propranolol.) In the
femoral vein (center) contracted with norepinephrine ($3 \times 10^{-7}M$), a
transient relaxation is induced by the aggregating platelets only in rings
with endothelium; in both rings with and without endothelium, the major
response is contractile. In the saphenous vein contracted with
norepinephrine ($10^{-7}M$), no appreciable endothelium-dependent relaxation
occurs; both rings with and without endothelium contract vigorously to
substances released from the platelets (bottom). Thus, the role of
endothelium in modulating responses to aggregating platelets is variable
and depends to large degree on the anatomic origin of the vessel.

In most isolated blood vessels that have been studied, substances released from aggregating platelets produce contraction of the smooth muscle. Commonly, this can be attributed to serotonin, though thromboxane A_2 and other vasoconstrictor prostanoids may also contribute (Ellis, et al., 1976). For example, the canine pulmonary artery contracts in response to platelets (McGoon and Vanhoutte, 1984; Fig. 4); this contraction, like that induced by serotonin, is antagonized dose-dependently by ketanserin, while the thromboxane analogue U46619 produces negligible change in tension. Contraction of the rat tail artery is also dose-dependently antagonized by ketanserin (De Clerck and Van Nueten, 1982). On the other hand, contraction of isolated human digital arteries to supernatant from aggregated platelets cannot be fully antagonized by ketanserin if the supernatant is withdrawn immediately after aggregation (within the short halflife of thromboxane A_2) whereas it can be fully antagonized if the platelets were aggregated in the presence of a drug to inhibit the formation of thromboxane A_2 (Moulds, et al., 1984).

In some vessels, the response to aggregating platelets is dependent on the presence or absence of endothelium. This is most dramatic in the canine coronary artery, where in the absence of endothelium, platelets induce a contraction while in its presence, a relaxation is evoked (Cohen, et al., 1983a, 1983b, 1984; Houston, et al., 1985; Figs. 5 and 6). Although serotonin does produce a modest endothelium-dependent relaxation in this vessel which can be blocked by the S_1-serotonergic antagonist, methiothepin, the major component of the relaxation seen to aggregating platelets is not blocked by methiothepin and probably is due to ATP and ADP (Houston, et al., 1985). In other vessels, such as the pulmonary artery (McGoon and Vanhoutte, 1984) or the saphenous vein (Fig. 5; Houston, et al., 1984) of the dog, platelets induce only constriction regardless of the presence of endothelium. Intermediate cases are the canine femoral artery and vein, in which transient endothelium-dependent relaxations may (not consistently) be seen in response to aggregating platelets (Figs. 5 and 7; Houston, et al., 1984). In the femoral artery, a relaxation can sometimes be unmasked by ketanserin (Fig. 7), indicating that it may be obscured by an overwhelming constrictor action of serotonin. The comparative difference between the coronary and femoral arteries with endothelium in response to platelet-released substances likely reflects the stronger contractile response to serotonin, and perhaps the lack of an endothelially mediated relaxation to the monoamine in the femoral artery (unpublished observations).

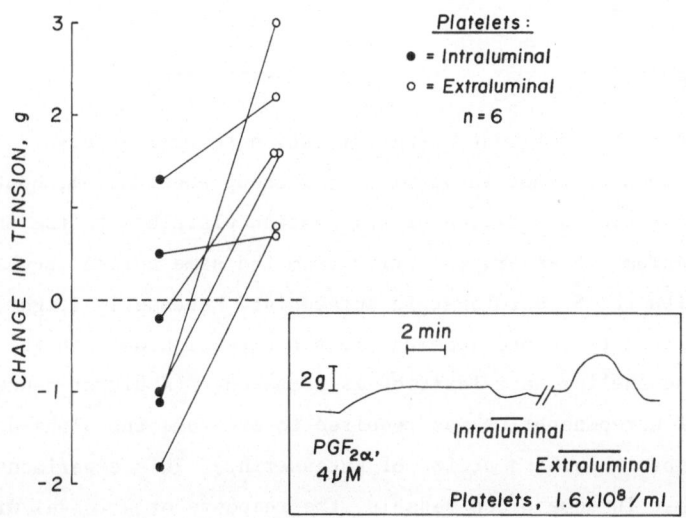

Fig. 6. Responses of perfused lengths of canine coronary artery to intraluminal or extraluminal exposure to aggregating platelets. Isometric tension was recorded by means of hooks placed in the vessel wall; to facilitate observing relaxations, tension was first induced with prostaglandin $F_{2\alpha}$ ($PGF_{2\alpha}$). Inset: tension recording of one of the experiments. In not all cases was a relaxation produced by intraluminal exposure to platelets, but in all cases a greater contraction was produced by extraluminal addition of the same concentration of platelets to the organ bath surrounding the outside of the vessel segment. (From Cohen, et al., 1984, by permission.)

5. IMPLICATIONS

The net effects of serotonin and other substances released by platelets in a local area of the vascular tree depends on a balance of actions. Vasodilation will be favored by endothelium-dependent relaxation, at least in some vessels; by presynaptic inhibition of norepinephrine release from vasoconstrictor sympathetic nerves; by

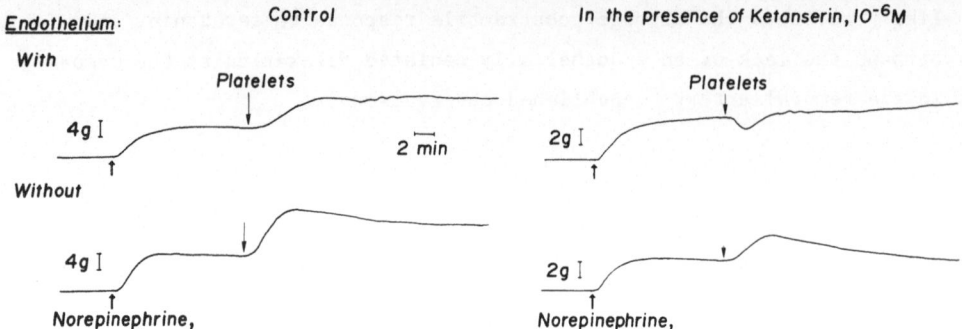

Fig. 7. In this experiment, rings of canine femoral artery (contracted
with norepinephrine), with or without endothelium, contract in
response to addition of aggregation platelets to the organ bath
medium. However, in a ring from the same animal incubated first
with the S_2-serotonergic antagonist, ketanserin (right),
platelets induce somewhat less contraction and, in the ring with
endothelium, a relaxation is unmasked. (A higher concentration
of norepinephrine was required to overcome the alpha-adreno-
ceptor-blocking action of ketanserin.) This experiment indicates
that in some blood vessels, the response of a vessel with
endothelium will be the sum of endothelially-mediated relaxation
and contractile effects directly on the smooth muscle; this net
effect can be shifted by selectively blocking one of the
component responses.

stimulation of inhibitory VIPergic nerves (at least in the gut); and by
acting directly on inhibitory receptors on the smooth muscle (Fig. 8).
Vasoconstriction will be favored principally by the direct contractile
action of serotonin on the vascular smooth muscle, but also by the
amplification effect of serotonin on other vasoconstrictor stimuli; by
displacement of norepinephrine from adrenergic nerve terminals; and by the
amplification effect on platelet aggregation (Fig. 9). Which response
predominates will depend on the specific blood vessel concerned, but may
be determined to large degree by the presence or absence of a functioning
endothelium. A number of pathological conditions to which platelets have
been linked are characterized by excessive contraction of the vascular
smooth muscle.

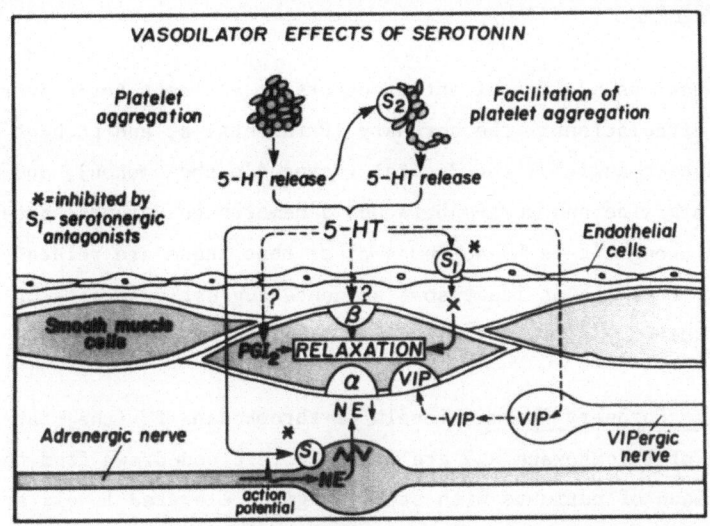

Fig. 8. Possible mechanisms by which serotonin may relax vascular smooth
muscle. See text for explanation. 5-HT = serotonin; NE =
norepinephrine; VIP = vasoactive intestinal polypeptide, or VIP
receptor; X = uncharacterized endothelium-derived relaxing
factor; PGI_2 = prostacyclin; S_1 = S_1-serotonergic receptor; S_2 =
S_2-serotonergic receptor; α= alpha-adrenoceptor; β=
beta-adrenoceptor. (from Vanhoutte, 1985, by permission.)

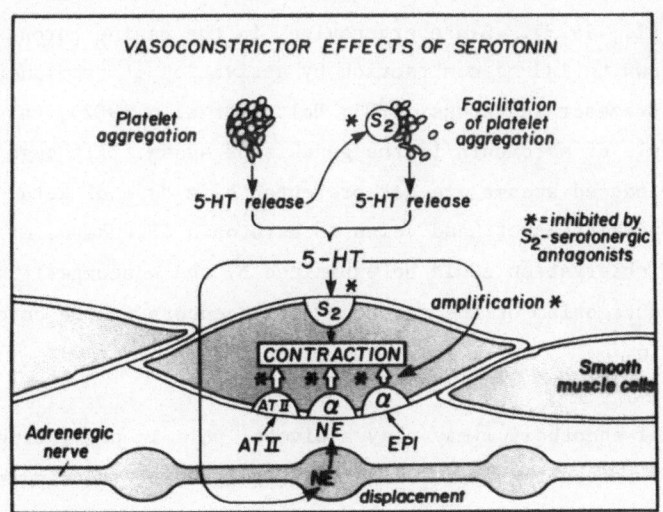

Fig. 9. Possible mechanisms by which serotonin may induce
vasoconstriction. see text for explanation. 5-HT = serotonin;
NE = norepinephrine; EPI = epinephrine; AT II = angiotensin II,
or angiotensin II receptor; S_2 = S_2-serotonergic receptor; α =
alpha-adrenoreceptor. (From Vanhoutte, 1985, by permission.)

5.1. Vasospasm

Vasospasm as a clinical entity occurs in at least three anatomically disparate circulations: the coronary (Prinzmetal's, and perhaps also effort-induced, angina), the digital (Raynaud's phenomenon), and the cerebral (migraine and post-subarachnoid hemorrhage cerebral arterial spasm). Although it is by no means clear that these are related conditions, there is at least some evidence suggesting a role of platelets in all of them.

5.1.1. Coronary Spasm. Levels of thromboxane B_2 (the stable metabolite of thromboxane A_2) are elevated in blood drawn from the coronary sinus of patients with stable angina; elevated levels are also found in patients with Prinzmetal's angina (Cortellaro, et al., 1983; Tada, et al., 1981). Since the probable source of thromboxane is from platelets, these findings suggest that platelet activation occurs during passage through the diseased coronary circulation -- likely upon contact with atherosclerotic or dysfunctional endothelium. Both the thromboxane A_2 and the serotonin released from platelets could contribute to contraction (spasm) of the coronary smooth muscle.

Ergonovine is commonly used during angiography to evoke spasm, which can mimic the spontaneous spasm of patients with Prinzmetal's angina (Curry, et al., 1979). Since ergonovine, in the canine coronary at least, has been shown to induce contraction by activation of serotonergic receptors (Brazenor and Angus, 1981; Holtz, et al., 1982), this would support a role of serotonin in the genesis of spasm. Although in man, ergonovine-induced spasms are not prevented by a dose of ketanserin that inhibits constriction of hand veins to serotonin (Freedman, et al., 1984), this latter observation could be explained by the noncompetitive and incomplete antagonism of the serotonergic receptors in the coronary artery by ketanserin.

A normal endothelium may play a pivotal role in preventing spasm -- not just by preventing platelet aggregation, but by inducing relaxation of the smooth muscle in response to serotonin and other platelet-released products. The aortas, limb arteries, and coronary arteries of rabbits, monkeys, and dogs respectively subjected to high cholesterol diet (and balloon-catheter denudation of the endothelium in the case of the dogs) are hypersensitive to serotonin and ergonovine (Henry and Yokoyama, 1980;

Heistad, et al., 1984; Kawachi, et al., 1984); this may reflect a loss of normal endothelially generated inhibition.

5.1.2. Raynaud's Phenomenon. Although structural changes in the digital vessels may occur in advanced stages of Raynaud's phenomenon, the episodic and cold-sensitive features of the condition indicate an active spastic process. Both serotonin and supernatant from aggregated platelets contract isolated human digital arteries (Moulds, et al., 1984). In support of the notion that serotonin-induced constriction is of primary importance in Raynaud's phenomenon are the observations that constrictions caused by serotonin and by platelets in isolated canine cutaneous blood vessels are augmented by cooling (Fig. 10; Vanhoutte and Shepherd, 1970; Lindblad, et al., 1984) and that intravenous ketanserin raises blood flow to the digits in patients suffering from Raynaud's syndrome (Stranden, et al., 1982). Although endothelium-dependent relaxations have not yet been reported in the digital vessels, it is reasonable to imagine that endothelial dysfunction or damage again could contribute to this disorder by enabling platelet activation and facilitating contraction of the smooth muscle cells.

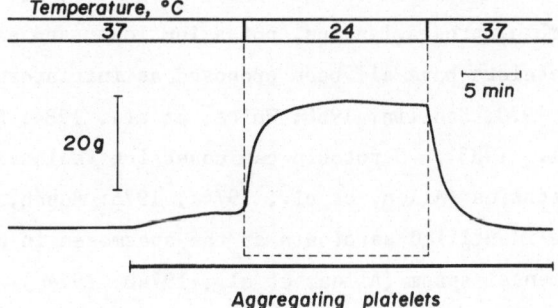

Fig. 10. Isometric tension recording of a canine saphenous vein ring. The ring contracts on exposure to aggregating canine platelets; this contraction is markedly augmented by cooling the organ bath from 37 to 24°C and reverses on rewarming to 37°.

149

5.1.3. **Migraine**. Since the observation by Sicuteri, et al. (1961)
of increased excretion of the metabolite of serotonin, 5-hydroxyindole-
acetic acid, during attacks of migraine, a role for serotonin has been
sought in this disorder. Concentrations of serotonin in the blood fall
during an attack (Somerville, 1976; Rydzewski, 1976; Anthony, et al.,
1967), suggesting that it is released from platelets and then cleared from
the plasma. Both vasoconstriction and vasodilation probably occur in
migraine; the former has been held responsible for the classical
neurological prodromal symptoms, and the latter for the throbbing pain.
Both may occur together (Raskin, 1981), and serotonin may be involved in
either or both phases.

More convincing, although not altogether straightforward, is the
effectiveness of various serotonergic and antiserotonergic drugs in
migraine. Antagonists such as methysergide, cyproheptadine, and pizotifen
are useful in migraine prophylaxis (Saper, 1978; Speight and Avery, 1972),
whereas the agonists ergotamine and ergonovine can abort acute attacks.
Perhaps the simplest interpretation is that the spasm phase is due to
transiently elevated free plasma serotonin levels after platelet release,
which would be combatted by the antagonists, whereas the pain is due to a
"rebound" vasodilation which would be combatted by the powerful
vasoconstrictive action of the serotonergic agonists. More complicated
interpretations exist, however: for example, some antagonists may have their
effect by inhibiting pain-sensitive afferent nerve endings (Fozard, 1984b).

5.1.4. **Cerebral spasm**. Although it has long been believed that the
spasm of the cerebral arteries that commonly occurs following subarachnoid
hemorrhage is caused or started by elements in the extravasated blood,
agreement has not been reached on what blood element(s) are responsible.
Hemoglobin, thrombin, prostaglandins, potassium ions, and serotonin
released from platelets have all been proposed as initiators of the spasm
(Zervas, et al., 1973; Boullin, 1980; White, et al., 1984; Sasaki, et al.,
1982; Ohta, et al., 1983). Serotonin can constrict isolated canine and
human cerebral arteries (Allen, et al., 1974a, 1976; Rusch, et al., 1985);
some studies have identified serotonin as the spasmogen in patients and
models of experimental spasm (Allen, et al., 1974b, 1974c). Others,
however, have found no increase in cerebrospinal fluid concentration of
serotonin in subarachnoid hemorrhage (Voldby, et al., 1982), and could not
inhibit with serotonergic antagonists the contraction of cerebral arteries
to cerebrospinal fluid obtained from patients with subarachnoid hemorrhage
(Boullin, et al., 1976).

5.2. Hypertension

Under "normal" circumstances, the disposition mechanisms for serotonin are active enough, and serotonin concentrations low enough, that it probably exerts effects only in discrete regions or vascular beds. However, each of the various elements in the formula dictating net vascular responsiveness to serotonin (release and disposition of the monoamine; constrictor responses and dilator responses of the vessel wall) can be altered. If such an alteration were generalized, it is conceivable that serotonin could participate in a systemic pathological process. Intriguing, though by no means conclusive, evidence suggests that the elevated peripheral vascular resistance characteristic of chronic arterial hypertension may be one such process (Fig. 11; see Vanhoutte, 1982a, 1982b, 1985; Vanhoutte and Cohen, 1983).

5.2.1. Elevated serotonin levels. Despite the difficulties in directly measuring free plasma serotonin concentrations, an inference that an elevation exists can be made from the observation that urinary excretion of 5-hydroxyindoleacetic acid is increased in hypertensive patients and can be lowered by antihypertensive therapy (Dzurik, et al., 1985). This could be due to increased release from platelets: indeed, platelets from hypertensive patients are reported to be more responsive to aggregating agents, including serotonin (Affolter, et al., 1984). Other evidences for increased platelet turnover in hypertension are decreased platelet survival times and increased circulating levels of β-thromboglobulin (Mehta and Mehta, 1981; Symoens and Vanhoutte, 1984). Increased plasma serotonin concentrations could also be due to decreased reuptake by non-aggregated platelets or diminished clearance by endothelial cells (Roth and Wallace, 1980; Gillis, 1985). Either way, free serotonin might persist in plasma long enough to exert more than a local effect.

5.2.2. Increased vasoconstrictor response. In a number of animal models of hypertension (genetic, renal, and corticosteroid-induced), marked hypersensitivity of vascular smooth muscle to the contractile effects of serotonin has been noted (see Webb and Vanhoutte, 1985). The augmentation of the response to serotonin is greater than that to norepinephrine or angiotensin II. In addition, the decay or tachyphylaxis of the constrictor response to serotonin (but not angiotensin II) is diminished (Collis and Vanhoutte, 1977, 1981; De Mey and Vanhoutte, 1981). Similar effects occur progressively with aging in normotensive animals (De Mey and Vanhoutte, 1981), suggesting that the vascular wall ages

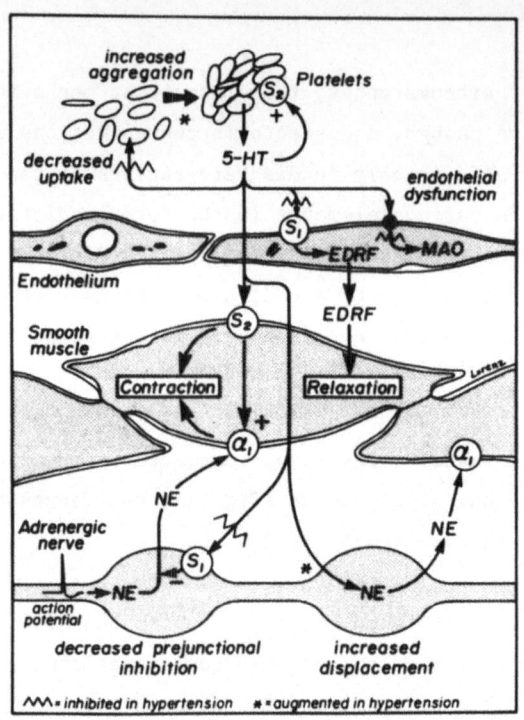

Fig. 11. Possible mechanisms by which serotonin may contribute to
hypertension. Serotonin uptake by platelets and elimination by
the endothelium are diminished; coupled with increased
aggregability of platelets, this may cause increased free
serotonin levels in plasma. Contractile effects of serotonin
are enhanced by: increased sensitivity to the direct
contractile action (mediated through S_2-serotonergic receptors);
reduced tachyphylaxis; amplification of other vasoconstrictor
responses, e.g. norepinephrine and angiotensin II; and
facilitated displacement of norepinephrine from adrenergic nerve
varicosities. The indirect relaxant effects of serotonin (by
presynaptic inhibition of norepinephrine release from adrenergic
nerve varicosities and by release of endothelium-derived
relaxing factor) are diminished. Increased serotonin
concentrations, increased contractile effects, and decreased
relaxant effects would all favour vasoconstriction and, if
generalized, hypertension. 5-HT = serotonin; MAO = monoamine
oxidase; S_1 = S_1-serotonergic receptor; S_2 = S_2-serotonergic
receptor; NE = norepinephrine; EDRF = endothelium-derived
relaxing factor; α= alpha-adrenoceptor. (From Vanhoutte, 1985,
by permission.)

prematurely when exposed to chronically increased distending pressure in hypertension. Thus hyperresponsiveness to serotonin could play a role in the maintenance, if not the initiation, of hypertension.

5.2.3. Decreased Vasodilator Responses. The prejunctional inhibitory effect of serotonin on norepinephrine release from adrenergic nerve varicosities is reduced in spontaneously hypertensive rats (Kubo and Su, 1983). Endothelial cells in various animal models of hypertension demonstrate morphological and functional abnormalities (Hüttner and Gabbiani, 1983) which could include diminished release of endothelium-derived relaxing factor in response to serotonin. Thus, there may be a diminished vasodilator effect of normal (or increased) concentrations of serotonin in hypertension to offset its constrictor effects, resulting in a greater contraction of the vessel wall.

5.2.4. Effects of Antagonists. Perhaps the most piquing point in the consideration of a role for serotonin in hypertension is the antihypertensive efficacy of ketanserin (De Cree, et al., 1981; see Schalekamp, 1985). Since this drug antagonizes the S_2-serotonergic receptor responsible for the contractile effects of serotonin in most blood vessels, the implication appears obvious that serotonergic tone plays a role in maintenance of elevated blood pressure. However, ketanserin also antagonizes $alpha_1$-adrenoceptors, like the commonly used antihypertensive agent, prazosin. Its hypotensive effect, at least in animal models of hypertension, may be due to alpha-adrenoceptor blockade (M.L. Cohen, et al., 1983; Fozard, 1982), although this question has not been settled in hypertensive patients (Wenting, et al., 1984). It must be noted that other commonly available serotonergic antagonists do not possess convincing antihypertensive properties, but this may relate to the fact that they are partial agonists at serotonergic receptors (Webb and Vanhoutte, 1985).

ACKNOWLEDGEMENTS

We would like to thank Mr. Robert Lorenz and Mrs. Helen Hendrickson for preparation of the illustrations, and Mrs. Janet Beckman for typing the manuscript. Dr. Houston is supported as a Fellow by the Medical Research Council of Canada.

REFERENCES

Affolter, H., Erne, P., Buhler, F.R. and Pletscher, A., 1984, Increased sensitivity of blood platelets to 5-hydroxytryptamine in essential hypertension, submitted for publication.

Ahlman, H., Bhargava, H.N., Donahue, P.E., Newson, B., Das Gupta, T.K. and Nyhus, L.M., 1978, The vagal release of 5-hydroxytryptamine from enterochromaffin cells in the cat, Acta Physiol. Scand. 104: 262-270.

Allen, G.S., Gold, L.H.A., Chou, S.N. and French, L.A., 1974c, Cerebral arterial spasm. Part 3: In vivo intracisternal production of spasm by serotonin and blood and its reversal by phenoxybenzamine, J. Neurosurg. 40: 451-458.

Allen, G.S., Gross, C.J., French, L.A. and Chou, S.N., 1976, Cerebral arterial spasm. Part 5: In vitro contractile activity of vasoactive agents including human CSF on human basilar and anterior cerebral arteries, J. Neurosurg. 44: 594-600.

Allen, G.S., Henderson, L.M., Chou, S.N. and French, L.A., 1974a, Cerebral arterial spasm. Part I: In vitro contractile activity of vasoactive agents on canine basilar and middle cerebral arteries, J. Neurosurg. 40: 433-441.

Allen, G.S., Henderson, L.M., Chou, S.N. and French, L.A., 1974b, Cerebral arterial spasm. Part 2: In vitro contractile activity of serotonin in human serum and CSF on the canine basilar artery, and its blockage by methylsergide and phenoxybenzamine, J. Neurosurg. 40: 442-450.

Anthony, M., Hinterberger, H., Lance, J.W., 1967, Plasma serotonin in migraine and stress, Arch. Neurol. 16: 544-552.

Apperley, E., Feniuk, W., Humphrey, P.P.A. and Levy, G.P., 1980, Evidence for two types of excitatory receptors for 5-hydroxytryptamine in dog isolated vasculature, Br. J. Pharmacol. 68: 215-224.

Boullin, D.J., 1980, Recent advances in vasospasm research, in: Cerebral Vasospasm, D.J. Boullin, ed., John Wiley and Sons, Chichester, pp. 293-299.

Boullin, D.J., Mohan, J. and Grahame-Smith, D.G., 1976, Evidence for the presence of a vasoactive substance (possibly involved in the aetiology of cerebral arterial spasm) in cerebrospinal fluid from patients with subarachnoid haemorrhage, J. Neurol. Neurosurg. Psychiat. 39: 756-766.

Brazenor, R.M. and Angus, J.A., 1981, Ergometrine contracts isolated canine coronary arteries by a serotonergic mechanism: no role for alpha adrenoceptors, J. Pharmacol. Exp. Ther. 218: 530-536.

Buccino, R.A., Covell, J.A., Sonnenblick, E.H. and Braunwald, E., 1967, Effects of serotonin on the contractile state of the myocardium, Am J. Physiol. 213: 483-486.

Burks, T.F., and Long, J.P., 1966, 5-Hydroxytryptamine release into dog intestinal vasculature, Am. J. Physiol. 211: 619-625.

Cocks, T.M. and Angus, J.A., 1983, Endothelium-dependent relaxation of coronary arteries by noradrenaline and serotonin, Nature 305: 627-630.

Cohen, M.L., Fuller, R.W. and Kurz, K.D., 1983, Evidence that blood pressure reduction by serotonin antagonists is related to alpha receptor blockade in spontaneously hypertensive rats, Hypertension 5: 676-681.

Cohen, R.A., Shepherd, J.T. and Vanhoutte, P.M., 1983a, Inhibitory role of the endothelium in the response of isolated coronary arteries to platelets, Science 221: 273-274.

Cohen, R.A., Shepherd, J.T. and Vanhoutte, P.M., 1983b, 5-Hydroxytryptamine can mediate endothelium-dependent relaxation of coronary arteries, Am. J. Physiol. 245: H1077-H1080.

Cohen, R.A., Shepherd, J.T. and Vanhoutte, P.M., 1984, Endothelium and asymmetrical responses of the coronary arterial wall, Am. J. Physiol. 247: H403-H408.

Collis, M.G. and Vanhoutte, P.M., 1981, Tachyphylaxis to 5-hydroxytryptamine in perfused kidneys from spontaneously hypertensive and normotensive rats, J. Cardiovasc. Pharmacol. 3: 229-235.

Collis, M.G. and Vanhoutte, P.M., 1977, Vascular reactivity of isolated perfused kidneys from male and female spontaneously hypertensive rats, Circ. Res. 41: 759-767.

Cortellaro, M., Boschetti, C., Antoniazzi, V., Moreo, G., Repetto, S., Verna, E., Boscarini, M., Limido, A., Binaghi, G. and Polli, E.E., 1983, Transcoronary platelet thromboxane A_2 formation without platelet trapping in patients with coronary stenosis - effect of sulphinpyrazone treatment, Thromb. Haemost. 50: 857-859.

Crawford, N.V., 1965, Systemic venous platelet-bound and plasma free serotonin levels in non-carcinoid malignancy, Clin. Chim. Acta 12: 274-281.

Curry, R.C., Pepine, C.J., Sabom, M.B. and Conti, C.R., 1979, Similarities of ergonovine-induced and spontaneous attacks of variant angina, Circulation 59: 307-312.

De Clerck, F. and Van Nueten, J.M., 1982, Platelet-mediated vascular contractions: inhibition of the serotonergic component by ketanserin, Thromb. Res. 27: 713-727.

De Clerck, F., David, J.-L. and Janssen, P.A.J., 1982, Inhibition of 5-hydroxytryptamine-induced and -amplified human platelet aggregation by ketanserin (R41468), a selective 5-HT$_2$-receptor antagonist, Agents and Actions 12: 388-397.

De Cree, J., Leempoels, J., Geukens, H., De Cock, W. and Verhaegen, H., 1981,

The antihypertensive effects of ketanserin (R 41468), a novel
5-hydroxytryptamine-blocking agent, in patients with essential
hypertension.

de la Lande, I.S., Cannell, V.A. and Waterson, J.G., 1966, The interaction of
serotonin and noradrenaline on the perfused artery, Br. J. Pharmacol.
28: 255–272.

De Mey, C. and Vanhoutte, P.M., 1981, Effect of age and spontaneous
hypertension on the tachyphylaxis to 5-hydroxytryptamine and angiotensin
II in the isolated rat kidney, Hypertension 3: 718–724.

Dzurik, R., Fetkovska, N., Brimichova, G. and Tison, P., 1985, Blood pressure,
5-OH indoleacetic acid, and vanilmandelic acid excretion and blood
platelet aggregation in hypertensive patients with ketanserin, J.
Cardiovasc. Pharmacol., in press.

Edvinsson, L., Deguerce, A., Duverger, D., Mackenzie, E.T. and Scatton, B.,
1983, Central serotonergic nerves project to the pial vessels of the
brain, Nature 306: 55–57.

Ellis, E.F., Oelz, O., Roberts, L.J. III, Payne, A.N., Sweetman, B.J., Nies,
A.S. and Oates, J.A., 1976, Coronary arterial smooth muscle contraction
by a substance released from platelets: evidence that it is thromboxane
A_2, Science 193: 1135–1137.

Engbaek, F. and Voldby, B., 1982, Radioimmunoassay of serotonin
(5-hydroxytryptamine) in cerebrospinal fluid, plasma, and serum, Clin,
Chem. 28: 624–628.

Engel, G., Göthert, M., Müller-Schweinitzer, E., Schlicker, E., Sistonen, L.
and Stadler, P.A., 1983, Evidence for common pharmacological properties
of ^{3}H-5-hydroxytryptamine binding sites, presynaptic 5-hydroxytryptamine
autoreceptors in CNS and inhibitory presynaptic 5-hydroxytryptamine
receptors on sympathetic nerves, Naunyn-Schmiedeberg's Arch. Pharmacol.
324: 116–124.

Erspamer, V. and Asero, B., 1952, Identification of enteramine, the specific
hormone of the enterochromaffin cell system, as 5-hydroxytryptamine,
Nature 169: 800–801.

Feniuk, W., Humphrey, P.P.A. and Watts, A.D., 1979, Presynaptic inhibitory
action of 5-hydroxytryptamine in dog isolated saphenous vein, Br. J.
Pharmacol. 67: 247–254.

Feniuk, W., Humphrey, P.P.A. and Watts, A.D., 1983, 5-Hydroxytryptamine-induced
relaxation of isolated mammalian smooth muscle, Eur. J. Pharmacol. 96:
71–78.

Fillion, G.M.B., Lluch, S. and Uvnas, B., 1971, Release of noradrenaline from
the dog heart in situ after intravenous and intracoronary administration
of 5-hydroxytryptamine, Acta Physiol. Scand. 83: 115–123.

Fozard, J.R., 1982, Mechanism of the hypotensive effect of ketanserin, <u>J. Cardiovasc. Pharmacol</u>. 4: 829-838.

Fozard, J.R., 1984a, MDL 72222: a potent and highly selective antagonist at neuronal 5-hydroxytryptamine receptors, <u>Naunyn-Schmiedeberg's Arch. Pharmacol</u>. 326: 36-44.

Fozard, J.R., 1984b, 5-Hydroxytryptamine in the pathophysiology of migraine. Proc. 5th International Symposium on Vascular Neuroeffector Mechanisms, Paris, August, in press.

Freedman, S.B., Chierchia, S., Rodriguez-Plaza, L., Bugiardini, R., Smith, G. and Maseri, A., 1984, Ergonovine-induced myocardial ischemia: no role for serotonergic receptors? <u>Circulation</u> 70: 178-183.

Furchgott, R.F., 1983, Role of endothelium in responses of vascular smooth muscle, <u>Circ. Res</u>. 53: 536-573.

Furchgott, R.F., 1984, The role of endothelium in the responses of vascular smooth muscle to drugs, <u>Ann. Rev. Pharmacol. Toxicol</u>. 24: 175-197.

Genefke, I.K., Garel, A. and Mandel, P., 1968, Factors influencing free serotonin in human plasma, <u>Clin. Chim. Acta</u>. 20: 61-67.

Gillis, C.N., 1985, Peripheral metabolism of serotonin, <u>in</u>: Serotonin and the Cardiovascular System, P.M. Vanhoutte, ed., Raven Press, New York, pp. 27-36.

Glusa, E. and Markwardt, F., 1984, Inhibition of 5-hydroxytryptamine-potentiated aggregation of human blood platelets by 5-hydroxytryptamine receptor-blocking agents, <u>Biomed. Biochim. Acta</u> 43: 215-220.

Griffith, S.G., Lincoln, J. and Burnstock, G., 1982, Serotonin as a neurotransmitter in cerebral arteries, <u>Brain Res</u>. 247: 388.

Heistad, D.D., Armstrong, M.L., Marcus, M.L., Piegors, D.J. and Mark, A.L., 1984, Augmented responses to vasoconstrictor stimuli in hypercholesterolemic and atherosclerotic monkeys, <u>Circ. Res</u>. 54: 711-718.

Henry, P.D. and Yokoyama, M., 1980, Supersensitivity of atherosclerotic rabbit aorta to ergonovine: mediation by a serotonergic mechanism, <u>J. Clin. Invest</u>. 66: 306-313.

Holmsen, H., 1975, Biochemistry of the platelet release reaction, <u>Ciba Symposia</u> 35: 175-205.

Holtz, J., Held, W., Sommer, O., Kuhne, G. and Bassenge, E., 1982, Ergonovine-induced constrictions of epicardial coronary arteries in conscious dogs: alpha-adrenoceptors are not involved, <u>Basic Res. Cardiol</u>. 77: 278-291.

Houston, D.S., Shepherd, J.T. and Vanhoutte, P.M., 1985, Adenine nucleotides, serotonin and endothelium-dependent relaxations to platelets, <u>Am. J. Physiol</u>. 248:H389-H395.

Houston, D.S., Shepherd, J.T. and Vanhoutte, P.M., 1984, Platelets cause endothelium-dependent relaxations in some but not all isolated canine vessels, Fed. Proc. 43: 1084 (abstract).

Hüttner, I. and Gabbiani, G., 1983, Vascular endothelium in hypertension, in: Hypertension, J. Genest, O. Kuchel, P. Hamet and M. Cantin, ed., McGraw-Hill, New York, pp. 473-488.

Imaizumi, Y., Baba, M., Imaizumi, Y. and Watanabe, M., 1984, Involvement of endothelium in the relaxation of isolated chick jugular vein by 5-hydroxytryptamine, Eur. J. Pharmacol. 97: 335-336.

Kawachi, Y., Tomoike, H., Maruoka, Y., Kikuchi, Y., Araki, H., Ishii, Y., Tanaka, K. and Nakamura, M., 1984, Selective hypercontraction caused by ergonovine in the canine coronary artery under conditions of induced atherosclerosis, Circulation 69: 441-450.

Kawasaki, H. and Takasaki, K., 1984, Vasoconstrictor response induced by 5-hydroxytryptamine released from vascular adrenergic nerves by periarterial nerve stimulation, J. Pharmacol. Exp. Ther. 229: 816-822.

Kubo, T. and Su, C., 1983, Effects of serotonin and some other neurohumoral agents on adrenergic neurotransmission in spontaneously hypertensive rat vasculature, Clin. Exp. Hyper.-Theory and Practice A5: 1501-1510.

Kuhn, D.M., Wolf, W.A. and Lovenberg, W., 1980, Review of the role of the central serotonergic neuronal system in blood pressure regulation, Hypertension 2: 243-255.

Larsson, I., Ahlman, H., Bhargava, H.N., Dahlstrom, A., Petterson, G. and Kewenter, J., 1979, The effects of splanchnic nerve stimulation on the plasma levels of serotonin and substance P in the portal vein of the cat, J. Neural. Transmission 46: 105-112.

Leysen, J.E., Awouters, F., Kennis, L., Laduron, P.M., Vandenberk, J. and Janssen, P.A.J., 1981, Receptor binding profile of R41468, a novel antagonist at 5-HT_2 receptors, Life Sci. 28: 1015-1022.

Lindblad, L.E., Shepherd, J.T. and Vanhoutte, P.M., 1984, Cooling augments platelet-induced contraction of peripheral arteries of the dog, Proc. Soc. Exp. Biol. Med. 176: 119-122.

Lingjaerde, O., 1977, Platelet uptake and storage of serotonin, in: Serotonin in Health and Disease, W.B. Essman, ed., Spectrum, New York, pp. 139-199.

McGoon, M.D. and Vanhoutte, P.M., 1984, Aggregating platelets contract isolated canine pulmonary arteries by releasing 5-hydroxytryptamine, J. Clin. Invest. 74: 828-833.

McGrath, M.A., 1977, 5-Hydroxytryptamine and neurotransmitter release in canine blood vessels: Inhibition by low and augmentation by high concentrations, Circ. Res. 41: 428-435.

Mehta, J. and Mehta, P., 1981, Platelet function in hypertension and effect of therapy, Am. J. Cardiol. 47: 331-334.

Moulds, R.F.W., Iwanov, V. and Medcalf, R.L., 1984, The effects of platelet-derived contractile agents on human digital arteries, Clin. Sci. 66: 443-451.

Ohta, T., Osaka, K., Siguma, M., Yamamoto, M., Shimizu, K. and Toda, N., 1983, Cerebral vasospasm following ruptured intracranial aneurysms, especially some contributions of potassium ion released from subarachnoid hematoma to delayed cerebral vasospasm, in: Vascular Neuroeffector Mechanism, 4th International Symposium, S.A. Bevan, et al., ed, Raven Press, New York, pp. 353-358.

Page, I.H., 1954, Serotonin (5-hydroxytryptamine), Physiol. Rev. 34: 563-588.

Page, I.H. and McCubbin, J.W., 1953, The variable arterial pressure response to serotonin in laboratory animals and man, Circ. Res. 1: 354-362.

Rapport, M.M., 1949, Serum vasoconstrictor (Serotonin) V: Presence of creatinine in the complex; a proposed structure of the vasoconstrictor principle, J. Biol. Chem. 180: 961-969.

Raskin, N.H., 1981, Pharmacology of migraine, Ann. Rev. Pharmacol. Toxicol. 21: 463-478.

Reinhard, J.F., Liebmann, J.E. and Schlosberg, A.J., 1979, Serotonin neurons project to small blood vessels in the brain, Science 206: 85-87.

Roth, R.A. and Wallace, K.P., 1980, Disposition of biogenic amines and angiotensin I by lungs of spontaneously hypertensive rats, Am. J. Physiol. 239: H736-H741.

Rusch, N.J., Chyatte, D., Sundt, T.M. and Vanhoutte, P.M., 1985, 5-Hydroxytryptamine: source of activator calcium in human basilar arteries Stroke, in press.

Rydzewski, W., 1976, Serotonin (5-HT) in migraine: levels in whole blood in and between attacks, Headache 16: 16-19.

Saper, J.R., 1978, Migraine: I. Classification and pathogenesis, J. Am. Med. Assoc. 239: 2380-2383.

Sasaki, T., Wakai, S., Asano, T., Takakura, K. and Sano, K., 1982, Prevention of cerebral vasospasm after SAH with a thromboxane synthetase inhibitor, OKY-1581, J. Neurosurg. 57: 74-82.

Schalekamp, M.A.D.H., 1985, Sertonergic blockade and hypertension, in: Serotonin and the Cardiovascular System, P.M. Vanhoutte, ed., Raven Press, New York, pp. 135-145.

Sicuteri, F., Testi, A. and Anselmi, B., 1961, Biochemical investigations in headache: increase in the hydroxyindoleacetic acid excretion during migraine attacks, Int. Arch. Allergy 19: 55-58.

Sjaastad, O., 1975, The significance of blood serotonin levels in migraine: a critical review, Acta. Neurol. Scand. 51: 200-210.

Somerville, B.W., 1976, Platelet-bound and free serotonin levels in jugular and forearm venous blood during migraine, Neurology 26: 41-45.

Speight, T.M. and Avery, G.S., 1972, Pizotifen (BC-105): a review of its pharmacological properties and its therapeutic efficacy in vascular headaches, Drugs 3: 159-203.

Stranden, E., Roald, O.K. and Krohg, K., 1982, Treatment of Raynaud's phenomenon with the 5-HT$_2$-receptor antagonist ketanserin, Br. Med. J. 285: 1069-1071.

Strum, J.M. and Junod, A.F., 1972, Radioautographic demonstration of 5-hydroxytryptamine-^{3}H uptake by pulmonary endothelial cells, J. Cell Biol. 54: 456-467.

Symoens, J. and Vanhoutte, P.M., 1984, Role of serotonin in blood pressure regulation, in: Care of the postoperative surgical patient, J.A.R. Smith and J. Watkins, eds., Butterworths and Co., London, in press.

Tada, M., Kuzuya, T., Inoue, M., Kodama, K., Mishima, M., Yamada, M., Inui, M. and Abe, H., 1981, Elevation of thromboxane B$_2$ levels in patients with classic and variant angina pectoris, Circulation 64: 1107-1115.

Thompson, J.H., 1971, Serotonin and the alimentary tract, Res. Comm. Chem. Pathol. Pharmacol. 2: 687-782.

Thompson, J.H., 1977, Serotonin (5-hydroxytryptamine) and the alimentary system, in: Serotonin in Health and Disease, vol. 4, Essman, W.B., ed., Spectrum, New York, pp. 201-392.

Vanhoutte, P.M., 1982a, Does 5-hydroxytryptamine play a role in hypertension? Tr. Pharmacol. Sci. 3: 370-373.

Vanhoutte, P.M., 1982b, 5-Hydroxytryptamine, vasospasm and hypertension, in: 5-Hydroxytryptamine in Peripheral Reactions, F. De Clerck and P.M. Vanhoutte, eds., Raven Press, New York, pp. 163-174.

Vanhoutte, P.M., 1985, Peripheral serotonergic receptors and hypertension, in: Serotonin and the Cardiovascular System, P.M. Vanhoutte, ed., Raven Press, New York, pp. 123-133.

Vanhoutte, P.M. and Cohen, R.A., 1983, The elusory role of serotonin in vascular function and disease, Biochem. Pharmacol. 32: 3671-3674.

Vanhoutte, P.M. and Shepherd, J.T., 1970, Effect of temperature on reactivity of saphenous, mesenteric, and femoral veins of the dog, Am. J. Physiol. 218: 187-190.

Vanhoutte, P.M. and Rimele, T.J., 1983, Role of the endothelium in the control of vascular smooth muscle function, J. Physiol. (Paris) 78: 681-686.

Van Nueten, J.M., Janssen, P.A.J., Van Beek, J., Xhonneux, R., Verbeuren, T.J. and Vanhoutte, P.M., 1981, Vascular effects of ketanserin (R41468), a novel antagonist of 5-HT$_2$ serotonergic receptors, J. Pharmacol. Exp. Ther. 218: 217-230.

Van Nueten, J.M., Janssen, P.A.J., De Ridder, W. and Vanhoutte, P.M., 1982, Interaction between 5-hydroxytryptamine and other vasoconstrictor substances in the isolated femoral artery of the rabbit; effect of ketanserin (R41468), Eur. J. Pharmacol. 77: 281-287.

Verbeuren, T.J., Jordaens, F.H. and Herman, A.G., 1983, Accumulation and release of [^{3}H]5-hydroxytryptamine in saphenous veins and cerebral arteries of the dog, J. Pharmacol. Exp. Ther. 226: 579-588.

Voldby, B., Engbaek, F. and Enevoldsen, E.M., 1982, CSF serotonin concentrations and cerebral arterial spasm in patients with ruptured intracranial aneurysm, Stroke 13: 184-189.

Webb, R.C. and Vanhoutte, P.M., 1985, Responsiveness of the hypertensive blood vessel wall to serotonergic stimuli. In: Serotonin and the Cardiovascular System, P.M. Vanhoutte, ed., Raven Press, New York, pp. 113-121.

Wenting, G.J., Woittiez, A.J.J., Man In't. Veld, A.J. and Schalekamp, M.A.D.H., 1984, 5-HT, alpha-adrenoceptors, and blood pressure: effects of ketanserin in essential hypertension and autonomic insufficiency, Hypertension 6: 100-109.

White, R.P., Shirasawa, Y. and Robertson, J.T., 1984, Comparison of responses elicited by alpha-thrombin in isolated canine basilar, coronary, mesenteric, and renal arteries, Blood Vessels 21: 12-22.

Wolf, W.A., Kuhn, D.M. and Lovenberg, W., 1985, Serotonin and central regulation of arterial blood pressure, in: Serotonin and the Cardiovascular System, P.M. Vanhoutte, ed., Raven Press, New York, pp. 63-73.

Zervas, N.T., Kuwayama, A., Rosoff, C.B. and Salzman, E.W., 1973, Cerebral arterial spasm: modification by inhibition of platelet function, Arch. Neurol. 28: 400-404.

TWO PRINCIPLES OF LARGE ARTERY DILATION: INDIRECT ENDOTHELIUM-MEDIATED AND
DIRECT SMOOTH MUSCLE RELAXATION

Eberhard Bassenge and Ulrich Pohl

Department of Applied Physiology
University of Freiburg, W-Germany

INTRODUCTION

Vasomotion of large arteries can be brought about by two different
mechanisms:

1.) Direct effects of various agonists or transmitters on the smooth
muscle cells, effecting either contractions or relaxation depending on the
induced level of myoplasmic free calcium (which in turn is largely
controlled by the concentration of cGMP and the activity of guanylate
cyclase).

2.) Vasomotion can also result from indirect effects on vascular
smooth muscle. Various agonists and physico-chemical stimuli act on the
endothelial surface to release, via the abluminal side, a vasomotor signal
to the adjacent vasculature. With regard to endothelium-mediated
vasomotion, there is convincing evidence for the presence and significance
of an endothelium-derived relaxant factor (EDRF). However there is much
less evidence for endothelium-mediated vasoconstriction.
Endothelium-mediated relaxation was first shown by Furchgott and Zawadzki
in 1980, using a sandwich technique in which EDRF could diffuse from a
segment with stimulated endothelium to a non-endothelium-containing
indicator vessel segment.

1. DIRECT EFFECTS OF AGONISTS ON VASCULAR SMOOTH MUSCLE

The main factor controlling the contractile force in vascular smooth
muscle is the level of myoplasmic free calcium. There is growing evidence

that the initiation of the contractile process, i.e., the cyclic interaction between myosin cross-bridges and the actin filament is dependent on the phosphorylation of the myosin light chain (MLC) (for review see Adelstein and Eisenberg, 1980; Aksoy, et al., 1983; Hartshorne and Mrwa, 1982). Myosin phosphorylation is achieved by myosin-light-chain kinase (MLCK). This specific enzyme is activated by the binding of the calcium-calmodulin-complex (Ca-CaM), which is formed at intracellular calcium concentrations above 1 μm. Reduction of myoplasmic calcium leads to an inactivation of the MLCK due to dissociation of the Ca-calmodulin-myosin-light-chain-kinase-complex (Adelstein, et al., 1980; Hartshorne and Mrwa, 1982). A complementary calcium-dependent MLC-phosphatase dephosphorylates the MLC resulting in a decrease of actin-myosin interaction: The smooth muscle relaxes.

The calcium dependence of contractile activity in vascular smooth muscle can be modulated by alteration of the level of MLCK activity: Phosphorylation of the MLCK by a cyclic AMP-dependent protein kinase inhibits the Ca-CaM-dependent activation of the enzyme. In the same manner CaM-antagonists can act as inhibitors of MLCK (Weiss, et al., 1982).

The increase in myoplasmic calcium which represents the final common step in the excitation-contraction coupling of a vascular smooth muscle is due to a transmembrane calcium influx and/or a release of calcium from the sarcoplasmic reticulum and other, membrane-bound intracellular stores (Morgan and Morgan, 1984). These functional extra- and intracellular calcium pools are utilized during excitation-contraction coupling in the various mammalian species, including man, to a different extent (Bolton, 1979; Casteels and Droogmans, 1984). The relative contribution may vary also within the same species along the vascular tree.

The main membrane pathway through which intracellular calcium penetrates into vascular smooth muscle is a voltage-dependent, gated channel (See Fig. 1A). Some findings support the hypothesis that there are also receptor-operated channels in vascular smooth muscle, which open by receptor occupation without depolarization as the initiating step (Droogmans, et al., 1977). However, a critical test to prove the possible

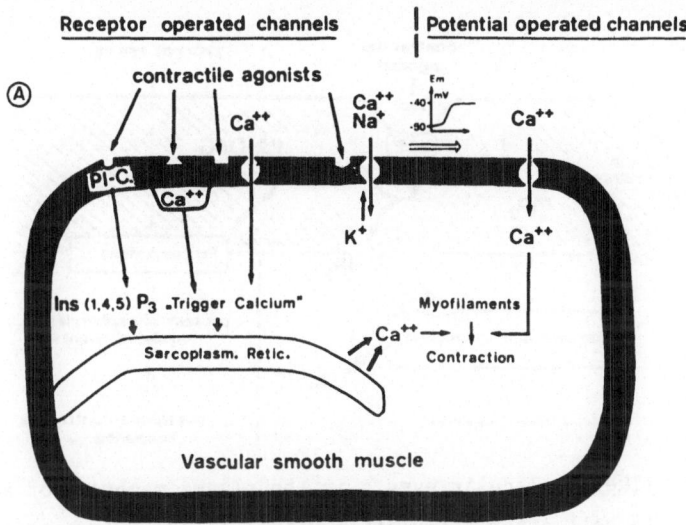

Fig. 1A: Diagrammatic representation of pathways of calcium delivery in excitation-contraction coupling in vascular smooth muscle. PIP_2 = phosphatidylinositol 4,5-bisphosphate, Ins $(1,4,5)P_3$ = D-inositol 1,4,5-trisphosphate. Depolarization opens potential operated calcium channels, increasing intracellular Ca-concentration, which in turn triggers contraction (right hand side). Evidence for voltage independent, receptor operated channels (by various agonists, left side) is still missing.

existence of receptor-operated channels by patch-clamp experiments has not yet been performed. The release of calcium bound to the inner surface of the cell membrane induced by receptor stimulation represents a further hypothetical possibility to increase myoplasmic calcium concentration without significant membrane depolarization (Casteels and Droogmans, 1984). It is thought that the interaction of an agonist with its membrane receptor promotes the hydrolysis of membrane phosphoinositides by activation of a phospholipase C. This leads to the enhanced production of 1,2-diacylglycerol and inositol-1,4,5-triphosphate (IP_3) (Fig. 1B). Recently it was shown that IP_3 induces a release of calcium from intracellular stores of vascular smooth muscle (Suematsu, et al., 1984). This strengthens the view that IP_3 can act as a messenger between receptor occupation and intracellular calcium release (Streb, et al., 1983).

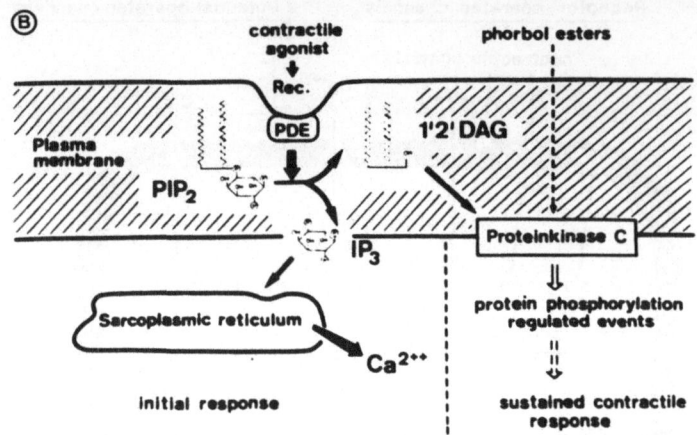

Fig. 1B: The likely molecular events in the plasma membrane of a smooth muscle cell after stimulation with a contractile agonist. DE = phosphodiesterase, probably activated after receptor occupation via a GTP-dependent protein mechanism. PIP_2 = phosphatidyl-inositol, IP_3 = inositol 1,4,5-trisphosphate, 1'2' DAG = 1,2-diacylglycerol. Tumor promoting phorbolesters, e.g., 12-o-tetra-decanoyl-phorbol-13-acetate will activate the C-kinase directly without causing an increase in Ca^{2+} inducing a sustained contractile response: Such drugs are, therefore, useful tools in assessing direct effects of protein kinase.

Although the decrease of myoplasmic calcium can be described as a basic principle of relaxation in vascular smooth muscle, other mechanisms have not yet been well defined. Vascular tone may be maintained even when the myoplasmic calcium declines (possibly by so-called "latch bridges") (Aksoy, et al., 1983). The role of cyclic nucleotides (c-AMP and c-GMP), which seem to be involved in the modulation of calcium sensitivity of the contractile apparatus (Ruegg and Paul, 1982; Hardman, 1984) rather than in the process of calcium sequestration, remains to be established (Ignarro and Kadowitz, 1985; Lincoln and Corbin, 1978).

2. INDIRECT, ENDOTHELIUM-MEDIATED EFFECTS ON VASCULAR SMOOTH MUSCLE

Smooth muscle relaxation can also be elicited by stimulation of the endothelium to release a not yet identified dilator substance (EDRF = endothelium derived relaxant factor) (Furchgott and Zawadzki, 1980). Various stimulants are effective to induce such a release: acetylcholine,

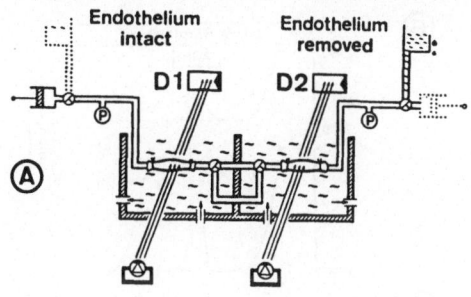

Fig. 2A: Experimental set-up applied for serial perfusion of
          endothelium-intact and endothelium-removed arterial segments.
          Intraluminal perfusion could be performed either from the
          de-endothelialized vessel to the endothelium-intact vessel or
          vice versa. External diameters of the arterial segments were
          recorded simultaneously by photo-electric devices (D1 and D2).
          (Modified from Busse and Bassenge, 1985).

norepinephrine, serotonin, histamine, ATP, ADP, thrombin, vasopressin,
bradykinin, substance P and a number of pharmacological compounds.
Reviews are given by Busse et al. (1985), Furchgott (1983), by Vanhoutte
and Rimele (1983) and in the chapter by Vanhoutte and Houston.

2.1. NATURE OF ENDOTHELIUM-DERIVED RELAXANT FACTOR

Although a great number of laboratories have engaged in the
investigation of EDRF for some years, the chemical nature of EDRF's has
still not been identified. To demonstrate, in principle, the humoral
nature of EDRF, a so-called "sandwich"-mount, which consists of a
longitudinal strip with intact endothelium suspended together with a
transverse strip without endothelium, was used in the early years of
EDRF-research (Furchgott and Zawadzki, 1980). In this way, Furchgott and
colleagues could demonstrate that during acetylcholine stimulation the
endothelium-intact strip must have released a relaxing factor(s) which
diffused to the smooth muscle cells of the endothelium-denuded strip.

A dilation of a de-endothelialized preconstricted rabbit coronary
artery by the effluent from an acetylcholine-stimulated rabbit thoracic
aorta with intact endothelium could be demonstrated by Griffith, et al. in

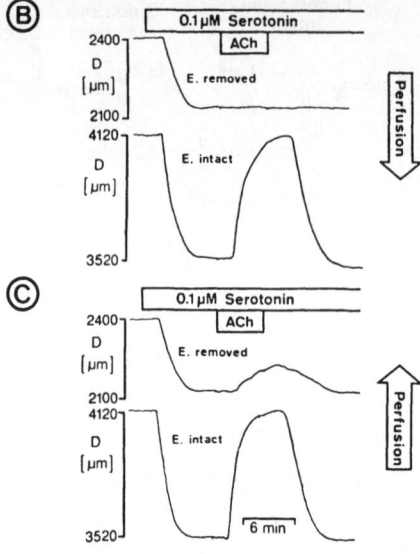

Fig. 2B: Recordings of external diameter (D) from a segment of rabbit thoracic aorta with intact endothelium and from a segment of canine coronary artery after removal of endothelium. The arteries precontracted by serotonin were perfused with Tyrode's solution (flow rate: 0.5 ml/min). Acetylcholine (ACh, 1 μM) was added to the perfusate to elicit endothelium-mediated dilation. Serial perfusion from the de-endothelialized coronary artery to the intact aorta elicits an ACh-induced dilation only in the aorta (with endothelial cells present).

2C: The direction of perfusion is reversed, the effluent from the ACh-stimulated aorta now elicits a dilation in the downstream de-endothelialized coronary segment.

1984. By varying the length of intervening perfusion tubing between the aortic and coronary preparations, the authors could determine the half-life of EDRF as amounting to about six seconds. They could furthermore demonstrate a basal continuous release of EDRF even from the unstimulated rabbit aorta. Both the continuous basal and the stimulated EDRF release could be maintained in good preparations for up to one hour. Using a newly developed bioassay system, which allows one to record directly the outer diameters of the EDRF donor and the EDRF detector vessel (Fig. 2A), the properties of EDRF released from the rabbit aorta and the canine femoral artery in response to acetylcholine were investigated (Busse, et al., 1985). An example of such a bioassay

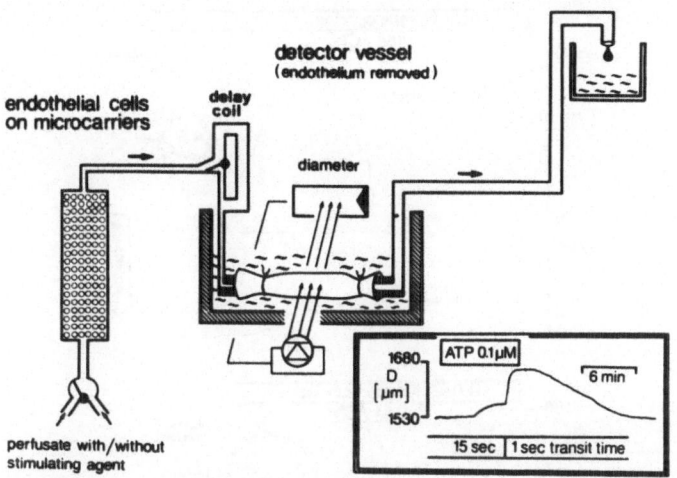

Fig. 3:   Bioassay system used for the detection of EDRF, released from
          cultured endothelial cells. Microcarrier beads (≈200 μm
          diameter) covered with calf aortic endothelial cells are packed
          into the barrel of a small syringe (about 0.3 ml volume) and
          perfused with oxygenased Tyrode's solution. The release of ERDF
          from these cells was assayed by passing the effluent through a
          precontracted endothelium-denuded detector vessel. Transit time
          from the column to the vessel could be altered by changing the
          length of the intervening tubing (delay coil). Insert: Example
          of EDRF-induced dilation in the de-endothelialized rabbit femoral
          artery with two different transit times. EDRF-release from the
          cultured endothelial cells was elicited by 0.1 m ATP. D =
          external diameter.

experiment is shown in Fig. 2B and 2C. It is obvious that EDRF is a very
unstable factor, which is inactivated in the perfusing blood by hemoglobin
and other proteins and exhibits a short half-life. Therefore it acts as
an autacoid, and will be largely degraded during the passage through the
downstream vessel sections.

     Bioassay experiments show that EDRF can also be released from
cultured calf aortic endothelial cells under appropriate conditions.
Microcarrier beads covered with endothelial cells were packed into columns
and perfused with oxygenated Tyrode's solution (see Fig. 3). The release
of vascular relaxing factor(s) from these endothelial cells was assayed by

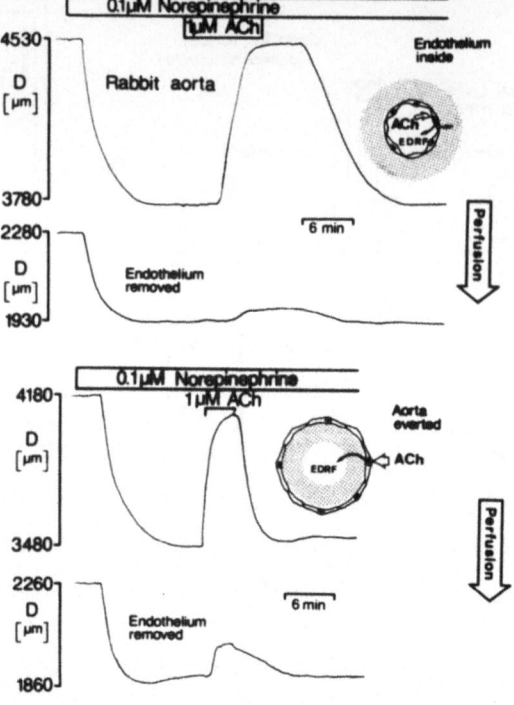

Fig. 4: Bioassay experiment (rabbit aorta – rabbit femoral artery)
demonstrating the greater <u>abluminal</u> release of EDRF. The
intraluminal flow-rate in these experiments was reduced from 0.5
ml/min to 0.1 ml/min. Upper panel: Control experiment.
Endothelium-mediated dilation and luminal EDRF-release was
elicited by intraluminal application of 1 μM acetylcholine (ACh).
Lower panel: After careful aversion of the aorta, the perfusate
reaching the detector vessel was now in contact with the
adventitia of the aorta. Acetylcholine was administered to the
outside (endothelium) of the aorta. The dilation in the detector
vessel was elicited by the abluminally released EDRF, which had
passed through the media and adventitia. Although the degree of
EDRF-inactivation might be much greater with this experimental
approach (diffusion time in the wall) the dilation in the
detector vessel is remarkably greater than that in the control
experiment (luminal EDRF-release). (Modified from Busse and
Bassenge, 1985.)

passing the effluent through precontracted endothelium–denuded arteries
(canine coronary or femoral arteries). Stimulation of the endothelial
cells by bradykinin (0.1 μM), ATP (0.5 μM) or the calcium ionophore A

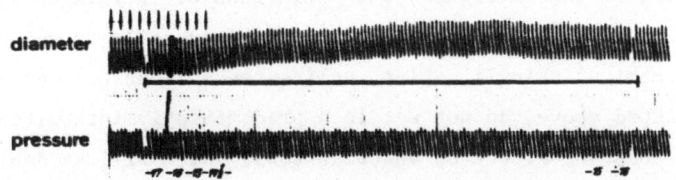

Fig. 5: Original tracings of Schretzenmayr's classical experiment:
During stimulation of the hindleg motor nerves (indicated by the
line between the tracing) the diameter of the femoral artery
increased (upper tracing), while arterial pressure remained
unchanged (lower tracing).

23187 (0.1 µM) induced a marked vasodilation, which was absent when the
stimulant containing perfusate bypassed the columns. This dilation was
not attenuated by pretreatment of the columns with cyclooxygenase
inhibitor (indomethacin 10 µM).

The demonstration of a luminal EDRF release from vessel segments
cannot be taken tacitly as evidence that a humoral EDRF acts on the
abluminal side of the vessel wall. In view of the marked polarity of
endothelial cells in metabolic and secretory processes, a possible
asymmetry in EDRF release should be taken into account. In this line of
reasoning, it was inferred that - independent from the luminal EDRF
release - the endothelium-dependent dilation is mediated by an unknown
(electrical ?) mechanism via myoendothelial junctions. However, bioassay
experiments performed in order to substantiate the existence of an
abluminal EDRF release revealed that this abluminal EDRF release (i.e.,
towards the media cells) is much greater than the luminal release (Fig.
4).

2.2 CONTINUOUS STIMULATION OF ENDOTHELIUM

In addition to the various endogenous and exogenous compounds a
number of physico-chemical stimuli seem to induce endothelium-mediated
dilation under various physiological conditions. Such stimuli are
increases in flow rate (accordingly reductions in flow rate result in
diameter reductions), presence and strength of pulsatility (rhythmic

changes in vessel circumference) and variations of intravascular pO_2 (e.g., localized selective hypoxia of the intima). The stimulater effects of physico-chemical stimuli, which, unlike the various hormonal-humoral compounds listed above, do not act in a predominantly intermittent manner, but exert a constant effect on vascular tone, which will be dealt with in this chapter.

2.2.1. FLOW-DEPENDENT DILATION (FDD)

During functional hyperemia, Schretzenmayr described as early as 1933 a dilation of the cat's femoral artery in situ. His original tracings are presented in Fig. 5. Fleisch in 1935 confirmed this observation. The mechanism for this dilation remained unclear. Fleisch (1935) and Hilton (1959) suggested as the mechanism the spreading of a relaxing signal from the arterioles upward along the vascular wall as a "cable". Consequently Fleisch introduced the name "ascending dilation". However, after the experimental dissection of the artery (Lie, et al., 1970) the "ascending dilation" could still be elicited and, therefore the name had to be corrected. In 1981 Gerova et al. demonstrated the presence of such a flow dependent dilation in coronary arteries of the dog (in situ preparation). In conscious animals such a dilation was observed by Hintze and Vatner (1984). In isolated, perfused canine coronary arteries in vitro, this flow dependent dilation was abolished by endothelial removal (Holtz, et al., 1983a). Therefore, the release of a dilator signal or compound from the endothelium resulting from increases in shear stress was suggested as a potential mechanism (Holtz, et al., 1983b). Thus, the vasomotion in response to changes in shear stress acting on the luminal surface of the endothelium would imply a mechano-transducer function of the endothelial cells (Fig. 6).

2.2.1.1. ENDOTHELIAL CELLS AS FLOW SENSORS

In 1975, Rodbard published a prediction of such an endothelial transducer function. This was a rather remarkable prediction, since at that time no experimental hints of any endothelium-mediated vasomotion existed. He summarized reflections on the mechanism of changes in arterial caliber induced by chronic changes in flow and proposed four steps (Fig. 7). Starting from the control situation with a certain

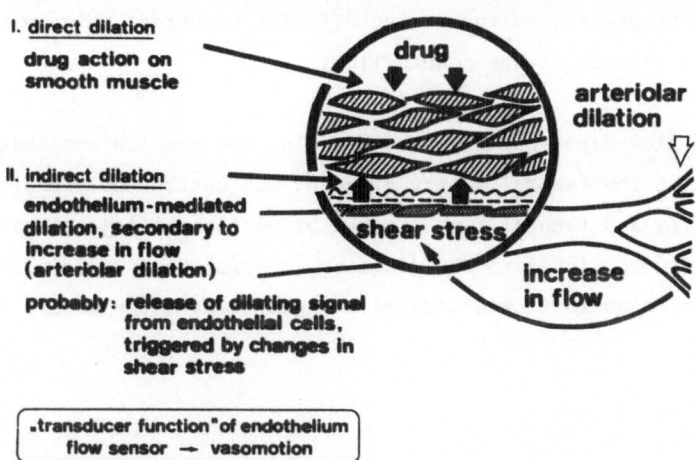

I. **direct dilation**

drug action on smooth muscle

drug

arteriolar dilation

II. **indirect dilation**

endothelium-mediated dilation, secondary to increase in flow (arteriolar dilation)

probably: release of dilating signal from endothelial cells, triggered by changes in shear stress

shear stress

increase in flow

„transducer function" of endothelium flow sensor → vasomotion

Fig. 6: Scheme of the mechano-transducer function of endothelial

cells modifying smooth muscle tone in addition to direct effects

of drugs, transmitters or hormones.

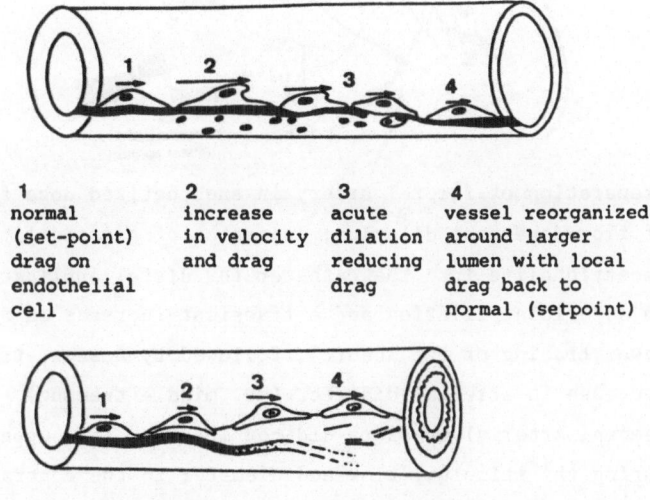

1	2	3	4
normal (set-point) drag on endothelial cell	increase in velocity and drag	acute dilation reducing drag	vessel reorganized around larger lumen with local drag back to normal (setpoint)

Fig. 7: Rodbard's prediction of "flow-dependent, endothelium-mediated

dilation" (step 3 in the upper panel) and of a dual influence of

flow on arterial diameter (which is included in step 3 of the

lower panel). Both predictions could be verified in the coronary

system of healthy dogs 10 years later.

viscous drag acting on the endothelium (1), any increase in flow would

increase this drag (2), this increase triggering an acute dilation (3), pressure, a progressively increasing dilation occurred, which was abolished by removal of the endothelium.

Such a flow-dependent dilation was also observed in anesthetized dogs, in which the femoral artery diameter was continuously monitored by sonomicrometry and femoral flow altered by peripheral injections of vasoactive drugs (distal to the diameter sensors (Fig. 8A)). The resultant increase in flow triggered a significant large artery dilation,

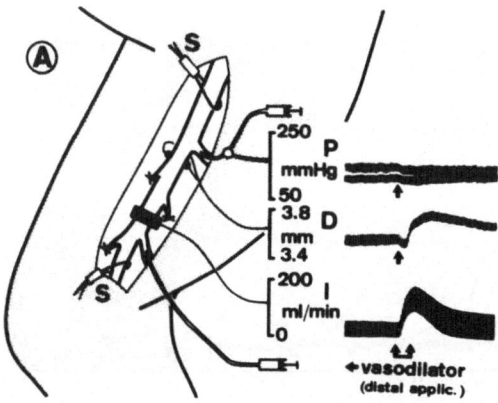

Fig. 8A: Preparation of femoral artery in anesthetized dogs for analysis of flow-dependent dilation: Injection of a vasodilator (acetylcholine into the catheter for distal application) induced an arteriolar dilation and a transient increase in flow (=I, lower tracing of the insert), followed by a slow, transient increase in arterial diameter (=D, middle tracing), while femoral arterial pressure did not increase (=P, upper tracing). During the stimulus, flow and diameter in the contralateral leg did not change.

which reduces this drag and finally the vessel is reorganized around this larger lumen (4) if the elevated flow persists chronically.

Thus, step 3 is a prediction of the "flow-dependent endothelium-mediated dilation." The same sequence of events occurring with chronic reductions in flow was also proposed. Step 3 (bottom part) predicts a constriction of the artery, when flow is reduced. Both predictions require the assumption of the potential existence of a dual influence of

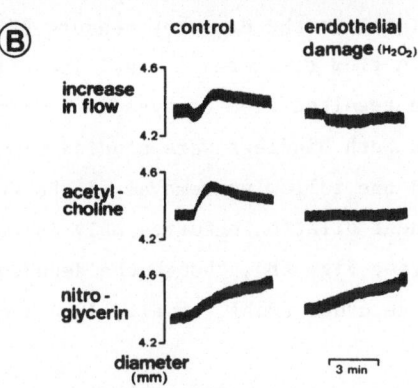

Fig. 8B: Diameter tracings from a typical experiment according to Fig.
8A. Left hand tracings: Femoral artery with intact endothelium
(control). Right hand tracings: Femoral artery with endothelium
removed by rubbing. Stimuli were applied prior to and following
rubbing. Stimuli from top to bottom: —increase in flow
(induced by distal application of a vasodilator) induces
arterial dilation only in the artery with intact endothelium.
Note that the vasodilator during distal application did not
reach the site of diameter recording (see Fig. 8A). —acetylcholine applied luminally (while flow was kept constant
mechanically) induced arterial dilation in the control artery,
demonstrating functional integrity of the endothelium. —nitroglycerin applied from the adventitial side (by superfusing the
artery) dilated both arteries, demonstrating well preserved
dilatory responsiveness of the denuded artery. (norepinephrine
applied from the adventitial side (superfusion) induced
constrictions in both arteries, not shown in this tracing.)

flow on arterial diameter. This assumption was tested under well
controlled in vitro conditions (Bassenge, et al., 1984). With
augmentation of the perfusion rate and compensation of transmural
pressure, a progressively increasing dilation occurred, which was
abolished by removal of the endothelium.

Such a flow-dependent dilation was also observed in anaesthetized
dogs, in which the femoral artery diameter was continuously monitored by
sonomicrometry and femoral flow altered by peripheral injections of

vasoactive drugs (distal to the diameter sensors (Fig. 8A)). The resultant increase in flow triggered a significant large artery dilation, which could not have resulted from changes in transmural pressure (Pohl, et al., 1986). When both hindlegs were studied simultaneously, one with an intact artery and one following removal of the endothelium by rubbing, then the flow dependent dilation occurred only in the artery with the intact endothelium (see Fig. 8B), though the denuded vessel responded properly to vasoactive drugs (Pohl, et al., 1984 and 1986).

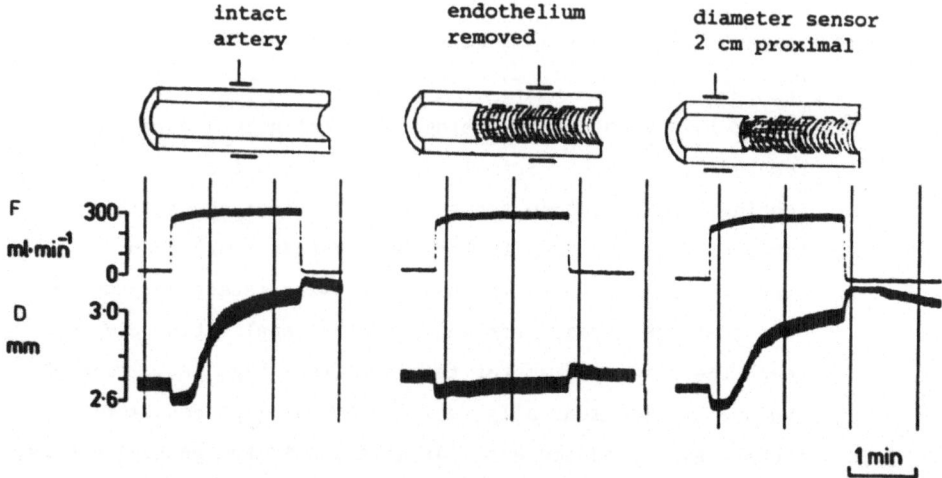

Fig. 9: The endothelium-mediated, flow-dependent dilation is strictly localized: Left hand panel: Increase in flow (F) by opening a shunt induced an increase in arterial diameter (D); Middle panel: Localized de-endothelialization by rubbing abolished the dilation; Right hand panel: Removing the diameter sensor 2 cm proximal to the denuded area demonstrates a well maintained dilation in response to augmented flow. This means that the flow-dependent, endothelium-mediated dilation is not transmitted downstream by a humoral factor. (Modified according to experiments from Smiesko, et al., 1985.)

The flow-dependent dilation was shown by Smiesko, et al. (1985) to be a strictly localized response, and this is true also for the role of the endothelium. When the endothelium underneath the diameter sensor was removed by rubbing, no flow-dependent dilation could be elicited (Fig. 9). Furthermore, it could be demonstrated, that myogenic autoregulation in response to the decline in transmural pressure plays only a negligible role in larger arteries. When the diameter sensor was shifted only 2 cm proximal to the area of endothelial destruction, the dilatory response was

well preserved, while it was abolished 2 cm more distally in the denuded
section (Fig. 9). Therefore, a transfer of the dilating endothelial
signal downstream does obviously not occur, even over distances of

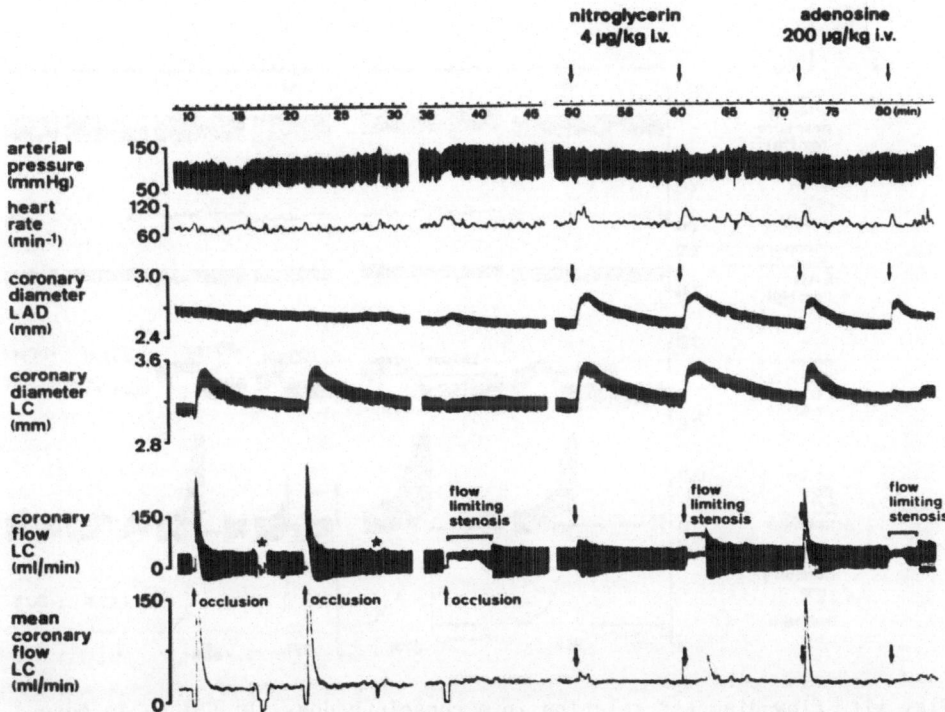

Fig. 10:   Flow-dependent and flow independent epicardial coronary artery
dilations in a conscious dog. Minutes give the time of rest on
the the experimental table; stars indicate electronic zero
balance control of the flowmeter. Abbreviations: LAD = left
anterior descending; LC = left circumflex. Repeated provocation
of reactive hyperemia by temporal (30 s) coronary occlusion
induced reproducible dilations in the affected epicardial
coronary artery (LC branch), but not in the unaffected control
branch (LAD). Prevention of the postocclusion reactive
hyperemia by a flow-limiting stenosis abolished the epicardial
artery dilation. Nitroglycerin induced dilations of both
epicardial arteries, demonstrating responsiveness of the control
branch; a flow-limiting stenosis did not modify this
flow-independent dilation induced by nitroglycerin. Adenosine
induced a short-lasting increase in coronary flow and dilations
in both epicardial branches; prevention of the increase in flow
by a stenosis abolished the epicardial artery dilation only in
this branch. (From Holtz, et al., 1984, by permission.)

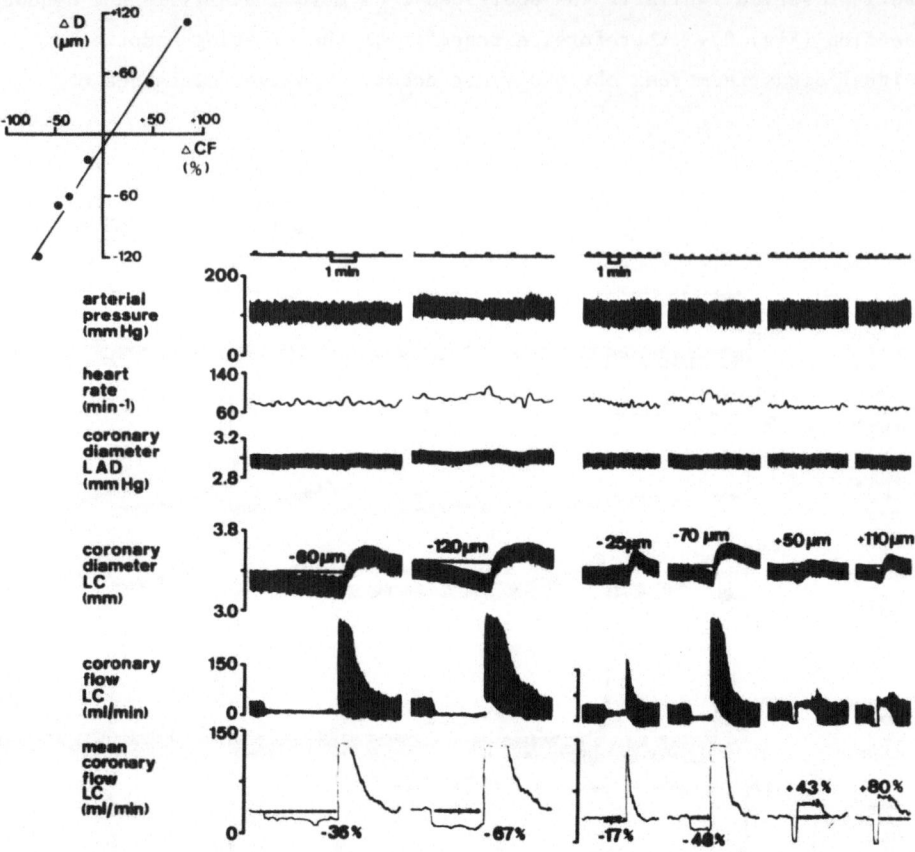

Fig. 11: Flow-diameter relation in a conscious dog. D: Change in mean
diameter (recorded proximally to the site of the implanted
occlusion cuff); CF: Change in mean coronary flow; other
abbreviations as in Fig. 10. The changes in flow were induced
either by partially inflating the occlusion cuff so that mean
coronary flow was kept rather constant at a lower level for at
least 2 min, or by limiting postocclusion (30 s) reactive
hyperemic flow to a rather constant level above preocclusion
flow by only partially deflating the cuff (see two panels on the
right hand side). Ninety seconds after lowering or elevating
mean coronary flow, the percent change in mean coronary flow and
the change (in m) in mean coronary diameter were read and
yielded the flow-diameter relation depicted in the right-hand
insert. Diameter readings were obtained from the electronically
filtered signal, but in this example, the unfiltered, phasic
signal is shown as illustration. Note that the reactive
hyperemias following temporal reductions in flow induced the
typical flow-dependent dilations of the artery. (From Holtz, et
al., 1984, by permission.)

centimeters. This is compatible with a suggested augmentation of EDRF release during enhanced flow rates since EDRF has a short life and is rapidly inactivated in the blood by hemoglobin and other proteins. The flow-dependent dilation was also studied in the intact, undisturbed organism, in the coronary circulation of the conscious, chronically instrumented dog.

The external diameter of two epicardial arteries was continuously monitored, one vessel serving as control. Distal to the sites of diameter recording, a pneumatic cuff served as an experimental stenosis for the induction of various defined flow rates.

Any increase in coronary flow during reactive hyperemia or by adenosine will induce an increase in diameter of the corresponding epicardial artery as shown in Fig. 10. As Rodbard predicted in his hydromechanical deliberations, there should be a dual effect of flow on the vessel diameter via the endothelium. Figure 11 demonstrates the immediate decrease in diameter, when the flow rates are experimentally reduced. On the other hand, doubling the flow rate will increase the diameter by 6%, halving it will decrease the diameter by approximately 6%. A shear stress of 0.8 N/m^2 can be calculated for the flow-dependent dilation of epicardial coronary arteries in Fig. 11. Doubling or halving the flow rate will increase or reduce this value by a factor of 2. Endothelial cells can "sense" shear stress levels of this or slightly lower magnitudes and consequently undergo structural changes, when exposed to shear stress.

Application of various shear stresses will modify cellular functions of endothelial cells substantially. Three different observations document hydromechanical impacts on endothelial structure and cell function:

a) induction of intracellular endothelial stress fibers, elements of the cytoskeleton (Franke, et al., 1984),

b) modifications of endothelial pinocytosis, a reaction of potential significance for the metabolism of plasma constituents (lipids),

c) changes in the shape of the cells and in the orientation of the stress fibres.

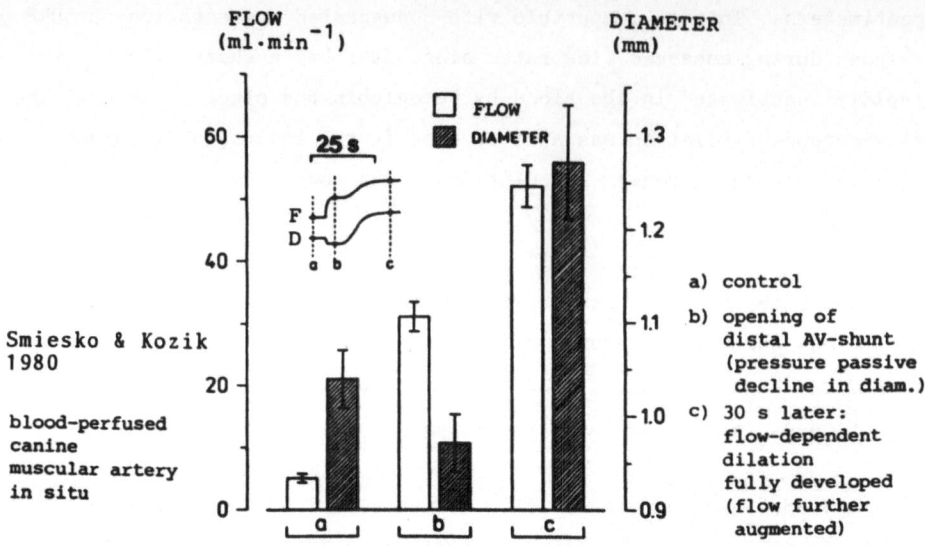

Fig. 12: Flow-dependent dilation (FDD) in a small muscular artery (canine
gracilis muscle preparation) contributes to overall vascular
conductance of the preparation. Mean values ± SD are given of
flow and diameter obtained at three different time points of the
experiments (opening of AV-shunts, see insert): a) prior to
shunt opening; b) 5 s following shunt opening; note that
diameter is now reduced (pressure passive decline in diameter);
c) 30 s following shunt opening; the FDD has now fully developed
and, due to this decline in proximal vascular resistance, flow
has augmented further. (Modified according to Smiesko and Kozik,
1980.)

2.2.1.2. CONTRIBUTION OF FLOW-DEPENDENT-DILATION TO OVERALL CONDUCTANCE

The second implication, resulting from the proposal of shear stress
induced release of EDRF is as follows: the dilatory reaction should be
larger and more rapidly developing in smaller vessels, since an augmented
endothelial/smooth muscle ratio will result in improved diffusion
conditions. The experimental evidence presented so far originated from
large arteries, several mm in diameter, and the flow-dependent dilations
amounted to 5 - 10%, taking 60 - 90 s to reach a maximum response. In
smaller vessels (approx. 1.0 mm) the flow-dependent dilation seems to
reach a significantly higher contribution, as shown in Fig. 12. Thirty s
following a sudden increase in perfusion rate (opening of the shunt) the

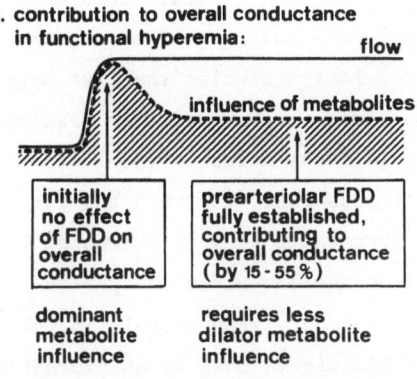

I. adjustment of serial prearteriolar resistances

II. contribution to overall conductance
    in functional hyperemia: flow

influence of metabolites

| initially no effect of FDD on overall conductance | prearteriolar FDD fully established, contributing to overall conductance (by 15-55%) |

dominant metabolite influence

requires less dilator metabolite influence

Fig. 13: Scheme of potential contributory role of flow-dependent dilation
          to the regulation of organ perfusion during functional
          hyperemia: Once flow is augmented due to accumulation of
          metabolites in the tissue, the slowly developing flow-dependent
          dilation of the prearteriolar vessels as well as some myogenic
          autoregulatory dilation of preterminal arterioles will
          contribute to the lowering of overall conductance of the bed,
          thus allowing a lower metabolic concentration for maintaining
          the functional hyperemia during persistence of the metabolic
          stimulus. The quantification (15-55%) of that contribution is
          rather tentative, resulting from experiments shown in Fig. 14.

flow-dependent dilation has already fully developed, eliciting an increase
in internal diameter by up to 45% (Smiesko and Kozik, 1980). Thus, the
dilation is much more pronounced and more rapidly achieved in smaller
arteries. In addition, this example illustrates another aspect.
Immediately following the opening of the shunt, the flow rate is augmented
5-fold, while the diameter is reduced due to the sudden decline in
transmural pressure (state b). Within the next ten seconds, the
flow-dependent dilation of the artery develops, which results in a
superimposed doubling of the flow rate (state c).

Under these experimental conditions, the flow-dependent dilation
contributes substantially to the overall conductance of the preparation,
leading to the more general questions: Is there such a contribution also
under less artificial conditions? What is the physiologic significance of
this dilation? The contribution of flow-dependent dilation apparently

Fig. 14: Effect of repeated coronary occlusions on peak active hyperemic
flow. Redrawings from a typical experiment in a conscious dog.
Abbreviations: CD = coronary diameter; CF = coronary flow.
Upper tracings: Post-occlusion reactive hyperemia induced a
flow-dependent large artery dilation lasting roughly 10 min.
Lower tracings: If the stimulus (occlusion) is repeated when
flow is back to baseline, but when the large artery is still
dilated, peak hyperemic flow is augmented by 15% (as is peak
diastolic conductance). Since changes in arterial capacity are
not taken into account, when calculating peak diastolic
conductance, this figure must be an overestimate of the real
change in conductance. However, in the smaller arteries the
flow-dependent dilation is appearing and disappearing more
rapidly, thus, the contribution of flow-dependent dilation to
overall conductance might be underestimated by this experiment.
When the same experiment is repeated under nitroglycerin
infusion (0.5 µg/kg/min) at a dosage not altering arterial
pressure of coronary flow, the large arteries are dilated and
peak postocclusion conductance is augmented by 55%, but repeated
occlusions no longer modify the peak reactive hyperemic flow
further. The value of 55% must be an overestimate of the
potential contribution of flow-dependent artery dilation to
overall conductance, since capacity effects are not taken into
account. These considerations assume without definite evidence,
that a constant stimulus (constant time of occlusion) will yield
a constant metabolic signal, and this again will result in a
constant dilation of the resistance vessels. Because of these
unproven assumptions, these quantifications must remain
tentative and speculative.

increases from large to medium sized arteries. However, there is little
evidence on its potential relevance within the important terminal
arterioles.

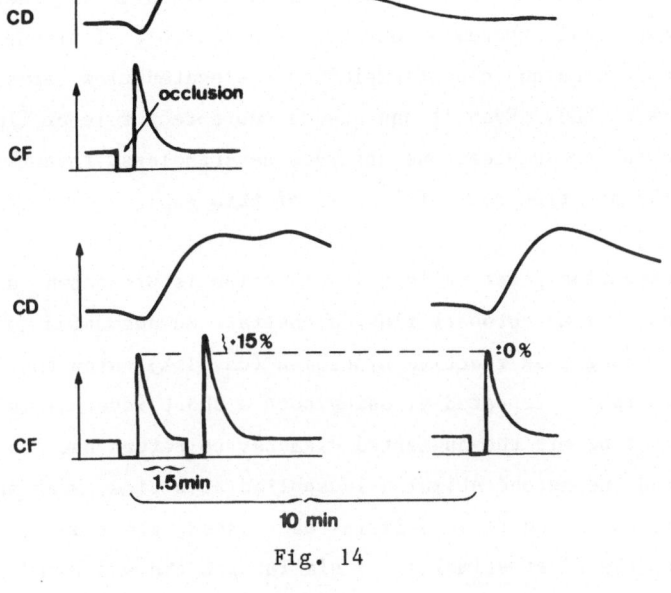

CD

CF

occlusion

CD

CF

·15%

:0%

1.5 min

10 min

Fig. 14

2.2.1.3. POTENTIAL REGULATORY FUNCTION OF FLOW-DEPENDENT DILATION

The flow-dependent dilation may be a tuning factor adjusting the conductance of the prearteriolar vessels during sudden changes in distal arteriolar resistance in such a way, that locally arising alterations in wall shear stress are minimized. In addition, the flow-dependent dilation may contribute to overall conductance during functional hyperemia in such a way, that two phases of functional hyperemia are superimposed (Fig. 13). If one assumes, that the metabolic rate of a particular organ increases to a new, elevated steady state, then this is associated with an appropriate elevation of flow to a new steady state, according to the local metabolic regulation of flow during functional hyperemia. Initially, this increase is completely due to metabolic dilation of the arterioles, since the flow-dependent dilation of the larger vessels occurs later. In the following second phase, however, when these prearteriolar vessels are already dilated, the effect of the vasoactive metabolites on the arterioles must be reduced.

Interestingly, this is exactly what integrated investigations of local metabolic vasodilation describe. Detailed analyses of metabolic dilation reveal that only the initial dilation might be explained by accumulating tissue metabolites, but leave open the particular question, by which mechanisms the elevated flow is maintained during the steady state of functional hyperemia, during which the role of tissue metabolites seems to be reduced and cannot explain the elevated flow rates (Shepherd, 1985; Sparks, 1980). Even if one assumes no direct role of flow-dependent dilation on the arterioles, one observes nevertheless, that the prearteriolar dilation could fill part of this gap.

An observation pointing in this direction is presented in Fig. 14. The tracings of mean coronary flow demonstrate an augmented coronary conduction during peak reactive hyperemia (ca. 15%), when the dilating stimulus is applied repeatedly, using such a short interval in between, which does not permit the augmented diameter to return back to control. However, when the second stimulus is applied at a time, when the large vessel dilation of the first stimulus had ceased, the flow response is identical to the first stimulus. A similar pattern was found using a variety of dilating stimuli acting on resistance vessels, applied as supramaximal bolus injections, which again elicited an augmentation of 15% in peak vascular conductance (if simultaneous capacity effects are neglected). This figure should be an underestimate, since the flow-dependent dilation of the medium sized arteries is occurring more rapidly and has ceased earlier, than the shortest intermittent interval, depending on the return of coronary flow to prestimulus control. When the same stimuli with the same short intervals under the conditions of nitro-induced large vessel dilation were applied, no such augmentation in the response to the second stimulus (compared to that induced by the first stimulus) is observed. Both flow responses are larger than without the nirate. These experiments are in accordance with the hypothesis that the flow-dependent dilation is an important contributor to the physiological regulation of organ perfusion. This hypothesis does not require the assumption of a direct contribution of the FDD within the arterioles.

Rodbard's (1975) considerations included the concept that any alterations in wall shear stress resulting from the induction of new flow levels should be minimized by the consecutively induced caliber changes. This concept was tested experimentally in chronic animal preparations. Approximated shear stress values were obtained by calculations according to Hagen-Poiseuille's law. It became obvious that in the chronic adaptation state the wall shear stress was indeed completely normalized up to a maximum of a 4-fold increase in flow. Only when flow rates exceeded this limit the calculated shear stress became augmented chronically.

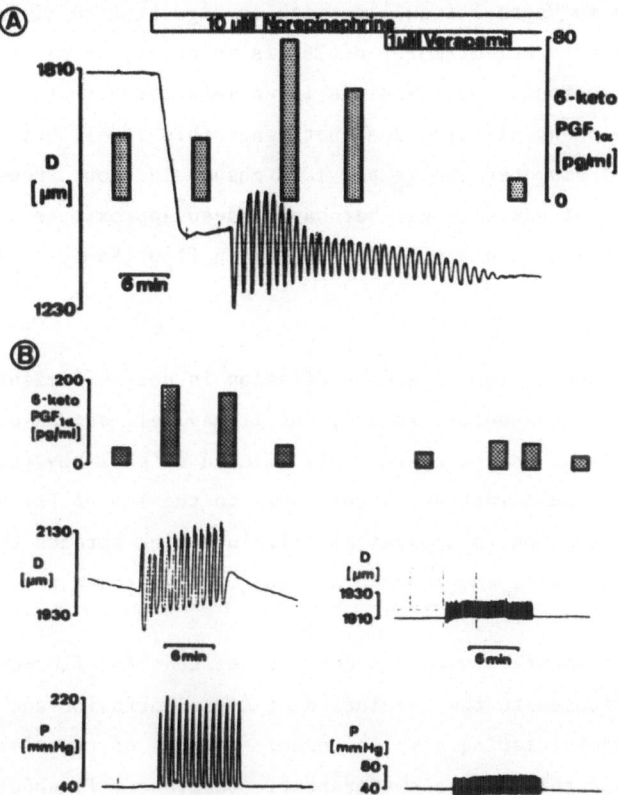

Fig. 15: Effect of rhythmic diameter oscillations on PGI$_2$-release A)
Diameter oscillations induced by rhythmic vascular activity:
activation of an isolated vessel segment (side branch of canine
femoral artery) by norepinephrine induces a contraction
(decrease in outer diameter (D)) with subsequent rhythmic
activity. During rhythmic diameter oscillations, PGI$_2$-release
(measured as 6-keto-PGF$_{1\alpha}$ concentration in the effluent) is
enhanced (hatched bars). Perfusion pressure virtually does not
change during contraction and rhythmic oscillation (recording
not shown). Administration of verapamil to the perfusate
inhibits both, rhythmic activity and enhanced PGI$_2$-release.
B) Diameter oscillations induced by pulsatile perfusion: in an
isolated vessel segment, (side branch of canine femoral artery,
relaxed) superimposition of large pressure pulses (P) induced
diameter oscillations (D) (left hand recordings). Fractionated
sampling of the effluent from the vessels reveals a significant
increase in 6-keto-PGF$_{1\alpha}$ concentration (bars, top panel), during
pulsatile perfusion. Smaller pressure and hence diameter
amplitudes (right hand recordings) induce only minor changes in
PGI$_2$-release. (For details see Pohl, et al., 1985b.)

According to Hagen-Poiseuille, with an augmentation of flow by 100% an increase in internal diameter of 26% is necessary to maintain wall shear stress constant. Observations in animals demonstrate that the acute flow-dependent dilation does not reach this level, but is only in the range of 6% diameter change per 100% change in flow. However, chronic reorganizations of vascular caliber have indeed approximated these levels of 26% diameter increase per 100% increase in flow (Kamiya and Togawa, 1980).

Thus, the acute flow-dependent dilation is not sufficient for a complete shear stress normalization, but it may well act as an initiating trigger mechanism. Once the artery is dilated slightly by this mechanism, the wall tension must increase accordingly to the law of LaPlace. An increased wall tension is a physical stimulus which spreads to all muscle cells within the wall, and therefore might induce growth processes.

From our present view we can demonstrate that the flow-dependent dilation contributes to the regulation of organ perfusion and that it might act as an initiating step in chronic changes of vascular caliber. Furthermore, and this may gain therapeutic relevance, it should modify the coronary vasomotor effects of drugs.

2.2.2. ENDOTHELIAL STIMULATION BY PULSATILE PERFUSION

Another physiochemical stimulus may also potentially affect vascular tone by an endothelium-mediated mechanism. From endothelial cells in culture it is well known that mechanical stimuli (pulsatile flow, flow changes) induce an enhanced PGI_2-release (Frangos, et al., 1985; Grabowski, et al., 1985; van Grondelle, et al., 1984). In the in situ perfused rat aorta, pulsatile perfusion at very high oscillation frequencies induced the release of prostacyclin like activity, which was not observed during non-pulsatile perfusion (Quadt, et al., 1982). When isolated vessel segments are perfused either in a pulsatile or non-pulsatile manner the release of prostacyclin is significantly enhanced during pulsatile perfusion (Pohl, et al., 1985b). Systematic investigations on the nature of the pulsatility related stimulus gave evidence, that the amplitudes of diameter oscillations may be a decisive factor for the enhanced release of PGI_2. Large diameter oscillations (experimentally induced by mechanically increased pulse pressure) result

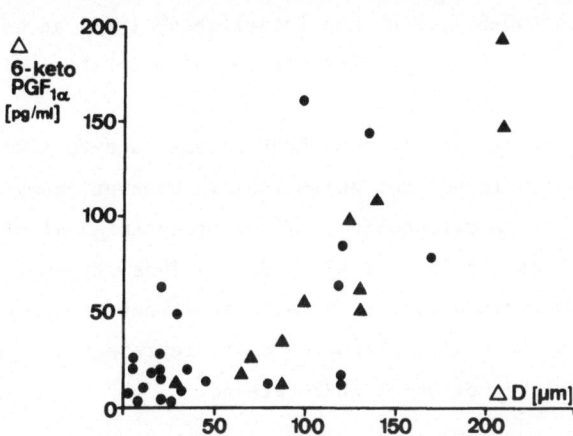

Fig. 16: Augmenting effect of rhythmic diameter oscillations, either elicited by agonist rhythmicity (Δ) or by pulsatile perfusion (o) on PGI_2-release (measured as 6-keto-$PGI_{1\alpha}$ concentration of the effluent). Each symbol represents a separate experiment. ΔD: Amplitude of diameter oscillation. Δ-6-keto-$PGF_{1\alpha}$: Increase in release compared to control (= steady nonpulsatile perfusion).

in a substantially enhanced PGI_2-release, while small oscillations hardly augment the basal release (Fig. 15A).

An enhanced PGI_2-release is also observed, when diameter oscillations result solely from vascular rhythmic activity induced by various agonists although - in contrast to pulsatile perfusion - the intravascular pressure is virtually constant in the vascular preparations. Upon the slow disappearance of these oscillations also the increased PGI_2-release ceases slowly (Fig. 15B). Thus, it appears that the PGI_2-release correlates in fact with the size of diameter amplitudes and not with the pulse pressure (Fig. 16). These findings suggest, that the rhythmic stretching of the vascular wall acts as adequate stimulus for PGI_2 production. It is well known that PGI_2 released in appropriate concentrations dilates the majority of vascular preparations. However, in the physiological range (below 50 μm diameter amplitudes) PGI_2 apparently is not released in vasoactive amounts by pulsatile stimulation, at least in great vessels. This does not exclude, that in the microcirculation (diameter amplitudes during vasomotion 40-80%; Funk and Intaglietta, 1983) an enhanced PGI_2-release contributes significantly to the modulation of vascular tone.

Whether or not a simultaneous EDRF release occurs also during pulsatile perfusion is not yet established. However, many stimulators of EDRF release (e.g., acetylcholine, ATP or bradykinin) simultaneously enhance PGI_2-release (Busse, et al., 1985). Thus, by analogy, it could be supposed, that pulsatile perfusion may also affect the EDRF release, which would imply, that pulsatile perfusion could represent a significant continuous stimulator of basal EDRF-release.

2.2.3. ENDOTHELIUM-DEPENDENT HYPOXIC DILATION

For many years it has been established both in vitro and in vivo experiments, that most vessels responded to hypoxia with a dilation (Chang and Detar, 1980; Bachofen, et al., 1971). These dilator responses have been explained either by the release of vasoactive metabolites from the hypoxic tissue or by direct effects of oxygen lack on vascular smooth muscle. It has been hypothesized that the hypoxic dilation of isolated vascular preparations may represent an in vitro artifact resulting from diffusion limitations within vascular smooth muscle layers under the specific conditions in a saline, non-blood perfused organ bath ("anoxic core hypothesis"; Pittman and Duling, 1973). However, in vivo there is evidence, that oxygen exerts a direct effect on vascular tone by acting on structures of the vascular wall far above hypoxic levels (Jackson and Duling, 1983; Pohl, et al., 1982). These findings imply the existence of specific "oxygen sensors" in the vascular wall. Although the vascular smooth muscle without any doubt shows a dependency of contractile state of pO_2, and thus may act as an oxygen sensor, from a functional point of view the vascular endothelium simply would represent the ideal sensing structure for rapid adaption of vascular tone to changes of intravascular plasma pO_2 (Busse and Bassenge, 1984).

We, therefore, investigated whether or not the endothelial cells react to changes in pO_2 by the release of vasodilator signals. For this purpose a model was developed, which allows the induction of intimal hypoxia, while the majority of the structures of the media can be maintained in a well oxygenated state (Busse, et al., 1983). Under the conditions of endothelial hypoxia in isolated vessels a dilator response

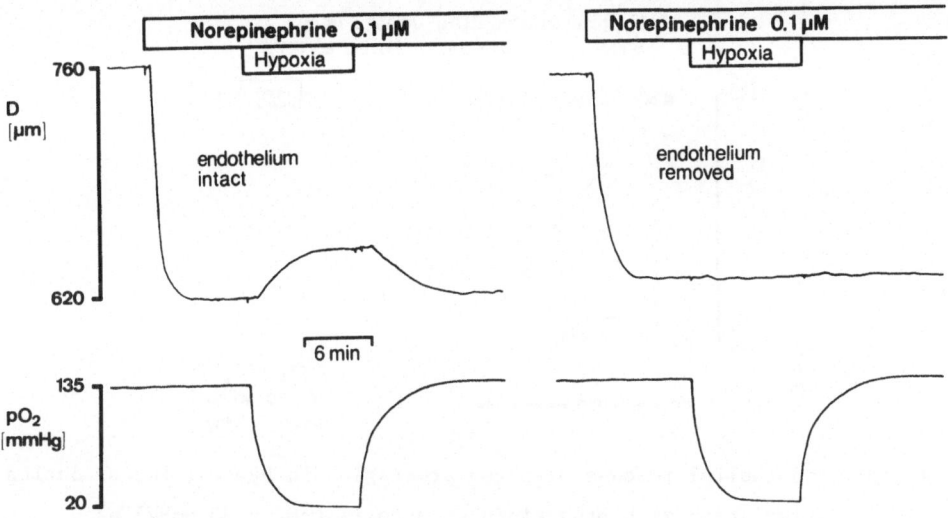

Fig. 17: Effect of intraluminal hypoxic perfusion on the tone of an
isolated vascular segment (rat tail artery). D = outer vascular
diameter, pO_2 = oxygen tension of the intraluminal perfusate
(the segment is mounted in an organ with an adventitial pO_2
constantly above 400 mmHg). With intact endothelium, lowering
of the intraluminal pO_2 elicits a vasodilation in the segment
activated by norepinephrine (left hand recordings). After
endothelial removal (rubbing), in the same segment no
vasodilation during intraluminal hypoxia occurs (right hand
recordings), which suggests that the hypoxic dilation needs the
presence of intact endothelial cells.

was observed, which was abolished after endothelial removal (Fig. 17).
Thus it appears, that the responses to hypoxia results from the
endothelial release of a vasodilator signal.

2.2.3.1. NATURE OF HYPOXIA INDUCED DILATOR COMPOUNDS

In the rat tail artery, enhanced PGI_2 production induced by
endothelial hypoxia apparently contributes to the endothelium-mediated
hypoxic dilation (Busse, et al., 1984). However, hypoxic dilation in the
presence of an intact endothelium, although attenuated, is also observed
in some vessels after indomethacin treatment (canine femoral artery) as
well as in vessels known to be relatively insensitive to PGI_2 (rabbit

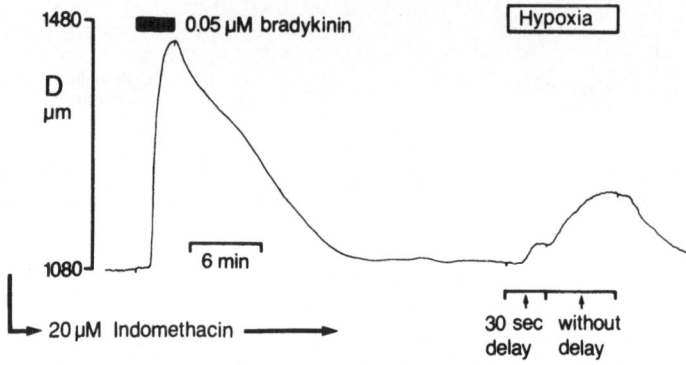

Fig. 18: Endothelial release of a non-prostaglandin humoral factor during
stimulation with bradykinin or hypoxia (pO_2 = 33 mmHg) as
demonstrated in a bioassay experiment. Donor = cultured
endothelial cells (bovine aorta); detector = isolated segment of
rabbit femoral artery (endothelium denuded) whose outer diameter
(D) is continuously recorded (see also Fig. 3). In the presence
of indomethacin, endothelial stimulation with bradykinin induces
a dilation of the detector perfused with the effluent of
cultured cells. Hypoxic perfusion of the endothelial cell
column also induces a vasodilation in the detector (hypoxic
perfusion of the detector alone had no effect). The humoral
factor(s) which is released from the endothelial cells during
hypoxic perfusion is obviously unstable as yielded from the
reduced dilation, when the effluent passes a delay coil.

femoral artery). Therefore besides PGI_2, other nonprostaglandin
vasodilator signals probably are released from the hypoxic endothelium
(Pohl, et al., 1985a).

Recently in bioassay experiments (cultured endothelial cells combined
with vessel segments denuded of endothelial cells as detector) we could
demonstrate the transfer of a non-prostaglandin humoral vasodilating
factor during endothelial cell hypoxia (Fig. 18).

2.2.3.2. PHYSIOLOGICAL SIGNIFICANCE OF ENDOTHELIUM-DEPENDENT HYPOXIC DILATION

From experiments using various intraluminal (endothelial) pO_2's (while adventitial pO_2 was maintained above 400 mmHg) there is evidence that oxygen exerts a gradual tonic influence on the endothelium-mediated vasodilator release. Over a wide range of pO_2-values (100-20 mmHg) a close correlation between dilator response and intimal pO_2 was observed. No evidence was found for a certain hypoxic threshold value necessary to trigger the hypoxic vasodilation.

The quantitative significance of the endothelium-mediated hypoxic vasodilation in the adjustment of vascular tone to changes in oxygen supply has not yet been established. Nevertheless, it is tempting to hypothesize, that in the area of resistance vessels, the endothelial cells act as oxygen sensors, allowing a rapid adjustment of organ flow under conditions of arterial hypoxemia. Thus part of the regulation of oxygen supply to tissue would be performed by structures, which do not need the occurrence of tissue ischemia and increased metabolites as necessary triggers. However, the role of the endothelium-dependent vasodilation has to be further clarified with respect to the in vivo situation. In peripheral segments of rabbit femoral vessels the endothelial component could not be clearly separated from concomitant hypoxia-induced effects on vascular smooth muscle (hypoxic vasodilation, albeit attenuated, also in the endothelium denuded control segments). In single preparations, very rarely, even a hypoxic contraction sensitive to indomethacin was observed. This indicates a rather delicate balance of different vasoactive signals originating from endothelium and vascular smooth muscle under hypoxic conditions, which may depend on the actual oxygen gradient over the vascular wall and the endothelium/smooth muscle ratio.

3. CONCLUSION

In summary there is good evidence that the actual large artery tone represents a net effect of direct actions on vascular smooth muscle and indirect actions mediated by stimulation of the endothelium to release vasomotor signals. This stimulation can result from various endogenous (humoral-hormonal factors, transmitters, a number of peptides and proteins) or exogenous compounds (mainly vasoactive drugs) or from the varying strength of stimulation by different physico-chemical phenomena like viscous drag, pulsatile stretching and intravascular hypoxia acting on the luminal endothelial surface. Since these latter stimuli exert a continuous influence on the vascular endothelium, their physiologic significance resides in the maintenance of a basal state of endothelial cell stimulation resulting in continuous release of vasoactive signals.

ACKNOWLEDGEMENTS

Supported by Deutsche Forschungsgemeinschaft, grant 408/12-5.

REFERENCES

Adelstein, R.S., and Eisenberg, E., 1980, Regulation and kinetics of the actin-myosin-ATP interaction, Ann. Rev. Biochem., 49:921-956.

Aksoy, M.O., Mras, S., Kamm, K.E., and Murphy, R.A., 1983, Ca^{2+}, cAMP, and changes in myosin phosphorylation during contraction of smooth muscle, Am. J. Physiol., 245:C255-C270.

Bachofen, M., Gage, A., and Bachofen, H., 1971, Vascular response to changes in blood oxygen tension under various blood flow rates, Am. J. Physiol., 220:1786-1792.

Bolton, T.B., 1979, Mechanisms of action of transmitters and other substances on smooth muscle, Physiol. Rev., 59:606-718.

Bassenge, E., Holtz, J., Busse, R., and Giesler, M., 1984, Nervous coronary constriction via α-adrenoreceptors: counteracted by metabolic regulation, by coronary β-adrenoreceptor stimulation or by flow-dependent, endothelium-mediated dilation?, in: Breakdown in Human Adaptation to "Stress", A. L'Abbate, ed., Martinus-Nijhoff Publishers, Boston-The Hague-Dordrecht-Lancester, pp. 949-960.

Busse, R. and Bassenge, E., 1984, Endothelium and hypoxic responses, Biblthca. Cardiol., 38:21-34, Karger, Basel-New York.

Busse, R., Förstermann, U., Matsuda, H., and Pohl, U., 1984, The role of prostaglandins in the endothelium-mediated vasodilatory response to hypoxia, Pflügers Arch., 401:77-83.

Busse, R., Pohl, U., Kellner, C., and Klemm, U., 1983, Endothelial cells are involved in the vasodilatory response to hypoxia, Pflügers Arch., 397:78-80.

Busse, R., Trogisch, G., and Bassenge, E., 1985, The role of endothelium in the control of vascular tone, Basic Res. Cardiol., 80: in press.

Casteels, R. and Droogmans, G., 1984, Cell membrane responsiveness and excitation-contraction coupling in smooth muscle, J. Cardiovasc. Pharmacol., 6:S304-S312.

Chang, A.E. and Detar, R., 1980, Oxygen and vascular smooth muscle contraction revisited, Am. J. Physiol., 238:H716-H728.

Droogmans, G., Raeymaekers, L., and Casteels, R., 1977, Electro- and pharmacomechanical coupling in the smooth muscle cells of the rabbit ear artery, J. Gen. Physiol., 70:129-148.

Fleisch, A., 1935, Les reflexes nutritifs ascendants producteurs de dilatation arterielle, Arch. Int. Physiol., 41:141-167.

Frangos, J.A., Eskin, S.G., McIntire, L.V., and Ives, C.L., 1985, Flow effects on prostacyclin production by cultured human endothelial cells, Science, 227:1477-1479.

Franke, R.-P., Gräfe, M., Schnittler, H., Seiffge, D., Mittermayer, C., and Drenckhahn, D., 1984, Induction of human vascular endothelial stress fibres by fluid shear stress, Nature, 307:648-649.

Funk, W. and Intaglietta, M., 1983, Spontaneous arteriolar vasomotion, in: Vasomotion and Quantitative Capillaroscopy, Messmer, Hammersen, eds., Karger, Basel, pp. 66-82.

Furchgott, R.F., 1983, Role of endothelium in responses of vascular smooth muscle, Circ. Res., 53:557-573.

Furchgott, R.F. and Zawadzki, J.V., 1980, The obligatory role of endothelial cells in the relaxation of arterial smooth muscle by acetylcholine, Nature, 288:373-376.

Gerova, M., Gero, J., Barta, E., Dolezel, S., Smiesko, V., and Levicky, V., 1981, Neurogenic and myogenic control of conduit coronary artery: A possible interference, Basic Res. Cardiol., 76:503-507.

Grabowski, E.F., Jaffe, E.A., and Weksler, B.B., 1985, Prostacyclin production by cultured endothelial cell monolayers exposed to step increases in shear stress, J. Lab. Clin. Med., 105:36-43.

Griffith, T.M., Hughes Edwards, D., Lewis, M.J., Newby, A.C., and Henderson, A.H., 1984, The nature of endothelium-derived vascular relaxant factor, Nature, 308:645-647.

Hardman, J.G., 1984, Cyclic nucleotides and regulation of vascular smooth muscle, J. Cardiovasc. Pharmacol., 6:S639-S645.

Hartshorne, D.J. and Mrwa, U., 1982, Regulation of smooth muscle actomyocin, Blood Vessels, 19:1-18.

Hilton, S.M., 1959, A peripheral arterial conducting mechanism underlying dilatation of the femoral artery and concerned in functional vasodilatation in skeletal muscle, J. Physiol. (London), 149:93-111.

Hintze, T.H. and Vatner, S.F., 1984, Reactive dilation of large coronary arteries in conscious dogs, Circ. Res., 54:50-57.

Holtz, J., Busse, R., and Giesler, M., 1983a, Flow-dependent dilation of canine epicardial coronary arteries in vivo and in vitro: Mediated by the endothelium, Naunyn-Schmid. Arch. Pharmacol., 322:R44.

Holtz, J., Forstermann, U., Pohl, U., Giesler, M., and Bassenge, E., 1984, Flow-dependent, endothelium-mediated dilation of epicardial coronary arteries in conscious dogs: Effects of cyclooxygenase inhibition, J. Cardiovasc. Pharmacol., 6:1161-1169.

Holtz, J., Giesler, M., and Bassenge, E., 1983b, Two dilatory mechanisms of anti-anginal drugs onepicardial coronary arteries in vivo: indirect, flow-dependent, endothelium-mediated and direct smooth muscle relaxation, Z. Kardiol., 72, Suppl. 3, 98-106.

Ignarro, L.J. and Kadowitz, P.J., 1985, The pharmacological and physiological role of cyclic GMP in vascular smooth muscle relaxation, Ann. Rev. Pharmacol. Toxicol., 25:171-191.

Jackson, W.F. and Duling, B.R., 1983, The oxygen sensitivity of hamster cheek pouch arterioles: In vitro and in situ studies, Circ. Res., 53:515-525.

Kamiya, A. and Togawa, T., 1980, Adaptive regulation of wall shear stress to flow change in the canine carotid artery, Am. J. Physiol., 239:H14-H21.

Lie, M., Sejersted, O.M., and Kiil, F., 1970, Local regulation of vascular cross section during changes in femoral arterial blood flow in dogs, Circ. Res., 27:727-737.

Lincoln, T.M. and Corbin, J.D., 1978, On the role of the cAMP and cGMP-dependent protein kinases in cell function, J. Cycl. Nucl. Res., 4:3-14.

Morgan, J.P. and Morgan, K.G., 1984, Stimulus-specific patterns of intracellular calcium levels in smooth muscle of ferret portal vein, J. Physiol., 351:155-167.

Pittman, R.N. and Duling, B.R., 1973, Oxygen sensitivity in vascular smooth muscle. I. In vitro studies, Microvasc. Res., 6:202-211.

Pohl, U., Busse, R., and Bassenge, E., 1985a, Endothelium-dependent hypoxic vasodilation: endothelium derived relaxant factor (EDRF) as mediator? Circulation, 72, II-...

Pohl, U., Busse, R., and Kessler, M., 1982, Vascular resistance and tissue pO_2 in skeletal muscle during perfusion with hypoxic blood, in: Cardiovascular System Dynamics, T. Kenner, R. Busse, eds., Plenum Press, New York and London, pp. 521-530.

Pohl, U., Forstermann, U., Busse, R., and Bassenge, E., 1985b, Endothelium-mediated modulation of arterial smooth muscle tone and PGI_2-release: Pulsatile versus steady flow. In: "Prostaglandins and Other Eicosanoids in the Cardiovascular System", K. Schror (Ed.), 553-558, Karger, Basel.

Pohl, U., Holtz, J., Busse, R., and Bassenge, E., 1984, Dilation of large arteries in response to increased flow in vivo: an endothelium-dependent reaction, Circulation, 70:II-123.

Pohl, U., Holtz, J., Busse, R., and Bassenge, E., 1986, Crucial role of endothelium in the vasodilator response to increased flow in vivo, Hypertension, 8:33-41.

Quadt, J.F.A., Voss, R., and Hoor, F.T., 1982, Prostacyclin production of the isolated pulsatingly perfused rat aorta, J. Pharmacol. Methods, 7:263-270.

Rodbard, S., 1975, Vascular caliber, Cardiology, 60:4-49.

Ruegg, J.C. and Paul, R.J., 1982, Vascular smooth muscle: Calmodulin and cyclin AMP-dependent protein kinase alter calcium sensitivity in porcine carotid skinned fibers, Circ. Res., 50:394-399.

Schretzenmayr, A., 1933, Über kreislaufregulatorische Vorgänge an den gro Ben Arterien bei der Muskelarbeit, Pflügers Arch. Ges. Physiol., 232:743-748.

Shepherd, J.T., 1985, Circulation to skeletal muscle, in: Handbook of Physiology, Sect. 2, Vol. III, Pt. 1, J.H. Shepherd, F.M. Abboud, eds., Am. Physiol. Soc., Bethesda, pp. 319-370.

Smiesko, V. and Kozik, J., 1980, Dilation response of a small artery of muscular type to increased blood flow, Bratisl. Lek. Listy 73:727-733.

Smiesko, V.,Kozik, J., and Dolezel, S., 1985, Role of endothelium in the control of arterial diameter by blood flow, Blood Vessels, in press.

Sparks, H.V. Jr., 1980, Effect of local metabolic factors on smooth muscle, in: Handbook of Physiology, Sect. 2, Vol. II, D.F. Bohr, A. P. Somlyo, H.V. Sparks, Jr., eds., Am. Physiol. Soc., Bethesda, pp. 475-513.

Streb, H., Irvine, R.F., Berridge, M.J., and Schulz, I., 1983, Release of Ca^{2+} from a nonmitochondrial intracellular store in pancreatic acinar cells by inositol-1,4,5-trisphosphate, Nature, 306:67-69.

Suematsu, E., Hirata, M., Hashimoto, T., and Kuriyama, H., 1984, Inositol 1,4,5-triphosphate releases Ca^{2+} from intracellular store sites in skinned single cells of porcine coronary artery, Biochem. Biophys. Res. Comm., 120:481-485.

van Grondelle, A. van, Worthen, G.S., Ellis, D., Mathias, M.M., Murphy, R.C., Strife, R.J., Reeves, J.T., and Volkel, N.F., 1984, Altering hydrodynamic variables influences PGI_2 production by isolated lungs and endothelial cells, J. Appl. Physiol. Respirant Environ. Exercise Physiol., 57:388-395.

Vanhoutte, P.M. and Rimele, T.J., 1983, Role of the endothelium in the control of vascular smooth muscle function, $\underline{J}$. $\underline{Physiol}$. (Paris), 78:681-686.

Weiss, B., Prozialeck, W.C., and Wallace, T.L., 1982, Interaction of drugs with calmodulin: Biochemical, pharmacological and clinical implications, $\underline{Biochem}$. $\underline{Pharmacol}$., 31:2217-2226.

ANTIHYPERTENSIVE DRUGS: CURRENT AND FUTURE THERAPIES

Harvey R. Kaplan and Michael J. Ryan

Department of Pharmacology
Warner-Lambert/Parke-Davis Pharmaceutical Research
Ann Arbor, Michigan 48105

INTRODUCTION

It is well accepted that persons with an elevated blood pressure, even mild hypertension, are at a greater health risk than those with normal blood pressure. In the United States alone, high blood pressure afflicts an estimated 38 million adults (American Heart Association, 1985). Over 5 million are achieving control of their blood pressure with medication; however, for most hypertensives their elevated pressure remains either undetected, untreated, or inadequately controlled. While our understanding of the pathophysiology of hypertension is incomplete there are several targets for therapeutic intervention. Many of these are illustrated in Fig. 1.

Table 1 presents an historical perspective on therapies used in the twentieth century for the treatment of hypertension and the blood pressure factor they modify. Many very safe and effective antihypertensive drugs are now available as a result of the remarkable advances in efficacy and reduction in side effects and toxicity.

Fig. 2 depicts the incremental advances in antihypertensive therapy over the past three decades. The ultimate therapeutic goal as we approach the "ideal" treatment is to eliminate hypertension-related morbidity and mortality.

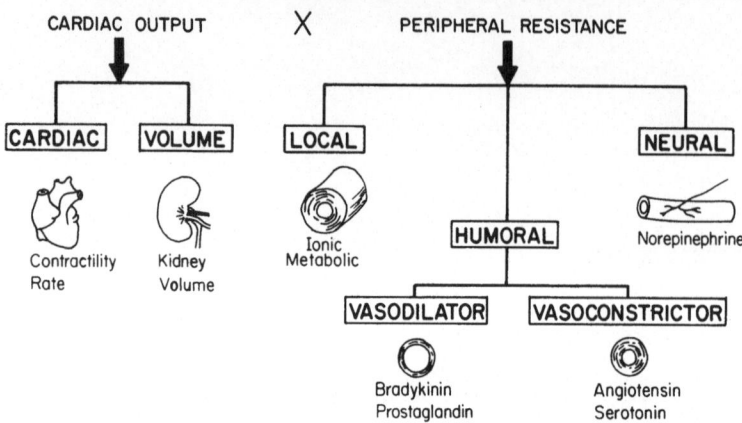

Fig. 1. Some factors involved in the control of arterial blood pressure.
Blood pressure is related to the product of cardiac output and
peripheral resistance. Cardiac output is dependent on the
performance of the heart which is regulated by contractility and
heart rate. The kidney is the principle organ regulating blood
volume. Peripheral resistance is dependent on local, humoral,
and neural factors. Control mechanisms normally regulate each
factor to maintain arterial blood pressure within a narrow range.
The role of the central nervous system while not illustrated
plays an important role in blood pressure regulation.

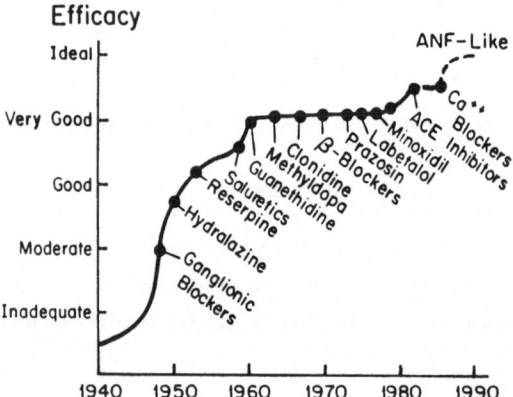

Fig. 2. Past, current and future antihypertensive therapies, incremental
advances as we approach the "ideal." (Adapted from Gross, 1982.)

TABLE 1

Therapies Used in The Treatment of Essential Hypertension

Year of Introduction	Therapy	Factor Modulated
1903	Thiocyanate	Direct Vascular
1925	Surgical Sympathectomy	Neural
1940	Veratrum Alkaloids	Neural
1947	Ganglionic Blockade	Neural
1949	Hydralazine	Direct Vascular
1949	Rauwolfia	Neural
1952	Phenoxybenzamine	Neural
1957	Thiazide Diuretic	Volume
1959	Spironolactone	Volume
1959	Guanethidine	Neural
1960	Alpha-Methyldopa	Neural
1960	MAO Inhibitors	Neural
1962	Diazoxide	Direct Vascular
1964	Furosemide	Volume
1964	Propranolol	Multiple
1964	Bethanidine	Neural
1965	Debrisoquin	Neural
1966	Clonidine	Neural
1969	Guancydine	Direct Vascular
1970	Minoxidil	Direct Vascular
1972	Prazosin	Neural
1973	Guanabenz	Neural
1977	Captopril	Humoral
1984	Labetalol	Neural
1985	Enalapril	Humoral
198?	Nitrendipine	Direct Vascular
1990's	Endogenous Substance eg ANF-like	Multiple

SURVEY OF ANTIHYPERTENSIVE DRUGS - 1985

This chapter surveys current antihypertensive drug therapy and presents perspectives on what we might expect in the pharmacologic management of this disease. Antihypertensive drugs can be categorized

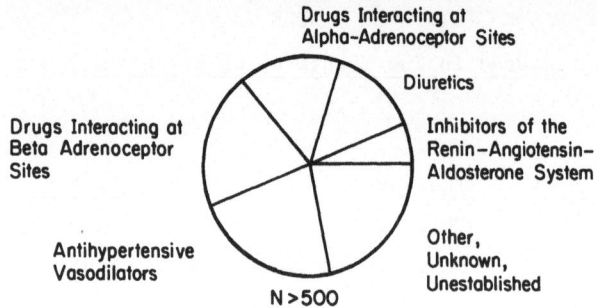

Fig. 3. Antihypertensive drug survey, 1985. Categorization of antihypertensive drugs reported in the world literature (see text for details).

based on their mechanisms and/or sites of drug action. Fig. 3 graphically depicts a classification scheme based on the principal mechanisms of action. The major drug classes include: 1) drugs interacting at alpha adrenoceptor sites, 2) drugs interacting at beta adrenoceptor sites, 3) inhibitors of the renin-angiotensin-aldosterone system, 4) antihypertensive vasodilators, 5) diuretics, and 6) other, unknown, or unestablished sites and/or mechanisms of action. Each section of the pie is proportional to the number of compounds cataloged from the world literature. Over 500 compounds were identified in this survey based on reviews of the world literature in 1981 (Kaplan and Smith, 1981), 1982 (Kaplan, 1983, or Kaplan, et al., 1982), and 1985 (Evans, Bristol, and Kaplan, 1985). Many of these agents may have already been proven unfit for further development because of adverse side effects, toxicity, undesirable pharmaceutical properties, undesirable pharmacokinetics, lack of clinical efficacy or perhaps for commercial reasons. As this figure suggests, the pharmacologist can readily discover compounds that lower blood pressure by a variety of diverse mechanisms. However, very few compounds survive the rigorous scrutiny of preclinical and clinical evaluations to become useful antihypertensive agents. The major classes listed in Fig. 3 are reviewed in this chapter with the exception of diuretics which have been reviewed elsewhere (Evans, Bristol, and Kaplan, 1985).

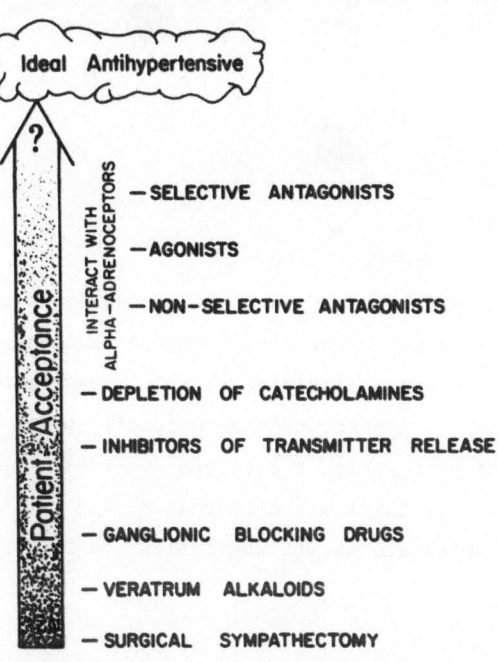

Fig. 4. Relative patient acceptance of various therapeutic interventions designed to interrupt or modify output of the sympathetic nervous system. The ideal antihypertensive agent would be effective and have universal patient acceptance. The apparent relationship between mechanism and side effects makes it uncertain as to whether this can be achieved by drugs interacting at alpha adrenoceptors.

DRUGS INTERACTING AT ALPHA ADRENOCEPTOR SITES

Drugs that interact with alpha adrenoceptors to interrupt or modify output of the sympathetic nervous system have already proven to be effective antihypertensives. In general, drugs that interact with alpha adrenoceptors enjoy better patient acceptance than the earlier approaches for interrupting or modifying output of the sympathetic nervous system (Fig. 4). Alpha adrenoceptors can be divided into $alpha_1$ and $alpha_2$ subtypes based on responses to selective agonists and antagonists. Figure 5 illustrates conceptually some of the possible combinations by which an antihypertensive drug might interact with alpha-adrenoceptor subtypes to reduce blood pressure. Prazosin is a prototype antagonist which is highly

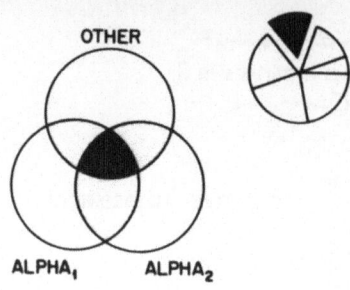

Fig. 5. Possible interactions of an antihypertensive drug with multiple receptor subtypes. Prazosin is highly selective for $alpha_1$ adrenoceptors while clonidine is selective for $alpha_2$ adrenoceptors. Agents, such as urapidil, which appear to interact at both receptor subtypes are represented by the shaded area. Finally, there are compounds having additional components of action which are as yet unestablished.

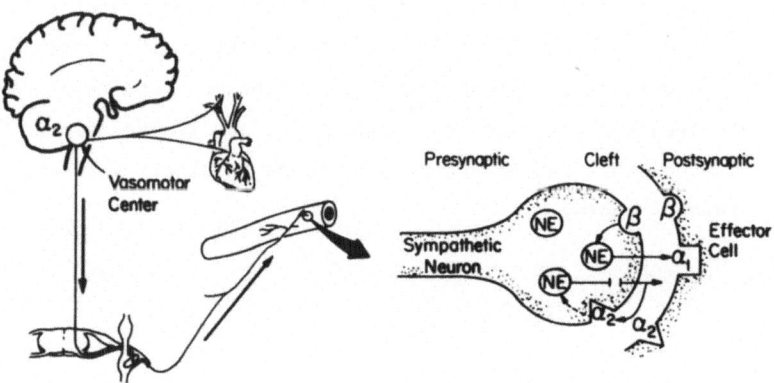

Fig. 6. Central and peripheral alpha-adrenoceptors at which antihypertensive drugs may interact.

selective for peripheral postsynaptic $alpha_1$ receptors which when activated causes contraction of vascular smooth muscle (Fig. 5,6). Stimulation of central $alpha_2$ receptors with clonidine causes an inhibition of sympathetic outflow resulting in a blood pressure reduction. Older alpha blockers such as phentolamine and phenoxybenzamine are non-selective and block both $alpha_1$ and $alpha_2$ adrenoceptor sites. Each prototype has

TABLE 2

Advantages and Limitations of Antihypertensive Drugs which Interact with
Alpha-Adrenoceptors

PROTOTYPE:	ADVANTAGES:	LIMITATIONS:
PHENOXYBENZAMINE non-selective antagonist	-effective treatment for pheochromocytoma	-limited BP effect -postural hypotension -tachycardia -sedation -sexual dysfunction -nasal stuffiness
CLONIDINE central $alpha_2$ agonist	-effective in mild to severe hypertension -reduction in sympathetic outflow	-sedation -dry mouth -rebound hypertension
PRAZOSIN selective $alpha_1$ antagonist	-effective in mild to moderate hypertension -consistent decrease in peripheral resistance -modest effects on heart rate	-postural hypotension -first dose phenomenon -sedation -dry mouth -sexual dysfunction -urinary incontinence -headache

limiting side effects, e.g., sedation and orthostatic hypotension, likely
an extension of their interaction with specific alpha adrenoceptor
subtypes. Their advantages and limitations are summarized in Table 2.

Interest in new approaches has been stimulated based on the clinical
experience that antihypertensives with a mechanism dependent solely on
either $alpha_1$ adrenoceptor blockade or central stimulation of $alpha_2$
adrenoceptors are likely to have limiting side effects. Urapidil is an
example of a new agent under development suggested to have a mixed
clonidine-prazosin type profile (Schoetensack, Bruckschen, and Zech,
1983). Another approach is the development of a post-synaptic $alpha_2$

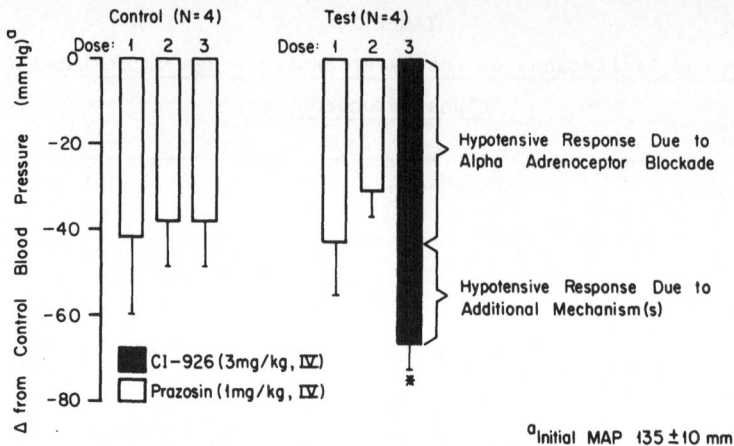

Fig. 7. Vasodepressor effects of CI-926 in the presence of complete
alpha$_1$ adrenoceptor blockade in anesthetized dogs. To determine
if CI-926 has in vivo actions in addition to alpha$_1$ adrenoceptor
blockade, CI-926 was given to anesthetized dogs during the peak
response to super maximal hypotensive doses of prazosin. In
control experiments (left panel), a 1 mg/kg dose of prazosin
lowered blood pressure approximately 40 mm Hg. A second and
third 1 mg/kg dose of prazosin caused no additional reduction in
blood pressure indicating the maximal hypotensive effect obtained
with alpha$_1$ adrenoceptor blockade. In a second group of dogs
(right panel), two 1 mg/kg doses of prazosin were followed by a
single 3 mg/kg dose of CI-926. Under these conditions, CI-926
produced an additional reduction in blood pressure. Asterisk
indicates significantly different from dose 2 (p 0.05) as
determined by one tailed Student's t-test.

antagonist. Such an antagonist might provide an opportunity to antagonize
the contractile effects of circulating epinephrine (e.g., from the
adrenal) acting on alpha$_2$ adrenoceptors while having no interaction with
alpha$_1$ receptors. The later receptor subtype, in close proximity to the
sympathetic neuron (Fig. 6), is presumably the principle receptor activated to maintain blood pressure during postural changes. SKF-86466, a
selective alpha$_2$ antagonist, lowers blood pressure in a dose-dependent
manner in experimental hypertension and has been reported to have less
orthostatic liability than prazosin (Roesler, Kopaciewicz, and Hieble,
1983; McCafferty et al., 1983). Others have coupled alpha$_1$ adrenoceptor

TABLE 3

Antihypertensive Drugs Interacting at Alpha-Adrenoceptor Site(s)

NEWER AGENTS:	COMMENTS
FLA 136 (HASSLE)	A SELECTIVE CENTRAL $ALPHA_2$ AGONIST
SKF 86466 TRIPAMIDE (ADRIA)	$ALPHA_2$ RECEPTOR ANTAGONISTS
D 2343 (DRACO)	$ALPHA_1$ ANTAGONIST-$BETA_2$-AGONIST
ALFLUOSINE (SYNTHELABO)	$ALPHA_1$ ANTAGONIST/CALCIUM ANTAGONIST COMPONENT
CI-926 (WARNER-LAMBERT/ PARKE-DAVIS)	$ALPHA_1$ ANTAGONIST + OTHER MECHANSIM(S)

NEW DIRECTIONS

LINKAGE BETWEEN ENDOGENOUS OPIATES, ALPHA-ADRENOCEPTORS AND THE HYPOTENSIVE PROCESS

ROLE AND PHARMACOLOGICAL IMPLICATIONS OF CENTRAL $ALPHA_1$ AND PERIPHERAL $ALPHA_2$ ADRENOCEPTORS IN HYPERTENSION

PHYSIOLOGICAL AND PATHOPHYSIOLOGICAL SIGNIFICANCE OF ALPHA-ADRENOCEPTOR MEDIATED, ENDOTHELIUM-DEPENDENT RELAXATION OF VASCULAR SMOOTH MUSCLE

BIOCHEMICAL BASIS (TRANSDUCTION) OF EVENTS BEYOND THE ALPHA-RECEPTOR

blockade with an additional blood pressure lowering mechanism such as beta blockade (labetalol), antagonism of serotonin (ketanserin), a calcium antagonist component (alfluosine), $beta_2$ agonist activity (D2343), or other unknown mechanisms (several).

CI-926 is another example of a new drug under development interacting at $alpha_1$ adrenoceptor sites and having an additional as yet undefined component of action (Ryan et al., 1984). Functionally this can be shown (Fig. 7) in an experiment in which CI-926 was given to anesthetized dogs during the peak response to super maximal hypotensive doses of the $alpha_1$ antagonist prazosin. The mechanism of this additional blood pressure lowering for CI-926 remains to be established.

Table 3 summarizes many of the emerging new developments in the area of alpha-adrenoceptor pharmacology. New opportunities for antihypertensive drugs might occur as we begin to understand the linkage between endogenous opiates and alpha-adrenoceptors, the role of central $alpha_1$ and peripheral $alpha_2$ receptors in the pathophysiology of hypertension, the role of alpha receptors in endothelium-dependent relaxation or contraction of vascular smooth muscle, and biochemical events beyond the alpha receptors, i.e., the biochemical basis for the transduction of extracellular signals into intracellular events. There are many recent and interesting laboratory observations in these areas. The opiate receptor antagonists naloxone and naltrexone block the antihypertensive and bradycardic effects of clonidine and clonidine releases an endogenous opiate from SHR but not WKY rats suggesting interactions between central opiates and alpha receptor systems (Farsang et al., 1980 or Kunos, Farsang, and Ramirez-Gonzales, 1981). Norepinephrine is a more powerful vasoconstrictor of vascular smooth muscle in the absence of endothelium and releases a vasodilator substance from endothelial cells that can act as a physiological antagonist of the well known smooth muscle contractile responses (Cocks and Angus, 1983). This effect appears to be mediated in part through $alpha_2$ receptors. Neither the mechanism nor the physiological or pathological significance of these fascinating observations are known. However, these and other interesting laboratory findings will likely lead to more successful alpha adrenoceptor drug therapy.

BETA ADRENOCEPTOR ANTAGONISTS

Several beta blockers of diverse chemical structure are at various stages of preclinical or clinical development throughout the world and at least five are marketed in the US. These agents have been arbitrarily classified according to their secondary characteristics, i.e., cardio-selectivity, intrinsic sympathomimetic activity and membrane stabilizing activity. As newer beta blockers evolved, they have been further classified according to their ability to penetrate into the CNS, their "direct" vasodilator or alpha blocking component of action, their acute hemodynamic effects, their oral bioavailability, and their duration of action. Most evidence suggests that $beta_1$-adrenergic receptor blockade per se is the essential component of action for the reduction of arterial blood pressure by this class of drugs. The juxtaglomerular apparatus of the kidney, the heart, brain, postganglionic sympathetic nerve terminals and vascular smooth muscle have been postulated as target sites for beta blockade contributing to the antihypertensive effects of these agents. The fact that so many different sites have been proposed provides an

TABLE 4

Beta-Adrenoceptor Antagonists

NEWER AGENTS:	COMMENTS
DTI (MERCK), CGP 20712A	HIGHLY CARDIOSELECTIVE
PRIZIDILOL (SKF), ISF 3382	CARDIOSELECTIVE AND VASODILATOR
SCH 19,927 (R,R ISOMER OF LABETALOL)	SELECTIVE VASCULAR (BETA$_2$) INTRINSIC SYMPATHOMIMETIC ACTIVITY
BEVANTOLOL (WARNER-LAMBERT/PARKE-DAVIS)	ALSO AN INTERACTION AT ALPHA ADRENOCEPTOR SITES
ESMOLOL (ACC)	ULTRA SHORT ACTING BETA BLOCKER

NEW DIRECTIONS

- CARDIO-PROTECTION
- INFLUENCE ON LIPOPROTEIN METABOLISM
- TISSUE/SITE SELECTIVITY

opportunity for more selective agents in the future. A detailed
discussion on the sites and mechanisms of the antihypertensive effects of
beta receptor antagonists is beyond the scope of this review. The
interested reader is referred to recent editorials and clinical studies on
the subject (Cowley, 1984 or Korner, 1984 or Colfer et al., 1984). Table
4 lists some of the newer agents and new directions for this class of
compounds.

A very attractive feature of the beta blockers is that they may
reduce the risk associated with coronary heart disease and reinfarction in
hypertensive patients (Stewart, 1982). On the other hand, certain of the
beta adrenoceptor antagonists alter triglyceride and cholesterol metabolism to the extent that an "unhealthy" plasma lipoprotein profile may
result which could promote the atherogenic process. Data from the recent
NIH-sponsored multicenter trials which show that lowering of plasma
cholesterol reduces the risk of coronary heart disease (Lipid Research
Clinics Program, 1984, Brensike et al., 1984, and Levy et al., 1984),
support the need for careful attention to pharmacologic modification of
lipoprotein profiles. The rational use of drugs having positive and
negative risk potential will have to be based on a careful assessment of
their net effects. This clearly represents a major challenge for health
care professionals utilizing new antihyertensive therapy.

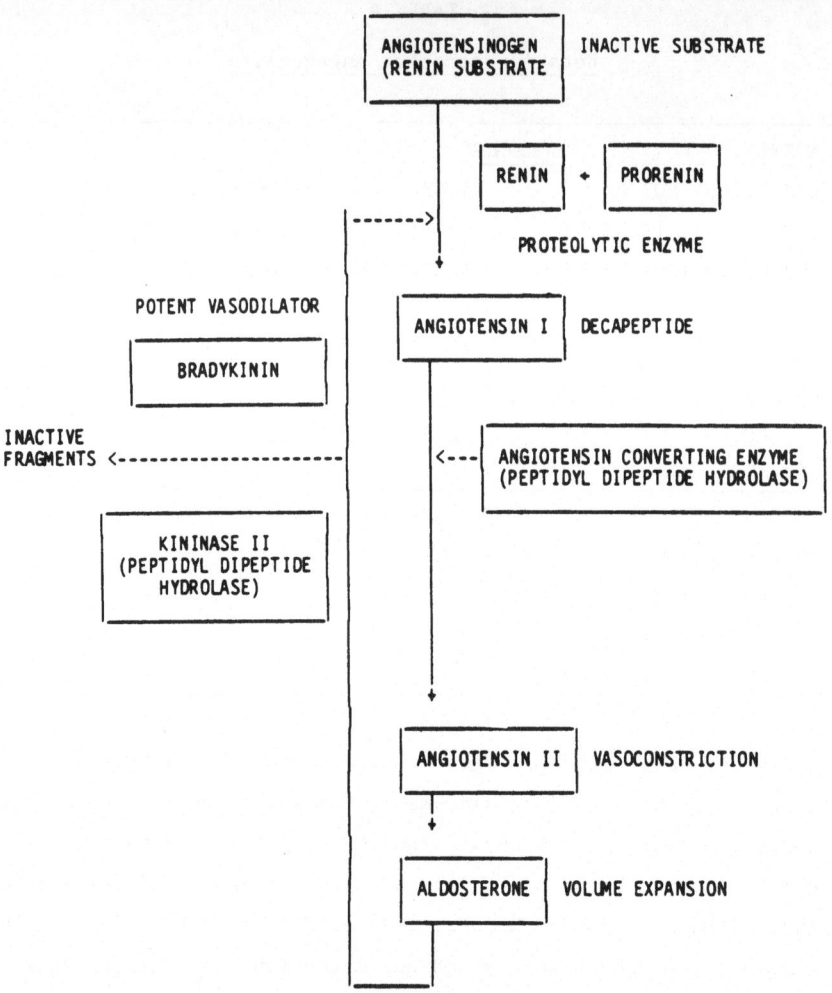

Fig. 8. The renin-angiotensin-aldosterone cascade. This diagram highlights the multiple points where a pharmacological agent might interrupt the renin-angiotension-aldosterone cascade to bring about a reduction in blood pressure.

RENIN ANGIOTENSIN ALDOSTERONE SYSTEM

The renin-angiotensin-aldosterone cascade is depicted in Figure 8 and mechanisms for pharmacologically manipulating this system are listed in Table 5. Angiotensin plays a fundamental physiological and pathophysio-logical role in blood pressure regulation, having both direct and indirect effects on vascular smooth muscle via peripheral and central mechanisms. In addition angiotensin influences determinants of blood volume. There-fore, it is apparent that a defect in this regulatory system might have a

TABLE 5

Drugs Modifying Renin-Angiotensin-Aldosterone System

---

1. RENIN INHIBITORS

    PHOSPHOLIPIDS (PE-104, URI, 74A)

    RENIN SUBSTRATE ANALOGS (H-142-DECAPEPTIDE INHIBITOR OF HUMAN RENIN)

    STATINE CONTAINING COMPOUNDS (PEPSTATIN, SCRIP, CGP-29287, R-PEP-27,
    ES-305)

2. MODULATION OF RENIN RELEASE (INCLUDING ACTIVATION OF PRORENIN)

    BETA BLOCKERS

    CLONIDINE-LIKE

3. ANGIOTENSIN CONVERTING ENZYME INHIBITORS[1]

SQ20881	USV 765	KETOACE	HOE 498	DITHIOLAN COMPOUNDS
CAPTOPRIL	SA-291	RHC 3659	RO 31-2848	INDOLAPRIL
ENALAPRIL	SA-446	CL-242817	CGS-14824A	QUINAPRIL
DU-1219				CI-925

4. ANGIOTENSIN II RECEPTOR ANTAGONISTS

    SARALASIN (NOT ORALLY ACTIVE)

5. ALDOSTERONE ANTAGONISTS

    SPIRONOLACTONE

NEW DIRECTIONS

NOVEL INHIBITORS OF RENIN RELEASE OR ACTIVATORS OF PRORENIN

SITE AND TISSUE SELECTIVE INHIBITORS OF THE RENIN ANGIOTENSIN SYSTEM

---

[1]SINCE THE ANNOUNCED DISCOVERY OF CAPTOPRIL IN 1977, AN ORALLY EFFECTIVE
INHIBITOR OF THIS ENZYME AND ITS THERAPEUTIC SUCCESS, THERE HAS BEEN AN
"EXPLOSIVE" DEVELOPMENT OF AGENTS OF THIS TYPE. A RECENT REVIEW (COHEN,
1985) OF THE AREA CATEGORIZED THESE AGENTS AS: 1) INHIBITORS FROM NATURAL
PRODUCTS, 2) SULFUR CONTAINING ACE INHIBITORS, 3) NON-SULFHYDRYL
CONTAINING ACE INHIBITORS, 4) PHOSPHOROUS CONTAINING ACE INHIBITORS, AND
5) MISCELLANEOUS TYPES. THERE ARE NO MORE THAN 32 AGENTS CITED IN HER
REVIEW.

profound effect on blood pressure. There are numerous points of regulation which could be pharmacologically controlled. These points provide
opportunities for new drug discovery, especially in the areas of orally
active inhibitors of renin and receptor antagonists of angiotensin II

(Antonaccio and Cushman, 1981). Drugs acting at these sites may provide greater selectivity and help to address some compelling questions, e.g., the role of bradykinin in the antihypertensive action of the angiotensin converting enzyme inhibitors. As nicely stated by Antonaccio and Cushman "...their discovery and development require only the foresight of their potential and the hindsight of former successes in this area" (Antonaccio and Cushman, 1981).

VASODILATOR-ANTIHYPERTENSIVES

Vasodilator-antihypertensives produce a generalized relaxation of vascular smooth muscle by acting on one or more components of the excitation-contraction coupling process. Hydralazine, diazoxide, nitro-prusside, minoxidil, and the calcium antagonists are in this class (Table 6). Vasodilator-antihypertensives share hemodynamic properties and side-effects largely as a consequence of their common site(s) and mechanism(s) of action. For example, they lower blood pressure independent of sympa-thetic tone, antagonize the vasoconstrictor effects of several endogenous agents, and usually have the side effects of reflex tachycardia, and salt and water retention. Several of these agents also have potentially-limiting side effects and toxicities which may be the result of mechanisms not common to the group, i.e., lupus-like syndrome (hydralazine), and hyperglycemia and hyperuricemia (diazoxide). In addition, both minoxidil and diazoxide can cause hirsutism.

Currently, the calcium antagonists represent one of the most exciting and mechanistically interesting approaches to vasodilator antihypertensive therapy. These agents, of which there are many, act presumably by inhib-iting the influx of extracellular calcium into the vascular smooth muscle cell. The utility of the calcium antagonists as antihypertensives over other available agents is not established. As vasodilators, there are several questions that need to be answered, e.g., effects on fluid reten-tion and renal blood flow, the relative degrees of venous and arteriolar vasodilation, effects on cardiac output and heart rate, and their relative selectivity for smooth muscle of vascular vs non-vascular sites. Calcium antagonists with a high degree of tissue selectivity are especially useful in view of the universal role of calcium in physiological functions (Rahwan, Piascik, and Witiar, 1979 or Smith, 1983). Certain of these agents appear to show remarkable specificity. Whether this is a function of dose, microdistribution of the drug to various tissues, or both is presently unclear. Future directions for new vasodilator anti-

TABLE 6

Vasodilator - Antihypertensives

PROTOTYPE(S)	HYDRALAZINE	MINOXIDIL	CALCIUM-ENTRY BLOCKERS NIFEDIPINE VERAPAMIL NITRENDIPINE	NITROPRUSSIDE
PHARMACEUTICAL CHARACTERISTICS	-	-	-	-
COMMON SIDE EFFECTS	HEADACHE, TACHYCARDIA, PALPITATIONS, NAUSEA	HYPERTRICHOSIS EDEMA		
SPECIAL COMMENTS	CORONARY DISEASE LUPUS-LIKE SYNDROME	CHD, CHF (EDEMA, FLUID RETENTION)	DECREASED ATHEROGENIC PLAQUE FORMATION	
POSSIBILITIES FOR IMPROVEMENT	INCREASED DURATION	SITE SELECTIVITY LACK OF TACHYCARDIA	SITE SELECTIVITY	ORAL ACTIVITY

NEWER AGENTS	COMMENTS
SEVERAL	PRECISE MECHANISM(S) UNDEFINED
CARPRAZIDIL	LESS TACHYCARDIA, HDL-CHOLESTEROL ↑ VLDL AND LDL CHOLESTEROL ↓
BRL 34915	CAUSES MEMBRANE HYPERPOLARIZATION, ↑ K^+ CONDUCTANCE

NEW DIRECTIONS

ANTIATHEROGENIC EFFECTS OF CALCIUM ANTAGONISTS, "CALCINOSIS"

ANTIPLATELET/ANTITHROMBOTIC COMPONENT OF PROSTACYCLIN-TYPES

MEDICINAL CHEMICAL HYBRIDS WITH COMPLIMENTARY BIOLOGICAL ACTIVITIES (E.G., VASODILATOR + β-BLOCKADE)

INTRACELLULAR MODULATION OF VASCULAR CONTRACTILITY (MLCK, CALMODULIN, ETC.)

hypertensive therapy will depend on a better understanding of the ionic involvement in vascular smooth muscle contractile function, excitation-contraction coupling, tissue (vascular bed) selectivity of these agents, and the nature of the differences in the handling of calcium and other ions by cardiac and nonvascular smooth muscle. An important future direction for the calcium antagonists relates to their potential anti-atherogenic effects. For example, the dihydropyridine, nifedipine can prevent plaque formation in cholesterol-fed rabbits without significantly modifying circulating lipids (Henry, 1982). Furthermore, the non-dihydro-pyridine diltiazem can prevent vitamin D_3-induced arterial "calcinosis."

Key Issues:*

● In which vascular bed does the agent reduce resistance to blood flow ?

● Is there a reduction in arteriolar resistance ?, venous capacity ?, both ? Which effect predominates ?

● How much sodium and fluid retention does the vasodilator cause ?

● How strong are the reflex and humoral counterregulatory responses to vasodilation ?

● What is the side effects profile ?

Fig. 9.   Issues regarding current and future antihypertensive vasodilators. (After Fitzpatrick and Julius, 1985.)

Differences in the profiles of the calcium antagonists and their diverse chemistry make future opportunities for this pharmacological class quite promising. Figure 9 raises key questions regarding current and future antihypertensive vasodilators.

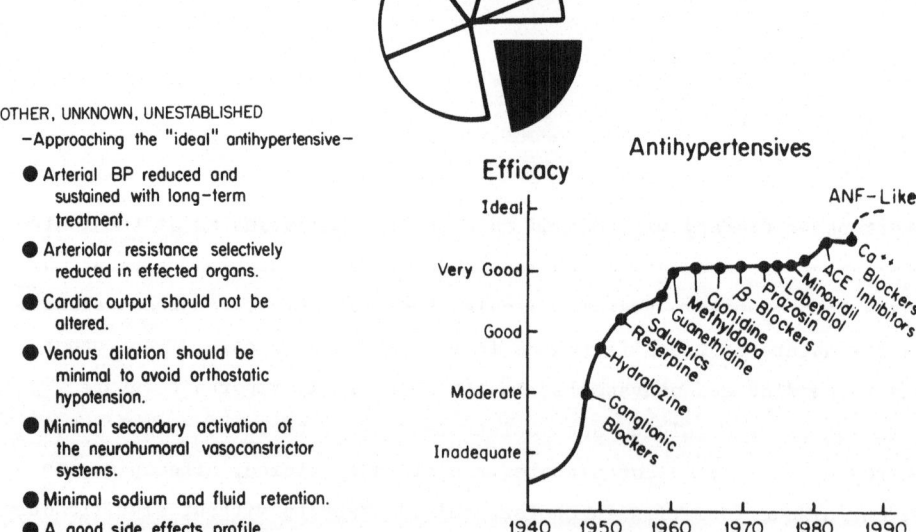

OTHER, UNKNOWN, UNESTABLISHED

—Approaching the "ideal" antihypertensive—

● Arterial BP reduced and sustained with long-term treatment.

● Arteriolar resistance selectively reduced in effected organs.

● Cardiac output should not be altered.

● Venous dilation should be minimal to avoid orthostatic hypotension.

● Minimal secondary activation of the neurohumoral vasoconstrictor systems.

● Minimal sodium and fluid retention.

● A good side effects profile.

Fig. 10.   Properties of an "ideal" antihypertensive.

TABLE 7

Open Questions - Antihypertensive Drugs and Lipoproteins

1. SHOULD LIPOPROTEIN PROFILES GUIDE SPECIFIC ANTIHYPERTENSIVE THERAPY?
    TO WHAT EXTENT SHOULD AGE AND RELATIVE RISK IMPACT?

2. WHAT ARE THE POTENTIAL DRUG INTERACTIONS WHICH MAY RESULT FROM
    COMBINED ANTIHYPERTENSIVE AND/OR COMBINED ANTIHYPERTENSIVE-
    HYPOLIPIDEMIC THERAPY?

3. IS THERE LONG-TERM BENEFIT FROM RISK OF CORONARY HEART DISEASE AND ITS
    COMPLICATIONS WITH LIPOPROTEIN MODIFICATION, PRAZOSIN VS BETA?
    BLOCKERS? DIURETICS?

4. SHOULD YOU TREAT UNDESIRABLE LIPID EFFECTS WITH COMBINATION
    ANTIHYPERTENSIVE THERAPY?

5. WHAT EFFECTS DO NEWER ANTIHYPERTENSIVE AGENTS (i.e., CONVERTING ENZYME
    INHIBITORS, CALCIUM ENTRY BLOCKERS, VASODILATORS, ALPHA-BETA BLOCKERS)
    HAVE ON LIPOPROTEIN METABOLISM?

6. IN NEW ANTIHYPERTENSIVE DRUG DISCOVERY, SHOULD AN ANIMAL MODEL BE USED
    (UNLIKE THE RAT) WHICH HAS A LIPOPROTEIN PROFILE MORE SIMILAR TO THE
    HUMAN?

7. WHAT ARE THE SHORT- AND LONG-TERM EFFECTS OF ALTERED PLASMA
    LIPOPROTEIN LEVELS ON VASCULAR REACTIVITY? WHAT ARE THE EFFECTS OF
    PHARMACOLOGIC MODIFICATION?

TRENDS AND PERSPECTIVES

A greater understanding of factors which regulate blood pressure,
will likely lead to improved antihypertensive drugs. New technology
permits the study of pathophysiological events at the organ, tissue,
cellular and subcellular levels. "Future directions" in areas which we
believe should provide opportunities for new drug discovery have already
been indicated in part in Tables 3, 4, 5, and 6. Figure 10 lists key
criteria for future antihypertensive agents.

TABLE 8

Emerging Concepts for New Antihypertensive Drugs

- BIOACTIVE ATRIAL PEPTIDES (ATRIOPEPTINS), SELECTIVE ORALLY ACTIVE
"MIMICS" (NON-PEPTIDE?)
- NATRIURETIC FACTOR (HUMORAL SODIUM-POTASSIUM PUMP INHIBITOR) - SELECTIVE
"DYSINHIBITORS" FOR VASCULAR SMOOTH MUSCLE
- SELECTIVE INHIBITORS OF THE RENIN-ANGIOTENSIN ALDOSTERONE SYSTEM (ORALLY
ACTIVE, NON-PEPTIDE?)
- MIMICS OR ANTAGONISTS OF NON-ADRENERGIC NEUROTRANSMITTERS IN THE CNS
(e.g. GABA, AcH, GLUTAMATE)
- MORE SELECTIVE MODULATORS OF SYMPATHETIC OUTFLOW
- DRUGS INTERACTING VIA THE ARACHIDONIC ACID CASCADE (e.g.
PROSTAGLANDIN/PROSTACYCLIN-TYPE)
- MODULATORS OF NEUROTRANSMITTER BIOSYNTHESIS (e.g. ORGAN SPECIFIC PNMT
INHIBITION)
- MODULATORS OF NA^+, K^+ AND CA^+ PERMEABILITY AND TRANSPORT (e.g. DRUGS
INCREASING MEMBRANE K^+ CONDUCTANCE)
- IMPROVEMENTS WITHIN A CLASS OR SUBCLASS OF CURRENTLY AVAILABLE
ANTIHYPERTENSIVE DRUGS (e.g. ACE INHIBITORS, CALCIUM ANTAGONISTS)
- MEDICINAL CHEMICAL HYBRIDS - ANTIHYPERTENSIVE DRUGS WITH COMPLEMENTARY
BIOLOGICAL ACTIVITIES (VASODILATOR-BETA BLOCKER, ACE INHIBITOR-DIURETIC
ETC.)
- PHARMACOLOGICAL IMPLICATIONS OF POTENTIAL ANTIHYPERTENSIVE DRUGS ACTING
ON THE HEART
- THE ROLE OF THE VASCULAR ENDOTHELIUM - PHARMACOLOGICAL MODUALTION OF
VASOPRESSOR AND DEPRESSOR SIGNALS
- INTRACELLULAR MODULATION OF VASCULAR CONTRACTILITY (REGULATION OF MYOSIN
LIGHT CHAIN KINASE, CALMODULIN)
- RENAL CORTICAL AND/OR MEDULLARY ANTIHYPERTENSIVE LIPIDS AND DERIVATIVES
ORALLY ACTIVE "ANRL-LIKE" AGENTS)
- DRUGS ALTERING VASCULAR COMPLIANCE (e.g. POLYETHER IONOPHORES -
BROMOLASALOCID)
- ANTIHYPERTENSIVE DRUGS HAVING A FAVORABLE LIPOPROTEIN PROFILE (e.g.
$ALPHA_1$ ADRENOCEPTOR ANTAGONISTS?)
- SELECTIVE ADENOSINE RECEPTOR AGONISTS
- ROLE OF SECOND MESSENGERS IN THE REGULATION OF VASCULAR SMOOTH MUSCLE
TONE AND HYPERTENSION

Alterations of plasma lipids by antihypertensive drugs is an
important future research area. Many of the issues related to this topic
are listed in Table 7. Table 8 identifies several concepts for new
antihypertensive drugs where there are intensive research efforts. It is
beyond the scope of this review to detail any single item. The table is
neither complete nor does it attempt to prioritize. Hopefully it will
serve to promote useful discussions. Future emphasis will likely be
directed towards highly selective, mechanistic approaches aimed at both
treatment and prevention of hypertension. Ideally, underlying patho-
physiologic mechanisms will be identified and selective pharmacologic
agents will be developed. We are in an era where major advances in the
treatment of hypertension is realistic and we might expect the continued
decline in hypertensive disease-related morbidity and mortality and
overall improvement in the quality of life for the hypertensive patient.

REFERENCES

American Heart Association, 1985, Heart Facts, Dallas, Texas.

Antonaccio, M.J., and Cushman, D.W., 1981, Drugs inhibiting the
        renin-angiotensin system, Fed. Proc., 40:2275-2284.

Brensike, J.F., Levy, R.I., Kelsey, S.F., Passamani, E.R., Richardson,
        J.M., Loh, I.K., Stone, N.J., Aldrich, R.F., Battaglini, J.W.,
        Moriarty, D.J., Fisher, M.R., Friedman, L., Friedewald, W., Detre,
        K.M., and Epstein, S.E., 1984, Effects of therapy with
        cholestyramine on progression of coronary arteriosclerosis:
        Results of the NHLBI Type II Coronary Intervention Study,
        Circulation 69:313-324.

Cocks, T.M., and Angus, J.A., 1983, Endothelium-dependent relaxation of
        coronary arteries by noradrenaline and serotonin, Nature
        305:627-630.

Cohen, M.L., 1985, Synthetic and fermentation-derived angiotensin
        converting enzyme inhibitors, Ann. Rev. Pharmacol. Toxicol.
        25:307-323.

Colfer, H.T., Cottier, C., Sanchez, R., and Julius, S., 1984, Role of a
        cardiac factor in the initial hypotensive action by beta
        adrenoceptor blocking agents, Hypertension 6:145-151.

Cowley, A.W., 1984, Assessment of the contributions of autoregulatory
        mechanisms to the antihypertensive actions of beta adrenergic
        therapy, Hypertension 6:137-139.

Evans, D.B., Bristol, J.A., and Kaplan, H.R., 1985, Antihypertensive
        Agents, in: "Handbook of Medicinal Chemistry," M. Verderame, ed.,
        CRC Press Inc. Boca Raton, Florida. In press.

Farsang, C., Ramirez-Gonzales, M.D., Mucci, L., and Kunos, G., 1980,
        Possible role of an endogenous opiate in the cardiovascular effects
        of central alpha adrenoceptor stimulation in spontaneously
        hypertensive rats, J. Pharmacol. Exp. Ther. 214:203-208.

Fitzpatrick and Julius, S., 1985, Hemodynamic effects of
        angiotensin-converting enzyme inhibitors in essential hypertension:
        A review. J. Cardiovasc. Pharmacol. 7:535-539.

Gross, F., 1982, Present concepts and perspectives of antihypertensive
        therapy, Clin. and Exper. Hyper. - theory and Practice
        A4(1-2):1-25.

Henry, P.D., 1982, Antiatherogenic effects of nifedipine in
        cholesterol-fed rabbit: Decreased vascular injury without
        concomitant reduction in circulating lipids, in: "Calcium
        Modulators, T. Godfraind, A. Albertini, and R. Paoletti, eds.,
        p. 233-245, Elsevier Biomedical Press.

Kaplan, H.R., and Smith, R.D., 1981, Proposed sites and mechansims of
        action, Fed. Proc. 40:2268-2274.

Kaplan, H.R., 1983, New antihypertensive drugs - Hypertension theme
        Symposium, In Fed. Proc. 42:153-210.

Kaplan, H.R., Ryan, M.J., Singer, R.M., Cohen, D.M., and Cygan, R.M.,
        1983, Survey of new antihypertensive drugs: 1982, Fed. Proc.
        42:154-156.

Korner, P.I., 1984, Beta blockers, autoregulation, and experimental
        design, Hypertension 6:140-144.

Kunos, G., Farsang, C., and Ramirez-Gonzales, M.D., 1981, beta-Endorphin:
        Possible involvement in the antihypertenisve effect of central
        alpha-receptor activation, Science 211:82-84.

Levy, R.I., Brensike, J.F., Epstein, S.E., Kelsey, S.F., Passamani,
        E.R., Richardson, J.M., Loh, I.K., Stone, N.J., Aldrich,R.F.,
        Battaglini, J.W., Moriarty ,D.J., Fisher, M.L., Friedman, L.,
        Friedewald, W., and Detre, K.M., 1984, The influence of changes in
        lipid values induced by cholestyramine and diet on progression of
        coronary artery disease: Results of the NHLBI Type II Coronary
        Intervention Study, Circulation 69:325-337.

Lipid Research Clinics Program, 1984, The lipid research clinics coronary
        primary prevention trials results, I. Reduction in incidence of
        coronary heart disease, J. Am. Med. Assoc. 251:351-364.

Lipid Research Clinics Program, 1984, The lipid research clinics coronary primary prevention trial results, II. The relationship of reduction in incidence of coronary heart disease to cholesterol lowering, <u>G. Am. Med. Assoc.</u> 251:365-374.

McCafferty, J.P., Roesler, J.M., Kopaciewicz, L.J., and Hieble, J.P., 1981, Orthostatic liability of antihypertensive drugs in the conscious rabbit, <u>Pharmacologist</u> 25:102.

Rahwan, R.G., Piascik, M.F., and Witiar, D.T ., 1979, The role of calcium antagonism in the therapeutic action of drugs, <u>Can. J. Physiol. Pharmacol.</u> 57:443-460.

Roesler, J.M., Kopaciewicz, L.J., and Hieble, J.P., 1983, Antihypertensive activity of SKF 86466: A selective alpha$_2$ antagonist, <u>Pharmacologist</u> 25:242.

Ryan, M.J., Cohen, D.M., Bjork, F.A., Mathias, N.P., Randolph, A.E., Evans, D.B., and Kaplan, H.K., 1984, CI-926 (3-[4-[4-(3-methyl-phenyl)-1-piperazinyl]butyl]-2,4-imidazolinedione): Antihypertensive (AH) profile and pharmacology, <u>Pharmacologist</u> 26:237.

Schoetensack, W., Bruckschen, E.G., and Zech, K., 1983, Urapidil, <u>in</u>: "New Drugs Annual: Cardiovascular Drugs," A. Scriabine, ed., p. 19-48, Raven Press, New York.

Smith, R.D., 1983, Calcium entry blockers: Key issues, <u>Fed. Proc.</u> 42:201-206.

Stewart, I., 1982, Beta-adrenoceptor blockade and the incidence of myocardial infarction during the treatment of severe hypertension, <u>Brit. J. Clin. Pharmacol.</u> 13:91-93.

THE FAWN-HOODED RAT AS A GENETIC HYPERTENSIVE ANIMAL MODEL

Albert Magro, Nisan Gilboa and Ulrich Rudofsky

Wadsworth Center for Laboratories and Research
Albany, N.Y. 12201

INTRODUCTION

Essential hypertension in its established phase is characterized by a raised total peripheral resistance which, in most cases, is accompanied by a normal cardiac output (Frohlich, Tarazi and Dustan, 1969). The disease in man has the property of multiplicity (Paul, 1976) in which hereditary factors play an important role (Page, 1961; Pickering, 1974). Probably over 95% of patients with hypertension have essential hypertension. By definition essential hypertension is a primary form of elevated blood pressure of undetermined causes. There seems to be a genetic predisposition and a slow and unrelenting progressiveness in the severity of the elevated blood pressure and increased vascular resistance. Therefore, in order to develop an experimental model that mimics aspects of the clinical conditions, it is necessary to have a naturally occurring form of hypertension that is predisposed genetically, that is progressive and that involves an adaptation of the circulatory system that is not rapid. The genetic component of essential hypertension is generally believed to be polygenic in nature, but by no means is this opinion unanimous (McKusich, 1960).

The development of animal models of essential hypertension (Folkow and Hallback, 1977) have promoted understanding. Of these models, the spontaneously hypertensive rat (SHR) developed by Okamoto and Aoki (1963) is the most widely used. Using this model progress has been made in determining the causes of increased peripheral resistance (Mulvany, 1983) and in the study of neurogenic mechanisms in essential hypertension (Antonaccio, 1977). However, it is clear that no model can be complete, even the SHR (McGiff and Quilley, 1981), to study the multiplicity of essential hypertension. Other strains of genetically susceptible hypertensive rats have provoked interest because of factors thought to be important to the pathogenesis of human hypertension. These include the Milan strain (Baer, Bianchi, and Liliana, 1978) which possesses a renal lesion related to the onset of its hypertension, and the Dahl strain (Dahl, Heine and Tassinari, 1962) which has a genetic susceptibility to develop hypertension when on a high salt diet. The New Zealand rat is another genetically hypertensive rat which may be relevant to human essential hypertension (Smirk, 1972), but it far less accessible than the SHR. Other rat strains, which are less well known but demonstrate high systolic blood pressure are, GH and the related AS strains (Simpson et al., 1973), NSD strain and the OM strain (Hansen, 1972). The AS and GH strains, as the SHR, were derived from Wistar rats. The genetic hypertension of the AS and GH strains may be associated with a defect in renal prostaglandin catabolism (Armstrong et al., 1976). All of these strains have the potential of being developed into useful models of hypertension.

The overuse of the SHR has promoted interest in the development of new genetic hypertensive models. There is activity in the development of new strains by introducing specific genes into spontaneously hypertensive rats. For example, the intercrossing of SHR and the Brattleboro rat has produced spontaneously hypertensive rats homozygous for hypothalamic diabetes insipidus (Ganten et al., 1983). This is a powerful method for increasing the number and scope of hypertensive strains which has long been advocated. Unfortunately the intercrossing gene insertion approaches which utilize only the SHR as the hypertensive strain are inherently limiting, for the hypertension aspects of the new strain will be dependent upon how the SHR strain relates to essential hypertension.

An additional strain which spontaneously develops elevated systolic blood pressure is the fawn-hooded (FH) rat. Although not widely known to the hypertension research community, there is a growing interest in using the FH rat as a model to investigate mechanisms which may be relevant to essential hypertension. It is the intent of this article to give an overview of the FH rat and to illustrate that the animals do have characteristics and varied accompanying phenotypic abnormalities relevant to aspects of essential hypertension.

Origins. The FH rat first appeared at the Department of Psychology of the University of Michigan, Ann Arbor, as a mutant from unplanned crosses in a colony consisting of randomly bred albino and pigmented rats. The rat appeared from crosses of various populations of laboratory rats of unknown geneology from colonies maintained by E. L. Walker and N. R. F. Maier. As far as can be determined, the rat spontaneously appeared from a combination of Long-Evans, Wistar and brown rats. The result is a rat strain unique in appearance, behavior and other characteristics.

In 1971 a stock of rats from the University of Michigan was introduced into the research facilities of the American Museum of Natural History, N.Y.C., N.Y., and has since that time been bred at random (Tobach, DeSantis and Zucker, 1984). In the early 1970's W. Jean Dodds introduced the rat from the University of Michigan into Griffin Laboratory Farms, Guilderland, N.Y., which is the animal breeding facilities of the New York State Department of Health. Since that time the Department of Health has provided the rat, upon request, to a number of research institutions. As a consequence, several of other institutions possess small colonies of the rat. Currently there are several lines of FH rats at Griffin Laboratory Farms. One of the colonies is a line which has a specific designation (FH/Wjd). The FH rat is currently listed in the ILAR Directory.

Background. The appearance of the FH rat is distinctive, particularly its "fawn" color of hoodedness and its ruby colored eyes. Initially, breeding selection was performed visually based on these two physical characteristics. Interest in the FH rat began with its very interesting behavior, in that they are very calm and unaggressive. This led to work on the 5HT storage pool defect that the rat manifests (Tschopp and Zucker, 1972; Raymond and Dodds, 1975) which is a defect similar to platelet storage pool disease in man. In addition to platelet secretory abnormalities, both sexes manifest a number of cellular secretory defects which include mast cell release mechanisms (Magro, 1981) and the release of renin from the JGA (Magro and Rudofsky, 1982). The FH rats, particularly males, begin to develop a low-renin (PRA) form of elevated systolic blood pressure by as early as 2-3 months of age (Rudofsky and Magro, 1982). The renin substrate, angiotensinogen, is not limiting in the low PRA form of elevated blood pressure the FH rat manifests (Magro and Rudofsky, 1982). The elevated systolic blood pressure appears spontaneously and may be labile. In addition to a low PRA, there is an associated low urinary kallikrein output (Gilboa, Rudofsky and Magro, 1984a). There is also the development of focal glomerular sclerosis at an early age and this occurs with a pronounced proteinuria (Kreisberg and Karnovsky, 1978; Magro and Rudofsky, 1982; Gilboa et al., 1984a). A longitudinal study comparing the renal morphologic changes with hemostatic defects indicates that the hemostatic defects are congenital in that they do not become more severe in older animals and seem unrelated to nephropathy (Urizar et al., 1984). Also, the FH rats demonstrate significantly higher urinary outputs of the catecholamines, norepinephrine, dopamine, and the dopamine metabolite homovanillic acid (Magro et al., 1986).

Lack of a Control Strain. Any discussion of abnormalities in the FH rat requires that the control be defined. Most of the information about phenotypic abnormalities in the FH rat was accumulated using the ancestral Wistar (W) strain as a control. There are values which may be clearly higher or lower than most other laboratory rats (e.g. the FH rat's systolic blood pressure). However, this clear distinction does not apply to all of the phenotypic abnormalities. The phenotypic abnormalities discussed above, and their relationship to hypertension, should be verified and further investigated after the proper controls are developed. Some investigators advocate using a variety of control rats (Clineschmidt et al., 1970). Unfortunately, the FH rat totally lacks any control strain. In addition, the FH rat was never selectively bred for hypertension. Although there is clear heritability, very little is known about the genetic control of the phenotypic abnormalities or the FH rat's elevated systolic blood pressure. The comparison of a number of physiological and biochemical characteristics of the hypertensive FH rat, with the ancestral normotensive W rat, is an obvious first step in identifying factors which indicate the desirability of developing the FH rat as a hypertensive model. It is acknowledged that evidence which will be presented here for some of the phenotypic abnormalities is preliminary and that each individual study may not be conclusive. However, cumulatively the evidence indicates that the FH rat has a combination of unique characteristics relevant to hypertension even if all these characteristics should be reinvestigated after the proper control strains are developed.

The following is a review of some of the phenotypic traits of the FH rat which may be relevant to some clinical forms of hypertension.

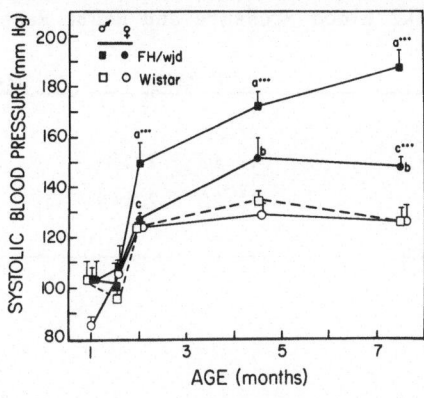

Fig. 1. Systolic blood pressure in FH and W rats at various ages (means +
SEM). Letters indicate statistically significant differences
between (a) FH and W males, (b) FH and W females, and (c) FH
males and FH females. p<.05 (no asterisk); p <.001 (***).

PHENOTYPIC TRAITS

FH Rats Have Elevated Systolic Blood Pressures. When measured by the
tail cuff method both male and female FH rats demonstrate an elevated
systolic blood pressure. The systolic blood pressure is higher in the
male FH rats than in the female FH rats but both are elevated relative to
the Wistar controls. The systolic blood pressures of the FH rat are
fairly normal over the first 2 months of life (Fig. 1), then elevate
intermittantly, and then slowly increase to hypertensive levels which
maximize at about 6-9 months of age. Although systolic blood pressures in
all of the FH rats are elevated relative to controls, the rats in our
colony demonstrate a range of blood pressures. Male FH rats demonstrate a
range of about 145-215 mm Hg for tail cuff measurements of indirect
systolic pressure. This range of blood pressure manifests itself even
after the rats have been acclimated to being confined in the blood
pressure measuring holder. Despite the large range of blood pressure it
can be seen from the deviations that most of the rats blood pressures fall
with + 15 mm Hg of the mean (Fig. 1). For mean arterial pressures
(carotid artery cannulations) at 5 months of age male FH rats show a range
of about 110 - 180 mm Hg (Table 1). At 5 months of age the values for
mean arterial blood pressures show a greater deviation than for indirect
systolic blood pressures which raises the question of whether the FH rats,
at this age, have a component of labile (borderline) hypertension which

Table 1. Mean Arterial Blood Pressure and Heart Rate of FH and W Rats

	BP mmHg	Heart Rate beats/min.
FH (6) mean ± std. dev.	147.17 ± 22.89	353.33 ± 44.12
W (6) mean ± std. dev.	91.25 ± 22.40	315.0 ± 50.0

can be exacerbated by stress. The FH rat colony has never been
selectively bred for high or low blood pressure. This in part could
account for the large deviation of the mean arterial pressures at 4-5
months age. Kuijpers and deJong (1982), have shown that at 1 year of age
the male FH rats mean arterial blood pressure is consistently elevated
with a value over 190 mm Hg with a relatively small deviation (198 ± 6.8
mm Hg).

The Reproduction and Development of the FH Rat is Healthy and
Reliable. The FH rat does not demonstrate any serious reproductive or
developmental disturbances. During breeding the rats in our colony are
maintained on tap water and regular rat chow. The animals are housed in
plastic cages on wood shavings in no special environment. Usually one
male is accompanied by 3 or 4 sibling females during mating and the
pregnant females are removed from the breeding cage prior to giving birth.

We have compared a number of reproductive parameters between an
outbred W rat colony and the inbred FH rat colony. The data of Table 2
are a compilation from the breeding records of the two colonies. Table 2
illustrates the FH rats show a high degree of fertility with about 95% of
the females becoming pregnant. The number of pups/litter, for the FH rat
averages around 8.5 which is a little higher than average for all inbred
strains at NIH (C. Hansen, personal communication). The survivability to
weaning is a little lower, in the FH strain, as compared to the outbred W
rats. This was primarily due to a higher rate of cannibalism in the FH
colony. In part this could be due to the FH mothers being on average a
little younger than the W mothers, and there being more first time mothers
in the FH colony. We feel the degree of cannibalism can be reduced by
trying various conditions and environments for the mothers to give birth
in. Overall, the data of Table 2 indicate that the fertility and

Table 2. Comparison of the Breeding Capacity of Outbred W Rats and
Inbred FH Rats.

	W (38 litters)	FH (31 litters)
% females which become pregnant	95%	95%
Range of pups/litter	6-16	4-15
Average number of pups/litter (mean $\pm$ std. dev.)	10.5 $\pm$ 1.4	8.3 $\pm$ 0.8
% pups surviving until weaned (culled to 10 pups)	94%	78%
Length of time to weaning	21 days	26 days

reproductive capacity of the FH rat are at a fairly high level for an
inbred strain and would present no problem to breeders.

Postweaning the development and growth of the FH rat proceeds similar
to that of the W rat. As in the W rats the degree of mortality in the FH
rat colony postweaning, up to the first 6 months of life, is negligible.

It can be seen from Fig. 2 that there is very little difference in
weight between W and FH rats over the first 5 months of life. After 6
months of age, the FH rats appear to be a little lighter than W rats and
this could be due to the development of severe proteinuria and glomerular
sclerosis in the male FH rats.

The FH Rat has an Interesting Genetic Makeup. In an attempt to
determine the linear order of the fawn [f] gene (Castle, 1919), the
retinal dystrophy [rdy] gene (LaVail, 1981a), and non-agouti [a] gene
(Robinson, 1965), LaVail (1981b) requested, from the New York State
breeding facilities, the FH rat. He selected the FH rat because he
believed it to be a source of ff. However his study (LaVail, 1981b)
demonstrated that the fawn color of the FH rat is not due to the f gene
(which the FH rat does not possess) but rather is the consequence of the
rat being homozygous for the red-eyed dilution [r] gene. This study also
demonstrated the FH rat is homozygous for the non-agouti [a] gene and the
non-pink eyed gene [P]. We have determined that the FH rat is of a RT1^u

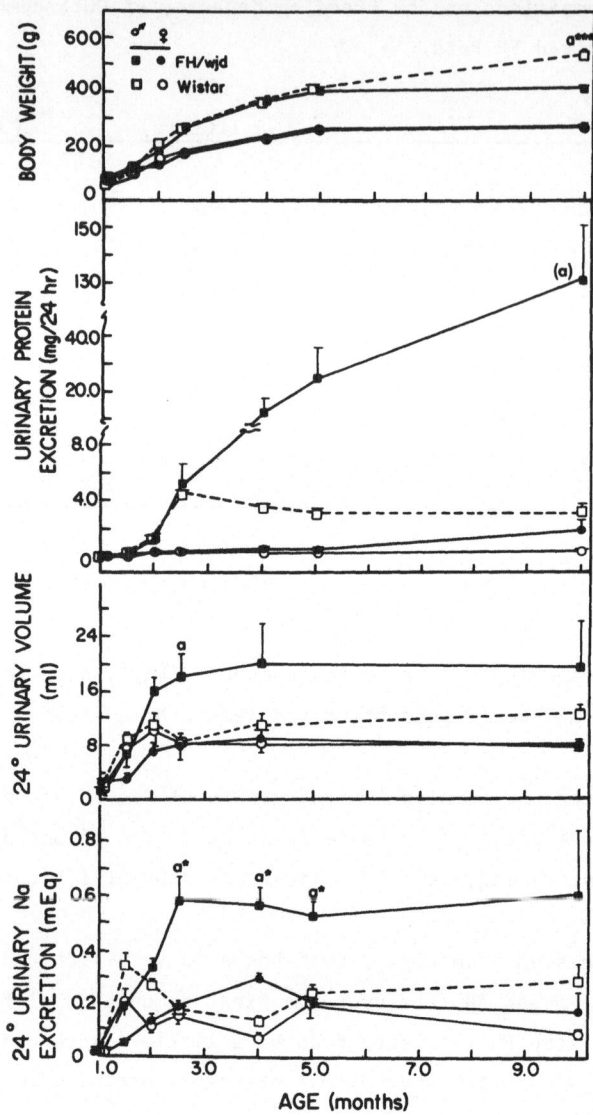

Fig. 2.   Comparison between male and female FH and W rats in weight,
urinary protein volume and Na excretion (means + SEM).   a,
Statistically significant difference between FH and W males.

haplotype (Table 3).   We have also determined (in collaboration with D.
Judge, Charles River Inc.) the allozyme pattern for the biochemical
markers Hbb, Pep-3, Es-1, Es-2, Es-6, Es-8 and Es-10 (Table 3).

Table 3. Genetic Makeup and Allozyme Pattern of SH, FH, WF, LE, WAG and
DA Rats

	SH	FH	WF	LE	WAG	DA	Linkage Group
Haplotype	k	u[+]	u	u	u	a	IX
Albino [c]	yes	no	yes	no	yes	no	I
Non-agouti [a]	no	yes	no	yes	yes	no	IV
Hooded [h]	no	yes	no	yes	yes*	no	VI
Agouti [A]	no	no	no	no	no	yes	
Non-pink eyed [P]	no	yes	no	no	yes	no	I
Non-anaphylactoid reaction to dextran [dx]	no	no	no	no	yes	no	
Red-eyed dilution [r]	no	yes	no	no	no	no	II
β-hemoglobin [Hbb]	a	b**	a	b	a	b	I
Peptidase [Pep-3]	a	b	a	a	a	b	
Esterase [Es-1]	a	b	a	b	a	b	V
Esterase [Es-2]	a	a	c	a	a	a	V
Esterase [Es-6]	a	b	a	a	a,b*	a	
Esterase [Es-8]	b	a	a	b	b,a*	b	V
Esterase [Es-10]	a	a	b	a	a,b*	a	V

See Robinson (1965); Altman and Katz (1979); and O'Brien (1984) for SH,
WF, LE, WAG and DA data.

*Depending on substrain. [+]Haplotype data for FH rat provided by R.
Guttmann. **Biochemical marker data for FH rat provided by D. Judge.

There are many interesting aspects of the data of Table 3. There is
a good deal of confusion about the ancestral origins of the FH rat. As
best as can be determined, it appeared spontaneously from crossbreedings
of albino and pigmented rats. The strains most often mentioned are
Wistar, Long Evans and "brown" or German brown. We feel that the albino
ancestral parent most likely was the WAG. Examination of Table 3
indicates that the RT1^{u} haplotype, the hooded [h] gene, non-agouti [a]
gene, the non-pink eyed dilution [P] gene which the FH rat possesses are
also present in the WAG.

Examination of the biochemical allozyme pattern (Table 3, 7 male FH rats all of which had the same pattern) reveals there is a high degree of correlation between the FH and DA. Since we have not been able to obtain solid breeding verification of the origins of the FH rat, it is tempting to speculate that the brown pigmented ancestral strain was possibly the agouti DA.

There are two places where mutations could have occurred in the FH rat. The red-eyed dilution [r] gene does not appear in any of the other strains. It is a rare mutation and LaVail (1981b) believes it appeared spontaneously in the FH rat. Also the Es-8, Es-10 allozyme pattern, which are both, a, in the FH rat, is extremely unusual (Table 3). In just about all other rat strains, the gel migratory patterns of these two esterases in combination are slow-fast [i.e. a,b or b,a]. In the FH rat however, the Es-8 and Es-10 migration patterns are slow-slow [a,a]. Since Es-2, Es-8 and Es-10 are very closely linked the unusual pattern of Es-2, Es-8, Es-10 in the FH rat [a,a,a] is most likely the result of a low probability recombination at Es-8 or Es-10. It is interesting to note, from the data of Table 3, that the SHR and the FH rat are different from each other in just about every genetic aspect listed in the Table.

The FH Rat has Low Plasma Renin Activity. The release of renin by the cells of the juxtaglomerular apparatus (JGA) (Cook, 1968; Robertson, et al., 1965) is a secretory process which can be induced systemically by β-adrenergic agonists (Lehr, Mallow and Krukoski, 1967; Reid, Schrier and Early, 1972; Meyer et al., 1973) and metabolites of arachidonic acid (Larsson, Weber and Änggård, 1974; Weber et al., 1976; Armstrong et al., 1978). The peripheral renin-angiotensin system can also be modulated by the stimulation of central α, 5HT and AII receptors. Plasma renin activity (PRA) can be increased by central stimulation of α_1 (Ganong, 1983) and 5HT receptors (Zimmerman and Ganong, 1980) and decreased by central stimulation of α_2 (Reid et al., 1975) and AII receptors (Malayan et al., 1979).

In rats, particularly males, PRA decreases with age. In FH rats the drop in PRA with age is more precipitous than other rat strains. As early as 3-4 month of age, FH rats have a PRA which is significantly lower than age and sex matched W rats (Fig 3). The bar diagram of Fig. 3 shows the PRA levels and the focal-glomerular-sclerosis (FGS)-histopathological-scores at various ages for male FH and Wistar rats. In the FH rat the PRA

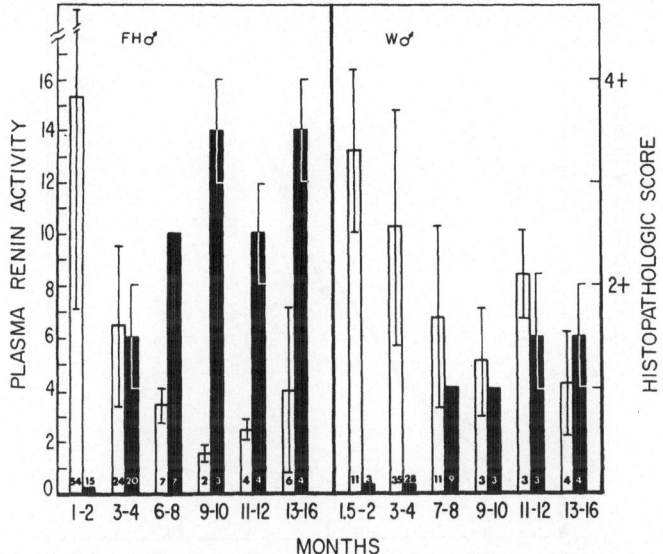

Fig. 3. Comparison of PRA and FGS severity in male and female FH and W rats. Light bars indicate PRA (mean ± SD); dark bars indicate approximate average of histologic score and range. Number of rats tested is shown in each bar.

drops before there is any significant decrease in the kidney renin content. Also the low PRA in the FH rat appears to be due to low plasma renin rather than a limitation of the renin substrate, angiotensinogen, which appears to be present in the FH rat in non-limiting quantities (Fig. 4).

β-adrenergic stimulation is one of the more effective ways of inducing the JGA to secrete renin. The capacities of male FH and W rats to secrete renin upon β-receptor stimulation were compared (Table 4). The rats were matched in age (3-4 months) and weight (250-300 g). The rats were injected intravenously with the β-agonist isoproterenol. After isoproterenol administration the rats were bled for plasmas which were then analyzed for PRA. At a 0.1 mg/kg i.v. dose of isoproterenol the PRA of both rat strains increased significantly but less so in the FH rat. The somewhat lower PRA response by the FH rat was also apparent at the 1 mg/kg dose of isoproterenol. In order to verify that the isoproterenol stimulation was β-adrenergic receptor-specific, the β-antagonist propranolol was administered 1-2 h before the isoproterenol injection. The propranolol was given orally in H_2O as a bolus (40 mg/kg). It can be seen from Table 4 that the propranolol was an effective antagonist of the isoproterenol PRA stimulation of renin secretion.

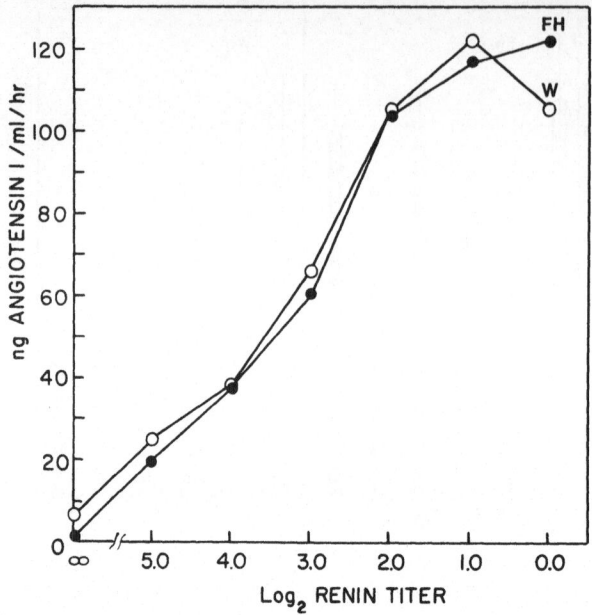

Fig. 4. Comparison of angiotensinogen substrate levels in male FH and W
plasmas. Constant amounts of renin-depleted plasma (0.2 ml) were
incubated for 1 h with doubling dilutions of renin-containing
kidney supernatant (0.05 ml). The zero point of the $\log_2$
dilution corresponds to the renin activity in the supernatant of
1.25 mg (wet weight) of homogenized whole kidney tissue added to
the 0.2 ml of plasma (6.25 mg kidney/ml plasma).

The data of Table 4, which show that the FH rat has less of a PRA
increase for a given degree of β-receptor stimulation and raise the
possibility that by 3-4 months of age the renin secretory capacity of the
JGA of FH rats is less than that of W rats.

It appears that the onset of spontaneous FGS in rats is preceded by a
drop in PRA. The correlation between low renin activity and the
appearance of FGS may indicate some connection between low activity of the
peripheral renin-angiotensin system (RAS) and FGS. Schmid (1962) first
postulated the RAS as a physiologic regulator of renal hemodynamics.
Since that time support has grown for the concept that the RAS plays an
important role in controlling postglomerular resistance through an
efferent arteriolar constrictor mechanism (Hall et al., 1977; Davis, 1977;
Ploth and Navar, 1979; Stowe, Schnermann and Hermle, 1979).

Table 4. Induction and Inhibition of Renin Release in FH and W Rats

Strain	n	Treatment	Dose mg/kg	Plasma Angiotensin I[1] ng/ml/h
FH	4	none	0	14.6 $\pm$ 7.8
	4	isoproterenol	0.1	33.7 $\pm$ 22.0
	3	isoproterenol	1.0	61.0 $\pm$ 21.3
W	4	none	0	19.2 $\pm$ 10.7
	4	isoproterenol	0.1	51.5 $\pm$ 7.0
	3	isoproterenol	1.0	94.0 $\pm$ 9.6
FH	1	none	0	2.4
	1	propranolol	40	2.7
	1	isoproterenol & propranolol	0.1 + 40	8.3
W	1	none	0	6.3
	1	propranolol	40	2.9
	1	isoproterenol & propranolol	0.1 + 40	2.4

[1]Samples were collected 20 min after isoproterenol administration.
Propranolol was administered 1.5 h before isoproterenol.

Increased Urinary Catecholamine Output in the FH Rat. Urinary
catecholamine levels, and their metabolites, are the result of a complex
series of events involving: rate of release, neuronal and extraneuronal
reuptake and metabolism, conjugation, storage, and renal handling.
(Please see chapter by W. Osswald, this volume.) These complexities, and
the heterogeneity of essential hypertension, have created inconsistencies
and difficulties in assessing the role of catecholamines in hypertension
in humans (Goldstein, 1983). Consequently, utilizing levels of plasma
catecholamines, and particularly levels of urinary catecholamines, as an
index of increased sympathetic nervous activity in essential hypertension,
remains a matter of controversy. However, it has been a consistent
finding that a high proportion of young patients with essential
hypertension show elevated circulating norepinephrine (NE) (Sever et al.,
1977; Goldstein, 1981) and that some patients with "borderline"
hypertension show increased urinary catecholamine output (Kuchel et al.,
1976). In addition, although free dopamine (DA) is extremely difficult to
detect in plasma, there are indications that essential hypertension

patients can have surges and high levels of conjugated DA in plasma
(Kuchel et al., 1982), as well as elevated urinary homovanillic acid (HVA)
excretion.

It has been proposed that high sympathetic output can play a role in
the development of high blood pressure in genetically hypertenisve rats
(Grobecker et al., 1975; Judy et al., 1976; Nagaoka and Lovenberg, 1976;
Nagatsu et al., 1976). In a homogenous (age matched population) of inbred
genetically hypertensive rats, slightly elevated levels and turnover of
catecholamines can be detected to a level of significance. Therefore
elevated NE and HVA output may serve as an indication of enhanced
sympathetic nerve activity. This was the rationale in obtaining the data
of Table 5 which demonstrate that the urinary outputs of NE, DA and HVA
are significantly higher in hypertenisve FH rats when compared to
normotensive, age matched, control rats.

Elevated plasma and urinary levels of NE, and its metabolites, are
probably better indicators of increased sympathetic output than elevated E
(Cuche et al., 1974; Kobayashi et al., 1979). The primary source of
plasma and urinary E is adrenomedullary, and therefore levels of E do not
closely reflect sympathetic output from neuronal varicosities. Although
renal nerves can release DA upon nerve stimulation (Kopp, Bradley and
Hjemdahl, 1983) and baroreflex activation (Morgunov and Baines, 1981)
urinary DA excretion, like E, is also a poor indicator of neuronal
activity (Kopp et al., 1983). A considerable portion of urinary DA is
produced by renal tubules (Christensen, Mathias and Frankel, 1976; Ball,
Lee and Oates, 1978; Unger, Buu and Kuchel, 1978; Buhler et al., 1978).
Therefore, high levels of urinary DA do not necessarily reflect elevated
sympathetic activity, but may be due to an increased synthesis of DA, from
dihydroxyphenylamine (DOPA) (Baines and Chan, 1980) and other DA
precursors (Baines and Drangova, 1984), by nonneuronal kidney tissue.
Pharmacologically, DA is a potent renal vasodialator (Goldberg, 1972) and
there are suggestions that physiologically DA possesses natriuretic
properties (Goldberg and Weder, 1980). Interestingly, high DA output in
the FH rat is accompanied by increased sodium output (Gilboa et al.,
1984a). Although, increased DA output is not in conflict with increased
sympathetic output, the significance of the high DA output relative to
natriuresis, kidney hemodynamics and the FH rats' hypertension, requires
investigation.

Table 5. 24 Hour Urinary Catecholamine Output of Five-month Old Untreated and Debrisoquin Treated W and FH Rats

Catecholamine Output	W	FH
Untreated		
DA µg/24 hr.	3.0 + 0.7 (12)	4.7[*] + 1.1 (12)
E ng/24 hr.	254 + 94 (12)	320 + 71 (12)
NE µg/24 hr.	1.3 + 0.3 (12)	1.9[*] + 0.3 (12)
Debrisoquin Treated		
DA µg/24 hr.	0.8 + 0.4 (9)	1.3[+] + 0.5 (8)
E ng/24 hr.	354 + 145 (9)	597 + 335 (8)
NE µg/24 hr.	0.5 + 0.2 (9)	1.1[*] + 0.5 (8)

Values given as mean + std. dev.

Parentheses indicate number of animals/group.

Denotes significant difference between values for Wistar and Fawn-hooded rats: [*]p <.001, [+]p <.05.

There is a question of whether the FH rats' increased NE, DA and HVA outputs are related to renal handling. It has been demonstrated that FH rats are similar to normotensive control rats in their ability to clear creatinine and in their renal blood flow (Urizar, et al., 1984). This would argue against abnormalities in renal clearance, being the cause of the high urinary NE, DA and HVA outputs in the FH rat. Also, these present data show that there is no significant difference in E output relative to controls, which also indicates that abnormalities in renal clearance of catecholamines is probably not the major cause of the FH rats' increased NE, DA and HVA output.

The mechanism of action of debrisoquin, as an antihypertensive, is due to the drug's ability to inhibit sympathetic nervous system activity at noradrenergic nerve terminals (Boura and Green, 1965; Orgain and Kern, 1970; Bauer, 1970). In addition to depleting and interfering with sympathetic neurons, debrisoquin inhibits monoamine oxidase (Luria and Freis, 1965; Kitchin and Turner, 1966; Solomon et al., 1968). A significant portion of NE is from the noradrenergic varicosities, which are affected by debrisoquin treatment, and most likely accounts for decreases in NE output following treatment. Unlike reserpine, debrisoquin does not readily cross the blood brain barrier (Medina, Giachetti and

Shore, 1969) and central catecholamines are not depleted. Adrenal medullary catecholamines also are not affected. Decreases in NE were not accompanied by decreases in E upon debrisoquin treatment (Table 5.) The source of E being primarily from the adrenal glands, which are not affected by debrisoquin, could account for the inability of debrisoquin to decrease E output.

The major metabolites of DA in the rat brain is DOPAC. However, the major metabolite of DA in the periphery of the rat is HVA (Austin, Livett and Chubb, 1967). The much higher HVA output found in the FH rat clearly indicates a greater DA turnover. It has been shown that a substantial portion of the plasma and urinary HVA originate from peripheral 6-hydroxydopamine sensitive sites (Bacopoulos, Hattox and Roth, 1979). Inhibition of monoamine oxidase systemically, but not centrally, with debrisoquin, shows that 33% of the HVA found in the urine of untreated animals originates in the brain (Hoeldtke, Rogawski and Wurtman, 1974). The decrease in HVA output upon debrisoquin treatment of the W rats are in good agreement with the 33% figure and also illustrates that upon debrisoquin treatment there is no significant difference in HVA production between the W rats and FH rats (Table 6). In addition the 33% figure suggests that the normotensive W rats, in terms of HVA production, behave as other normal rat strains while the FH rat does not. The urinary HVA being neuronally derived (Bacopoulos et al., 1979), and the FH rats demonstrating about twice the HVA output as do W rats, can also be considered evidence for a hyperactive sympathetic nervous system in the FH rat. Overall the data indicate that the high HVA observed in the hypertensive FH rats is due to a higher turnover of DA neuronally and systemically.

The major observations in the FH rat relative to catecholamine output are: (1) they have a high peripheral NE output; (2) they can be rendered normotensive, as control rats are, by inhibition of sympathetic nervous activity and (3) they have a higher HVA output that is primarily from increased systemic DA turnover. Overall the initial evidence is supportive of the idea that the FH rat possesses a hyperactive sympathetic nervous system which could play a role in the development and manifestation of the hypertension in these animals.

Low Urinary Kallikrein in the FH Rats. Urinary kallikrein (Ukal) is a proteolytic enzyme produced in the kidney and excreted in the urine. Ukal forms kallidin which is later converted to bradykinin by cleaving

Table 6.   24 Hour Urinary HVA Output in Five-Month Old Untreated and
            Debrisoquin Treated W and FH Rats

HVA Output (μg/24 hr.)	W	FH
Untreated	18 + 2 (9)	37[*] + 8 (9)
Debrisoquin Treated	7 + 2 (7)	10 + 4 (8)

Values given as mean + std. dev.

Parentheses indicate number of animals/group.

Denotes significant difference between value of Wistar and Fawn-hooded
rats: [*]p <0.001.

these peptides from a plasma globulin kininogen. Bradykinin and kallidin
promote vasodilation, diuresis, and natriuresis (Levinsky, 1979). The
kallikrein-kinin system appears to play a role in the regulation of
systemic blood pressure. There is evidence that renal kallikrein is
involved in the regulation of renal blood flow, sodium and water excretion
as well as in the pathogenesis of experimental and clinical hypertension
(Carretero et al, 1974, Carretero and Scicli, 1981a; Mills, Newport and
Obika, 1979; Shimamoto et al, 1981; Hattori, Hasumura and Takeuchi, 1984).

The renal kallikrein-kinin system is a multi-enzymatic system that
appears to be involved in complex intrarenal as well as extrarenal events.
Vio and Figueroa (1985) using ultrastructural immunocytochemistry showed
that kallikrein was present exclusively in the cells of collecting tubules
in the rat. Subcellularly, kallikrein is distributed in luminal
membranes, basal membranes, rough endoplasmic reticulum, Golgi apparatus,
and vesicles. The presence of kallikrein in both the luminal and basal
plasma membrane allows it to act on the substrate kininogen in either the
tubular fluid or the peritubular space generating kinins which in turn are
able to influence the renal function directly or via stimulation of
prostaglandin production (McGiff, Itskovitz, and Terragno, 1975).
Recently it has been shown that a stable analog of prostacyclin increases
Ukal excretion in patients (Ylitalo et al., 1985).

An additional regulatory role for glandular kallikrein in natriuresis
and diuresis has been recently suggested by Briggs et al. (1984). These
authors demonstrated that kallikrein inactivates the atrial natriuretic

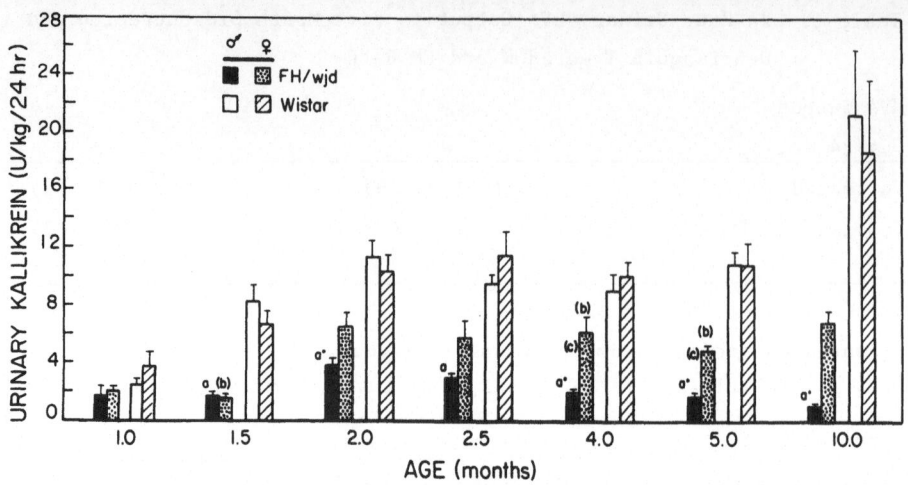

Fig. 5. UKal excretion by FH and W rats (means ± SEM). Letters indicate
statistically significant differences between (a) FH and W males,
(b) FH and W females, and (c) FH males and FH females. Only
significant differences are noted. No asterisk, $\alpha = 0.05$;
asterisk, $\alpha = 0.01$; brackets, analysis performed using log
transformation.

peptide in vitro. Thus, inactivation of the natriuretic peptide by renal
kallikrein is a possible mechanism for in vivo regulation of the
natriuretic activity.

The central nervous system plays a significant role in the
pathogenesis of hypertension. Chao et al. (1983), have shown the
existence of an endogenous tissue kallikrein in the central nervous system
using a monoclonal antibody against rat Ukal. Recently Kamitani, Little
and Ellis (1985) demonstrated that bradykinin produced from endogenous
brain kininogen by kallikrein causes a dose dependent vasodilation.
Kallikrein-like activity was also found in the neurointermediate lobe of
the rat pituitary (Powers, 1985). This activity was increased by blockade
of dopamine receptors.

Recently we found lower Ukal and a lower renal kallikrein in the FH
rat as compared to normal sex and age matched W rats (Gilboa, Rudofsky,
Magro, 1984a). The decrease in Ukal in the FH rats precedes the
appearance of hypertension (Figs. 1 and 5). No inhibitors of Ukal were
found in the urine of FH rats. These findings suggest that the

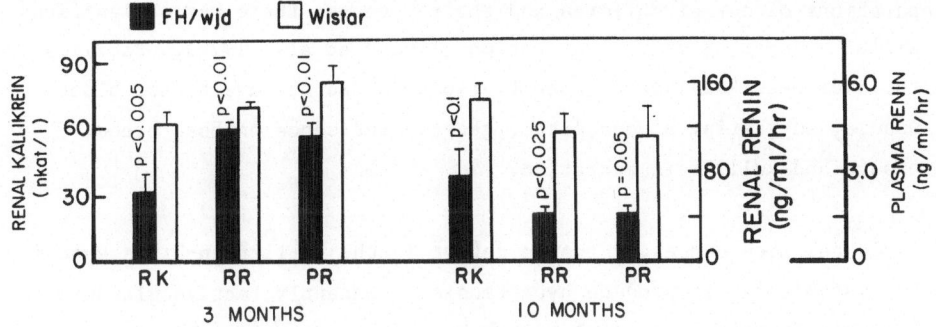

Fig. 6. Renal kallikrein (RK), renal renin (RR), and plasma renin (PR)
activities in FH and W males (means $\pm$ SEM). p values indicate
the differences between the FH and W rats.

abnormality in the renal kallikrein system may be involved in the
pathogenesis of hypertension in FH rats.

Fig. 6 shows a comparison of the renal kallikrein (RK), renal renin
(RR), and plasma renin (PR) of FH and W rats. A comparison between FH and
W rats is made at both 3 months and 10 months of age. At ages 3 and 10
months male W rat kidneys contained about twice as much kallikrein as male
FH rat kidneys. Renal renin activity and plasma renin activity were also
significantly lower in FH rats as compared to W rats. Renal renin and
urinary kallikrein both decrease with age in FH rats. However, renal
kallikrein does not change with age in either FH or W rats. The reason
for decreases in urinary kallikrein without decreases in renal kallikrein
is unknown, but a possible explanation is that intracellular stores of
renal kallikrein may remain stable despite the changes in activation and
excretion of this protein.

Studies by Carretero et al. (1976, 1977, 1978) have reported
decreased Ukal in Dahl salt-sensitive and New Zealand genetically
hypertensive rats at the time when they were still normotensive.
Normotensive children of patients with essential hypertension also have
reduced Ukal (Zinner et al., 1978). Low kallikrein excretion and
hypertension in humans may also be regulated by genetic factors (Carretero
and Scicli, 1981b, Sinaiko et al., 1982, Zinner, et al., 1978).
Normotensive black people excrete less Ukal than normotensive white people
and essential hypertensive white or black individuals less than their
respective controls (Shimamoto et al., 1981). Significantly lower mean
Ukal activity was found in children from the upper 0.26% of blood pressure
distribution than in the lower 5% (Sinaiko et al., 1982). In large

populations of normal children and their mothers, there was a significant familial clustering of Ukal excretion (Zinner et al., 1978). Families with lower mean kallikrein concentrations tended to have higher blood pressure, suggesting a concomitant genetic influence on both blood pressure and kallikrein excretion.

It has been suggested that a defect in the kallikrein-kinin system acts systemically to produce hypertension. Recently, active glandular kallikrein identical to renal and Ukal has been found in the plasma (Geiger et al., 1980; Lawton et al., 1981; Rabito et al., 1982, Masferrer et al., 1985). It has been known for a long time that it is present in salivary and sweat glands, pancreas, and intestine and in their exocrine secretions and may participate in the local regulation of blood flow in these organs (Orstavik, Carretero, and Scicli, 1982). There is apparently an organ specific control of glandular kallikrein, suggesting that they are under different electrolyte and endocrine control, and may perform different functions at each anatomical site (Van Leeuwen, Grinblat and Johnson, 1984). Isolated perfused rat kidneys release kallikrein into perfusate (Corthorn, Bio, Roblero, 1983). Proud, et al. (1984) found glandular kallikrein in lymph fluid. Orstavik, Carretero, and Scicli (1982) demonstrated that upon adrenergic stimulation of the submandibular gland in the rat, glandular kallikrein released into the vascular compartment produced kinins in the peripheral circulation. These kinins produced hypotension when their breakdown was prevented by the kininase II (angiotension I – converting enzyme) inhibitor, captopril. In addition, Nolly and Lama (1982) showed that arterial tissue contains an enzyme with physicochemical characteristics resembling those of kallikreins of glandular origins and capable of generating kinins.

Kallikrein is excreted in the urine as a zymogen and as an active enzyme (Corthorn and Roblero, 1981). As we did not detect an inhibitor of kallikrein in the urine of FH rats, we compared the urinary excretion of active Ukal and of prekallikrein (Pkal) in 5-months-old FH and W males (Fig. 7) (Gilboa, Sidoti, and Magro, 1985). Ukal activity was measured before and after activation of Pkal by trypsin as described by Corthorn and Roblero (1981). There was no significant difference between Pkal in the two strains, but significantly lower Ukal ($p < 0.005$) was present in the FH rats. In W rats the Ukal/Pkal ratio was 1.2. The Ukal/Pkal ratio in the FH was 0.8, suggesting that in FH rats the larger part of Ukal is inactive, whereas in Wistar rats the opposite is correct. In other words,

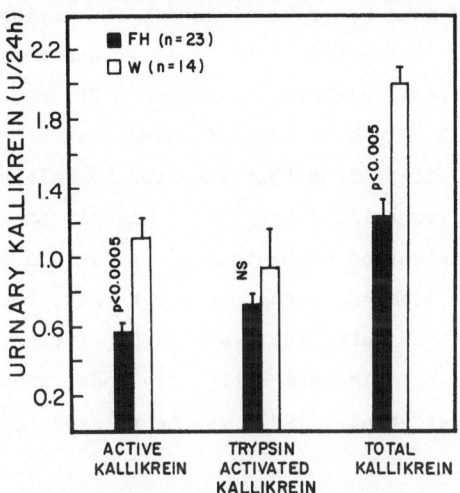

Fig. 7.   Excretion of activated, trypsin-activated (Prekallikrein) and
          total kallikrein in the urine of 5 month old FH and W male rats.
          NS - No significant difference.

FH and Wistar rats excrete similar amounts of Pkal but FH excrete less
active Ukal which may indicate a decrease in Pkal to Ukal conversion by
the FH rat.

     Carretero et al. (1978) found a decreased total and active Ukal in
normotensive Dahl salt-sensitive rats. The decrease in Ukal activity was
far greater than the decrease in total Ukal, suggesting that Ukal in these
rats may be inhibited, abnormal or inactive. In a recent study Arbeit and
Serra (1985) demonstrated that both 7 week and 12 week old Dahl
salt-sensitive rats excreted significantly less active Ukal and total
kallikrein than the age matched Dahl salt resistant counterparts.
Normotensive Dahl salt-sensitive rats failed to increase maximally
kallikrein activity or total kallikrein when the diet was switched from
normal salt diet (0.7% Na$^+$) to low salt diet (0.004%Na$^+$). Low sodium diet
did not increase total Ukal in salt sensitive rats but increased in the
salt resistant rats. The authors suggested that the decrease in Ukal,
seen in Dahl salt-sensitive rats is an inherited defect and not a result
of the hypertension. Koolen, Daha and van Brummelen (1985) found that
patients whose blood pressure increased with high salt diet had lower
levels Ukal excretion and a blunted kallikrein response to dietary sodium
restriction and furosemide (Grunfeld et al., 1984). A blunted response in
Ukal was also noted in hypertensive children following diuretics.   Ader

et al. (1985) showed that under basal conditions active Ukal is lower in SHR compared to WKY. Excretion of both active and total kallikrein was directly related to renal perfusion pressure (RPP) and less responsive to changes in RPP in SHR. In hypertensive one-kidney pole-ligated rats the total Ukal does not differ from that excreted by solely uninephrectomized rats, though the active kallikrein shows a significant drop, suggesting an alteration in the activation mechanisms of kallikrein (Vergara and Corthorn, 1983). We also found that unlike W rats, the excretion of Ukal did not change in the FH rats regardless of Na^+ intake. This could suggest that failure to stimulate kallikrein under normal physiologic conditions (Lieberthal et al., 1983) may be related to the pathogenesis of hypertension.

In a recent study Takaoka et al. (1985) demonstrated that rat kidney cortex contains a protease catalyzing conversion of urinary inactive kallikrein into its active form and this protease has properties of a thiol protease but not of trypsin. This protease may be one of the regulators of the kallikrein-kinin system in the kidney.

Renal Abnormalities in the FH Rat. FH males develop progressive proteinuria at the age of 4 months (Gilboa, Rudofsky, and Magro, 1984a, Urizar et al., 1984; Kreisberg and Karnovsky, 1978) (Fig. 2). Although FH females in our study did not develop severe proteinuria, they nevertheless excreted 2-4 fold more protein than normal W females from age 4 to 10 months. Magro and Rudofsky (1982) reported focal glomerular sclerosis (FGS) in FH males as early as 3 months of age. It is not clear whether the proteinuria in FH males is solely attributed to the FGS or the combination of FGS and hypertension or other physiologic or renal abnormalities.

Although FH females develop mild FGS and moderate hypertension they do not excrete more than 10 mg of protein/24 hr. SHR male rats develop proteinuria but at a much older age than FH males. It is of interest that when the blood pressure in the SHR rats has been normalized, at young age, by antihypertensive medications they nevertheless developed considerable proteinuria associated with glomerular pathologic changes after 11 months of age (Feld et al., 1981). This suggests that proteinuria and glomerular pathology in the SHR rat may develop as in normotensive rats. Sustarsic, McPartland, and Rapp (1981) found slightly increased protein excretion in 1.5 month-old Dahl salt-sensitive rats which were still normotensive. More recently Sterzel et al. (1985) found significant proteinuria associated with glomerular changes in prehypertensive Dahl salt-sensitive rats. They also found that normal NaCl intake delays but does not prevent hypertension and progressive renal disease.

It has been suggested that renal parenchymal damage in rats decreases kallikrein excretion (Glasser and Michael, 1976). The renal damage in this study was induced by antiglomerular basement membrane nephritis and aminonucleoside nephrosis. However, we found that in FH rats the decrease in Ukal preceeds the development of FGS by at least 2 months. As compared to normal controls, Cumming and Robson (1985) found that Ukal was decreased in patients with glomerulonephritis without nephrotic syndrome, but was markedly increased in patients with nephrotic syndrome.

Due to the fact that the hypertension, renal pathology, and proteinuria are much more severe in the FH males than FH females we postulated that male hormones may play a role in this difference between the sexes. We therefore orchiectomized or sham operated young (5-8 week old) FH males and followed them until they were 9 months old (Gilboa, Rudofsky, Magro, 1984b). During this period 7 twenty-four hour urine collections were obtained. The urine was analyzed for Na^+, K^+, protein and kallikrein. Blood pressure was obtained before each urine collection. We found no significant differences in the excretion of Na^+, K^+, Ukal or urine volume between the sham or orchiectomized rats. Blood pressure after about 20 weeks of age stabilized around mean systolic pressure of approximately 160 mmHg in the orchiectomized rats, but continued to increase in the controls. This divergence, however, did not achieve statistical significance. Between 20 and 38 weeks of age the body weight of control rats was significantly higher than of the orchiectomized rats.

The most striking differences between the orchiectomized and the control rats were in the protein excretion and the pathologic changes of the glomeruli. Control rats started excreting significantly more protein than the orchiectomized rats as early as 6 weeks after the gonadectomy. This difference increased with time until prior to the rats' sacrifice at age 40 weeks when the control rats excreted an average of 100 mg of protein/24 hr whereas the orchiectomized rats lost only 30 mg/24 hr ($p<0.005$). Examination of the kidneys revealed that they were significantly ($p<0.0005$) heavier in the control rats than in the orchiectomized rats. It was also found that 25% of glomeruli were sclerosed in the control rats, but only 10% in the orchiectomized animals ($p<0.001$).

The direct reason for the reduced FGS and proteinuria in the orchiectomized rats is not clear, but it is obviously related to the male hormones. It is hard to attribute the difference to a change in blood pressure for 2 reasons. First, although the blood pressure was lower in the older orchiectomized rats than in the age matched controls, the difference was not statistically significant. Second, protein excretion was already significantly lower in the orchiectomized rats several weeks before any difference in the blood pressure had been noted.

Jams and Wexler (1979) reported that gonadectomy in young SHR males and females retards the increase in the blood pressure, although the ovariectomized females "escaped" from this inhibiting effect. Treatment with testosterone or estradiol demonstrated that estradiol was particularly effective in inhibiting the usual rise in blood pressure in sham-operated or gonadectomized males and females. As in our study, they also found that the gain in body weight was lower in the castrated males.

It is possible that while removal of testosterone in the FH males has a protective effect, presence of estradiol in the females reduces the severity of hypertension, FGS, and proteinuria. It is of interest that pregnancy has a profound antihypertensive effect in the SHR rats (Lorenz et al., 1984). Recently Matsudaira, Kogo, and Satoh (1985) found that orchiectomy in male SHR rats retards the development of hypertension. Because urinary PGE_2 and electrolyte excretion tended to be higher in the castrated group they postulated that renal PGE_2 may participate in the gonads-mediated blood pressure regulation.

It is also of interest that Meyer et al. (1983) demonstrated that ablation of the pituitary in rats with reduced nephron number prevents the glomerular injury usually caused by the glomerular capillary hypertension and hyperperfusion associated with increase in glomerular volume. The pituitary ablation reduced body weight, prevented the pathologic changes, and maintained glomerular volume and morphology near normal.

Renal Function and Excretion of Electrolytes. We found no significant difference in creatinine clearance in 5-month old FH and W males. Similar findings have been shown by Kreisberg and Karnovsky (1978) in 12-month old rats despite the progressive FGS. With the exception of 1.0 and 1.5 months old rats, FH males persistently excreted more urine than W males; however this difference was significant only at the age of 2.5 months (Fig. 2). It has been suggested by some authors that Ukal excretion is directly related to the volume of the excreted urine (Croxatto et al., 1976; Mills et al., 1979). This however does not seem to be the case in the FH rats, particularly the males, who from the age of 2 months had higher urine output but lower Ukal excretion than FH females or W rats.

FH males, 2 months and older excreted more sodium, but there was no difference in the excretion of K^+. The reasons for the higher urine volume and Na^+ excretion in FH males are not yet clear. These may be related to the hypertension that FH males develop by 2 months of age. Pressure diuresis has been also proposed in the genetically hypertensive mice (Sustarsic et al., 1980). Rosas, Albertini, and Croxatto (1977) found that a decrease in renal kallikrein is associated with an increase in plasma volume and in doubling of urine volume in renal hypertensive rats. Analysis of serum for the levels of creatinine, BUN, Na^+, K^+, Cl^-, glucose, CO_2 and Ca^{++} did not reveal any difference between the FH and W rats.

The Pathology of Focal Glomerular Sclerosis in FH Rats. The etiopathology of focal glomerular sclerosis (FGS) is not known although hormonal, dietary, and immunologic factors have been postulated (Bolton et al., 1976; Couser and Stillmant, 1975; Elema and Arends, 1975). The male FH rat is therefore an interesting model because FGS as well as tubulointerstitial lesions occur as early as 2 months of age (Magro and Rudofsky, 1982; Kuijpers and Gruys, 1984). Both FGS (Kreisberg and Karnovsky, 1978) and hypertension (Rudofsky and Magro, 1982) have persisted as phenotypic traits. FGS and hypertension have also been described in FH rats from other laboratories (Kuijpers and Gruys, 1984) suggesting that these genetic traits have been in this breed for many years.

The onset of FGS in male FH rats in our colony occurs between 1.5 and 2.5 months of age involving a small number of glomeruli (2-5% per kidney cross-section on PAS-stained slides); female FH rats have a milder form starting at 4-6 months of age (Magro and Rudofsky, 1982; Kuijpers and Gruys, 1984). It is noteworthy that both segmentally as well as globally sclerosed glomeruli are present even at this early age and it is the percent of involved glomeruli which increases with age. For example, male FH kidneys examined at ages 2,5,6, and 7 months may have 5%,20%,25%, and 40% of affected glomeruli respectively; maximal renal disease (>50% glomerular involvement) manifests itself by 9 months of age. Thus there is an age dependent increase in segmentally as well as globally sclerosed glomeruli. A second renal lesion in FH rats which increases in severity with age occurs in the renal proximal tubules and interstitium.

By light microscopy the earliest glomerular abnormalities appear to consist of swollen endo- and -epithelial cells some of which contain PAS positive, silver negative droplets (Magro and Rudofsky, 1982). Whether these glomeruli become progressively sclerotic is not known. Secondly, a few histologically normal glomeruli are surrounded by plasma protein which has leaked into Bowman's space and hyaline casts are seen in adjacent proximal tubules. The characteristic glomerular lesions (Fig. 8) vary in degree of basement membrane thickness and splitting, as determined by electron microscopy (Urizar, et al., 1984), and there are increases in collagen deposition and hyperplasia of the mesangial areas; sclerosis may occur in a single capillary loop or mesangial axis or it may involve one or all glomerular tufts. Occasionally there are adhesions of glomerular tufts to Bowman's capsule and there are structures resembling crescents. A few glomeruli show membranoproliferative changes. Exudative, fibrinoid, necrotic, hemorrhagic or inflammatory changes have not been observed in the glomeruli.

Rare focal tubulointerstitial lesions are seen in the renal outer cortex as early as 1 month of age i.e. at a time when there are virtually no histologic glomerular changes or blood pressure elevations. These consist of small areas of interstitial fibrosis, peritubular mononuclear cell infiltration and thickening, deformation, and wrinkling and duplication of the tubular basement membranes (Urizar, et al., 1984); occasionally hyaline casts are also seen. Extensive disruption of the cortical architecture occurs by 7-9 months-of-age. There is dilitation and atrophy of tubules, and the presence of numerous large hyaline casts obstructing tubular lumens giving rise to a "thyroid-like" appearance to the renal cortex; there are cystic glomeruli with greatly extended Bowman's spaces which appear to be the result of plugged tubules (Magro and Rudofsky, 1982; Urizar, et al., 1984).

Immunohistopathologic analyses of renal tissues of FH rats at various ages have demonstrated mesangial and global deposition of serum proteins (IgG, IgM, C3, fibrin and fibronection) (Kuijpers and Gruys, 1984; Urizar, et al., 1984). The granular and amorphous patterns of immunofluorescence are suggestive of deposition of immune complexes and are similar to those seen in most mice and guinea pigs (Markham, Sutherland and Mardiney, 1973;

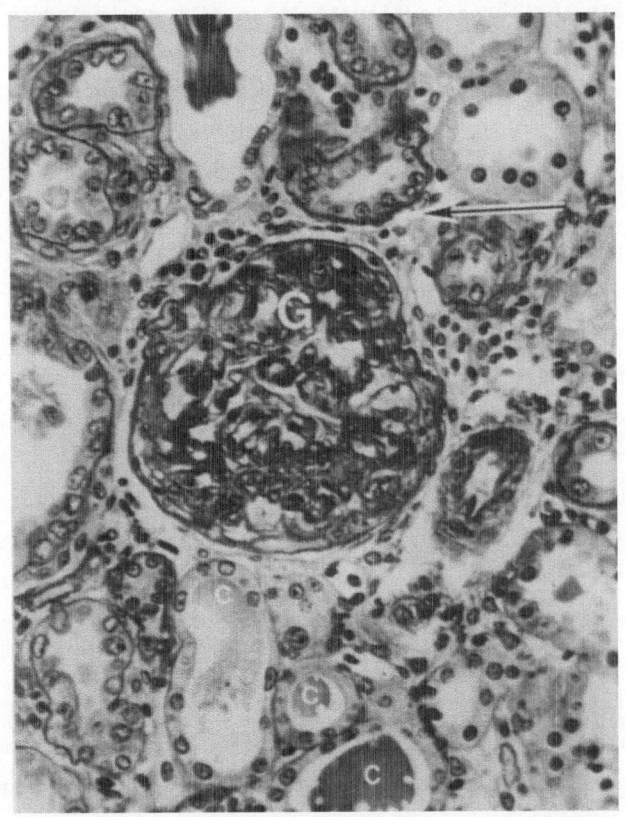

Fig. 8.   Severely sclerotic glomeruli and tubulointerstitial lesions as
present in the kidneys of male FH rats. Glomerular (G) changes
include expansion of mesangial areas, obsolescence of capillary
tufts, and adhesions to Bowman's capsule. Tubulointerstitial
lesions consist of mononuclear cell infiltrates, dilation and
atrophy of tubules, thickening of basement membranes (arrow), and
formation of hyaline casts (C). (PAS-stain, X420).

Steblay and Rudofsky, 1971), but the evidence for an immunopathologic
mechanism of renal disease is insufficient. Furthermore, a search for
circulating immune complexes or electron-dense glomerular deposits in FH
rats was inconclusive (Urizar, et al., 1984); moreover, autoantibodies to
renal or nuclear antigens could not be detected in the sera of FH rats
(Urizar, et al., 1984 and unpublished observations). Thus, the evidence
for an immunopathogenesis of FGS has not been established.

The relationship between FGS and hypertension is not known. It does not seem likely that hypertension in these rats could induce FGS since these lesions are present in some glomeruli before the onset of elevated blood pressures. Conversely, since FGS is present in non-hypertensive aging rats of other strains, it does not appear that there is an important contribution of FGS to the development of hypertension.

The FH rat appears to offer an opportunity to perform both biochemical, pharmacologic and genetic studies on the pathogenesis of FGS not available in any other rat strains. However, we can only speculate about the cause-and-effect relationships among the numerous presently known abnormalities in FH rats (see above) and the pathogenesis of FGS. The local nature of FGS, i.e., the heterogeneity of morphologic changes among glomeruli suggests that the lesions may arise from local disturbances in biochemical and hemodynamic mechanisms in glomeruli or perhaps the juxtaglomerular apparatus. Indeed, we have noted that there is a decline in plasma renin activity before the appearance of diffuse FGS in young FH as well as aging W rats (Magro and Rudofsky, 1982). This evidence could be interpreted as an indication that there may be angiotensin-related glomerular or arteriolar vasomotor abnormalities or disturbances in filtration which may be involved in the sclerotic changes. Certainly, the fact that up to 25% of glomeruli remain histologically normal even in aging FH rats suggests a local rather than a systemic insult such as circulating immune complexes. It is also apparent that no single metabolic abnormality may explain this pathogenesis and that a combination of several factors may influence the development and progression of FGS of these FH rats. The accelerated development of FGS in the FH rat may, therefore, be an excellent model for FGS in man.

Impaired Baroreflex Function in the FH Rat. The SHR demonstrates abnormalities in baroreceptor function including an impaired baroreflex (Casto and Phillips, 1985). In part abnormalities in baroreceptor function are due to a resetting of the baroreceptors (Thant, Yamori and Okamoto, 1969; Nosaka and Okamoto, 1970; Nosaka and Wang, 1972; Andresen, Kuraoka and Brown, 1980) and may also be related to an increase in vessel wall stiffness (Andresen and Brown, 1980). The FH rat also demonstrates an impaired baroreflex cardio response, which could in part be centrally mediated and due to a lack of increase in cardiac vagal activity.

Cardio baroreflex responses were measured in conscious and freely moving FH and control W rats (Table 7). The slopes of the heart period (mSec/beat) vs. blood pressure (mmHg) were calculated from the least square regression line of the data points for pressures between 145–160 mmHg. By looking at the cardio baroreflex at pressures above 150 mmHg, it was possible to distinguish between a resetting (shift to higher pressure for a change in heart rate but with a similar slope) and an actual impairment of the response (lower slope). The average slope of heart period vs. pressure for the FH rat was 1.27 mSec/beat/mmHg $\pm$ .41 and for the W rat was 0.44 mSec/beat/mmHg $\pm$ .26 (Table 7).

In essential and other forms of human hypertension, baroreceptor reflexes, with respect to control of heart rate, show a change in sensitivity (slope of heart period vs. pressure) in addition to a resetting of the reflex at a higher level of blood pressure (Bristow, et al., 1969).

Table 7. Average Slope of Heart Period vs. Pressure for FH and W Rats

Rat	# of Data Points	Average Slope	std. dev.
W	11	1.27 mSec/beat/mmHg	$+0.41$
FH	20	0.44 mSec/beat/mmHg	$+0.26$

Besides being relevant to essential hypertension, an additional significance of the FH rat having an increased sympathetic output, low PRA, and impaired baroreflex response is that they can all be increased or decreased by central α_1, α_2, AII, 5HT, and $5HT_2$ receptors (Casto and Phillips, 1985; Brody, Haywood and Touw, 1980; Ganong, 1983; Zimmermann and Ganong, 1980; Reid, et al., 1975; Finch and Haeusler, 1973; Kobinger and Pichler, 1975; Kobinger and Walland, 1972; Wing and Chalmers, 1974; Kuhn, Wolf, and Lovenberg, 1980; Head and Korner, 1982; Korner, et al., 1981; Hokfelt, Hedeland and Dymling, 1970; Ferrario, Gildenberg and McCubbin, 1972; Fukiyama, McCubbin and Page, 1971; Ferrario and Barnes, 1981).

Considering the abnormalities in PRA, sympathetic output and baroreflex responses, the FH rat has the potential to be used as a model to make advances in understanding how central α, AII and 5HT receptors, interact and exercise neural control over the peripheral renin-angiotensin system, sympathetic output and cardiovagal activity.

Table 8. Effect of Antihypertensive Drugs on SBP of FH and W Rats

Treatment	Dose	Length of Treatment	SBP mmHg FH	SBP mmHg W
None	----	----	173 + 9 (9)	128 + 7 (9)
Nifedipine[+]	1.0 mg/kg	2 hr.	130 + 8 (5)	116 + 10 (5)
Clonidine[+]	70 µg/kg	3 hr.	111 + 5 (5)	101 + 4 (5)
Debrisoquin	60 mg/kg/day	2 days	104 + 9 (8)	91 + 7 (9)
Furosemide[*]	10 mg/kg	2.5 hr.	168 + 11 (5)	120 + 6 (5)
Captopril	300 mg/kg/day	3 days	126 + 13 (9)	94 + 8 (9)

Data shown as Mean + std. dev., parenthesis indicate number of animals, male rats 4.5 - 5.0 months old. [*]Reflex constriction of capacitance vessels may mask diuretic effects. [+]Effects of clonidine and debrisoquin prove conclusively only that the FH rat's sympathetic system is intact.

Some Evidence the FH Rat has Increased Vasoconstriction Tone. It is most likely that the FH rat has increased vasoconstrictor tone with an increased total peripheral resistance. As a first estimation of whether the rat is vasoconstricted, we observed the effect upon blood pressure of blood volume changes (low Na^+, high Na^+, diuretics) and of vasodilation (Nifedipine). For 3 month old FH rats, where the elevated blood pressures can be intermittent, low Na^+ intake exacerbates the hypertension and high Na^+ alleviates it. This may be due to sensitivity to AII, for we have shown that as expected there is an inverse relationship between PRA and blood pressure at this age in the FH rat. Low salt intake exacerbating low PRA forms of human essential hypertension has been documented (Laragh, et al., 1982).

In itself, drug responsiveness is not conclusive when deducing a mechanism of hypertension. Also, we are aware that in comparing effects to normotensive controls, it is easier to lower blood pressure when blood pressures are elevated. However, we do feel drug responsiveness can be relevant and indicative.

The Ca^{2+} channel blocker nifedipine can have cardiac effects but its primary mode of action, as an antihypertensive, is vasodilation (Kazda, Garthoff and Thomas, 1982; Nayler and Poole-Wilson, 1981). It can be seen from Table 8 that an I.P. injection of nifedipine effectively reduces the FH rats blood pressure while showing very little effect on the W rats. Also an I.M. injection of an acute dose of the diuretic furosemide was

ineffective in relieving the FH rats' hypertension even though there was
an average of 15 ml urine output over the first three hours after
furosemide treatment. The effects of long term diuretic treatment have
not been looked at in the FH rat as yet. However, we were interested in
the effects of reduced plasma volume and long term diuretic treatment has
other effects in addition to reducing plasma volume. The response to
nifedipine and furosemide is another indication the animal may be
vasoconstricted and not volume expanded.

All forms of human essential hypertension manifest an increased total
peripheral resistance (Frohlich, 1982). Elevated total peripheral
resistance is an important aspect of an animal model (Pfeffer and
Frohlich, 1977; Ferrone, et al., 1979; Smith and Hutchins, 1979), if it is
to be relevant to essential hypertension. Conclusive evidence for
elevated peripheral resistance may require cardiac output measurements
(i.e., normal or reduced cardiac output with elevated pressures implies
elevated total peripheral resistance) in the FH rat, but initial evidence
indicates that the FH rat does have increased vasoconstrictor tone.

Cellular Activation Abnormalities in the FH Rat. Mast cells from
normal rats isolated from the peritoneal cavity, and challenged with
dextran in vitro, release histamine. The process is an energy dependent
one which is not fatal to the mast cell and is very similar to antigen
induced histamine release. Both antigen induced histamine release, and
dextran induced histamine release from rat mast cells are greatly
potentiated in the presence of phosphatidylserine. However, IgE mediated
histamine release from mast cells isolated from FH rats is not potentiated
by phosphatidylserine (Fig. 9). In the FH rat, dextran mediated histamine
is also not potentiated (Fig. 10). It appears that the FH rat possesses a
non-potentiation response to phosphatidylserine which may be genetically
controlled.

The data of Table 9 show how the mast cells of 72 FH and 65 W rats
respond to the in vitro challenge of 15 µg/ml of ovalbumin in the presence
and absence of phosphatidylserine. The phosphatidylserine potentiation
test is a very reliable in vitro test. The lack of phosphatidylserine
potentiation may be related to some of the platelet activation
abnormalities that the FH rat manifest. Platelets of the FH rat do
not aggregate or secrete 5HT normally upon stimulus by collagen and
other agonists as platelets from most other rat strains do (Raymond
and Dodds, 1975; Urizar et al., 1984). The FH rat also shows a

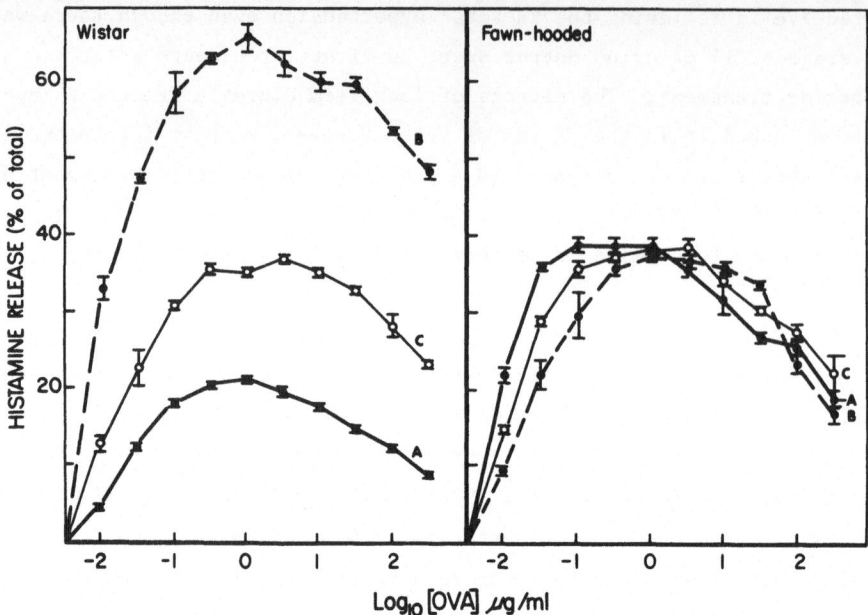

Fig. 9.   Histamine released from mast cells of ovalbumin-immunized Wistar
and Fawn-hooded rats at varying doses of ovalbumin. Curves A: no
exogenous phospholipid. Curves B: 1 µg/ml of exogenous
phosphatidylserine. Curves C: 0.1 µg/ml of exogenous
Lyso-phosphatidylserine.

platelet 5HT storage defect similar to that found in man (O´Brien, 1984)
and a reduced 5HT uptake (Raymond and Dodds, 1975). Also, there appears
to be an association of the red-eyed dilution gene [r] of the FH rat with
some platelet disorders the rat manifests (Prieur and Meyers, 1984;
Tobach, DeSantis and Zucker, 1984). It has been reported that the 5HT
uptake in human platelets of hypertensive subjects is reduced (Ahtee et
al., 1974; Bhargava et al., 1979). The blood platelet has pharmacological
and kinetic properties similar to central serotoninergic neurons (Stahl
and Meltzers, 1978). It is not clear whether production of 5HT uptake in
platelets in humans is the result of the hypertension or indicates a
central dysfunction in central serotoninergic neurons which may be causing
or exacerbating the hypertension. In themselves, the platelet defects in
the FH rat are significant for we feel the rat has the potential to
provide information as to whether central and/or platelet 5HT binding and
uptake sites are modified by elevated blood pressures (Kamal et al., 1984;
DeWardener and Clarkson, 1982; Hamlyn et al., 1982; Kamal et al., 1983),
or whether some basic genetic defect in 5HT uptake, storage and release
precedes and is etiologically related to the appearance of hypertension.

Table 9. Mean Value and Std. Dev. of Histamine Release from Peritoneal
          Mast Cells of W and FH Rats

| | % Histamine Release | |
| | No PS | 1 µg/ml PS |
Rat	Average ± std. dev.	Average ± std. dev.
W (65)	16 ± 16	56 ± 26
FH (72)	20 ± 19	24 ± 19

In an investigation into 5HT uptake and imipramine binding in blood
platelets and brain synaptosomes of FH and Sprague-Dawley rats, Arora et
al. (1983) found that 5HT and imipramine binding sites may be
independently regulated and that platelets do not always manifest
abnormalities present in brain 5HT neurons and vice versa. Arora et al.
(1983) found that the B_{max} values for ^{3}H-imipramine binding in both
platelets and brain of FH rats was significantly lower than those of
Sprague-Dawley rats. On the other hand, Ieni et al. (1985) found no
significant differences in high- and low-affinity ^{3}H-imipramine binding in
brain homogenates of FH rats as compared to Long Evans rats.

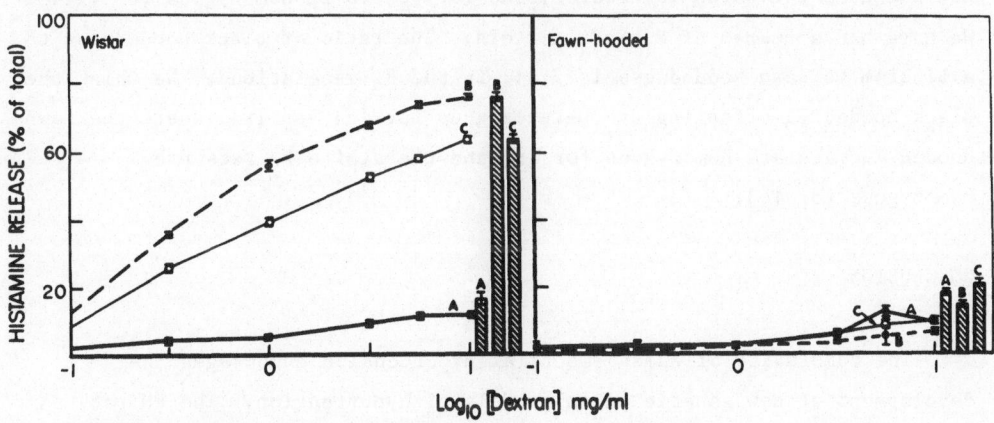

Fig. 10. Histamine released from mast cells in vitro of
          ovalbumin-immunized W and FH rats at varying doses of dextran
          500. Curves A: no exogenous phospholipid. Curves B: 1 µg/ml of
          exogenous phosphatidylserine. Curves C: 0.1 µg/ml of exogenous
          lyso-phosphatidylserine. Bar diagrams show the quantity of
          histamine released for the same cell populations by 15 µg/ml of
          ovalbumin.

<u>The F_1 Generation of FHXW and WXFH are all Black Hooded and</u>
<u>Intermediate in Blood Pressures</u>. The data of Table 10 illustrate the
blood pressure and color of litters of FHxW. All of the offspring (male
and female) of the FHxW were black hooded. The indirect systolic blood
pressures shown in Table 10 are for male F_1 hybrids. These data are
preliminary but indicate the regulation in blood pressure of the FH rat is
either polygenic or autosomal codominant. If the inheritance of blood
pressure in the FH rat was autosomal recessive the blood pressure would be
closer to 125 mmHg which is about the average of the 4 month old W males.
On the other hand, if the inheritance of blood pressure in the FH rat was
autosomal dominant, one would expect the F_1 generation to be closer to 170
mmHg in blood pressure which is about the average of the 4 month old FH
males. It will be necessary to look critically at the blood pressure
distribution frequency of the F_2 generation to determine conclusively
whether inheritance of blood pressure in the FH rat is polygenic,
autosomal codominant, autosomal recessive, or autosomal dominant.

The fact that all of the F_1 hybrids are black hooded is consistent
with the fawn color in the FH rat being the result of the red-eyed
dilution [r] gene (LaVail, 1981b). The albino W rats being [cc] and the
FH rats being [r,r], and since [r] and [c] are tightly linked with a
recombination value of only 0.3 + 0.1 percent (Robinson, 1965), one would
not expect a rat which is heterozygous for [c] to be homozygous for [r].
We have had a number of F_1 X F_1 litters. The ratio of black hoodedness to
albinoism to fawn hoodedness is 2:1:1 in the F_2 generations. We think the
black hooded F_2 offspring are heterozygous for [c] and [r], where the fawn
hooded F_2 rats are homozygous for [r] and the albino F_2 rats are
homozygous for [c].

CONCLUSION

The complexity of human essential hypertension encourages the
development of new genetic animal models of hypertension. The FH rat, as
a hypertensive strain, should offer a model which can be used to study
facets of essential hypertension. The rat demonstrates a highly heritable
hypertension which is readily maintained and which appears spontaneously.
Besides a genetic predisposition, the severity of the hypertension in the
FH rat increases with age. The blood pressures are fairly normal over the
first 2 months of life, then elevate intermittently, and then slowly
increase to hypertensive levels which maximize at about 5-9 months of age.
The severity of the hypertension and the fact that the animals were never

Table 10. Systolic Blood Pressure of Male F_1 Offspring of
FH X W Crossings at 4 Months of Age

Rats	Designation	No.	Color	SBP (mmHg) $\pm$ S.D.
RH	ESP 10F5	4	FH	161 $\pm$ 5
W	none	5	Albino	125 $\pm$ 5
(FH X W)F_1	Line 1	10	BH	147 $\pm$ 9
	Line 2	5	BH	152 $\pm$ 5*
	Line 3	5	BH	145 $\pm$ 4
	Line 4	9	BH	142 $\pm$ 5**
	Line 5	5	BH	149 $\pm$ 2

*Significant decrease compared to FH (p=<0.05)

**Significant increase compared to W (p= <0.001)

selectively bred for hypertension indicates that it is the result of a
precipitous combination or mutation of genes. This is quite different
from most other hypertensive rat strains where the elevated blood
pressures were intensified by breeding selection from moderate levels over
a period of time.

The combination of physiological, enzymatic and endocrine
abnormalities accompanying the elevated blood pressures are unique to the
FH rat. Most of the phenotypic abnormalities expressed by the FH rat are
expressed in patients with essential hypertension. Because of these
circumstances, the FH rat provides a model in which to study the etiology
of vascular, endocrine-enzymatic abnormalities and their relationship to
elevated blood pressure. In addition there is every indication that the
FH rat is uniquely suited to be developed into a model which can be used
to study neurogenic aspects and interactions in the central and peripheral
regulation of vascular resistance.

The long-term breeding record of the FH rat indicates that the
hypertension in these animals will be maintained and most likely
intensified by selective breeding. This and the fact that the animals are
easily bred, and require no special diet or care, is significant for it
indicates there is a high probability that the animal model can be

further developed in a manner useful to the research community.

Based on our experience with the FH rat, we hope that the developed
FH rat, with the proper control strains, will take investigators into
directions where basic questions will be answered that other models are
not presently addressing.

REFERENCES

Adler, J.L., Pollock, D.M., Butterfield, M.I., and Arendshorst, W.J.,
    1985, Abnormalities in kallikrein excretion in spontaneously
    hypertensive rats, Am. J. Physiol., 248 (Pt 2):F396–F403.

Ahtee, L., Pentikainen, L., Pentikainen, P.J., and Paasonen, M.K., 1974,
    5–hydroxytryptamine in the blood platelets of cirrhotic and
    hypertensive patients, Experientia, 30:1328–1329.

Altman, P.I. and Katz, D.D., 1979, Inbred and genetically defined strains
    of laboratory animals. Federation of American Societies for
    Experimental Biology, Bethesda, Maryland.

Andresen, M.C. and Brown, A.M., 1980, Baroreceptor function in
    spontaneously hypertensive rats. Effect of preventing hypertension,
    Circ. Res., 47:829–834.

Andresen, M.C., Kuraoka, S., and Brown, A.M., 1980, Baroreceptor function
    and changes in strain sensitivity in normotensive and spontaneously
    hypertensive rats, Circ. Res., 47:821–828.

Antonaccio, M.J., 1977, Neuropharmacology of central mechanisms governing
    the circulation, in: "Cardiovascular Pharmacology", M. Antonaccio,
    ed., Raven Press, New York.

Arbeit, L. A., and Serra, S.R., 1985, Decreased total and active urinary
    kallikrein in normotensive Dahl salt susceptible rats, Kidney Int,.
    28:440–446.

Armstrong, J.M., Blackwell, G.J., Flower, R.J., McGiff, J.C., Mullane,
    K.M., and Vane, J.R., 1976, Genetic hypertension in rats is
    accompanied by a defect in renal prostaglandin catabolism, Nature,
    260:582–586.

Armstrong, J.M., Lattimer, N., Moncada, S., and Vane, J.R., 1978,
    Comparison of the vasodepressor effects of prostacyclin and
    6–oxo–prostaglandin $F_{1\alpha}$ with those of prostaglandin E_2 in rats and
    rabbits, Br. J. Pharmacol., 62;125–130.

Arora, R.C., Tong, C., Jackman, H.L., Stoff, D., and Meltzer, H.Y., 1983,
    Serotonin uptake and impiramine binding in blood platelets and brain
    of Fawn–hooded and Sprague Dawley rats, Life Sci., 33:437–442.

Austin, L., Livett, B.G., and Chubb, I.W., 1967, Increased synthesis and release of noradrenaline and dopamine during nerve stimulation, Life Sci., 6:97-104.

Bacopoulos, N.G., Hattox, S.E., and Roth, R.H., 1979, 3,4 Dihydroxyphenylacetic acid and homovanillic acid in rat plasma. Possible indicators of central dopaminergic activity, Eur. J. Pharmacol., 56:225-236.

Baer, P.G., Bianchi, G., and Liliana, D., 1978, Renal micropuncture study of normotensive and Milan hypertensive rats before and after development of hypertension, Kidney Int., 13:452-466.

Baines, A.D. and Chan, W., 1980, Production of urine free dopamine from DOPA; a micropuncture study, Life Sci., 26:253-259.

Baines, A.D. and Drangova, R., 1984, Dopamine production by the isolated perfused rat kidney, Can. J. Physiol. Pharmacol., 62:272-276.

Ball, S.G., Lee, M.R., and Oates, N.S., 1978, The effect of inorganic salts on renal tissue dopamine levels in the rat, (Proc.) Br. J. Pharmacol., 63(2):343P.

Bauer, G.E., 1970, Debrisoquin. A five-year study of a new hypotensive agent, Med. J. Aust., 2:911-915.

Bhargava, K.P., Raina, N., Misra, N., Shanker, K., and Vrat, S., 1979, Uptake of serotonin by human platelets and its relevance to CNS involvement in hypertension, Life Sci., 25:195-199.

Bolton, W.K., Benton, F.R., Maclay, J.S., and Sturgill, B.C., 1976, Spontaneous glomerular sclerosis in aging Sprague-Dawley rats. I. Lesions associated with mesangial IgG deposits, Am. J. Pathol., 85:277-300, 1976.

Boura, A.L. and Green, A.F., 1965, Adrenergic neuron blocking agents, Ann. Rev. Pharmacol., 5:183-212.

Briggs, J.P., Marin-Grez, M., Steipe, B., Schubert, G., and Schnermann, J., 1984, Inactivation of atrial natriuretic substance by kallikrein, Am. J. Phys., 247:F480-F484.

Bristow, J.D., Honour, A.J., Pickering, G.W., Sleight, P., and Smyth, H.S., 1969, Diminished baroreflex sensitivity in high blood pressure, Circulation, 39:48-54.

Brody, M.J., Haywood, J.R., and Touw, K.B., 1980, Neural mechanisms in hypertension, Ann. Rev. Physiol., 42:441-453.

Buhler, H.U., da Prada, M., Haefely, W., and Picotti, G.B., 1978, Plasma adrenaline, noradrenaline and dopamine in man and different animal species, J. Physiol. (London), 276:311-320.

Carretero, O.A., Oza, N.B., Scicli, A.D., and Schork, A., 1974, Renal tissue killikrein, plasma renin and plasma aldosterone in renal hypertension, Acta Physiol. Lat. Am., 24:448.

Carretero, O. A., Polomski, C., Hampton, A., and Scicli, A. G., 1976, Urinary kallikrein, plasma renin and aldosterone in New Zealand genetically hypertensive (GH) rats, Clin. Exp. Pharmacol. Physiol., 3 (Suppl):55.

Carretero, O. A., Scicli, A. G., Piwonska, A., and Koch, J., 1977, Urinary kallikrein in rats bred for susceptibility and resistance to the hypertensive effect of salt and in New Zealand genetically hypertensive rats, Mayo Clin. Proc., 52:465.

Carretero, O. A., Amin, V. M., Ocholik, T., Scicli, A. G., and Koch, J., 1978, Urinary kallikrein in rats bred for their susceptibility and resistance to the hypertensive effect of salt: A new radioimmunoassay for the direct determination, Circ. Res., 42:727.

Carretero, O. A., and Scicli, A. G., 1981a, The renal kallikrein-kinin system: implications for renal function regulation, Proc. 8th International Cong Nephrol, Athens, p. 1037, Basel, Karger.

Carretero, O. A., and Scicli, G., 1981b, Possible role of kinins in circulatory homeostasis. State of the art review, Hypertension 3 (Suppl 1):I-4.

Casto, R. and Phillips, M.I., 1985, Neuropeptide action in nucleus tractus solitarius: Angiotensin specificity in hypertensive rats, Am. J. Physiol., 249:R341-R347.

Castle, W.E., 1919, Observations on the occurrence of linkage in rats and mice, Carnegie Inst. Wash. Pub., 288:29-36.

Chao, J., Woodley, C., Chao, L., and Margolius, H. S., 1983, Identification of tissue kallikrein in brain and in the cell-free translation product encoded by brain mRNA, Biol. Chem., 258:15173-15178.

Christensen, N.J., Mathias, C.J., and Frankel, H.L., 1976, Plasma and urinary dopamine: Studies during fasting and exercise and in tetraplegic man, Eur. J. Clin. Invest., 6(5):403-309.

Clineschmidt, B.V., Geller, R.G., Govier, W.C., and Sjoerdsma, A.,1970, Reactivity to norepinephrine and nature of the alpha adrenergic receptor in vascular smooth muscle of a genetically hypertensive rat, Eur. J., Pharmacol., 10:45-50.

Cook, W.F., 1968, The detection of renin in juxtaglomerular cells, J. Physiol. (London), 194:73p-74p.

Corthorn, J., Bio, L., and Roblero, J., 1983, Release of activatable kallikrein by isolated rat kidneys, Hypertension 3 (Suppl 11):II39-II41.

Corthorn, J. and Roblero, J., 1981, Release of activatable kallikrein by isolated rat kidneys, _Hypertension_ 3:II39-II41.

Couser, W.G. and Stillmant, M.M., 1975, Mesangial lesions and focal
glomerular sclerosis in the aging rat, _Lab. Invest._, 33:491-501.

Croxatto, H. R., Albertini, R., Arriagada, R., Roblero, J., Rojas, M., and
Rosas, R., 1976, Renal urinary kallikrein in normotensive and
hypertensive rats during enhanced excretion of water and
electrolytes, _Clin. Sci. Mol. Med._, 51 (Suppl 3):259s.

Cuche-J.L., Kuchel, O., Barbeau, A., Langlois, Y., Boucher, R., and
Genest, J., 1974, Autonomic nervous system and benign essential
hypertension in man. I. Usual blood pressure, catechlamines, renin
and their interrelationships, _Circ. Res._, 35:281-289.

Cumming, A. D., and Robson, J. S., 1985, Urinary kallikrein excretion in
glomerulonephritis and nephrotic syndrome, _Nephron_ 39:206-210.

Dahl, L.K., Heine, M., and Tassinari, L., 1962, Effects of chronic salt
ingestion: Evidence that genetic factors play an important role in
susceptibility to experimental hypertension, _J. Exp. Med._,
115:1173-1190.

Davis, J.O., 1977, Advances in our knowledge of the renin-angiotensin
system, _Fed. Proc._, 36:1753-1754.

DeWardener, H.E. and Clarkson, E.M., 1982, The natriuretic hormone: Recent
developments, _Clin. Sci._, 63:415-420.

Elema, J.D. and Arends, A., 1975, Focal and segmental glomerular
hyalinosis and sclerosis in the rat, _Lab. Invest._, 33:554-561.

Feld, L. C., van Liew, J. B., Brentjens, J. R., and Boylan, J.W., 1981,
Renal lesions and proteinuria in the spontaneously hypertensive rat
made normotensive by treatment, _Kidney Int._, 20:606.

Ferrario, C.M. and Barnes, K.L., 1981, Inactivation of either the area
postrema or the nucleus tractus solitarii produces opposite changes
in the hemodynamic characteristic of conscious trained dogs. In:
"Central Nervous System Mechanisms in Hypertension", Buckley,
Ferrario, eds., Raven Press, New York, p. 61.

Ferrario, C.M., Gildenberg, P.L., and McCubbin, J.W., 1972, Cardiovascular
effects of angiotensin mediated by the central nervous system, _Circ.
Res._, 30:257-262.

Ferrone, R.A., Walsh, G.M., Tsuchiya, M., and Frohloch, E.D., 1979,
Comparison of hemodynamics in conscious spontaneous and renal
hypertensive rats, _Am. J. Physiol._, 236:H403-H408.

Finch, L. and Haeusler, G., 1973, Further evidence for a central
hypotensive action of alpha-methyl-dopa in both the rat and cat, _Br.
J. Pharmacol._, 47:217-228.

Folkow, B. and Hallback, M., 1977, Physiopathology of spontaneous hypertension in rats, In: "Hypertension", J. Genest, E. Koiw, O. Kuchel, eds., McGraw-Hill, New York, pp. 507-529.

Frohlich, E.D., 1982, Hemodynamic factors in the pathogenesis and maintenance of hypertension, Fed. Proc., 41:2400-2408.

Frohlich, E.D., Tarazi, R.C., and Dustan, H.P., 1969, Re-examination of the hemodynamics of hypertension, Am. J. Med. Sci., 257:9-23.

Fukiyama, K., McCubbin, J.W., and Page, I.H., 1971, Chronic hypertension elicited by infusion of angiotensin into vertebral arteries of unanaesthetized dogs, Clin. Sci., 40:283-291.

Ganong, W.F., 1983, Relation of central α adrenoreceptor and other receptors to the control of renin secretion, Chest, 83:299-302.

Ganten, U., Rascher, W., Lang, R.E., Dietz, R., Rettig, R., Unger, T., Taugner, R., and Ganten, D., 1983, Development of a new strain of spontaneously hypertensive rats homozygous for hypothalmic diabetes insipidus, Hypertension, 5:1119-1128.

Geiger, R., Clausnitzer, B., Fink, E., and Fritz, H., 1980, Isolation of an enzymatically active glandular kallikrein from human plasma by immunoaffinity chromatography, Hoppe Seylers Z Physiol. Chem., 361:1975.

Gilboa, N., Rudofsky, U., and Magro, A., 1984a, Urinary and renal kallikrein in hypertensive fawn-hooded (FH/Wjd) rats, Lab. Invest., 50:72-78.

Gilboa, N., Rudofsky, U., and Magro, A., 1984b, Effect of Na$^+$ intake and orchiectomy on the spontaneously hypertensive FH/Wjd rat, Proc IX Int Congress Nephrol, p. 290A.

Gilboa, N.. Sidoti, L., and Magro, A., 1985, Urinary kallikrein and prekallikrein in the FH/Wjd spontaneously hypertensive rat. Effects of low sodium diet, Clin. Res.,.

Glasser, R. J., and Michael, A. F., 1976, Urinary kallikrein in experimental renal disease, Lab. Invest., 34:616-621.

Goldberg, L.I., 1972, Cardiovascular and renal actions of dopamine: Potential clinical applications, Pharmacol. Rev., 24:1-29.

Goldberg, L.I. and Weder, A.B., 1980, Connections between endogenous dopamine, dopamine receptors and sodium excretion: Evidence and hypothesis, Recent Adv. Clin. Pharmacol., 3:149-166.

Goldstein, D.S., 1981, Plasma norepinephrine in essential hypertension. A study of the studies, Hypertension, 3:48-52.

Goldstein, D.S., 1983, Plasma catecholamines and essential hypertension. An analaytical review, Hypertension, 5:86-104.

Grobecker, H., Roizen, M.F., Weise, V., Saavedr, J.M., and Kopin, I.J., 1975, Sympathoadrenal medullary activity in young spontaneously hypertensive rats, Nature London. 258:267-268.

Grunfeld, B., Gimenez, M., Simsolo, R., Mendilaharzu, F., and Becu, L., 1984, Urinary kallikrein in hypertension secondary to hemolytic uremic syndrome: Response to diuretic stimulus, Int. J. Pediatr. Nephrol., 5:205-208.

Hall, J.E., Guyton, A.C.,Jackson, T.E., Coleman, T.G., Lohmeier, T.E., and Trippodo, N.C., 1977, Control of glomerular filtration rate by the renin-angiotensin system, Am. J. Physiol., 233:F366-F372.

Hamlyn, J.M., Ringel, R., Schaeffer, J., Levinson, P.E., Hamilton, B.P., Kowarski, A.A., and Blaustein, M.P., 1982, A circulating inhibitor of $(Na^+ + K^+)$ATPase associated with essential hypertension, Nature, 300:650-652.

Hansen, C.T., 1972, A Genetic Analysis of Hypertension in the Rat with Spontaneous Hypertension - Its Pathogenesis and Complications, K. Okamoto, ed., Igaku Shoin LTD, Tokyo.

Hattori, K., Hasumura, J., and Takeuchi, J., 1984, Role of renal kallikrein in the derangement of sodium and water excretion in cirrhotic patients, Scand. J. Gastroenterol., 19:844-848.

Head, G.A. and Korner, P.I., 1982, Cardiovascular function of brain serotonergic neurons in the rabbit as analyzed from the acute and chronic effects of 5,6-dihydroxytryptamine, J. Cardiovasc. Pharmacol., 4:398-408.

Hoeldtke, R., Rogawski, M., and Wurtman, R.J., 1974, Effect of selective destruction of central and peripheral catecholamine-containing neurones with 6-hydroxydopamine on catecholamine excretion in the rat, Br. J. Pharmacol., 50:265-270.

Hokfelt, B., Hedeland, H., and Dymling, J.F., 1970, Studies on catecholamines, renin, and aldosterone following Catapresan (2-(2,6-dichloro-phenylamine)-2-imidazoline hydrochloride) in hypertensive patients, Eur. J. Pharmacol., 10:389-397.

Ieni, J.R., Tobach, E., Zukin, S.R., Barr, G.A., and Van-Praag, H.M., 1985, Multiple tritiated imipramine binding sites in brains of male and female fawn-hooded and Long-Evans rats, Eur. J. Pharmacol., 112:261-264.

Jams, S.G. and Wexler, B.C., 1979, Inhibition of the development of spontaneous hypertension in SH rats by gonadectomy or estradiol, J. Lab. Clin. Med., 94:608.

Judy, W.V., Watanabe, A.M., Henry, D.P., Besch, H.R., Murphy, W.R., and Hockel, G.M., 1976, Sympathetic nerve activity: Role in regulation of blood pressure in the spontaneously hypertensive rat, Circ. Res., 38:II21-II29.

Kamal, L.A., Cloix, J.F., Devynck, M.A., and Meyer, P., 1983, [3H]serotonin uptake in human blood platelets is reduced by ouabain and endogenous digitalis-like inhibitors of Na^+, K^+-ATPase, Eur. J. Pharmacol., 92:167-168.

Kamal, L.A., Raisman, R., Meyer, P., and Langer, S.Z., 1984, Reduced Vmax of 3H-serotonin uptake but unchanged 3H-imipramine binding in the platelets of untreated hypertensive subjects, Life Sci., 34:2083-2088.

Kamitani, T., Little, M. H., and Ellis, E. F., 1985, Evidence for a possible role of the brain kallikrein-kinin system in the modulation of the cerebral circulation, Circ Res., 57:545-552.

Kazda, S., Garthoff, B., nd Thomas, G., 1982, Antihypertensive effect of calcium agonists in rats differs from that of vasodilators, Clin. Sci., 63:363S-365S.

Kitchin, A.H. and Turner, R.W.D., 1966, Studies on debrisoquin sulphate, Brit. Med. J., 2:728-731.

Kobayashi, K., Kolloch, R., de Quattro, V., and Miano, L., 1979, Increased plasma and urinary normetanephrine in young patients with primary hypertension, Clin. Sci. Mol. Med., 57:1735-1765.

Kobinger, W. and Pichler, L., 1975, The central modulatory effect of clonidine on the carodiopressor reflex after suppression of synthesis and storage of noradrenaline, Eur. J. Pharmacol., 30:56-62.

Kobinger, W. and Walland, A., 1972, Facilitation of vagal reflex bradycardia by an action of clonidine on central receptors, Eur. J. Pharmacol., 19:203-217.

Koolen, M., Daha, M., and van Brummelen, P., 1985, Is the renal kallikrein system relevant to sodium sensitivity in patients with essential hypertension, Eur. J. Clin. Invest., 15:151-156.

Kopp, V., Bradley, T., and Hjemdahl, P., 1983, Renal venous outflow and urinary excretion of norepinephrine, epinephrine, and dopamine during graded renal nerve stimulation, Am. J. Physiol., 244:E52-E60.

Korner, P.I., Head, G.A., Angus, J.A., Oliver, J.R., Dorward, P.K., and Blombery, P.A., 1981, Neural mechanisms involved in the actions of clonidine on blood pressure, heart rate, and on the baroreceptor reflexes, In: "Central Nervous System Mechanisms in Hypertension", Buckley, J.P., Ferrario, C.M., eds., Raven Press, New York, pp. 191-202.

Kreisberg, J.I. and Karnovsky, M.J., 1978, Focal glomerular sclerosis in the fawn-hooded rat, Am. J. Pathol., 92:637.

Kuchel, O., Buu, N.T., Hamet, P., Larochelle, P., Baurque, M., and Genest,

J., 1982, Dopamine surges in hyperadrenergic essential hypertension, Hypertension, 4:845-852.

Kuchel, O., Cuche, J-L, Buu, N.T., and Genest, J., 1976, An increase in urinary catecholamines of renal origin in patients with "borderline" hypertension, Am. J. Med. Sci., 272:263-268.

Kuhn, E.M., Wolf, W.A., and Lovenberg, W., 1980, Review of the role of the central serotonergic neuronal system in blood pressure regulation, Hypertension, 2:243-255.

Kuijpers, M.H. and de Jong, W., 1982, The spontaneously hypertensive rat as a model to study human hypertension, In: Proceedings of the 4th International Symposium on Rats with Spontaneous Hypertension and Related Studies, W. Rascher, D. Clough, and D. Ganten, eds., Schattauer Verlag, Stuttgart.

Kuijpers, M.H.M. and Gruys, E., 1984, Spontaneous hypertensive renal disease in the fawn-hooded rat, Br. J. Exp. Pathol., 65:181-190.

Laragh, J.H., Sealey, J.E., Niarchos, A.P., and Pickering, T.G., 1982, The vasoconstriction-volume spectrum in normotension and in the pathogenesis of hypertension, Fed. Proc. 41:2415-2423.

Larsson, C., Weber, P., and Anggård, E., 1974, Arachiodonic acid increases and indomethacin decreases plasma renin activity in the rabbit, Eur. J. Pharmacol., 28:391-394.

LaVail, M.M., 1981a, Assignment of retinal dystrophy (rdy) to linkage group IV of the rat, J. Hered., 72:294-296.

LaVail, M., 1981b, Fawn-hooded rats, the fawn mutation and interaction of pink-eyed and red-eyed dilution genes, J. Hered., 72:286-287.

Lawton, W. J., Proud, D., Frech, M. E., Pierce, J. V., Keiser, H. R., and Pisano, J. L., 1981, Characterization and origin of immunoreactive glandular kallikrein in rat plasma, Biochem. Pharmacol., 30:1731.

Lehr, D., Mallow, J., and Krukowski, M., 1967, Copious drinking and simultaneous inhibition of urine flow elicited by beta-adrenergic stimulation and contrary effect of alpha-adrenergic stimulation, J. Pharmac. Exp. Ther., 158:150-163.

Lieberthal, W., Oza, N. B., Arbeit, L. A., Bernard, D. B., and Levinsky, N.G., 1983, Effect of alterations in sodium and water metabolism on urinary excretion of active and inactive kallikrein in man, J. Clin. Endocrinol. Metab., 56:513-519.

Levinsky, N. G., 1979, The renal kallikrein-kinin system, Circ. Res., 44:441-451.

Lorenz, P., Picchio, L. P., Weisz, J., and Lloyd, T., 1984, The relationship between reproductive performance and blood pressure in the spontaneously hypertensive rat, Am. J. Obstet. Gynecol., 150:519-523.

Luria, M.H. and Freis, E.D., 1965, Treatment of hypertension with
Debrisoquin Sulfate (Declinax), Curr. Ther. Res., 7:289-296.

Magro, A.M., 1981, Histamine release from fawn-hooded rat mast cells is
not potentiated by phosphatidylserine, Immunology, 44:1-10.

Magro, A. M., and Rudofsky, U. H., 1982, Plasma renin activity decrease
precedes spontaneous focal glomerular sclerosis in aging rats,
Nephron 31:245-253.

Magro, A.M., Rudofsky, U.H., Gilboa, N., and Seegal, R., 1986, Increased
catecholamine output in the hypertensive fawn-hooded rat, Lab. Anim.
Sci., in press.

Malayan, S.A., Keil, L.C., Ramsay, D.J., and Reid, I.A., 1979, Mechanism
of suppression of plasma renin activity by centrally administered
angiotensin II, Endocrinology, 104:672-675.

Markham, R.V., Jr., Sutherland, J.C., and Mardiney, M.R., Jr., 1973, The
ubiquitous occurrence of immune complex localization in renal
glomeruli of normal mice, Lab. Invest., 29, 111-120.

Masferrer, J., Albertini, R., Croxatto, H. R., Garcia, P., and Pinto, I.,
1985, Isolation and characterization of rat plasma glandular
kallikrein, Biochem. Pharmacol., 34:51-56.

Matsudaira, T., Kogo, H., and Satoh, T., 1985, A possible role of gonad
and renal prostaglandin E_2 on the development of hypertension in
spontaneously hypertensive rats, Jpn. J. Pharmacol., 37:51-57.

McGiff, J. C., Itskovitz, H. D., and Terragno, N.A., 1975, The actions of
bradykinin and eledoisin in the canine isolated kidney:
Relationships to prostaglandins, Clin. Sci. Mol. Med., 49:125-131.

McGiff, J.C. and Quilley, C.P., 1981, The rat with spontaneous genetic
hypertension is not a suitable model of human essential
hypertension, Circ. Res., 48:455-464.

McKusich, V.A., 1960, Genetics and the nature of essential hypertension,
Circulation, 22:857-863.

Medina, M.A., Giachetti, A., and Shore, P.A., 1969, On the physiological
disposition and possible mechanisms of the antihypertensive action
of debrisoquin, Biochem. Pharmacol., 18:891-901.

Meyer, D.K., Rauscher, W., Peskar, B., and Herting, G., 1973, The
mechanism of the drinking response to some hypotensive drugs.
Activation of the renin-angiotensin system by direct reflex-mediated
stimulation of β-receptors, Arch. Pharmacol., 276:13-24.

Meyer, T. W., Troy, J. L., Rennke, H. G., and Brenner, B.M., 1983,
Prevention of glomerular injury by pituitary ablation in rats with
reduced nephron number, Amer. Soc. Nephrol. Proc. p. 113A.

Mills, J. H., Newport, P. A., and Obika, F.O., 1979, Interaction between arterial pressure and vasoactive hormones in the release of renal kallikrein, Adv. Exp. Med. Biol., 120B:561.

Morgunov, N. and Baines, A.D., 1981, Renal nerves and catecholamine excretion, Am. J. Physiol., 240:E75-E81.

Mulvany, M.J., 1983, Do resistance vessel abnormalities contribute to the elevated blood pressure of spontaneously-hypertensive rats? A review of some of the evidence, Blood Vessels, 20:1-22.

Nagaoka, A. and Lovenberg, W., 1976, Plasma norepinephrine and dopamine-β-hydroxylase in genetic hypertensive rats, Life Sci., 19:29-34.

Nagatsu, T., Ikuta, K., Numata, Y., Kato, T., Sano, M., Nagats, H., Umezawa, M., Matsukaki, M., and Takeuchi, T., 1976, Vascular and brain dopamine β-hydroxylase activity in young spontaneously hypertensive rats, Science, 191:290-291.

Nayler, W.G. and Poole-Wilson, P.H., 1981, Calcium antagonists: Definition and mode of action, Basic Res. Cardiol., 76:1-15.

Nolly, H., and Lama, M. C., 1982, "Vascular kallikrein": A kallikrein-like enzyme present in vascular tissue of rat, Clin. Sci., 63 (Suppl 8):249s.

Nosaka, S. and Okamoto, K., 1970, Modified characteristics of the aortic baroreceptor activities in the spontaneously hypertensive rat, Jap. Circ. J., 34:685-693.

Nosaka, S. and Wang, S.C., 1972, Carotid sinus baroreceptor functions in the spontaneously hypertensive rat, Am. J. Physiol., 222:1079-1084.

O'Brien, S.J., 1984, Genetic Maps, Vol. 3: Cold Spring Harbor Laboratory Publications, Cold Spring Harbor, New York.

Okamoto, K. and Aoki, K., 1963, Development of a strain of spontaneously hypertensive rats, Jap. Circ. J., 27:82-93.

Orgain, E.S. and Kern, A., 1970, Debrisoquin in the therapy of hypertension, Arch. Intern. Med., 125:255-257.

Orstavik, T. B., Carretero, O. A., and Scicli, A. G., 1982, The kallikrein-kinin system in the regulation of submandibular gland blood flow, Am. J. Physiol., 242:H1010.

Page, I.H., 1961, Concepts of the etiology of arterial hypertension, Med. Clin. N. Amer., 45:235-238.

Paul, O., 1976, A review of the symposium on the epidemiology of hypertension, Chicago, September, 1974, In: "The Arterial Hypertensive Disease", G. Rorive, H. Van Couwenberge, eds., Masson, New York, pp. 129-139.

Pfeffer, M.A. and Frohlich, E.D., 1977, Electromagnetic flowmetry in anesthetized rats, J. Appl. Physiol., 33:137-140.

Pickering, G.W., 1974, "Hypertension: Causes, Consequences, and Management", Churchill Livingstone, Edinburgh.

Ploth, D.W. and Navar, L.G., 1979, Intrarenal effects of the renin-angiotensin system, Fed. Proc., 38:2280-2285.

Powers, C.A., 1985, Dopamine receptor blockade increases glandular kallikrein-like activity in the neurointermediate lobe of the rat pituitary, Biochem. Biophys. Res. Commun., 127(2):668-672.

Prieur, D.J. and Meyers, K.M., 1984, Genetics of the fawn-hooded rat strain. The coat color dilutions and platelet storage pool deficiency are pleiotropic effects of the autosomal recessive red-eyed dilution gene, J. Hered., 75:349-352.

Proud, D., Nakamura, S., Carone, F. A., Herring, P. L., Kawamara, M., Inagami, T., and Pisano, J. J., 1984, Kallikrein-kinin and renin angiotension systems in rat renal lymph, Kidney Int., 25:880-885.

Rabito, S. G., Scicli, A. G., Kher, V., and Carretero A. O., 1982, Immunoreactive glandular kallikrein in rat plasma: A radioimmunoassay for its determination, Am. J. Physiol., 242:H602.

Raymond, S.L. and Dodds, W.J., 1975, Characterization of the fawn-hooded rat as a model for hemostatic studies, Thromb. Haemost. (Stuttgart), 33:361-369.

Reid, I.A., Macdonald, D.M., Pachnis, B., and Ganong, W.F., 1975, Studies concerning the mechanism of suppression of renin secretion by clonidine, J. Pharmacol. Exp. Ther., 192:711-721.

Reid, I.A., Schrier, R.W., and Early, L.E., 1972, An effect of extrarenal beta-adrenergic stimulation on the release of renin, J. Clin. Invest., 51:1861-1869.

Robertson, A.L., Smeby, R.R., Bumpus, F.M., and Page, I.H., 1965, Renin production by organ cultures of renal cortex, Science, 149:650-651.

Robinson, R., 1965, Genetics of the Norway rat, Pergamon Press, Oxford.

Rosas, A., Albertini, R., and Croxatto, H. R., 1977, Renal kallikrein system, volemia and renal hypertension, Mayo Clin. Proc., 52:459.

Rudofsky, U.H. and Magro, A.M., 1982, Spontaneous hypertension in fawn-hooded rats, Lab. Anim. Sci., 32:389-391.

Schmid, H.E., 1962, Renin, a physiological regulator of renal hemodynamics?, Circ. Res., 11:185-193.

Sever, P.S., Birch, M., Osikowska, B., and Turnbridge, R.D.G., 1977, Plasma-noradrenaline in essential hypertension, Lancet, 1:1078-1081.

Shimamoto, K., Ura, N., Tanaka, S., Ogasawara, A., Nakao, T., Nakahashi, Y., Chao, J., Margolius, H. S., and Iimura, O., 1981, Excretion of human urinary kallikrein quantity measured by a direct radioimmunoassay of human urinary kallikrein in patients with essential hypertension and secondary hypertensive diseases, Jpn. Circ. J., 45:1092.

Simpson, F.O., Phelan, E.L., Clark, D.W., Jones, D.R., Gresson, C.R., Lee, D.R., and Bird, D.L., 1973, Studies on the New Zealand strain of genetically hypertensive rats, Clin. Sci. Mol. Med., 45:Suppl 1:15S-21S.

Sinaiko, A. R., Glasser, R. J., Gillum, R. F., and Prineas, R. J., 1982, Urinary kallikrein excretion in grade school children with high and low blood pressure, J. Pediatr., 100:938.

Smirk, F.H., 1972, Characterization of the New Zealand strain of genetically hypertensive rats considered in relation to essential hypertension in spontaneous hypertension: Its pathogenesis and complications, Okamto, ed., Tokyo, Igaku Shoin.

Smith, T.L. and Hutchins, P.M., 1979, Central hemodynamics in the development stage of spontaneous hypertension in the unanesthetized rat, Hypertension, 1:508-577.

Solomon, H.M., Ashley, C., Spirt, N., and Abrams, W.B., 1968, The influence of debrisoquin on the accumulation and metabolism of biogenic amines by the human platelet, in vivo and in vitro, Clin. Pharmacol. Ther., 10:229-238.

Stahl, S.M. and Meltzers, H.Y., 1978, A kinetic and pharmacologic analysis of 5-hydroxytryptamine transport by human platelets and platelet storage granules: Comparison with central serotonergic neurons, J. Pharmac. Exp. Ther., 205:118-132.

Steblay, R.W. and Rudofsky, U.H. 1971, Spontaneous renal lesions and glomerular deposits of IgG and complement in guinea pigs, J. Immunol., 107:1192-1196.

Sterzel, R. B., Gao, Y., Kriz, W., Waldherr, R., Sakai, T., Luft, F., Ganten, D., Briggs, J., and Schnermann, J., 1985, Renal disease precedes hypertension in salt-sensitive Dahl rats, Proc. Amer. Soc. Nephrol., p. 110A.

Stowe, N., Schnermann, J., and Hermle, M., 1979, Feedback regulation of nephron filtration rate during pharmacologic interference with the renin-angiotensin and adrenergic system in rats, Kidney Int., 15:473-486.

Sustarsic, D. L., McPartland, R. P., and Rapp, J. R., 1981, Developmental patterns of blood pressure and urinary protein, kallikrein and prostaglandin E_2 in Dahl salt-hypertension-susceptible rats, J. Lab. Clin. Med., 98:599.

Sustarsic, D. C., McPartland, R. P., Rapp, J. R., Schlager, G., and Tan, S. J., 1980, Urinary kallikrein and urinary prostaglandin E_2 in genetically hypertensive mice (40746), Proc. Soc. Exp. Biol. Med., 163:193.

Takaoka, M., Okamura, H., Kuribayashi, Y., Matsuoka, H., and Morimoto, S., 1985, Activation of urinary inactive kallikrein by an extract from the rat kidney cortex, Life Sci., 37:1015-1922.

Thant, M., Yamori, Y., and Okamoto, K., 1969, Baroreceptor function revealed by acute sinaortic denervation in spontaneously hypertensive rats, Jap. Circ. J., 33:501-507.

Tobach, E., DeSantis, J.L., and Zucker, M.B., 1984, Platelet storage pool disease in hybrid rats. F_1 fawn-hooded rats derived from crosses with their putative ancestors (Rattus Norvegicus), J. Hered., 75:15-18.

Tschopp, T.B. and Zucker, M.B., 1972, Hereditary defects in platelet function in rats, Blood, 40:217-220.

Unger, T., Buu, N.T., and Kuchel, O., 1978, Renal handling of free and conjugated catecholamines following surgical stress in the dog, Am. J. Physiol., 235(6):F542-F547.

Urizar, R. E., Cerda, J., Dodds, W. J., Raymond, S. L., Largent, J. A., Simon, R., and Gilboa, N., 1984, Age-related renal, hematologic, and hemostatic abnormalities in FH/Wjd rats, Am. J. Vet. Res., 45:1624-1631.

VanLeeuwen, B. H., Grinblat, S.M., and Johnson, C. I., 1984, Tissue-specific control of glandular kallikreins, Am. J. Physiol., 274:F760-F764.

Vergara, M., and Corthorn, J., 1983, Urinary excretion of activatable kallikrein in one-kidney pole-ligated hypertensive rats, J. Hypertens., 1:387-391.

Vio, C. P., and Figueroa, C. D., 1985, Subcellular localization of renal kallikrein by ultrastructural immunocytochemistry, Kidney Int., 28:36-42.

Weber, P.C., Larsson, C., Anggard, E., Hamberg, M., Corey, E.J., Nicolaou, K.C., and Samuelsson, B., 1976, Stimulation of renin release from rabbit renal cortex by arachidonic acid and prostaglandin endoperoxides, Circ. Res., 39:868-874.

Wing, L.M.G. and Chalmers, J.P., 1974, Participation of central serotonergic neurons in the control of the circulation of the unanesthetized rabbit. A study using 5,6 dihydroxytryptamine in experimental neurogenic and renal hypertension, Circ. Res., 35:504-513.

Ylitalo, P., Kankinen, S., Nurmi, A. K., Seppala, E., Passi, T., and Vapaatala, H., 1985, Effects of a prostacyclin analog iloprost on kidney function, renin-angiotensin and kallikrein-kinin systems, prostanoids and catecholamines in man, Prostaglandins 29:1063-1071.

Zimmerman, H. and Ganong, W.F., 1980, Pharmacological evidence that stimulation of central serotonergic pathways increases renin secretion, Neuroendocrinology, 30:101-107.

Zinner, S. H., Margolius, H. S., Rosner, B., and Kass, E. H., 1978, Stability of blood pressure rank and urinary kallikrein concentration in childhood: an eight year follow-up, Circulation 58:908.

ORGANIZATION OF DIFFERENTIAL SYMPATHETIC RESPONSES TO ACTIVATION OF VISCERAL RECEPTORS AND ARTERIAL BARORECEPTORS

Lynne C. Weaver, Robert L. Meckler, Jean C. Tobey, and
Reuben D. Stein

Department of Physiology, Michigan State University
111 Giltner Hall,
East Lansing, MI 48824-1101

INTRODUCTION

Cannon reasoned that the sympathetic nervous system is organized to produce "widespread changes in smooth muscle and glands throughout the organism" (Cannon, 1929; 1930). This diffuse action of the sympathetic system is possible because of its "arrangement for simultaneous and unified action." Specifically, the extensive divergence of preganglionic inputs to postganglionic neurons as well as to secretory cells of the adrenal medulla would seem to ensure concordant physiological responses required to homeostatically regulate mammals' internal environments during changes in the external environment. Cannon's postulated "en masse" action of the sympathetic nervous system was strictly interpreted until thirty years later, when reports of non-uniform sympathetic responses began to appear in the literature. The concept of sympathetic mass-activation was first challenged by Folkow, Johansson, and Lofving (1961) and Lofving (1961). These investigators made simultaneous measurements of the reflex changes in blood flow to hindlimb muscle and skin, kidney, and intestine that were initiated by stimulation of arterial chemoreceptors or unloading of carotid arterial baroreceptors. In these studies, renal and cutaneous blood flows were relatively unaffected, whereas skeletal muscle vessels constricted in response to these stimuli; intestinal blood flow decreased less than skeletal muscle blood flow. These results could be interpreted as an indication of reflex specificity among the sympathetic nerves. However, differences in regional distribution of blood flow may arise from heterogeneity of end organs'

sensitivities to equivalent magnitudes of sympathetic excitation. That
is, a mass-response of sympathetic outflow may have unequal effects on
blood flow to different regions due to inequalities of target organs'
responsiveness to a given amount of nerve activity Bevan (1979).
Therefore, simultaneous recordings of activity from different sympathetic
nerves are necessary to distinguish between target organ responsiveness
and specificity of sympathetic outflows per se.

More recently, electrophysiological studies have demonstrated
non-uniform reactions of different sympathetic nerves in a variety of
circumstances such as during chemoreceptor or baroreceptor stimulation or
during exercise (Folkow and Rubinstein, 1966; Iriki et al., 1971;
Ninomiya, Nisimaru, and Irisawa, 1971; Nisimaru, 1971; Iriki, Riedel, and
Simon, 1972; Ninomiya, Irisawa, and Nisimaru, 1973; Ninomiya and Irisawa,
1975; Kollai and Koizumi, 1977; Grignolo, Koepke, and Örbrist, 1982;
Jänig, 1982). Furthermore, differential responses of sympathetic nerve
activities are not due simply to segmental organization of afferent or
efferent components of the reflex arcs. For example, activity of single
neurons innervating blood vessels of skeletal muscle are inhibited by
stimulation of arterial baroreceptors and are excited by stimulation of
cutaneous nociceptors on the ipsilateral limb (Horeyseck and Jänig, 1974);
Blumberg et al., 1980). In contrast, vasoconstrictor neurons to cutaneous
vasculature are only weakly inhibited by baroreceptor stimulation and are
strongly inhibited by stimulation of the cutaneous nociceptors (Horeyseck
and Jänig, 1974; Blumberg, et al., 1980). Since both sets of
vasoconstrictor postganglionic nerves arise from similar lumbar spinal
segments, these non-uniform reflex responses are not due to a segmental
organization of the sympathetic system. These differential responses of
sympathetic neurons indicate that the sympathetic nervous system does not
always respond in a unitary fashion and therefore is organized more
complexly than Cannon had supposed. The regional redistributions of blood
flow usually are referred to by specifying particular vascular beds:
skeletal muscle, cutaneous, and visceral (Folkow et al., 1961; Lofving,
1961; Adams et al., 1969; Iriki et al., 1971). For example, during
fighting behavior in the cat, dilation of muscle blood vessels and
"visceral vasoconstriction" occur simultaneously (Adams et al., 1969).
However, this compartmentalization of vascular regions may also be
misleading, since blood flow to all viscera is not necessarily affected
identically (Folkow et al., 1961; Lofving, 1961; Iriki et al., 1972;
Grignolo et al., 1982). This homogeneity of visceral blood flow responses
has been questioned by our laboratory (Weaver et al., 1983; Weaver, Fry,

and Meckler, 1984; Calaresu et al., 1984) and by others (Folkow et al.,
1961; Lofving, 1961; Folkow and Rubenstein, 1966; Adams et al., 1969;
Iriki et al., 1971; Ninomiya et al., 1971; Nisimaru, 1971; Iriki et al.,
1972; Ninomiya et al., 1973; Horeyseck and Jänig, 1974; Ninomiya and
Irisawa, 1975; Kollai and Koizumi, 1977; Blumberg et al., 1980; Grignolo
et al., 1982; Jänig, 1982) who have demonstrated non-uniformity of
sympathetic outflow to different abdominal viscera in response to
stimulation of visceral or somatic receptors.

The purpose of several of the investigations in our laboratory was to
contrast reflexes affecting the kidney with those directed toward the
spleen, a major constituent of the "capacitive" compartment of the
circulation in cats and dogs (Barcroft and Stephens, 1927; Guntheroth and
Mullins, 1963; Greenway, Lawson, and Stark, 1968; Greenway and Lister,
1974; Carneiro and Donald, 1977; Greenway and Innes, 1980). In addition,
renal nerve responses were compared to those of mesenteric sympathetic
nerves innervating another significant capacitive vascular bed (Greenway
and Lister, 1974). If the kidney is viewed as a complex filter system
with ongoing excretory and endocrine activities, whereas an important
function of the spleen and mesenteric capacitive bed is to act as a
reservoir suited to meet sporadic demands for instant mobilization of
blood volume, the influences of the sympathetic nervous system on these
organs seem unlikely to be identical in all circumstances. The goal of our
research was to determine the extent to which these organs receive the
same or different input from the central nervous system.

GENERAL EXPERIMENTAL METHODS

Afferent Stimulation

Visceral receptors were stimulated in chloralose-anesthetized cats by
intra-arterial injections of bradykinin or capsaicin. These algogenic
substances excite groups III and IV afferent nerves (Longhurst et al.,
1984) and the receptors stimulated by these substances are thought to be
polymodal, i.e., reactive to mechanical, chemical and noxious mechanical
stimuli (Kumazawa and Mizumura, 1980; Malliani, 1982). The reflex
responses to these agents may resemble, in part, those to visceral pain.

Quantitation of Neural Responses

In these studies only magnitudes of change in multifiber sympathetic activity were compared and interpreted. Some responses were quantified by spike counting and others by voltage integration. Each technique has some limitations, which have been discussed in a previous paper (Weaver, et al., 1984). However, the observation of differences in mass responses to different target organs may be a first step toward understanding the cellular events and orgaization underlying discrete actions of the sympathetic nervous system. Changes in nerve activity from control were tested with an analysis of variance using a complete block design. Mean values were compared with a test of least significant differences (Sokol and Rohlf, 1969). In addition, to account for differences in baseline voltages of integrated nerve activity, the relationships between simultaneously recorded voltages on two different nerves were compared during the control state and during the period of maximum reflex response. Ratios were constructed and tested by a nonparametric Friedman test (Sokol and Rohlf, 1969) to compare the magnitudes of change in the two nerves. The Friedman test assessed the effect of a reflex response on the relationship between activities of the two nerves. All statistical analyses of resulted described in this paper were performed using integrated voltages or spike-counted discharge rates. For clarity of illustration, results in some text figures have been presented as percentage changes.

COMPARISON OF RENAL AND SPLENIC NERVE RESPONSES

Stimulation of Vagal and Spinal Afferent Nerves

Bradykinin (10 µg) was injected into the left atrial appendage to stimulate cardiac receptors innervated by vagal afferent nerves and also to stimulate abdominal, visceral, and skeletal muscle receptors innervated by lower thoracic and lumbar spinal afferent nerves. Sinoaortic baroreceptors had been denervated in these cats, and upper thoracic (T_1-T_6) sympathetic chains had been excised. A pattern of renal and splenic nerve responses to this stimulus is illustrated in Fig. 1, A and B. Renal nerve activity was first inhibited for approximately 30 s and then excited for 40 s. In contrast, splenic nerves responded with maintained excitation. Arterial pressure initially decreased, then

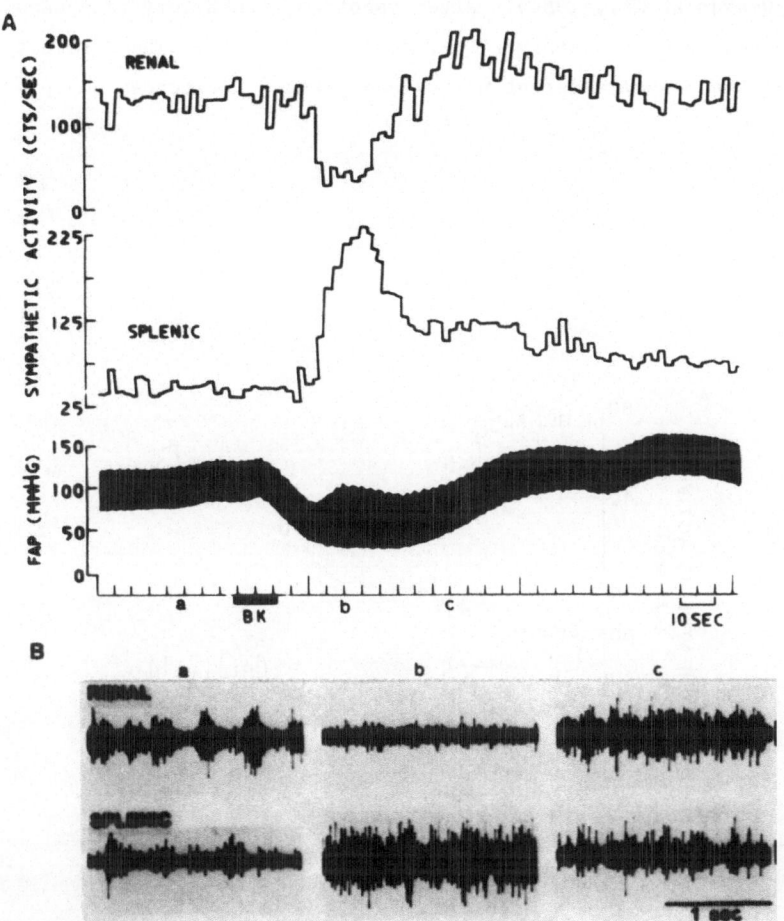

Fig. 1. Renal and splenic sympathetic responses to atrial injection of
bradykinin (BK). A: responses of renal and splenic efferent
nerve activity and systemic arterial pressure. Ordinate axes
indicate renal and splenic nerve activity (upper panels) and
femoral arterial pressure (FAP, lower panel). Abscissa indicates
times in 5 sec periods. Marker under time axis indicates period
of bradykinin injection (BK). B: oscillographic samples of
sympathetic responses at times a, b, and c indicated under time
axis of A. Vertical calibration is 20 μV. This stimulus caused
inhibition of renal nerve activity followed by excitation. In
contrast, excitation of splenic activity was initiated
immediately and maintained.

reversed toward an increase during the maximum splenic nerve excitation
and, approximately 1 min after the injection, increased to a value greater
than control. This approximate pattern occurred in each of six cats

tested (Weaver et al., 1984). After vagotomy, inhibitory responses were not observed, and excitatory reflexes were unchanged. These results suggested that vagal afferent nerves had greater influences on splenic

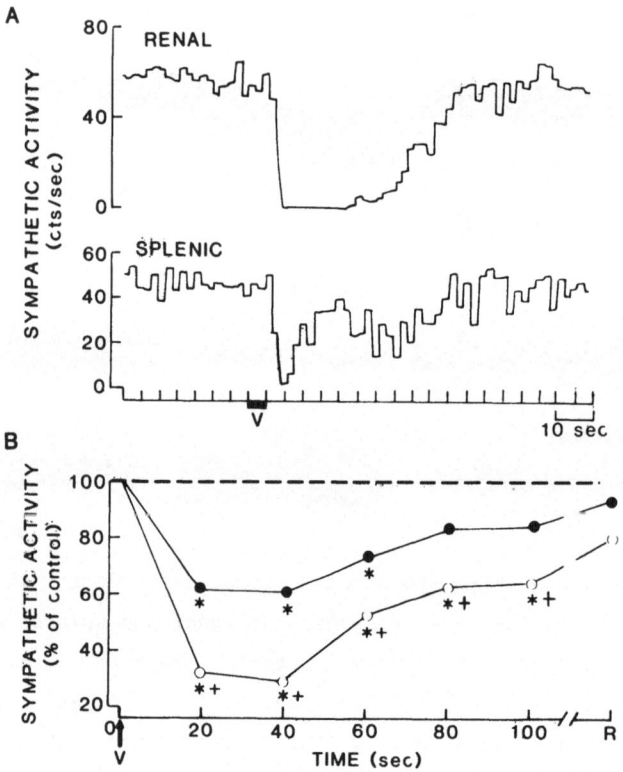

Fig. 2. Renal and splenic sympathetic responses to atrial injections of veratridine (v). A: sympathetic responses of 1 cat to veratridine. Lower axis is time in 5-s intervals. Time of injection is marked on time axis. B: mean sympathetic responses (5 cats) expressed as % of control discharge. Responses of renal (open circles) and splenic (closed circles) nerves are represented as 20-s averages of activity after veratridine injection (arrow). Recovery (R) is average activity of 60-s period, beginning 3 min after injection. *, Mean values significantly different from control. +, Renal responses significantly different from splenic. Renal nerve activity was inhibited significantly more than splenic nerve activity.

than renal nerves. Indeed, stimulation of vagally-innervated cardiac
receptors by veratridine caused significantly greater inhibition of
activity of renal than splenic nerves (Fig. 2). In addition, injections
of bradykinin into the thoracic aorta caused significantly greater
excitation of splenic than renal nerves (Fig. 3). These results
illustrated that simultaneous stimulation of different groups of receptors
can lead to widely divergent responses in renal and splenic nerves and
that abdominal visceral sympathetic outflow does not necessarily react
uniformly.

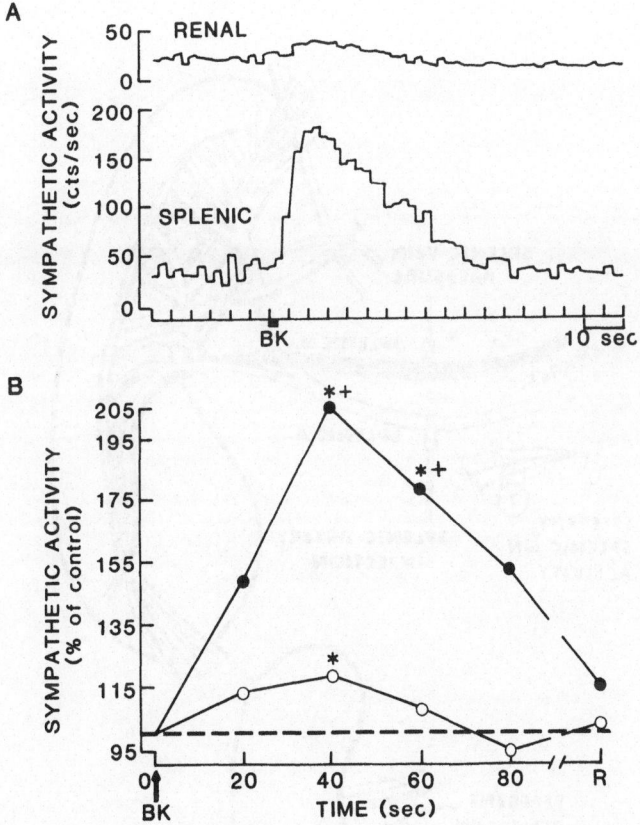

Fig. 3. Renal and splenic sympathetic responses to aortic injection of
bradykinin (BK). Format as in Fig. 2. Mean values in B derived
from 4 cats. Splenic nerve activity was excited significantly
more than renal nerve activity.

Because these experiments demonstrated generalized differential excitatory influences of spinal afferent nerves on renal and splenic nerves, the following experiments were designed to determine if specific visceral afferent nerves could produce this effect. Responses of renal and splenic nerves to stimulation of splenic receptors by the chemical stimulants capsaicin and bradykinin, and by congestion of the spleen with physiological saline, were investigated in vagotomized cats in which arterial baro- and chemoreceptors had been denervated to prevent buffering influences of these nerves on reflex responses. The spleen was vascularly isolated to ensure that only splenic receptors were stimulated. This isolation was accomplished (Fig. 4) by ligating and sectioning all

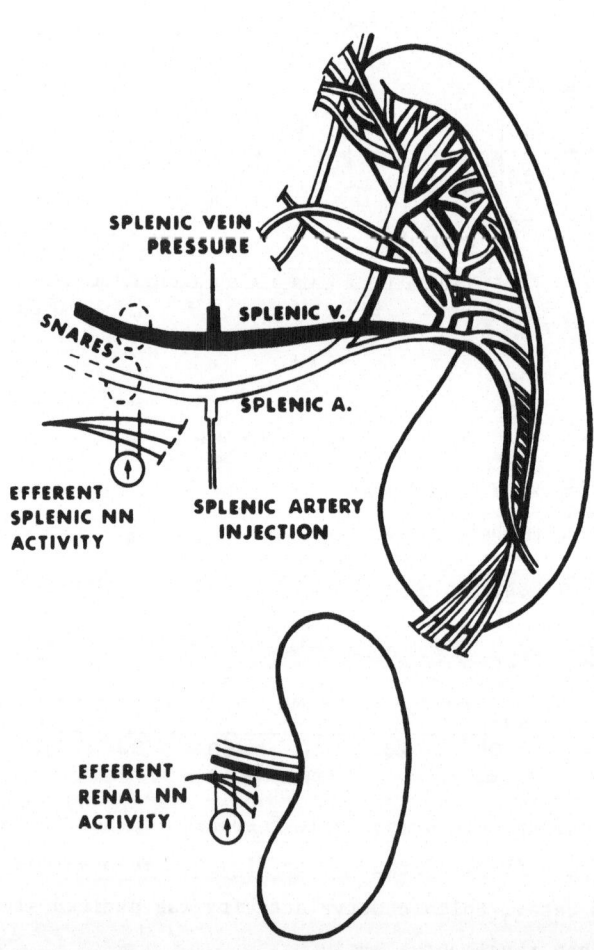

Fig. 4. Diagram of experimental preparation of spleen and kidney.

vascular connections to other organs while leaving the splenic artery, vein, and nerves intact. The splenic artery and vein were cannulated via small branches to allow arterial injections and measurements of venous pressure.

Injection of 5 μg capsaicin into the circulation of the vascularly isolated spleen caused excitation of splenic and renal nerves and increases in splenic venous pressure and systemic arterial pressure (Fig. 5). The mean increases in activity of splenic and renal nerves in five cats were 173% and 59%, respectively. The increase in splenic nerve activity was significantly greater in magnitude than the increase in renal nerve activity, as determined by the Friedman test (Fig. 6). The ratios of splenic to renal nerve activity were 0.16, 0.27, and 0.16 in the

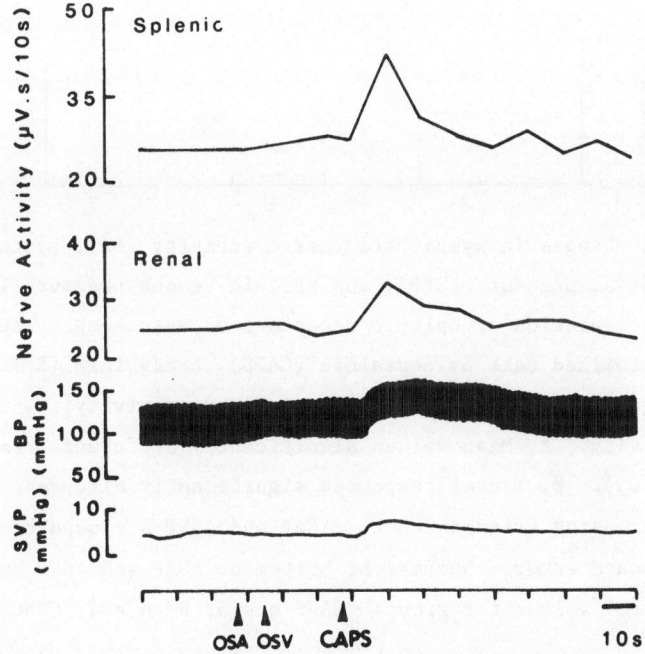

Fig. 5. Effects of injection of 5 μg capsaicin into splenic artery of baroreceptor-denervated vagotomized cat on integrated splenic and renal sympathetic nerve activity, blood pressure (BP), and splenic venous (SVP). Sympathetic activity is expressed as μV.s/10-s interval. The time-base marks indicate occlusion of splenic artery (OSA), occlusion of splenic vein (OSV), and injection of capsaicin (CAPS).

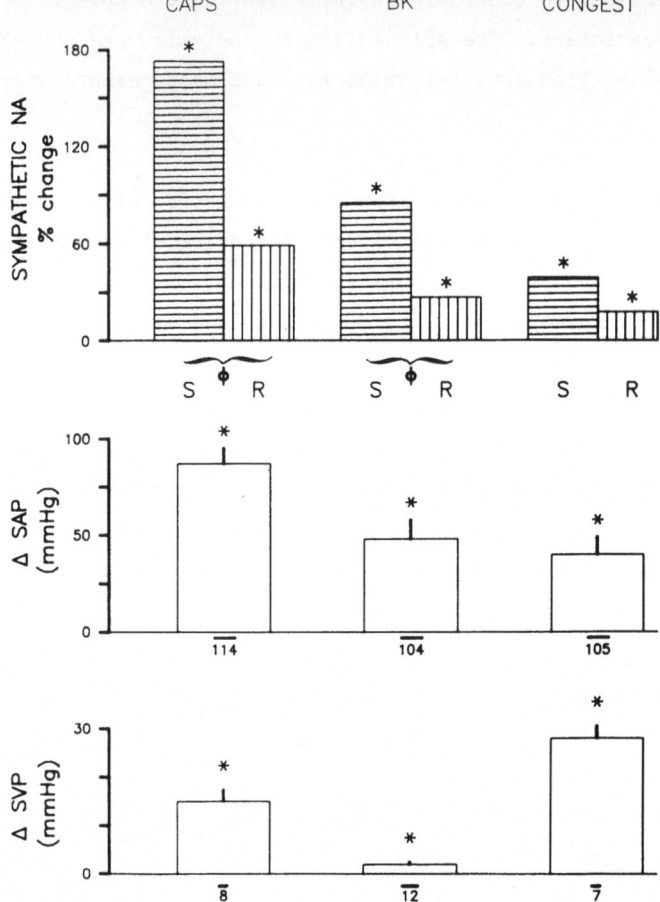

Fig. 6. Mean changes in sympathetic nerve activity (NA), systemic
arterial pressure (SAP), and splenic venous pressure (SVP) caused
by stimulation of splenic receptors of baroreceptor-denervated
vagotomized cats by capsaicin (CAPS), bradykinin (BK), and
congestion (CONGEST). S, splenic nerve activity; R, renal nerve
activity. *, Mean values significantly different from control (P
< 0.05). ∅, Neural responses significantly different from each
other using Friedman test. ∆SAP and ∆SVP are mean changes +
standard error. Numbers at bottom of ∆SAP and ∆SVP bars are mean
control values for group. CAPS n = 5, BK n = 4, CONGEST n = 6.

control, stimulation and recovery periods, respectively. Occlusion of the
splenic artery and vein without injection of capsaicin, or with injection

of 1.2-1.5 ml saline alone, caused no responses. Stimulation of splenic
receptors by injection of 1.0 µg bradykinin into the splenic arterial
circulation also caused sympathetic excitation in four cats (Fig. 6).
Again, splenic sympathetic excitation (85% increase) was greater than
renal (27% increase).

To provide a stimulus to splenic receptors more natural than the
chemical stimulation, pressure was increased within the spleen by infusing
3-20 ml physiological saline into the splenic artery. This splenic
congestion increased mean splenic venous pressure 27 $\pm$ 4 mmHg in six cats
and caused increases in activity of splenic and renal nerves and increases
in systemic arterial pressure (Fig. 6). The magnitudes of increase in
activity of splenic and renal nerves (39% and 18%, respectively) were not
significantly different from each other.

The reflex nature of responses to administration of capsaicin,
bradykinin, and physiological saline into the splenic arterial circulation
was verified by examining responses to these stimuli in the same cats
after splenic denervation. Sympathetic excitation and pressor and
chronotropic responses to the three stimuli were abolished by denervation.

Because these reflexes were accompanied by pressor responses,
activation of sino-aortic and cardiopulmonary baroreceptors would be
likely to suppress them in intact animals (Thoren, Donald, and Shepherd,
1976). However, equal suppression of splenic and renal sympathetic
responses was questionable, as intense stimulation of splenic receptors by
capsaicin and bradykinin caused significantly greater excitation of
splenic than renal nerves. In this situation, the integration of
baroreceptor-induced sympathoinhibition with different magnitudes of
reflex excitation in the two nerves may not lead to identical suppression
of responses. Therefore, splenic receptors were stimulated (as described
previously) in cats with intact vagi and arterial baroreceptor nerves.
Responses of one cat to the stimulation of splenic receptors by capsaicin
are illustrated in Fig. 7. The splenic nerve was excited, whereas the
renal nerve was inhibited, due to concomitant baroreceptor stimulation by
the pressor response. A multiphasic (excitation-inhibition) response
pattern in renal nerve occurred in three of the nine aminals, excitation
occurred in four of the nine, and inhibition, such as that illustrated in
Fig. 7, occurred in two. Mean responses to this stimulation are shown in
Fig. 8. Stimulation of splenic receptors by capsaicin, bradykinin, or

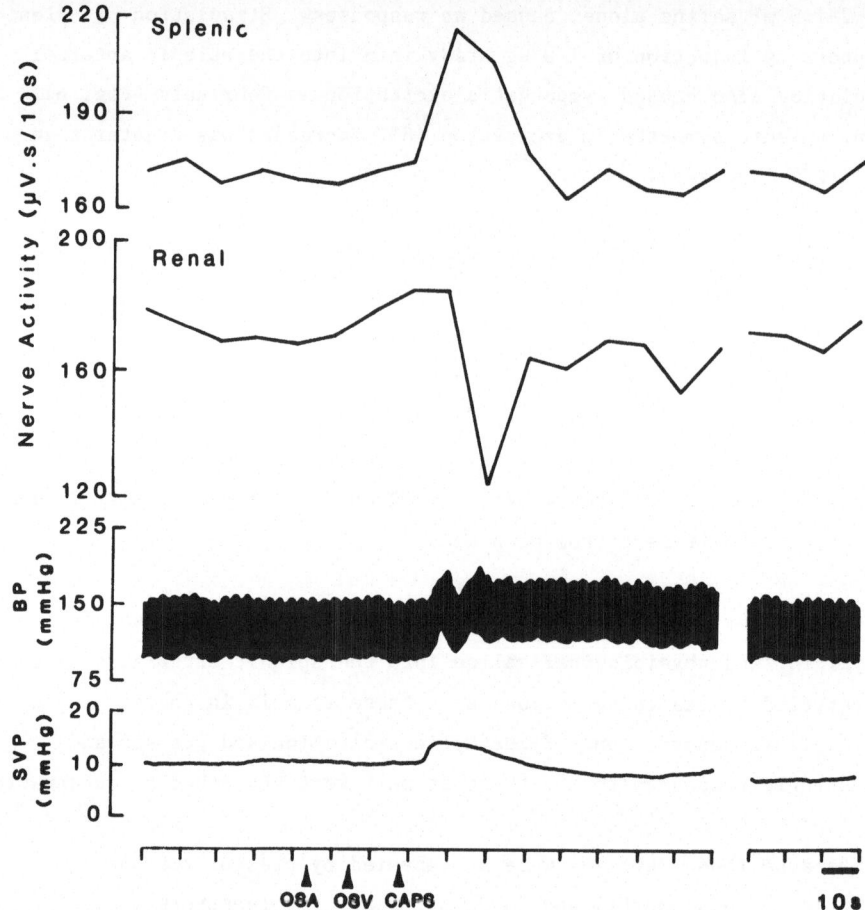

Fig. 7. Effects of injection of 5 μg capsaicin into splenic artery of baroreceptor-intact cat. Format is similar to that of Fig. 5. Recovery is shown at far right.

congestion caused excitation of splenic nerves but no significant mean renal nerve responses. Therefore, with baroreceptors present, each of the stimuli produced unequal responses in splenic and renal nerves. Pressor responses and contractions of the spleen or increases in splenic vein pressure (with congestion) accompanied the splenic reflexes (Fig. 8).

Stimulation of Baroreceptors

Because the baroreceptors appeared to suppress excitation of renal

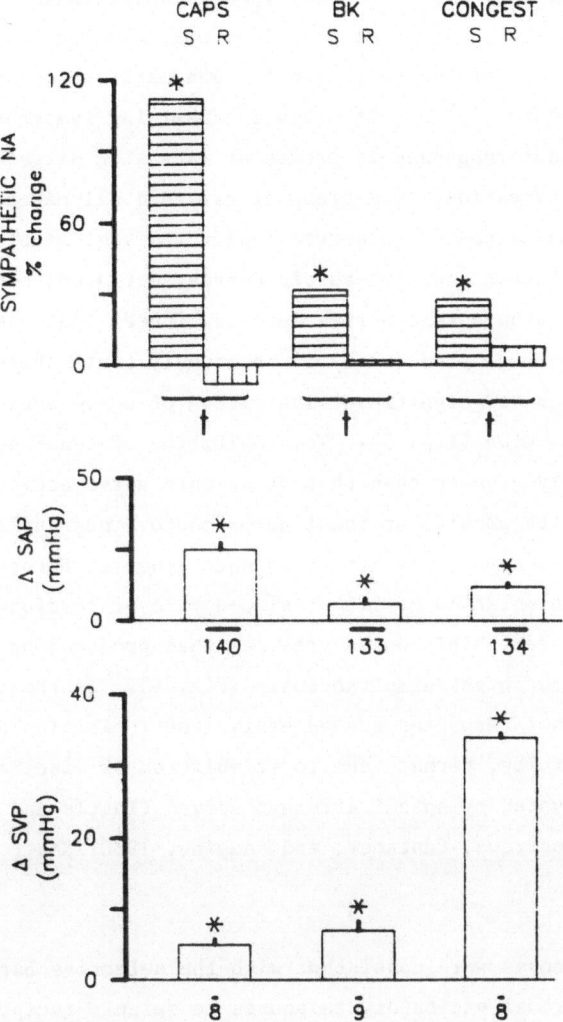

Fig. 8. Mean changes to stimulation of splenic receptors of baroreceptor-intact animals. Format is similar to that of Fig. 6. †, Neural responses significantly different from each other using Friedman test. CAPS n = 8, BK n = 8, and CONGEST n = 6.

nerves more than that of splenic nerves, the inhibitory influences of arterial baroreceptors and cardiopulmonary receptors on activity of these nerves were compared. This investigation was also prompted by observations of others demonstrating that arterial and cardiopulmonary baroreceptors can have differential influences on peripheral vascular resistance and sympathetic outflow (Ninomiya and Irisawa, 1969; Little, Wennergren, and Oberg, 1975; Mancia, Shepherd, and Donald, 1976). The relative inhibitory efficacy of arterial and cardiopulmonary baroreceptors was assessed separately in this study by examining sympathoinhibition induced by pressor responses in groups of cats with different states of baroreceptor innervation. One group of cats had all baroreceptors intact; another group had sino-aortic nerves intact and vagi sectioned; a third group had vagi intact and sino-aortic nerves sectioned; and in the final group, vagi and sino-aortic nerves were sectioned. Increases in arterial pressure (50-66 mmHg) produced by intravenously administered norepinephrine caused significant inhibition of nerve activity in all four states of innervation (Fig. 9). The inhibition of renal nerve activity was significantly greater than that of splenic nerve activity in each group of cats with partial or total baroreceptor innervation. These large pressor responses apparently activated each group of baroreceptors intensely, as stimulation of either sino-aortic or cardiopulmonary receptors produced inhibition as great as that produced by stimulation of both groups of receptors simultaneously (Fig. 9). In the denervated group, the pressor responses caused equivalent inhibition of splenic and renal nerve activity, perhaps due to stimulation of visceral low-pressure receptors innervated by spinal afferent nerves (Tuttle and McCleary, 1975; Weaver, 1977; Kostreva, Castaner, and Kampine, 1980; Kostreva et al., 1981).

These responses were consistent with the selective baroreceptor suppression of renal excitatory responses to splenic receptor stimulation observed previously. They also provided further documentation of selective cardiopulmonary afferent influences on renal nerves, as described previously in this paper (Fig. 2) and by others (Oberg and White, 1970; Mancia, Donald, and Shepherd, 1973; Weaver et al., 1984). In contrast, these findings were not consistent with those of Ninomiya et al., 1971; Ninomiya and Irisawa, 1975), which suggested greater arterial baroreceptor influences on splenic than renal nerves. A possible cause for the discrepancy between these two investigations is the use of chloralose anesthesia in our experiments and the use of pentobarbital in those of Ninomiya et al. (1971) and Ninomiya and Irisawa (1975). Previous studies in our laboratory, which will be described later, have demonstrated that renal nerves are more dependent upon supra-spinal excitatory drive than are splenic nerves (Weaver et al., 1983; Meckler and

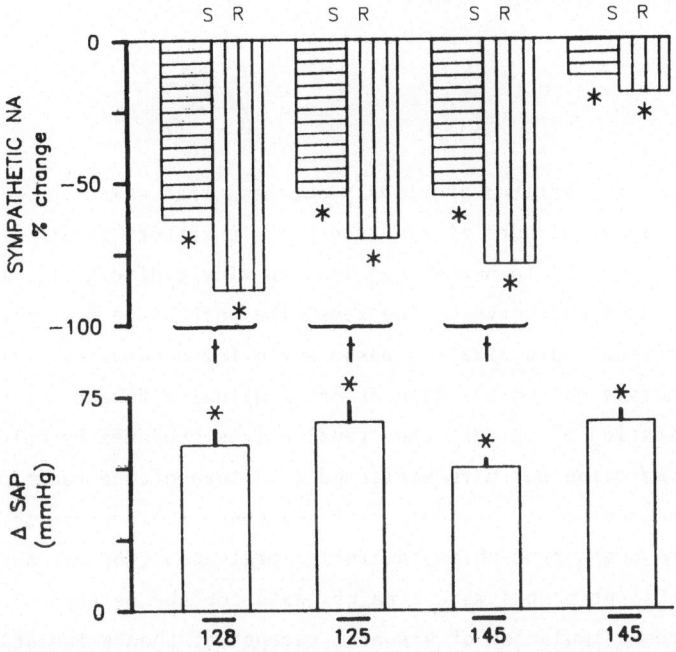

Fig. 9. Sympathetic nerve responses to large increases in blood pressure produced by injection of norepinphrine. Format is similar to that of Fig. 8. The four states of baroreceptor innervation are baroreceptor-intact (BARO INTACT), carotid sinus and aortic depressor nerves intact (CSN/ADN), vagi intact (VAGI INTACT), and carotid sinus nerves, aortic depressor nerves, and vagi sectioned (BARO DEN). BARO INTACT n = 7, CSN/ADN n = 5, VAGI INTACT n = 7, and BARO DEN n = 9.

Weaver, In press). As baroreceptors are thought to have supraspinal and spinal sites of inhibition (Gebber, Taylor, and Weaver, 1973), renal nerve activity may be affected by inhibition of brainstem or spinal neurons, whereas splenic nerves may be affected mostly by sympathoinhibition at spinal sites. If pentobarbital interferes with either descending excitatory drive from the brainstem or with supraspinal integration of baroreceptor-induced inhibition, then pentobarbital anesthesia may suppress actions of baroreceptors on renal nerves more than actions on splenic nerves. This conjecture requires further investigation.

NEURAL ORGANIZATION RESPONSIBLE FOR DIFFERENTIAL EXCITATION OF SYMPATHETIC NERVES BY SPINAL AFFERENT INPUTS

Local Sign

The studies described previously document that summation of subtle inequalities in magnitudes of excitatory or inhibitory reflex influences on different sympathetic nerves can lead to widely divergent, opposite responses among these nerves. The remaining work to be described concerns the specific neural organization responsible for unequal excitation among sympthetic nerves caused by stimulation of spinal afferent nerves. The greater excitation of splenic than renal nerves produced by splenic receptor stimulation may have reflected a feature of the functional central organization of autonomic reflexes that provides the largest output to the organ from which the reflex orginates (Koizumi and Brooks, 1972). If this principle were a major basis for the pattern of reflexes generated from stimulation of visceral receptors, then stimulation of spinal afferent nerves originating from another organ, such as the heart, should cause more intense responses in cardiac nerves than in either splenic or renal nerves. This hypothesis was tested in 16 vagotomized cats in which sinoaortic baroreceptor nerves had been severed. Cardiac receptors were stimulated in these cats by superfusion of the heart with 1.0 ml of saline containing 1.0–10.0 μg bradykinin, after creating a cradle of the opened pericardial sac. Multifiber activity was recorded from the left cardiac, splenic or renal sympathetic nerves. Activity was recorded simultaneously from two of the three nerves, and all possible combinations were compared. In these experiments activity was quantified by spike counting.

Reflex influences of cardiac spinal afferent nerves on renal, splenic, and cardiac nerves are illustrated in Fig. 10. All responses are grouped according to nerve in this analysis without consideration of the specific recording pairs used in the cats. Application of bradykinin produced excitation in the sympathetic efferent nerves after a latency of approximately 10 s. Splenic and cardiac nerve responses were significantly greater than those of renal nerves when responses of all animals were considered together. In addition, when only simultaneously recorded responses were considered, splenic nerve excitation was significantly greater than renal; and splenic and cardiac nerve responses were similar. That splenic nerves responded more than renal nerves to cardiac receptor stimulation suggested that "local sign" was not a major feature of the organization of differential excitatory sympathetic reflexes. Cardiac receptor stimulation caused splenic nerve responses which were equivalent to, rather than less than, those of cardiac nerves. Cardio-cardiac reflexes did not predominate; and, again excitation of renal nerves was less than that of cardiac of splenic nerves.

Segmental Organization of Spinal Reflexes

If the excitatory reflexes originating from spinal afferent nerves are organized solely within the spinal cord, segmental organization of these spinal reflexes also could contribute to the final response pattern. Early experiments of Beacham and Perl (1964) demonstrated that excitatory spinal sympathetic reflexes recorded in white rami decreased in magnitude at spinal segments distant from the site of dorsal root afferent stimulation. This organization suggests that spinal reflexes originating from visceral receptor stimulation would be greatest in sympathetic nerves originating from the segments of entry of the stimulated spinal afferent nerves. This hypothesis also was tested. After assessing cardiac, splenic, and renal sympathetic responses to cardiac receptor stimulation by bradykinin, the cervical spinal cord was transected and reflex responses were tested again in the same cats. As illustrated in Fig. 10, reflex responses to cardiac receptor stimulation were readily evoked in the spinal cats. Again, even in the spinal animals, cardio-cardiac reflexes did not predominate. Splenic sympathetic responses were equal to, or greater than, the cardiac sympathetic responses; and both were greater than renal reflex responses. The similarity of cardio-cardiac and cardio-splenic sympathetic reflexes suggested that segmentation of spinal reflexes within the cord also is not a major organizational feature underlying differential sympathetic responses. The consistently smaller excitatory responses of renal nerves may have reflected a unique characteristic of renal nerves. Perhaps they are generally less responsive to excitatory spinal afferent influences than are other sympathetic nerves.

A further test, which employed the concept of preferential distribution of a response back to the organ of origin, was applied to the renal and splenic nerves. Renal receptors were stimulated by injecting 10 μg bradykinin into the renal artery, and responses of splenic and renal nerves were compared. This intense stimulus to the kidney was delivered

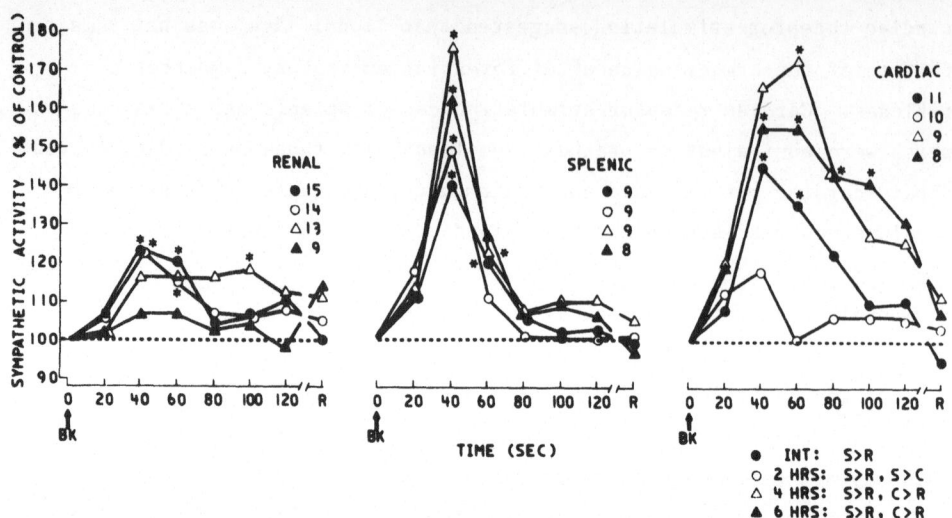

Fig. 10. Sypathetic responses to chemical stimulation of cardiac
sympathetic afferent nerves in intact cats and 2,4, and 6 h after
spinal cord transection. Responses are expressed as percent of
control discharge averaged during 20-s intervals after
administration of bradykinin (BK) at arrow. Recovery sample (R)
is 60-s average obtained 4 min after afferent stimulation.
Symbols identified in lower legend indicate statistically
significant comparisons of responses observed in intact state or
at 2,4, and 6 h after transection. Population sizes are noted
beside symbols above each panel. Asterisks indicate significant
differences from control determined by test of least significant
difference after analysis of variance of activity (counts/s).
Maximum responses (expressed as percent of control) were compared
between nerves at each time interval using a Wilcoxon 2-sample
test, and significant differences are indicated in lower legend.
Cardiac afferent stimulation caused excitation of all 3 nerves
before and after spinal transection. Splenic and cardiac nerve
responses generally were greater than renal responses.

in an attempt to provoke a major renal nerve response. Several
investigators have demonstrated excitatory reno-renal reflexes (Calaresu
et al., 1978; Recordati, Genovesi, and Cerati, 1982). Injections were
made through a small (PE 50) cannula which had been threaded into the

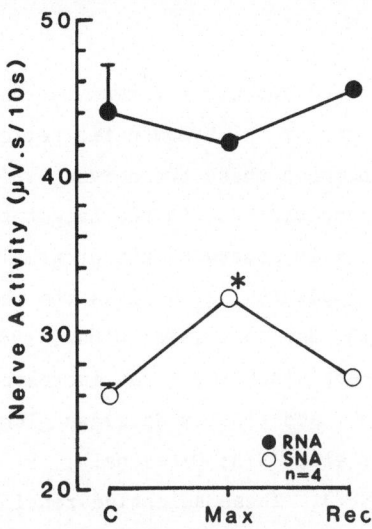

Fig. 11. Mean responses to stimulation of renal receptors with 5 µg
bradykinin in 4 vagotomized, sino-aortic denervated cats. The
ordinate indicates activity of renal (closed circles) and splenic
(open circles) nerves that was integrated in 10-s intervals. The
control period (C) and recovery period (Rec) represent average
nerve discharge accumulated during 1 min. The 10-s period of
maximum response after bradykinin injection is indicated as Max.
The variability in these responses is indicated by standard error
terms determined from an analysis of variance. Stimulation of
renal receptors caused significant excitation of splenic nerve
activity and no change in renal nerve activity.

renal artery from the femoral artery. This cannula did not obstruct renal
blood flow. Stimulation of renal receptors in four vagotomized,
sino-aortic denervated cats caused no changes in renal nerve activity but
produced significant excitation of splenic nerves (Fig. 11). Splenic
nerve activity increased by approximately 30%. Systemic arterial pressure
decreased by 10-30 mmHg in three cats and increased by 30 mmHg in one cat
in response to this stimulus. Intravenous injections of 10 µg bradykinin
caused similar decreases in arterial pressure and no reflex sympathetic
excitation. Therefore, even stimulation of receptors in the kidney, which
is adequate to cause significant sympathetic excitation, produces very
minor renal nerve responses.

The comparisons between splenic and renal nerve responses always yielded greater splenic than renal sympathetic reflexes. A potentially significant difference between these two nerves, which may mandate dissimilar response characteristics, is the target tissue each nerve innervates. The renal nerves innervate the proximal and distal tubules and the juxtaglomerular apparatus, as well as the renal vasculature (Barajas and Muller, 1973; Barajas, 1978; Dibon, 1982). Progressive increases in renal efferent discharge first increase renin release, then decrease sodium excretion, and finally decrease glomerular filtration rate and blood flow (Slick et al., 1975; Gottschalk, 1979; Osborn, Dibona, and Thames, 1981; Dibona, 1982). These selective renal responses seem particularly suited for specific, finely-graded neural inputs to the kidney and imply little need for massive excitation of real nerves or for mass action of the sympathetic nervous system. In contrast, the splenic innervation is distributed to splenic arteries inside and outside the spleen, to the capsule and trabecular smooth muscle, and to splenic veins (Utterback, 1944; Fillenz, 1970). The spleen is a reservoir for hemoconcentrated blood (Donald and Aarhus, 1974) and is a major constituent of the capacitive circulation in dogs and cats (Barcroft and Stephens, 1927; Guntheroth and Mullins, 1963; Greenway et al., 1968; Greenway and Lister, 1974; Carneiro and Donald, 1977; Greenway and Innes, 1980). Excitation of splenic nerves leads to contraction of the spleen and to constriction of the splenic arteries and veins, resulting in ejection of stored erythrocytes into the systemic circulation (Utterback, 1944; Boatman and Brody, 1964; Fillenz, 1970; Donald and Aarhus, 1974). The almost single purpose of splenic sympathetic innervation (to cause instant mobilization of blood) contrasts with the multiple actions of sympathetic nerves on functions of the kidney. Do splenic and renal nerves respond to excitatory stimuli unequally because splenic nerves are typical of splanchnic vasoconstrictor sympathetic outflow, and this outflow differs from that distributed to the kidney? Does splanchnic sympathetic outflow to the capacitive abdominal circulation react as a unit? If so, mesenteric vasoconstrictor nerves should respond similarly to splenic nerves, as the intestinal vasculature also is a major capacitive bed (Greenway and Lister, 1974). The differences between splenic and renal nerve responses may be paralleled by differences between mesenteric and renal nerve responses, suggesting that a principle underlying the organization of these reflexes relates to the function of the target organ innervated.

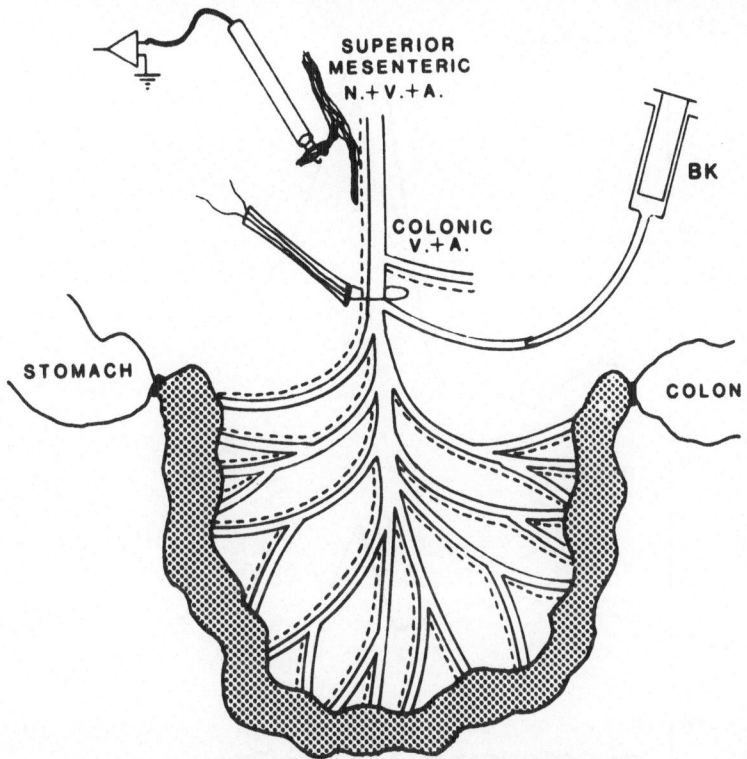

Fig. 12. Diagram of experimental preparation of the small intestine.

To test this possibility, responses of renal and mesenteric (vasoconstrictor) nerves to stimulation of visceral receptors were compared. In these experiments baroreceptor nerves were left intact to allow mesenteric sympathetic activity to be characterized. Only mesenteric nerve activity which could be inhibited by baroreceptor activation was studied in these experiments to ensure that sympathetic responses were recorded in vasomotor mesenteric sympathetic outflow. Baroreceptors were activated by increasing arterial pressure with intravenous injections of norepinephrine (0.2 μg/kg). In these experiments, receptors within the isolated vasculature of the small intestine were stimulated by intra-arterial injections of 1.0 μg bradykinin. The isolated intestinal loop preparation was similar to that described for the splenic circulation. Snares were placed around the superior mesenteric artery and vein, a small branch of the distal mesenteric artery was cannulated for injections of bradykinin, and collateral circulation from other vessels was eliminated by ligating the intestine at its gastro-duodenal and ileocolic junctions (Fig. 12).

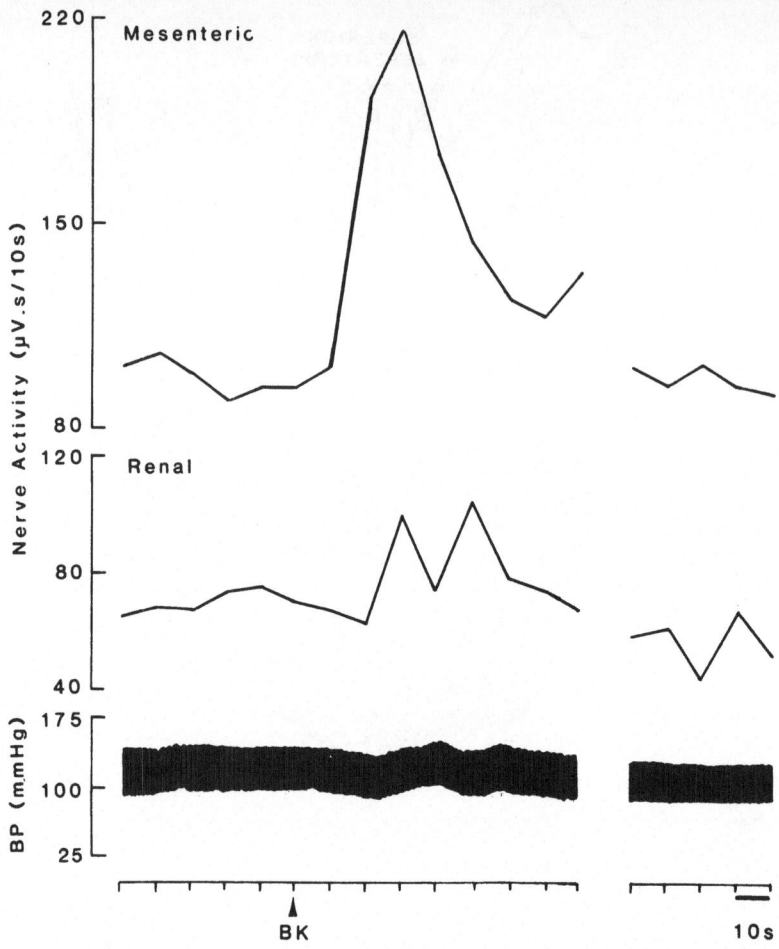

Fig. 13. Response of one cat to stimulation of mesenteric receptors.
Integrated activity of mesenteric and renal nerves is shown in
the top two panels. Systemic arterial pressure (BP) is
illustrated in the lower panel. Time in seconds and injection of
1.0 μg bradykinin (BK) are indicated beneath these panels.
Occlusion of the mesenteric artery and vein was completed
immediately before the time period of the illustrated samples of
sympathetic activity and arterial pressure. The recovery values
of these values 4 min after bradykinin injection and 1 min after
release of the occluded blood vessels are illustrated in the
group of panels on the right.

Stimulation of mesenteric receptors by 1.0 μg bradykinin caused intense
excitation of mesenteric sympathetic nerves and modest excitation of renal
nerves, as illustrated in Fig. 13. This sympathetic excitation sometimes

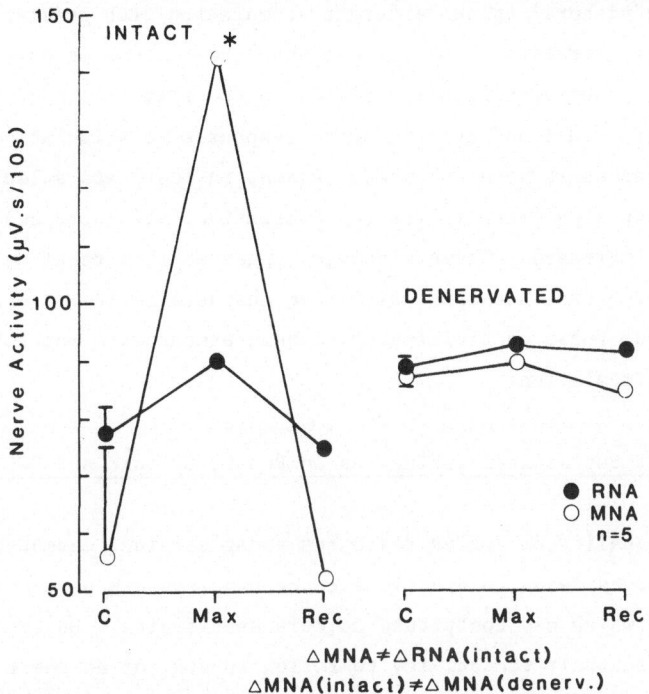

Fig. 14. Responses of sympathetic nerve activity to stimulation of
intestinal receptors with bradykinin before and after denervation
of the intestine in 5 cats. Format is similar to that of Fig.
11. The ordinate indicates activity of renal (closed circles)
and mesenteric (open circles) nerves that was integrated in 10-s
intervals. The control periods (C) and recovery periods (Rec)
represent average nerve discharge accumulated during 1 min, and
the 10-s periods of maximum response after bradykinin injection
(Max) are indicated. Left panel: intact intestinal innervation;
right panel: after denervation of the intestine.

was accompanied by small pressor responses which led to suppression of
renal responses by baroreceptor activation. Mean responses to this
stimulation are illustrated in Fig. 14. Mesenteric nerve excitation was
significantly greater than that of renal nerves in five cats, and these
responses were abolished by denervation of the intestine. The mean change
in systemic arterial pressure during this excitation was only 6 $\pm$ mmHg,
illustrating that the small magnitude of renal nerve responses could not
be attributed solely to activation of baroreceptors. These findings
illustrated that another component of splanchnic sympathetic outflow also

responded to visceral spinal afferent stimulation with greater intensity that did renal nerves. To be certain that the intense mesenteric nerve responses were not specifically related to the stimulation of receptors in the intestine, renal and splenic nerve responses to stimulation of mesenteric receptors were compared. Again, in four cats splenic nerves responded with significantly greater excitation (98% increase) than renal nerves (53% increase). These findings suggested that renal nerves are influenced by neural networks which have characteristics of organization different from those controlling splanchnic sympathetic outflow to the capacitive circulation.

Comparison of Supraspinal Influences on Renal, Splenic and Cardiac Nerves

The inequality of reflex responses among cardiac, mesenteric, splenic and renal nerves leads to questions concerning aspects of central neural organization which may contribute to this specificity. Do these nerves differ only in their receptivity to reflex inputs, or do their sources of ongoing drive differ, as well? Do they depend equally upon tonic supraspinal sources of excitation or inhibition? The following experiments were designed to compare the impact of high cervical spinal transection on basal discharge in cardiac, splenic, and renal nerves. Interruption of supraspinal inputs was proposed to reveal potential differences in their dependence upon descending excitation or inhibition which may be relevant to their responsiveness to reflexes. Sino-aortic baroreceptor nerves and vagi were severed in the cats used in these experiments.

Ongoing activity of splenic, renal, and cardiac nerves was monitored before and for 2 hr after transection of the spinal cord at the first cervical segment. As illustrated in Fig. 15, the discharge of these nerves before cord transection was characterized by synchronized bursts of activity. Following transection of the cord, the discharge rates of renal and cardiac nerves were significantly depressed by more than 50%, whereas splenic nerve activity was not decreased significantly (Fig. 15, Fig. 16). Splenic nerve activity was unchanged or actually increased after cord transection in 4 of 16 cats; an example from one of these cats is shown in Fig. 15. Discharge rates within each group (splenic, renal, or cardiac) did not change significantly during the 2 hr monitored after cord transection. Paired analyses also were done to allow comparisons of simultaneously recorded responses of pairs of nerves (i.e., splenic and

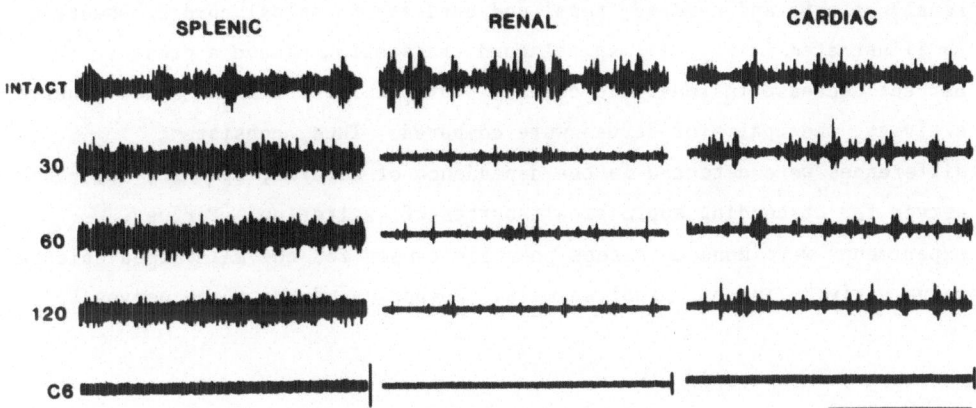

Fig. 15. Photographs of oscillograph tracings of sympathetic nerve
activity from 3 cats before and after spinal cord transection.
Activity of splenic, renal, and cardiac nerves was sampled when
the neuraxis was intact (INTACT), at 30, 60, and 120 min after
cord transection, and following administration of hexamethonium
(C6). Vertical calibrations are 30 μV; horizontal calibration,
1 s.

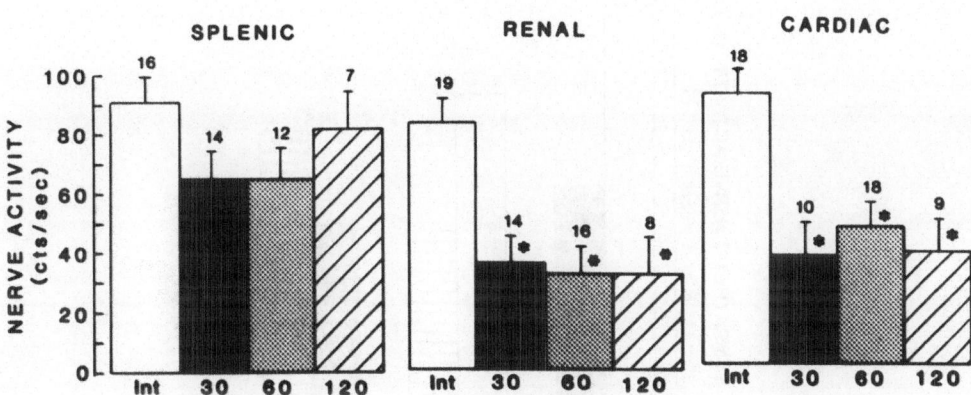

Fig. 16. Ongoing discharge of sympathetic nerves before and after high
cervical spinal cord transection. Bars represent averages of 1-2
min samples of splenic, renal, and cardiac nerve activity at the
4 sampling periods: intact neuraxis (Int.), 30, 60, and 120 min
following C1 spinal cord transection. The number of animals is
indicated above each bar. The thin lines on bars represent
standard error. Asterisks denote significant differences between
nerve discharge rates in the intact state and after spinal cord
transection. Discharge rates of splenic, renal, and cardiac
nerves were not significantly different from each other when the
neuraxis was intact.

renal; splenic and cardiac; renal and cardiac) to spinal cord transection. As illustrated in Fig. 17, spinal cord transection caused a greater percent decrease in renal and cardiac nerve activity than in splenic nerve activity, when pairs of nerves were compared. Thus, consistent differences were detected in the dependence of cardiac, splenic, and renal nerves for descending supraspinal sources of excitation. Further experiments were done to assess possible causes for the maintained splenic nerve activity in the spinal animals. Responses of the three nerves to

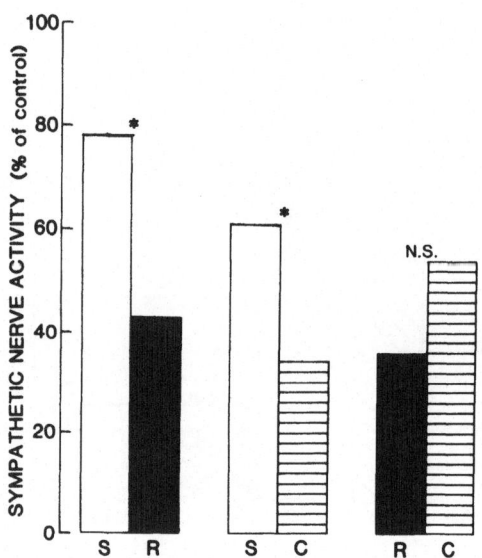

Fig. 17. Effects of spinal cord transection on simultaneously recorded
discharge rates of pairs of sympathetic nerves. Bars represent
rates of discharge of nerves expressed as a percent of control
(intact neuraxis) activity. Percentages were logarithmically
transformed prior to analysis of variance, because the
percentages were not normally distributed. Therefore,
coefficients of variation (C.V.) from the analyses of variance
are given (below) rather than the standard errors. S, splenic;
R, renal; C, cardiac; *, significant difference between the two
nerves; N.S., not significant; S/R (n = 6) C.V.: 74%, S/C (n = 6)
C.V.: 28%; R/C (n = 9) C.V.: 15%.

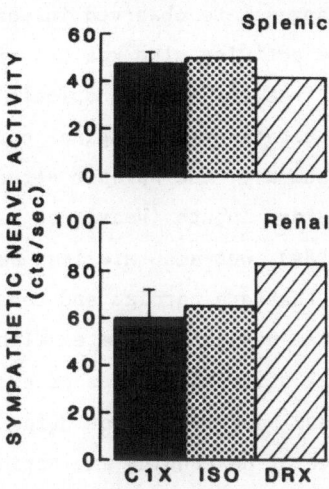

Fig. 18. Ongoing discharge of sympathetic nerves following dorsal
rhizotomy in 6 cats. Bars represent averages of 1-2 min samples
of activity of splenic and renal nerves following C1 spinal cord
transection (C1X), after transection of the spinal cord at T4 and
L5 (ISO), and following dorsal rhizotomy from T4 to L5 (DRX).
The thin lines on bars represent standard error. Activity of
both nerves was unaffected by dorsal rhizotomy. Mean arterial
pressure was 92 ± 4, 91 ± 4, and 90 ± 4 mmHg following C1X, ISO,
and DRX, respectively.

potential external sources of excitation, such as hypercapnia and
hypotension, were compared to determine possible hypersensitivity of
splenic sympathetic nerves. Neither hypotension, hypertension, nor
hypercapnia produced responses in the spinal animals which could be
associated with the maintained splenic sympathetic discharge. To evaluate
the possibility that activity of spinal afferent nerves tonically or
preferentially contributed to the ongoing discharge of splenic or renal
nerves in the spinal cat, an isolated spinal cord T_4-L_5 segment was
prepared in situ in six cats. The spinal cord was severed at the first
cervical segment, as usual. Thirty minutes later, the cord was transected
at L_5 and T_4. The paravertebral sympathetic chains also were transected
between T_4 and L_5. Isolation of this T_4-L_5 segment of spinal cord did not
cause changes in splenic or renal nerve activity (Fig. 18). Fifteen to 60
min after isolation of the spinal cord segment, the dorsal roots from
T_4-L_5 were severed, depriving this section of spinal cord of spinal

afferent input. No changes were observed in the rate of splenic nerve discharge. Renal nerve activity also was not affected significantly. These findings suggest that the ongoing splenic nerve activity maintained in spinal animals is generated within spinal cord circuits, or possibly at ganglionic sites. In summary, the splenic nerves are not only more reactive to many excitatory inputs (Weaver et al., 1983; Calaresu et al., 1984; Weaver et al., 1984), but also are less dependent upon tonic supraspinal excitation that are cardiac and renal nerves. This difference between splenic and renal nerves correlates with greater receptivity of splenic nerves to spinal afferent sources of excitation. Spinal afferent excitatory inputs to renal nerves may not bring spinal cord neurons to threshold in the absence of descending excitation. In addition, spinal afferent inputs may be distributed only to a limited "renal" population within the spinal cord, resulting in minor reflexes in the presence or absence of supraspinal excitatory drive. The large magnitude of spinal cardio-cardiac reflexes, despite the low discharge rates of cardiac nerves in spinal cats, may be attributed to especially powerful synaptic efficacy of spinal afferent inputs to cause this segmental spinal reflex.

SUMMARY AND CONCLUSIONS

This research has consistently illustrated that renal nerves are less responsive to excitatory spinal afferent influences than are nerves innervating the capacitive circulation or the heart. In addition, renal nerves appear to be particularly sensitive to inhibitory influences of afferent nerves which enter the neuraxis at brainstem levels. Each group of experiments designed to investigate a specific aspect of differential reflex patterns demonstrated the uniqueness of renal nerve reactions. The comparatively small magnitudes of renal excitatory responses could not be solely attributed to features of neural organization such as "local sign" or segmentation of spinal reflexes. The observation that renal nerves were especially sensitive to supraspinal reflexes generated from arterial and cardipulmonary baroreceptors and that renal nerves are dependent upon supraspinal sources of excitation for basal ongoing activity suggests that control of renal nerves may involve highly integrated neural mechanisms at supraspinal and spinal levels. Spinal afferent excitatory inputs may have minimal effects on renal nerves, because a major portion of the neural circuits controlling renal nerves are unavailable to these inputs. Alternatively, excitation of spinal afferent origin may be directly suppressed in renal nerves by a supraspinal source of inhibition.

These potentially complex control mechanisms are consistent with the fine control of renal function which can be accomplished by sympathetic nerves. Major perturbations requiring changes in renal function, such as changes in blood pressure and volume, changes in blood composition (pH, pCO_2, pO_2), and changes in osmolality of extracellular fluid, all can alter sympathetic outflow via supraspinal pathways. The spinal afferent neurons were usually stimulated by noxious substances in these studies; and reflexes were caused, at least in part, by activation of visceral nociceptors. Such stimulation can cause intense sympathetic excitation, but the utility of massive renal vasocontriction in response to visceral pain is not obvious. Hence, the exclusion of renal neurons from this major sympathetic excitation may be useful. In contrast, the mobilization of blood from the splanchnic circulation and increases in heart rate are well-known responses to pain. The neural circuits needed to cause these responses may be more simple than those regulating renal function and involve direct spinal pathways in many cases. The fact that splenic nerves depend minimally upon supraspinal influences, even for basal ongoing activity, suggests more primitive organization of this component of sympathetic outflow. Recent experiments from our laboratory have shown that mesenteric sympathetic nerves also have minimal dependence upon supraspinal excitation for maintenance of basal activity. The splanchnic circulation appears to have neural control circuitry which differs significantly from that directed to the kidney. The neural organization of different components of sympathetic outflow may be well suited to the widely different functions of different target organs.

ACKNOWLEDGMENTS

This research was supported by grant HL21436 from the National Heart, Lung, and Blood Institute. The authors are indebted to Drs. Simonetta Genovesi and Andrea Stella for their assistance with the design of some of these experiments, to Dr. Franco Calaresu for his critical evaluation of this manuscript, to Mr. James Hayner and Mr. P. O. Eric Mueller for expert technical assistance, and to Ms. Lynne Woods for excellent editorial and secretarial assistance.

REFERENCES

Adams, D. B., Baccelli, G., Mancia, G., and Zanchetti, A., 1969, Cardiovascular changes during naturally elicited fighting behavior in the cat, Am. J. Physiol. 216:1226-1235.

Barajas, L., 1978, Innervation of the renal cortex, Fed. Proc. 37:1192-1201.

Barajas, L., and Muller, J., 1973, The innervation of the juxtaglomerular apparatus and surrounding tubules: A quantitative analysis by serial section electron microscopy, J. Ultrastruc. Res. 43:107-132.

Barcroft, J., and Stephens, J. G., 1927, Observations upon the size of the spleen, J. Physiol. London 64:1-22.

Beacham, W. S., and Perl, E. R., 1964, Background and reflex discharge of sympathetic preganglionic neurones in the spinal cat, J. Physiol. London 172:400-416.

Bevan, J. A., 1979, Some bases of differences in vascular response to sympathetic activity, Circ. Res. 45:161-171.

Blumberg, H., Janig, W., Rieckmann, C., and Szulczyk, P., 1980, Baroreceptor and chemoreceptor reflexes in postganglionic neurones supplying skeletal muscle and hairy skin, J. Auton. Nerv. Syst. 2:223-240.

Boatman, D. L., and Brody, M. J., 1964, Analysis of vascular responses in the spleen, Am. J. Physiol. 207:155-161.

Calaresu, F. R., Kim, P., Nakamura, H., and Sato, A., 1978, Electrophysiological characteristics of renorenal reflexes in the cat, J. Physiol. London 283:141-154.

Calaresu, F. R., Tobey, J. C., Heidemann, S. R., and Weaver, L.C., 1984, Splenic and renal sympathetic responses to stimulation of splenic receptors in cats, Am. J. Physiol. 247 (Regulatory Integrative Comp Physiol 16):R856-R865.

Cannon, W. B., 1929, The sympathetic division of the autonomic system in relation to homeostasis, Arch. Neurol. Psychiat. 22:282-294.

Cannon, W. B., 1930, The Linnare Lecture on the autonomic nervous system: an interpretation, Lancet 1:1109-1115.

Carneiro, J. J., and Donald, D. E., 1977, Blood reservoir function of dog spleen, liver, and intestine, Am. J. Physiol. 232 (Heart Circ Physiol 1):H67-H72.

Dibona, G. F., 1982, The functions of the renal nerves, Rev. Physiol. Biochem. Pharmacol. 94:75-181.

Donald, D. E., and Aarhus, L. L., 1974, Active and passive release of blood from canine spleen and small intestine, Am. J. Physiol. 227:1166-1172.

Fillenz, M., 1970, The innervation of the cat spleen, Proc. Roy. Soc. Lond. B 174:459-468.

Folkow, B., Johansson, B., and Lofving, B., 1961, Aspects of functional differentiation of the sympatho-adrenergic control of the cardiovascular system, Med. Exp. 4:321-328.

Folkow, B., and Rubinstein, E. H., 1966, Cardiovascular effects of acute and chronic stimulations of the hypothalamic defence area in the rat, Acta. Physiol. Scand. 68:48-57.

Gebber, G. L., Taylor, D. G., and Weaver, L. C., 1973, Electrophysiological studies on organization of central vasopressor pathways, Am. J. Physiol. 224:470-481.

Gottschalk, C. W., 1979, Renal nerves and sodium excretion, Ann. Rev. Physiol. 41:229-240.

Greenway, C. V., and Innes, I. R., 1980, Effects of splanchnic nerve stimulation on cardiac preload, afterload, and output in cats, Circ. Res. 46:181-189.

Greenway, C. V., Lawson, A. E., and Stark, R. D., 1968, Vascular responses of the spleen to nerve stimulation during normal and reduced blood flow, J. Physiol. London 194:421-433.

Greenway, C. V., and Lister, G. E., 1974, Capacitance effects and blood reservoir function in the splanchnic vascular bed during non-hypotensive haemorrhage and blood volume expansion in anesthetized cats, J. Physiol. London 237:279-294.

Grignolo, A., Koepke, J. P., and Obrist, P. A., 1982, Renal function, heart rate and blood pressure during exercise and avoidance in dogs, Am. J. Physiol. 242 (Regulatory Integrative Comp Physiol 11):R482-R490.

Guntheroth, W. G., and Mullins, G. L., 1963, Liver and spleen as venous reservoirs, Am. J. Physiol. 204:35-41.

Horeyseck, G., and Janig, W., 1974, Reflexes in postganglionic fibres within skin and muscle nerves after noxious stimulation of skin, Exp. Brain Res. 20:125-134.

Iriki, M., Riedel, W., and Simon, E., 1972, Patterns of differentiation in various sympathetic efferents induced by changes of blood gas composition and by central thermal stimulation in anesthetized rabbits, Jap. J. Physiol. 22:585-602.

Iriki, M., Walther, O.-E., Pleschka, K., and Simon, E., 1971, Regional cutaneous and visceral sympathetic activity during asphyxia in the anesthetized rabbit, Pflügers Arch. 322:167-182.

Janig, W., 1982, An idea about the organization of the lumbar sympathetic outflow supplying skeletal muscle and skin, Proc. Austral. Physiol. Pharmacol. Soc. 13:47–58.

Koizumi, K., and Brooks, C. M., 1972, The integration of autonomic system reactions: a discussion of autonomic reflexes, their control and their association with somatic reactions, Ergeb. Physiol. 67:1–68.

Kollai, M., and Koizumi, K., 1977, Differential responses in sympathetic outflow evoked by chemoreceptor activation, Brain Res. 138:159–165.

Kostreva, D. R., Castaner, A., and Kampine, J. R., 1980, Reflex effects of hepatic baroreceptors on renal and cardiac sympathetic nerve activity, Am. J. Physiol. 238 (Regulatory Integative Comp Physiol 7):R390–R394.

Kostreva, D. R., Seagard, J. L., Castaner, A., and Kampine, J. P., 1981, Reflex effects of renal afferents on the heart and kidney, Am. J. Physiol. 241 (Regulatory Integrative Comp Physiol 10):R286–R292.

Kumazawa, T., and Mizumura, K., 1980, Chemical responses of polymodal receptors of the scrotal contents in dogs, J. Physiol. London 299:219–231.

Little, R., Wennergren, G., and Oberg, B., 1975, Aspects of the central integration of arterial baroreceptor and cardiac ventricular receptor reflexes in the cat, Acta. Physiol. Scand. 93:85–96.

Lofving, B., 1961, Differential vascular adjustments reflexly induced by changes in the baro- and chemoreceptor activity and by asphyxia, Med. Exp. 4:307–312.

Longhurst, J. C., Kaufman, M. P., Ordway, G. A., and Musch, T. I., 1984, Effects of bradykinin and capsaicin on endings of afferent fibers from abdominal visceral organs, Am. J. Physiol. 247 (Regulatory Integrative Comp Physiol 16):R552–R559.

Malliani, A., 1982, Cardiovascular sympathetic afferent fibers, Rev. Physiol. Biochem. Pharmacol. 94:11–74.

Mancia, G., Donald, D. E., and Shepherd, J. T., 1973, Inhibition of adrenergic outflow of peripheral blood vessels by vagal afferents from the cardiopulmonary region in the dog, Circ. Res. 33:713–721.

Mancia, G., Shepherd, J. T., and Donald, D. E., 1976, Interplay among carotid sinus, cardiopulmonary, and carotid body reflexes in dogs, Am. J. Physiol. 230(1):19–24.

Meckler, R. L., and Weaver, L. C. (In Press, 1985), Splenic, renal, and cardiac nerves have unequal dependence upon tonic supraspinal inputs, Brain Res.

Ninomiya, I., and Irisawa, H., 1975, Non-uniformity of the sympathetic nerve activity in response to baroreceptor inputs, Brain Res. 87:313–322.

Ninomiya, I., and Irisawa, H., 1969, Summation of baroceptor reflex effects on sympathetic nerve activities, Am. J. Physiol. 216(6):1330-1336.

Ninomiya, I., Irisawa, H., and Nisimaur, N., 1973, Non-uniformity of sympathetic nerve activity to the skin and kidney, Am. J. Physiol. 224:256-264.

Ninomiya, I., Nisimaru, N., and Irisawa, H., 1971, Sympathetic nerve activity to the spleen, kidney, and heart in response to baroceptor input, Am. J. Physiol. 221(5):1346-1351.

Nisimaru, N., 1971, Comparison of gastric and renal nerve activity, Am. J. Physiol. 220:1303-1308.

Oberg, B., and White, S., 1970, The role of vagal cardiac nerves arterial baroreceptors in the circulatory adjustments to hemorrhage in the cat, Acta. Physiol. Scand. 80:395-403.

Osborn, J. L., DiBona, G. F., and Thames, M. D., 1981, Beta-1 receptor mediation of renin secretion elicited by low-frequency renal nerve stimulation, J. Pharmacol. Exp. Ther. 216:265-269.

Recordati, G., Genovesi, S., and Cerati, D., 1982, Renorenal reflexes in the rat elicited upon stimulation of renal chemoreceptors, J. Autonomic Nervous System 6:127-142.

Slick, G. L., Aguilera, A. J., Zambraski, E. J., Dibona, G. F., and Kaloyanides, G. J., 1975, Renal neuroadrenergic transmission, Am. J. Physiol. 299:60-65.

Sokol, R. R., and Rohlf, F. J., 1969, "Biometry. The Principles and Practice of Statistics in Biological Research," W. H. Freeman and Co., San Francisco, CA.

Thoren, P. N., Donald, D. E., and Shepherd, J. T., 1976, Role of heart and lung receptors with nonmedullated vagal afferents in circulatory control, Circ. Res. 38:II2-II9.

Tutle, R. S., and McCleary, M., 1975, Mesenteric baroreceptors, Am. J. Physiol. 229(6):1514-1519.

Utterback, R. A., 1944, The innervation of the spleen, J. Comp. Neurol. 81:55-68.

Weaver, L. C., 1977, Cardiopulmonary sympathetic afferent influences on renal nerve activity, Am. J. Physiol. 233 (Heart Circ Physiol 2):H592-H599.

Weaver, L. C., Fry, H. K., and Meckler, R. L., 1984, Differential renal and splenic nerve responses to vagal and spinal afferent inputs, Am. J. Physiol. 246 (Regulatory Integrative Comp Physiol 15):R78-R87.

Weaver, L. C., Fry, H. K., Meckler, R. L., and Oehl, R. S., 1983, Multisegmental spinal sympathetic reflexes originating from the heart, Am. J. Physiol. 245 (Regulatory Integrative Comp Physiol 14):R345-R352.

SPINAL FACTORS IN SYMPATHETIC REGULATION

Lawrence P. Schramm

Departments of Biomedical Engineering and Neuroscience
The Johns Hopkins University School of Medicine
Baltimore, MD 21205

INTRODUCTION

The spinal cord represents the final central neural site for the generation of patterns of sympathetic activity. It is the locus of the sympathetic preganglionic neurons (SPNs), whose axons project to the sympathetic ganglia and the adrenal medulla, transmitting these patterns to final sites of neural or humoral integration. It is the locus of pathways descending upon SPNs and their associated interneurons from supraspinal systems, and it is the site of termination of visceral and somatic primary afferents which affect sympathetic activity. Finally, the spinal cord contains extensive intrinsic or propriospinal systems which may play significant roles in the regulation of sympathetic activity.

Except for the final section of this review, which deals with sympathetic regulation in animals with transected spinal cords, much of what will be described as "spinal systems" are, in fact, the spinal portions of systems which are organized at many levels of the neuroaxis: spinal cord, brainstem, hypothalamus, and even cerebral cortex. I will view the spinal sympathetic columns, their SPNs and interneurons, as the sites of convergence of information from all of these levels. Spinal sympathetic reflexes are thoroughly covered by Dr. Weaver in this volume, and I will have little to say about them here. Barman's (1984) excellent review of spinal sympathetic systems should be consulted for a somewhat different outlook on spinal sympathetic regulation.

This review will cover recent anatomical, neurophysiological and pharmacological findings relevant to spinal sympathetic regulation. In addition, I will try to extract from the literature data which are relevant to the functional specialization of SPNs. Finally, I will briefly review sympathetic regulation after spinal transection in animals (including humans).

ANATOMY OF THE SPINAL SYMPATHETIC COLUMNS

Locations of Sympathetic Preganglionic Neurons

In order to be able to electrophysiologically record from SPNs, in order to be able to study their associated circuitry, in order to be able to predict the effect of spinal cord lesions on sympathetic regulation, we must know the loci of these neurons. After 90 years of research, it is clear that the SPNs are located in the spinal cord between the caudal cervical and caudal lumbar segments. The lateral horn location of SPNs was first clearly stated by Gaskell (1886), based largely on what would now be considered the circumstantial evidence that the longitudinal distribution of the white rami corresponded best to the longitudinal distribution of the neurons of the intermediolateral column. Biedl (1895) noticed that section of the splanchnic nerves caused degeneration of neurons in the lateral horn, but he also found degeneration in the ventral horn and concluded that SPNs resided there as well.

In the twentieth century, nearly all investigators have concluded that the SPNs reside in the intermediate zone of the spinal gray matter (lamina VII of Rexed), extending, at some levels, from the region of the central canal into the lateral funiculus (see, for instance, Bok, 1928; Gagel, 1932; Petras and Cummings, 1972). The report by Lebedev, Petrov, and Skobelev (1976) of antidromic activation of neurons in the T3 and L2 ventral horns by stimulation of white rami is provocative. It suggests that SPNs may reside outside of the intermediate zone. However, this observation has not been replicated. Furthermore, the embryological work of Altman and Bayer (1984) suggests that the SPNs may arise from a generative region somewhat more dorsal than that which generates the somatic motoneurons. The presence of SPNs in the ventral horn would require a ventral migration to that region during development.

Each investigator has tended to coin his own terminology for the sympathetic columns. However, the most commonly-used terminology, and that used in this review, is that of Petras and Cummings (1972). Columns of SPNs have been identified in the lateral funiculus (ILf), in the lateral horn (ILp), in the intermediate zone between the lateral horn and the central canal (IC), and in the vicinity of the central canal (IM). The longitudinal range of the SPNs is usually thought to be from C8 through L2 or L3. However, Baron, Jänig, and McLachlan (1985a) have recently demonstrated that SPNs whose axons course in the hypogastric nerve of the cat may originate from as far caudal as the fifth lumbar segment.

Soma-dendritic Morphology of Sympathetic Preganglionic Neurons

The morphology of putative and identified SPNs has been thoroughly studied with the light microscope and to a lesser extent with the electron microscope. Most early investigators were forced to identify SPNs by their locations within the intermediate zone of thoracic or rostral lumbar spinal cord (Bok, 1928; Gagel, 1932). More recently, investigators have used the chromatolytic reaction following transection of the axons of SPNs or the retrograde transport of stainable materials from axons to SPN somas to unambiguously identify SPNs (for example Young, 1939; Petras and Cummings, 1972; Chung, Chung, and Wurster, 1975; Schramm, Adair and Stribling, 1975; Chiba and Murata, 1981). Several recent morphological studies have used electrophysiological techniques (Dembowsky, Czachurski and Seller, 1985b; Cabot, Bogan and Reid, 1985). Intracellular recordings were made from the somas of SPNs, identified by antidromic activation of their axons from white rami. Then, horseradish peroxidase (HRP) was injected intracellularly through the recording electrode to fill the cell with enzyme. Subsequent reaction of this material to stain for HRP provided a detailed, Golgi-like manifestation of the neuron.

Taken together, these studies provide a fairly complete picture of the morphology of SPNs. The neurons may be spherical or spindle-shaped. Their major dimension is rarely greater than 40 microns, and their minor dimension may be as small as 7 microns. Their Nissl substance does not usually stain as darkly as that of somatic motoneurons, but their nuclei contain prominent nucleoli. Most SPNs have between four and eight prominent dendrites which may extend over 1000 microns from the soma (Dembowsky et al., 1985b). the dendrites of SPNs in ILp (just medial to the lateral funiculus) and those dorsal to the central canal have a strong longitudinal orientation (Laruell, 1936; Rethelyi 1972; Schramm, Stribling and Adair, 1976; Barber et al., 1984; Dembowsky et al., 1985b). The rostrally- and caudally-directed dendrites of SPNs in ILp are of approximately equal length, but the former are more heavily branched (Dembowsky et al., 1985b), suggesting a preferentially heavy input to those processes.

Dembowsky et al. (1985b) reported that, in rostral thoracic cord, dendrites of SPNs tend to be restricted to the intermediolateral column. This feature originally caused Rethelyi (1972) to refer to the intermediolateral column as a "closed" nucleus, meaning that SPNs were affected only by projections which entered the column. There are several reasons to question this as a universal property of the organization of sympathetic columns. Rethelyi and many subsequent observers substantially underestimated the extent of the dendritic trees of SPNs. Certainly, the presence of a strong longitudinal orientation of some dendrites does not preclude the mediolateral orientation of others. Indeed the lateral horn neurons illustrated by Barber et al. (1984) exhibit dendrites with strong mediolateral orientations which extend into the lateral funiculus as well as to the central canal. Therefore, it would appear that most SPNs may be in a position to receive information which enters any portion of the intermediate zone as well as information conveyed by pathways coursing in the lateral funiculi. However, it does appear to be true that the dendrites of the SPNs extend very little into the dorsal or ventral horns.

Schramm et al. (1976) observed that the strong longitudinal orientation of adrenal SPNs in the thoracic spinal cord of the rat did not become fully established until well after weaning, 45 to 65 days postnatal. Previously, Rethelyi (1972) had noted that transversely oriented dendrites were more apparent in kittens than in cats, and subsequently, Phelps et al. (1984) have published photographs of horizontal sections of the spinal cord of neonatal rats stained for choline acetyltransferase in which the transverse orientation of the SPNs of the ILp is more prominent than in adult rats (Barber et al., 1984). If this adolescent reorientation is real (rather than being an artifact of plane of section), it represents a very late neural morphogenesis and occurs long after most autonomic regulation is established. Unfortunately, simple methods do not exist to correlate soma-dendritic morphology with function. Recent successes in morphological characterization of SPNs by intracellular injection of HRP raise hopes that this problem may be solved by determining the functional properties of SPNs prior to injection (see below).

Axonal Morphology

The axons of SPNs may originate either from somas or from proximal dendrites. It is particularly interesting that dendrites from which axons originate exhibit spines very near the point of origin of the axon. These spines may indicate the presence of inputs which could have a preferential effect on the excitability of the neuron (Dembowsky et al., 1985b). The axons of laterally-located SPNs, in ILf and ILp, course ventrad at the border between the lateral funiculus and the ventral horn before turning to leave with the lateral portion of the ventral root. SPNs located near the central canal often send their axons laterally where they turn ventrad near the intermediolateral column (Hancock and Peveto, 1979). The course of these axons may trace the course of a relatively late, "reverse", lateral-to-medial migration of these SPNs during development (Altman and Bayer, 1984).

Nearly all SPNs project to white rami ipsilateral to the side of the spinal cord in which they reside (Petras and Cummings, 1972; Schramm et al., 1975; Rubin and Purves, 1980; Chiba and Murata, 1981). However, at lumbar levels, some axons are derived from SPNs on the contralateral side of the spinal cord. Mediolaterally, somas of SPNs may be found from the lateral funiculus to the central canal.

As reviewed below, most sympathetic nerves receive axons from SPNs residing in a wide longitudinal range of spinal cord segments. This broad source of SPN innervation has suggested to some that these neurons might have long intraspinal courses or even branch intraspinally, and both neurophysiological and anatomical data have been presented which support these hypotheses (Deuschl and Illert, 1978; Faden, Petras and Woods, 1978; Faden, Jacobs and Woods, 1979). It is now clear from a number of studies that the axons of SPNs leave only from the segment in which their somas reside and that intraspinal branching of axons is rare or nonexistent (see, for example, Rubin and Purves, 1980; Oldfield and McLachlan, 1980; McKenna and Schramm, 1983).

There appear to be considerable species differences in the degree of myelination of the axons of SPNs. In man, apparently the axons of most SPNs are myelinated (Pick, 1970). Early electrophysiological studies in cats indicated that the majority of the SPNs had myelinated axons (Bishop and Heinbecker, 1932; Fernandez de Molina, Kuno and Perl, 1965). More recent physiological studies in the cat show axonal conduction velocities which are consistent with the conclusion that many SPNs have unmyelinated axons (McLachlan and Hirst, 1980; Jänig and Szulczyk, 1981). Recent anatomical studies suggest that as many as 40% of the axons of SPNs in the cat may be unmyelinated (Coggeshall, Hancock and Applebaum, 1976; Coggeshall and Galbraith, 1978). As discussed below, a crude relationship may exist in the cat between axon diameter or degree of myelination and the function of SPNs (Jänig and Szulczyk, 1981). In rat, it would appear that even more, perhaps most, SPNs have unmyelinated axons (Gilbey, Peterson and Coote, 1982).

There has been a long-standing interest in the possible existence of axon collaterals on SPNs. This interest has developed since the discovery of such collaterals on axons of somatic motoneurons and because some SPNs exhibit long after-spike-hyperpolarizations which resemble those mediated by collateral inhibition (see below). Rethelyi (1972) was the first to look carefully for collaterals on SPNs in well-stained Golgi preparations. He was unable to find any. Similar searches have been conducted by most subsequent investigators, usually in material in which the SPNs were visualized by the retrograde transport of HRP. For instance, McKenna and Schramm (1983) were able to place a ventral root of the isolated spinal cord of the neonatal rat directly in a concentrated HRP solution and obtain excellent uptake in axons and somas of SPNs. Nevertheless, they were unable to observe any axon collaterals. Dembowsky et al. (1985b) were unable to demonstrate collaterals on axons of SPNs which had been intracellularly injected with HRP in the cat. However, axons in their preparations did exhibit clusters of "spines." They suggested that if these spines were not artifactual, they might represent "synapses at nodes of Ranvier" (Bodian and Taylor, 1963), which could function as collaterals. Of course, confirmation that these structures are "synapses at nodes of Ranvier" will require electron microscopic studies. Also using the intracellular injection of HRP, Cabot et al. (1985) have recently reported that axon collaterals are visible in some SPNs which have been intracellularly injected with HRP in the pigeon. Whether these true collaterals are specific to avian SPNs, which differ from SPNs in other species in being located very medially in the nucleus of Terni, will need to be examined further.

The Neuropil of the Sympathetic Columns

The SPNs and their associated interneurons reside in a dense network of ascending, descending, and transversely-oriented fibers. Since the efferents of the sympathetic columns are the axons of the SPNs, and since the trajectory of these axons is relatively straightforward (see above), most of this fiber network represents the supraspinal afferents, the intraspinal afferents, and, perhaps, the primary sensory afferents to the sympathetic columns.

The neuropil of ILp and, to a lesser extent, of IM, have been studied with both the light and the electron microscope (Rethelyi, 1972; Galabov and Davidoff, 1976; Chung, Lavelle and Wurster, 1980; Tan and Wong, 1975; Wong and Tan, 1980; Chiba and Murata, 1981; Hwang and Williams, 1982). Data from these anatomical studies may be combined with those from physiological studies to form a picture of the composition of the processing network within the columns.

On anatomical grounds, Bok (1928) stressed the importance of afferent projections to ILp from interneurons near the central canal. Indeed, excellent neurophysiological evidence has been obtained for such interneurons by McCall, Gebber, and Barman (1977), who attributed to them a role in baroreceptor inhibition of the SPNs. Important as these interneurons and their projections might be, they may not be numerous, at least in the cat. McCall et al. (1977) did not find many of these neurons, and Rethelyi (1972) saw relatively few fibers entering the ILp from medial regions of the cord. Instead, Rethelyi felt that the majority of input to ILp came from the lateral funiculus. These fibers stream into the column and rapidly turn longitudinally to course parallel to the longitudinally-oriented dendrites in what Chiba and Murata (1981) referred to as a "gelatinous core" of largely unmyelinated fibers. Most authors agree that the ILp contains few myelinated fibers, and that the "gelatinous core" contains many axons which make "en passant" synapses with dendrites of SPNs. Since the conduction velocities of many propriospinal fibers (Kirchner, Wyszogrodski and Polosa, 1975; Laskey, Schondorf and Polosa, 1979) and descending fibers (Morrison and Gebber, 1985; Barman and Gebber, 1985; and see Dembowsky et al., 1985a for review) suggest that they are myelinated, these fibers must lose their myelin as they turn into the ILp.

What are the sources of these lateral funicular fibers? Conceivably, some of them are primary afferents which course ventrad at the border of the dorsal horn and the lateral funiculus (see below). Many more are likely to be collaterals of ascending secondary and tertiary sensory neurons, propriospinal fibers (Chung and Coggeshall, 1983), and fibers

descending from supraspinal systems. These latter projections are of
particular importance in the normal regulation of sympathetic activity
(Loewy, Wallach and McKellar, 1981), and their electrophysiological
characterization is well under way. Caverson, Ciriello and Calaresu
(1983) have mapped the ventral medullary nuclei of origin of some of these
axons, and they have estimated their conduction velocities to be around 19
m/s. Barman and Gebber (1985) have described the origins, funicular
trajectories and collateralization of other, much more slowly conducting
(3.5 m/s) medullospinal pathways to the ILp. Dembowsky et al. (1985a)
have described, on the basis of conduction velocities, five pathways in
the dorsolateral funiculus of the cat which, when electrically stimulated,
mediate EPSPs in SPNs.

The controversy over the "correct" conduction velocities for
medullospinal sympathomodulatory pathways has been resolved only partially
by Dembowsky et al. (1985a). The **range** of pathway conduction velocities
reported by them corresponded well to the **range** of conduction velocities
reported by laboratories for descending pathways exciting SPNs. In fact,
using intraspinal stimulation, they described a pathway whose conduction
velocity was similar to that reported by Caverson et al. (1983). However,
Dembowsky et al. (1985a) describe unpublished preliminary experiments in
which latencies for EPSPs in SPNs, elicited by stimulation of the rostral
ventral medulla itself, correspond to conduction velocities intermediate
between those reported by Caverson et al. (1983) and Barman and Gebber (1985).

Do SPNs receive direct input from primary sensory neurons? Until
recently, the consensus has been that they did not, and that primary
inputs projected via "secondary visceral gray" or related portions of the
dorsal horn in Rexed's lamina VI. Recent studies, using the
transganglionic transport of HRP, have shown that primary afferents may
well project to within the range of dendrites of SPNs in both the
intermediolateral column and the central autonomic nucleus (Ciriello and
Calaresu, 1983; Kuo et al., 1983; Cervero and Connell, 1984).

Sympathetic Interneurons among Preganglionic Neurons

To what extent do the spinal sympathetic nuclei contain interneurons?
This very difficult question has probably been most directly approached by
Oldfield and McLachlan (1980) in the intermediolateral column of the cat.
They demonstrated HRP in 90% of the neurons in that column after the white
ramus of that segment had been exposed to the enzyme. Since even this 90%

probably underestimated the SPNs projecting through the ramus, it would seem that interneurons represent a relatively small proportion of the total number of neurons in the intermediolateral column.

Nevertheless, electrophysiological studies suggest that the ILP contains neurons the activity of which is strongly related to sympathetic activity, which cannot be antidromically activated from the appropriate white ramus, and which discharge in bursts at frequencies above those attainable by SPNs (Gebber and McCall, 1976). These neurons are likely candidates to be sympathetic interneurons. Their repetitive discharge may be responsible for the large summation EPSPs observed in response to some inputs by Dembowsky et al. (1985a). The scarcity of these interneurons may not be relevant.

Synaptic Input

Several investigators have used the electron microscope to examine the synaptic input to SPNs in detail (Rethelyi, 1972; Chung et al., 1980; Wong and Tan, 1980; Chiba and Murata, 1981; Hwang and Williams, 1982). Only an overview of the major observations can be given here. The heterogeneity of synaptic vesicles in the intermediolateral column, first observed by Rethelyi (1972), has been amply confirmed by all subsequent workers. However, not all have found the same distribution of vesicle types among the same types of synaptic endings. All observers have found spherical-agranular vesicles, large dense-cored vesicles, and flattened vesicles. Also, all authors have observed both axosomatic and axodendritic synapses containing all three types of vesicles. A particularly notable species difference is that axoaxonal synapses are said to be absent in the ILp of the monkey, although they are present in the rat (Wong and Tan, 1980; Tan and Wong, 1976). One wonders whether such a fundamental morphological difference could occur without manifesting fundamentally different spinal processing of sympathetic activity in the two species.

Chung et al. (1980) reported that 39.6% of dendritic surface and 12.3% of somatic surface was invested with synaptic endings. Endings with spherical vesicles were more likely to be found on dendrites, and endings with flattened vesicles were more likely to be found on soma. If, as is commonly supposed, flattened vesicles manifest inhibitory synapses, this observation would suggest that inhibitory synapses are more likely on soma than on dendrites.

Synaptic glomeruli appear to be common phenomena in the intermediolateral column in both rat and monkey (Tan and Wong, 1976; Wong and Tan, 1980). These glomeruli contain one or several dendrites and multiple axons with synaptic endings of various types. This synaptic complex is encapsulated by astrocytic processes. The functional significance of these glomeruli is not known. However, it seems clear that they form the substrate for interactions between multiple synaptic inputs to the SPNs.

A curious attribute of the neuropil of the ILp is the close apposition of processes which course dorsad from the ventral horn to the somas and dendrites of SPNs (McKenna and Schramm, 1983). These processes could be ventral root afferents. However, ventral root afferents in more caudal portions of the spinal cord are thought to course fairly directly to the dorsal horn, where they terminate on secondary sensory neurons (Light and Metz, 1978). Alternately, these dorsally-coursing processes could be dendrites of somatic motoneurons. Perhaps dendrites of motoneurons and SPNs are closely apposed in the intermediate zone because they share inputs which terminate there. Perhaps dendrites of motoneurons and SPNs have synaptic relations in the intermediate zone whose functions are, as yet, unknown.

Perspective on Spinal Sympathetic Anatomy

About the time that we felt that the "gross" anatomy of the SPNs had been worked out, investigators began to intracellularly inject these neurons with HRP, and new and important features became evident. Previously invisible axon collaterals, axonal "spines" and periaxonal dendritic spines all enriched the SPNs' repertoire of potential processing mechanisms. It is probably characteristic of neuroscience today that these anatomical discoveries were made by investigators who would have been considered "physiologists" a few years ago. Hopefully, this new information will revitalize the study of the sympathetic columns by "anatomists", especially at the electron microscopic level. The new features demonstrated by intracellular injection must be confirmed at an ultrastructural level which will permit an analysis of synaptic relationships.

The morphogenesis of SPNs raises a number of possibilities for future research. Most regulations involving the sympathetic nervous system are established, at least in rats, during the preweaning, postnatal 21 days.

What are the morphological and pharmacological concomitants of this maturation? Neurophysiological and, especially, neuroanatomical techniques can be applied to animals this age. Indeed, the young rat's resistance to hypoxia recommends it as an excellent experimental subject for such studies. The apparent late morphogenesis of SPNs, lasting until the onset of sexual maturity, is a great puzzle because it does not correlate with the maturation of any well-defined metabolic process. Nevertheless, it should be possible to design experiments which might demonstrate such processes.

Although the sympathetic ganglia represent an integral part of the sympathetic nervous system and appear to be capable of considerable processing, they also prevent localization of SPNs which affect given organs or tissues because neither electrical activity nor tracers can move antidromically across them. Perhaps the greatest boon to the study of functional sympathetic anatomy would be the development and application of retrograde, trans-synaptic, transganglionic transport techniques. We need to find readily-stainable materials which will be transported from the periphery to the SPNs. A further improvement on this (as yet unavailable) technique would be materials whose transport was proportional to the activity of the neurons. Such materials would permit functional as well as anatomical mapping.

NEUROPHYSIOLOGY OF SYMPATHETIC PREGANGLIONIC NEURONS

Technical Considerations

It has proven difficult to record the electrical activity of either the somas or the axons of sympathetic preganglionic neurons (SPNs). The somas of the neurons, because of their small size and (often) spindle shape, make difficult targets for microelectodes. To make matters worse, the spinal cord is difficult to stabilize mechanically. Thus, these small targets are often small moving targets. Compared to the long, accessible route of axons of somatic motoneurons, the course of the axons of the SPNs through the white rami is inconvenient, either for stimulation of antidromic potentials (for the identification of SPN somas within the spinal cord) or for direct axonal recording. Finally, it now seems clear that recordings both from axons of SPNs (Jänig and Szulczyk, 1981) and from somas of SPNs (Dembowsky et al., 1985a) favor larger neurons and may, therefore, bias interpretations. Nevertheless, a substantial number of extracellular recordings have now been made from SPNs, both within the spinal cord and in the periphery (see, for example, Polosa, 1968; Jänig and Schmidt, 1970; Seller, 1973; Gilbey et al., 1982; McKenna and Schramm, 1983; Araki, Ito, Kurosawa and Sato, 1984). At the time of writing, there have been only six complete reports on **intracellular** recordings from SPNs (Fernandez de Molina et al., 1965; Coote and Westbury, 1979; McLachlan and Hirst, 1980; Yoshimura and Nishi, 1982; Dembowsky et al., 1985a,b).

Ongoing Activity of Sympathetic Preganglionic Neurons

Most authors report that between 10 and 30% of SPNs exhibit ongoing activity, depending on the animal, the anesthetic, or the preparation. There is considerable agreement that the ongoing discharge rate of uninjured SPNs is low, ranging between 0.1 and 6 Hz. In two series of experiments, SPNs exhibited no ongoing activity (Fernandez de Molina et al., 1965; Yoshimura and Nishi, 1982). The experiments of Fernandez de Molina et al. (1982) were conducted in acutely spinalized cats. The spinal shock exhibited by this animal may account for the lack of spontaneously active SPNs in that study, although several studies have subsequently been conducted in cats after acute spinal transection in which SPNs have exhibited substantial ongoing activity (see below). The failure of Yoshimura and Nishi (1982) to observe spontaneously active SPNs may be explained by the extreme isolation of their spinal cord slice preparation. It seems likely that insufficient afferents to SPNs remained in their thin transverse slices of spinal cord to sustain ongoing discharge.

In this context, it is interesting to note that, although McKenna and Schramm (1983) found that 18% of the SPNs in their isolated spinal cord preparation were spontaneously active, the average discharge rate of these SPNs was lower than that reported by many investigators in more intact preparations (1 Hz compared to approximately 3 Hz), and the maximum discharge rate was much lower than that reported in other preparations (2 Hz compared to approximately 6 Hz.). As discussed below, ongoing discharge rates are probably determined both by synaptic input and the inherent properties of SPNs. Maximum ongoing discharge rates would appear to be determined more by synaptic inputs than by inherent properties of the SPNs since many of these neurons can be driven by antidromic stimulation to discharge at rates of 20 to 50 Hz.

Intracellular recordings show no evidence of ongoing depolarizations or "pacemaker" potentials (Coote and Westbury, 1979; McLachlan and Hirst, 1980; Yoshimura and Nishi, 1982: Dembowsky et al., 1985a). Instead, SPNs are bombarded at irregular intervals with many EPSPs and occasional IPSPs. The average rate of arrival of identifiable, single-step EPSPs in intact anesthetized cats ranged from 0.9 to 8.5 IPSPs/s (Dembowsky et al., 1985a). Not only did single-step EPSPs occur in bursts, but summation EPSPs, consisting of up to 7 steps, were common. These multiple-step EPSPs could have resulted from the near coincidence of several excitatory inputs or from the multiple discharge of an excitatory interneuron (Gebber and McCall, 1976). EPSPs had durations which ranged from 25 to 45 ms and amplitudes which approached 20 mV for summation EPSPs. The threshold for discharge of SPNs appeared to range between 3 and 10 mV in different SPNs (Mclachlan and Hirst, 1980).

The occurrence of IPSPs appears to be rare in SPNs, and only a few have been tested for reversal by increasing intracellular chloride (Dembowsky et al., 1985a). The IPSPs which have been convincingly demonstrated so far have been evoked by electrical stimulation of either visceral or somatic afferents or by intraspinal stimulation. the amplitudes of these IPSPs ranged from 1 to 9 mV, and their durations ranged from 40 to 150 ms.

Does the relative rarity of IPSPs mean that the inhibition of discharge of SPNs is mediated largely by disfacilitation (reduction in the arrival of EPSPs)? Certainly hyperpolarization of SPN membranes can be observed as an appropriate concomitant of baroreceptor excitation (McLachlan and Hirst, 1980). There is still a chance that failure to observe IPSPs may be a consequence of experimental methodology. Dembowsky et al. (1985a) noted that stimulation which led to generation of IPSPs did not reduce either arterial pressure or overall sympathetic activity. Perhaps their electrode locations or stimulation parameters were not appropriate for inhibitions mediated by IPSPs. Further, most intracellular studies in which IPSPs have been sought have been conducted under chloralose anesthesia, which may diminish IPSPs. On the other hand, Yoshimura and Nishi (1982) conducted their experiments in a spinal cord slice preparation with no permanent anesthesia, and they too failed to see IPSPs. Finally, it is possible that IPSPs occur at a rate much greater than they appear to and that they are simply obscured by the bombardment of EPSPs. One of the more exciting directions for future research will be to determine the relative importance of direct inhibition and disfacilitation of SPNs in the normal regulation of sympathetic activity.

Characteristics of Action Potentials of Preganglionic Neurons

The earliest intracellular recordings from SPNs exhibited a somewhat long-duration (7 ms) action potential (Fernandez de Molina et al., 1965). Action potentials reported by McLachlan and Hirst (1980) had similar durations of 8 ms. On the other hand, Coote and Westbury (1979) and Yoshimura and Nishi (1982) observed action potentials with approximately half these durations. Perhaps these differences simply represent a sampling problem since Dembowsky et al. (1985a), in a very large sample of SPNs, reported durations which ranged between 3 and 12ms.

All workers using intracellular recordings have reported after-spike-hyperpolarizations (AHs) ranging in duration from 8ms to 7s! The amplitudes of the AHs may be as large as 17 mV, more than doubling the maximum depolarization necessary for generating an action potential and preventing the generation of high frequency, repetitive action potentials. In fact, AHs appear to be a primary factor in regulating the frequency of ongoing sympathetic activity (Mannard, Rajchgot and Polosa, 1977; Polosa, Schondorf and Laskey, 1982).

A short AH, lasting between 8 and 200 ms in the experiments of Yoshimura and Nishi (1982), was not sensitive to cholinergic blocking agents. These data can be taken as evidence that the short AH is not generated by a conventional Renshaw-type collateral inhibition. Yoshimura and Nishi also observed a long AH which lasted between 1 and 7 **seconds**. This AH occurred after a soma-dendritic action potential but not after an initial segment potential. The long AH was associated with a decrease in membrane resistance, and its polarity reversed if the membrane potential was driven more than 90 mV negative. Of greatest interest was the observation that the long AH was abolished during perfusion with calcium-free solutions. Based on the 90 mV reversal potential, these authors conclude that the long AH may be mediated by a calcium-dependent change in the potassium conductance.

Dembowsky et al. (1985a) described a long-lasting AH which, with a duration of 200 to 500 ms, falls at the upper range of the fast AHs observed by Yoshimura and Nishi. However, unlike the AHs observed by earlier workers, these required only a preceding EPSP rather than a soma-dendritic action potential. The magnitudes of these AHs depended on the magnitudes of the preceding EPSPs, and they ranged from 1 to 5 mV. Neither membrane hyperpolarization (effected by passing current through the recording electrode) nor increasing intracellular chloride concentration altered the magnitudes of these AHs. Dembowsky et al. argue convincingly that these AHs represent disfacilitation rather than inhibition, although they point out that conductance measurements would be necessary to answer this question with certainty. Nevertheless, the normally high intensity of ongoing excitatory input to SPNs, documented by these workers, lends credibility to the potential efficacy of disfacilitation as an inhibitory process.

Collateral Inhibition of Sympathetic Preganglionic Neurons

Are SPNs subject to collateral inhibition as are some of their somatic sisters? The evidence for and against collateral inhibition of SPNs has recently been carefully reviewed by Lebedev, Petrov and Skobelev (1980) and by Polosa et al. (1982), and earlier studies will not be considered again here. Despite thorough descriptions of families of long and short after-spike-hyperpolarizations (see especially McLachlan and Hirst, 1980; Yoshimura and Nishi, 1982; and Dembowsky et al., 1985a) which could account for discharge behavior attributable to collateral inhibition, the issue remains alive due to several recent observations. As described above, axon collaterals have been observed on some SPNs in the pigeon using state-of-the-art intracellular HRP injection techniques (Cabot et al., 1985). In cats, similar intracellular injection techniques demonstrate clusters of spines on axons which are candidates for "synapses at nodes of Ranvier" (Bodian and Taylor, 1963). These synapses are functionally identical to axon collaterals. Thus, one of the major objections of previous critics of axon collateral hypotheses (that these collaterals had not been demonstrated) has been rebutted. Further, McKenna and Schramm (1985) presented evidence consistent with a collateral mechanism for postexcitatory sympathoinhibition (the "sympathetic silent period"), and they suggested that a co-transmitter in SPNs, not acetylcholine, might mediate this effect.

Collateral inhibition, if it functions in the sympathetic neuropil, could accomplish more than simple regulation of the ongoing discharge rate of SPNs. As pointed out by Gebber and Barman (1979), it could play an important role in interactions between functional groups of SPNs. With the advent of studies combining intracellular recording and the intracellular injection of HRP, it should now be possible to detect actual or functional collaterals, determine their incidence, and relate their occurrence to the functional properties of SPNs.

Perspective on Sympathetic Preganglionic Neurophysiology

Basic somato-sympathetic and viscero-sympathetic "reflexes" are now adequately described. The key word for the future must be "regulation." More emphasis is now being placed on the analysis of sympathetic responses to ongoing conditions or states rather than responses to transient, often artificial, stimuli, and this is appropriate.

We need to have twenty laboratories, not five, recording intracellularly from SPNs. Where are the IPSPs in SPNs? Do they mediate only certain kinds of sympathoinhibition? Are they necessary at all? What mechanisms generate the various after-spike-hyperpolarizations (AHs) in SPNs? Are some of them mediated by collateral inhibition? Why do some appear after depolarization of the initial axon segment and others only after a complete soma-dendritic action potential? Is there a relationship between these AHs and the sympathetic silent period? Surely, we do not want for work!

PHARMACOLOGY OF SYMPATHETIC PREGANGLIONIC NEURONS

In this section, I will review some of the evidence for the physiological functions of epinephrine and norepinephrine, substance P, serotonin, and a small assortment of amino acids and peptides. This review cannot be exhaustive, but it will serve to demonstrate some of the kinds of questions being asked by central autonomic pharmacologists and physiologists. It will also demonstrate some of the techniques being used to answer these questions.

Epinephrine and Norepinephrine

Probably no class of putative transmitters has attracted the attention of more central autonomic pharmacologists than the catecholamines. There are several reasons for this interest. Catecholamines and their receptors are particularly dense within the sympathetic columns (Carlson et al., 1964; Dahlstrom and Fuxe, 1965;

Glaser and Ross, 1980; Hwang and Williams, 1982; Unnerstall et al., 1984; Seybold and Elde, 1984; Dashwood et al., 1985). Both catecholaminergic terminals and receptors are localized among the clusters of SPNs in the ILp (Ross et al., 1984a: Unnerstall et al., 1984), and the persistence of heavy binding of an alpha-2-adrenergic ligand around SPNs after chronic spinal transection in the rat suggests that the associated receptors are postsynaptic (Unnerstall et al., 1984). Electron microscopic data are also consistent with heavy catecholaminergic innervation of SPNs.

Functional data also point to an important role for catecholamines in sympathetic regulation. Coote and MacLeod (1974, 1975) presented evidence that monoaminergic pathways are important in mediating supraspinal sympathoinhibitory effects, including those elicited by baroreceptor activation (Coote et al., 1981). In a recent series of studies, reviewed in this volume, Reis and coworkers have shown that adrenergic neurons in the C1 region of the rostral ventrolateral medulla project heavily upon SPNs. They have also shown that activation of these neurons **excites** sympathetic activity, and that destruction of this region produces **hypotension** (Ross et al. , 1981; Reis et al., 1984; Ross et al., 1984a,b). Noradrenergic projections exist from pontine nuclei (Loewy, McKellar and Saper, 1979; Neil and Loewy, 1982), and stimulation of these neurons with glutamate apparently inhibits ongoing sympathetic activity.

If the data from studies using stimulation and lesions of supraspinal systems are conflicting as to whether catecholamines excite or inhibit sympathetic activity, electrophysiological and pharmacological studies are, for the most part, in agreement that epinephrine, norepinephrine, and clonidine inhibit the activity of SPNs (DeGroat and Ryall, 1967; Hongo and Ryall, 1966; Coote et al., 1981a; Kadzielawa, 1983b; Guyenet and Cabot, 1981). These data are consistent with the hypothesis that supraspinal adrenergic and noradrenergic systems generally inhibit sympathetic activity. How can they be reconciled with an excitatory role for these compounds? One explanation might be that receptors mediating inhibition reside near the soma, and receptors mediating excitation reside more distally on dendritic trees. Since recording electrodes probably must be near soma to record action potentials, release of catecholamines by the nearby iontophoretic electrode would have a primarily inhibitory effect. This hypothesis is supported by the general feeling among neurobiologists that inhibitory synapses are abundant on soma. It is also supported by the observation of Yoshirmura and Nishi (1982) that superfusion of slices of thoracic spinal cord with norepinephrine, presumably affecting both dendritic and somatic receptors, depolarized (in other words, excited) SPNs. On the other hand, intravenous clonidine might also be expected to reach receptors on both somas and dendrites. Yet, clonidine has been found to inhibit both ongoing and evoked sympathetic activity at a spinal level _via_ all routes of delivery (see, for example, Guyenet and Cabot, 1981; Franz and Madsen, 1982; Franz, Hare, and McClosky, 1982; Kurosawa, Minami, Togashi and Sato, 1985; Baum and Shropshire, 1977; Krier, Thor and DeGroat, 1979).

Another potential resolution of the discrepancy between the local action of epinephrine on the activity of SPNs and that predicted by the effects of activation of brainstem catecholaminergic nuclei is provided by the recent work of Lorenz _et al_. 1985). These investigators used immunohistochemical techniques to co-localize phenylethanolamine-N-methyltransferase (PNMT), a synthetic enzyme for epinephrine, and substance P within the same neurons in the rostral ventrolateral medulla. As discussed below, substance P excites SPNs. These authors suggest that the excitatory effect of these "adrenergic" neurons might be through the release of their co-transmitter, substance P.

Substance P

Substance P is heavily localized in the sympathetic columns (see, for example, Hancock, 1982; Takano and Loewy, 1984; Davis _et al_., 1984; Oldfield, Sheppard and Nilaver, 1985; Charlton and Helke, 1985). Substance P appears to excite SPNs. Direct application by iontophoresis of substance P increases the firing rate of discharging SPNs or of previously inactive SPNs which have been caused to discharge by application of glutamate (Gilbey _et al_., 1983; Backman and Henry, 1984). However, substance P is not capable of exciting silent SPNs to discharge.

If the effect of substance P on the activity of SPNs seems fairly clear, the source of the substance P-containing terminals which so heavily invest the SPNs has been controversial. Because primary afferents, especially small caliber ones, are known to contain substance P, it might be expected that dorsal rhizotomy would reduce the levels of this material in the intermediolateral column. However, Takahashi and Otsuka (1975) showed that dorsal rhizotomy had little effect on substance P levels in lateral horn, although it reduced dorsal horn levels of substance P by 90%.

Most authors now agree that the neurons of the ventrolateral medulla represent a major source of the substance P innervation of SPNs in the intact spinal cord. Helke et al. (1982) greatly reduced spinal levels of substance P with ventral medullary lesions. That excitatory, substance P-mediated responses may be elicited from these same regions is strongly supported by the fact that cardiovascular responses to the excitation of ventral medullary regions can be blocked by spinal administration of substance P antagonists (Loewy and Sawyer, 1982; Keeler and Helke, 1985).

Although the brainstem source and functional role of substance P seem to agree well, there are still a number of mysteries to resolve. The recent work of Davis et al. (1984) addresses some of these. These investigators examined the concentration of substance P-like material in the intermediolateral cell column after rostral cervical hemisections and after complete caudal cervical or thoracic transections. Confirming the work of Helke et al. (1982), they found that cervical hemisections reduced the substance P in the intermediolateral column. However, complete transection either had no effect on substance P concentrations in the sympathetic nuclei or it increased these concentrations. They concluded that some form of intraspinal regulation maintains substance P levels after transection. This intriguing observation raises the question of why this regulatory mechanism did not maintain substance P levels after the less complete cervical lesions. Nevertheless, the data suggest that the concentration of this excitatory putative transmitter may not decrease (and may even increase) after spinal cord transection. Such an increase in substance P might provide one mechanism for the hyper-reactivity of spinal sympathetic systems after spinal cord injury.

Serotonin

The role of serotonin in regulating sympathetic activity has been controversial, but this controversy appears to be yielding to newer techniques and to more sophisticated conceptions of the nervous system. Two points of apparently universal agreement are that SPNs, or at least the sympathetic nuclei, receive a dense serotonergic input (Dahlstrom and Fuxe, 1965; Loewy, 1981; Loewy and McKellar, 1981; Holets and Elde, 1982; Howe et al., 1983; Hadjiconstantinou et al., 1984), and that iontophoretically-applied serotonin is excitatory to SPNs (DeGroat and Ryall, 1967; Ryall, 1967; Coote et al., 1981a; Kadzielawa, 1983a; McCall, 1983; Yoshimura and Nishi, 1982). In fact, iontophoretic application of serotonin antagonists not only blocked the excitatory effect on SPNs of iontophoretically-applied serotonin (Kadzielawa, 1983a; McCall, 1983), but it inhibited the ongoing activity of SPNs in intact (although not in spinal) cats (McCall, 1983).

Despite unanimity about the direct effects of serotonin on SPNs, there has been confusion about the functional significance of serotonergic systems. Much of this confusion has arisen from conflicting results of experiments which have studied the sympathetic or cardiovascular effects of electrical stimulation or destruction of putative serotonergic nuclei or pathways. Often, stimulation or destruction of these systems has suggested a sympathoinhibitory role for descending serotonergic systems (Neumayr et al., 1974; Adair et al., 1977; Cabot, Wild and Cohen, 1979; Coote and Gilbey, 1980; Gilbey et al., 1981). These results conflict with the excitatory effects of iontophoretic application of serotonin to SPNs. They also conflict with the sympathoexcitatory effects of stimulation of the same medullary regions to which others attribute sympathoinhibition (see, for example, Kuhn, Wolf and Lovenberg, 1980; Howe et al., 1983). However, the careful pharmacological studies of McCall (1983, 1984) show that sympathoexcitation produced by stimulation of putative serotonergic brainstem nuclei is blocked by serotonin antagonists while sympathoinhibitions elicited from these same nuclei are not.

The clearest explanation of these conflicting results is that the nuclei containing the serotonergic neurons are heterogeneous with respect to their transmitters, and that experiments which have demonstrated sympathoinhibition from these nuclei have done so by exciting non-serotonergic sympathoinhibitory systems.

Another explanation is suggested by the work of Hokfelt et al. (1978). They found substance P co-localized with serotonin in brainstem nuclei, much the way Lorenz et al. (1985) found substance P co-localized with PNMT. It is tempting to hypothesize that serotonin and substance P could inhibit and excite SPNs respectively, and the net effect of activating these multi-transmitter neurons would depend on the relative amounts of serotonin and substance P released or on the properties of the receptors on the SPNs. However, the consistent finding that iontophoretic application of serotonin excites SPNs makes this hypothesis very unlikely.

One must also consider the sources of sympathetic activity in different preparations. Some of the studies cited above examined the effect of putative serotonergic systems on sympathetic activity elicited by stimulation of somatic or visceral afferents. In other cases, effects were studied on "spontaneous" sympathetic activity, which may, to varying extents, have been generated by nociceptors in these surgically-prepared animals. The regions stimulated are known to be the origins of powerful **sensory** modulatory systems (Basbaum, Clanton and Fields, 1978; Basbaum and Fields, 1979), and some of these systems are certainly serotonergic (see Duggan, 1985, for review). Therefore, sympathoinhibition could be produced by stimulation of serotonergic systems which diminish afferent input, input which drives sympathetic activity.

Other Putative Transmitters

Pharmacological evidence exists for actions of a sizable number of substances on sympathetic activity at spinal levels. However, the site of action of some of these substances is uncertain, and the following discussion will be limited to substances which have been applied iontophoretically to identified SPNs.

Not surprisingly, glutamate excites most SPNs (Gilbey et al., 1982, 1983; Yoshimura and Nishi, 1982; Backman and Henry, 1983a). On the other hand, several investigators have noticed that some SPNs originally excited by glutamate become silent with continued application of the amino acid, and some SPNs will not discharge at all in response to iontophoresis of glutamate. Backman and Henry (1983a) found that 30% of the SPNs they observed could not be excited by glutamate. However, glutamate did decrease the magnitude of the antidromically-excited action potential in these neurons. Therefore, glutamate probably caused depolarization block in these SPNs instead of causing them to discharge. Similar effects were seen upon iontophoretic application of aspartate to SPNs (Backman and Henry, 1983a; Yoshimura and Nishi, 1982).

Both GABA and glycine dramatically inhibit the ongoing activity of SPNs, and these effects are antagonized by bicuculline and strychnine respectively (Backman and Henry, 1983b). That the effects of these two putative transmitters were postsynaptic was strongly suggested by the observation that each was able to prevent the soma-dendritic invasion of an antidromically-conducted action potential.

Modern reports of direct projections from the hypothalamus to the thoracic spinal cord (Smith, 1965; Kuypers and Maiskey, 1975; Saper et al., 1976) have raised interest in the possibility that hypothalamic hormones might act as transmitters upon SPNs (see Swanson and Sawchenko, 1983, for review). Several recent studies have pursued this possibility by iontophoretically applying some of these substances to SPNs. Gilbey et al. (1982) showed that electrical stimulation in the paraventricular nucleus and iontophoretic application of both oxytocin and vasopressin inhibited the ongoing activity of SPNs. Vasopressin had a more variable effect in that it also excited about one third of the neurons to which it was applied. Backman and Henry (1983c) found that thyrotropin-releasing hormone had a mild excitatory effect on SPNs, similar to that exhibited by substance P (see above).

Ever since Beattie, Brow, and Long (1930) suggested (on the basis of no-longer-acceptable evidence) the existence of direct hypothalamo-sympathetic projections, this possibility has fascinated physiologists. Part of this fascination is derived from the apparently unorthodox projections directly from the diencephalon to the spinal cord without synapse in the pons and medulla. Probably even more fascinating than their unconventionality are the possibilities which they suggest for interaction between neuroendocrine and sympathetic regulatory mechanisms. That the transmitters mediating these interactions appear to be substances which previously have been thought to subserve primarily neuroendocrine function makes these pathways especially intriguing.

Perspective on Central Sympathetic Pharmacology

One of the most perplexing problems in determining the action of putative transmitters on SPNs has been differences between the effects of direct iontophoresis onto SPNs and the effects of manipulations (stimulation, lesions) of nuclei or pathways thought to use those transmitters. Some of these problems are being solved. The limitations of iontophoresis are now better recognized. We also know that just because a nucleus contains neurons which manifest one transmitter, this does not preclude the presence of other neurons which use other transmitters in that same nucleus. What should we think about neurons with multiple putative transmitters? This is a question being asked throughout neurobiology, and, judging from past experience, it will have different answers from different subdisciplines. We can hope that these answers reduce, rather than compound, our confusion.

Certainly, central sympathetic pharmacology shares with all of neuropharmacology the quest for mechanisms of transmitter action. Only by understanding these mechanisms can we truly understand similarities and differences between actions of putative transmitters. This understanding will also provide us with a rationale for the design of compounds which might affect central sympathetic processing.

FUNCTIONAL SPECIFICITY OF PREGANGLIONIC NEURONS

Few would deny that the sympathetic nervous system plays a role in metabolic regulation. Few would deny that the sympathetic nervous system shares this role with local metabolic regulation. However, few would venture to estimate the relative importance of sympathetic regulation and local regulation in many metabolic states. If we wanted to test what is, for the autonomic neurobiologist, the most optimistic hypothesis, i.e.; that the sympathetic nervous system plays a constant role in what Bell (1830) called "those thousand secret operations of a living body which may be called constitutional," then our data would be woefully inadequate.

We are severely constrained by both our methods and by our outlook. We savor observations that, in response to some manipulation or other, sympathetic activity to the kidney changes in one direction while that to another organ changes in the opposite direction, or does not change at all. In other words, we are still at the level of **organ** specificity, not **metabolic** specificity. We know, for instance, that both the renal blood vessels and the renal tubules are innervated, and that renal circulatory and renal tubular effects can be separately elicited by altering the parameters with which we stimulate the renal nerves (see DiBona, 1982, for review). Are there separate SPNs exciting the renal circulatory and renal tubular innervation? Under physiological conditions, are these two categories of SPNs separately excited depending on metabolic needs?

In the material that follows, I have tried to cull from the literature some of the data which bear on the existence of functionally specific SPNs. Like the investigator trying to impale a SPN with a microelectrode, I too am trying to hit a moving target. The term "function" might sometimes refer to a mass sympathoadrenal response. At other times, it might refer to the constellation of cardiovascular adjustments which precede exercise. Differential changes in sympathetic activity to several organs in response to an experimental stimulus might constitute a "function," although we might not know if these differential changes occur normally. Our most serious conceptual limitation is that we

often find ourselves thinking in 3-dimensional, **anatomical** space instead
of n-dimensional, **metabolic** space. This limitation will be clear in the
discussion which follows.

Loci of SPNs as Manifestations of Functional Specificity

Many of us who have studied the longitudinal distribution of SPNs
projecting to a peripheral nerve or organ (such as the cervical
sympathetic chain or the adrenal gland) have hoped to be able to discern
evidence of functional specificity in our data. In retrospect, it is easy
to see that the interpretation of these results would be limited. For
instance, the cervical sympathetic trunk and the greater splanchnic nerve
are each functionally heterogeneous. What could we conclude about
functional specificity by tracing their cells or origin? Furthermore,
these, and most other preganglionic sympathetic nerves, contain axons of
SPNs which project through the sympathetic chain from many spinal levels.
Usually, tracing the origins of a given sympathetic nerve results in only
a crude mapping. For instance, axons in the cervical sympathetic chain
are derived form C8 to T5 (Rando, Bowers and Zigmond, 1981). Those
innervating the stellate ganglion are derived from C8 to T8 (Chung et al.,
1979; Dalsgaard and Elfin, 1981; Pardini and Wurster, 1984). The greater
splanchnic nerve contains a mixture of preganglionic and postganglionic
axons (Celler and Schramm, 1981), and the preganglionic axons originate
from T1 to T13 (Kuo et al., 1980; Torigoe et al., 1985). Axons of the
hypogastric nerve are derived from T10 to L5 (Neuhuber, 1982; McLachlan,
Baron and Jänig, 1984; Baron, Jänig and McLachlan, 1985a; McLachlan,
1985).

Few of us are surprised to learn that cerebral vessels receive their
sympathetic input from caudal cervical or rostral thoracic spinal cord and
that colonic vessels receive their sympathetic input from lumbar spinal
cord. Discouraging as these experiments are in the information they have
yielded on the **functional** properties of SPNs, they form an important
anatomical basis for future experiments which will use neurophysiological
or metabolic mapping techniques to characterize functionally-specific
SPNs. They will tell investigators where to look in order to have the
greatest probability of finding the appropriate neurons.

Somewhat more has been learned about functional specificity by
studying the destinations of SPNs located mediolaterally, in different
sympathetic columns. Both Petras and Faden (1978) and Hancock and Peveto

(1979) presented evidence that the SPNs innervating paravertebral ganglia are likely to be located more laterally, in the ILp of the intermediate zone. Those SPNs innervating the prevertebral ganglia are more likely to be located medially, in the IC or IM columns. Similarly, SPNs projecting to paravertebral ganglia which innervate limb vasoconstrictors are located more laterally in lumbar ILp (McLachlan, Baron and Jänig, 1984), whereas those coursing in the hypogastric nerve are located just medial to the main portion of the ILp in cat (Baron et al., 1985a) or very medially, just dorsal to the central canal, in the rat (Hancock and Peveto, 1979).

The adrenal medulla is the only organ which receives a significant direct projection of SPNs. Although the functions of these SPNs are probably heterogeneous, corresponding to the functional heterogenity of the adrenal medulla itself, they have long been the targets of those seeking clues to the functional organization of sympathoadrenal output (Young, 1939; Cummings, 1969; Schramm et al., 1975; Haase, Contestabile and Flumerfelt 1982; Holets and Elde, 1982; Seybold and Elde, 1984). There seems to be almost universal agreement that the adrenal SPNs are located ipsilateral to the innervated adrenal gland. There is also agreement that SPNs from between T1 and T13 innervate the adrenal gland. However, the majority of the adrenal SPNs reside between T8 and T10. This broad source of innervation could represent the broad range of functions subserved by the adrenal medulla. Unfortunately, it tells us little about the functional organization of the sympathetic nervous system.

Transmitter Specificity and Functional Specificity

Functionally heterogeneous as the adrenal SPNs may be, a logical next step in seeking functional specificity is to ask whether other CNS neurons establish specific connections with them. In other words, do adrenal SPNs receive inputs not received by other SPNs? In the first study designed to answer this question, Holets and Elde (1982) identified adrenal SPNs by retrograde transport of fluorescent dyes (Fast Blue or True Blue) from injections into the adrenal gland. Then, using immunohistochemical techniques, they localized in the spinal cord endings which stained for antibodies to met-enkephalin, serotonin, substance P, neurophysin, oxytocin, and somatostatin. By studying the relationships between stained endings and SPNs which had transported dye, they were able to determine whether there was any selectivity in the putative transmitters which appeared to impinge upon adrenal vs nonadrenal SPNs. Fibers which stained with the antibody to somatostatin seemed to selectively impinge upon adrenal SPNs. Fibers which stained for met-enkephalin, serotonin, and substance P impinged on both adrenal and nonadrenal SPNs. Fibers staining with antibodies to neurophysin and oxytocin showed no special relationship to SPNs.

A complementary study was conducted by Seybold and Elde (1984). Quantitative receptor autoradiography was used to look for relationships between receptors for putative transmitters and the locations of identified adrenal SPNs. Serotonergic, mu-opiate, alpha-2-adrenergic and muscarinic cholinergic receptors were visualized. Adrenergic and serotonergic receptors were found in greater abundance in areas containing adrenal SPNs.

These studies show that a particular class of SPNs, those which project to the adrenal medulla, probably receive a spectrum of inputs which differs from that of nearby, nonadrenal SPNs. These results are extremely encouraging, for their only major conceptual drawback is the lack of functional specificity of adrenal SPNs. The results suggest that, when it is possible to functionally identify SPNs, perhaps with 2-deoxy-glucose or perhaps by intracellular recording followed by injection of a marking substance, it will be possible to test the pharmacological specificity of their afferents.

Neurophysiological Indications of Functional Specificity

A number of neurophysiological studies bear on the issue of functional specificity of SPNs. Gebber and Barman (1979) showed an orderly arrangement of interactions, possibly mediated by collateral inhibition, between species of SPNs. SPNs with small, slowly-conducting axons inhibited the discharges of SPNs with larger axons. If, as has been supposed throughout the peripheral nervous system, a relationship exists between fiber size and function, this interaction may manifest a reciprocal connection between SPNs of different functions.

In more recent experiments, Morrison and Gebber (1985) and Barman and Gebber (1985) have studied the specificity, with respect to spinal level of termination, of medullospinal projections of neurons whose discharge patterns are correlated with sympathetic activity. They found descending neurons which had both generalized and level-specific projections to the region of the sympathetic columns. Some neurons projected only as far caudal as T2. Others projected to T12, and these neurons also had collaterals which projected to the ILp at T2. Barman and Gebber suggested that these two classes of neurons, those with only T2 projections and those with both T2 and T12 projections, might mediate specific and general sympathetic regulation respectively. It remains to be seen whether brainstem neurons exist which project specifically to restricted levels of the sympathetic columns other than the rostral thoracic level. That such projections do exist was suggested by the observation of Morrison and Gebber (1985) that some neurons which projected to caudal thoracic spinal cord did not seem to have collaterals at T2.

Dembowsky et al. (1985a) have generated important new data with respect to functional specificity from recordings of EPSPs elicited in SPNs by stimulation of primary afferents, propriospinal pathways, and putative descending pathways. By an extremely detailed analysis of conduction times between stimulus onset and the arrival of EPSPs in SPNs, they detected five distinct descending pathways. The strength of the input conveyed by different pathways (i.e; at different latencies) varied considerably from one SPN to another. Thus, one SPN might be strongly depolarized by one of the more rapidly-conducting pathways, but not by one of the slower pathways. The reverse might be true for another SPN. They suggest that differential input by different pathways may manifest different functions for these SPNs.

The next steps will be difficult ones. First, it will be necessary to determine whether natural sympathoexcitatory stimuli which elicit different patterns of sympathetic output elicit different spectra of EPSPs in the same SPN. Second, it will be necessary to see if SPNs of the same functional category (perhaps using criteria comparable to those described by Jänig, see below) are excited by the appropriate stimuli. Whether a functional specificity can be found for SPNs to match specificities of input remains to be seen. However, it is clear that such experiments are now possible.

Probably the greatest progress in identifying characteristics of functionally-identified SPNs has been made by Jänig and co-workers (see Jänig and Szulczyk, 1981, for review). In an extensive series of studies, they have identified discharge patterns in postganglionic sympathetic cutaneous vasoconstrictor fibers, muscle vasoconstrictor fibers, sudomotor fibers, and piloerector fibers. These fibers may be identified by their distinct differences in ongoing activity or by their responses to baroreceptor, chemoreceptor, and nociceptor stimuli.

Recording from SPNs which were known to impinge on the ganglia providing the previously-characterized postganglionic neurons, Jänig and coworkers were able to identify species of SPNs whose discharge characteristics matched those of the postganglionic neurons. Indeed, when they then looked at the conduction velocities of SPNs of each modality, they found, despite the expected overlap, substantial differences in the speed of transmission, and presumably axon diameters, for the four modalities. Pilomotor-related axons had the highest conduction velocities, and muscle vasoconstrictor-related axons had the lowest.

The technique of determining discharge "signatures," first for postganglionic fibers of known destination or (from electrical stimulation studies) of known function, and then looking for preganglionic neurons with similar "signatures," is ingenious. It could probably be used profitably by others. Its major technical difficulty is the requirement for recording from single sympathetic postganglionic fibers, a difficult task for most investigators. Its major conceptual difficulty is, again, defining the "functions" of these fibers.

Perspective on Sympathetic Functional Specificity

The anatomical localization of SPNs projecting via various sympathetic nerves has provided little information which aids a modern understanding of metabolic regulation. On the other hand, neurophysiologists have laid the groundwork for more detailed functional studies. Intracellular recording from SPNs which are being activated (or inhibited) during well-defined regulatory processes, followed by intracellular injection of HRP will probably be most useful. However, progress will be limited by the difficulty of the recording techniques. The use of patterns of postganglionic sympathetic activity to functionally identify SPNs has already proven very successful, and it should be emulated by more laboratories.

By way of dreaming, we might hope that SPNs involved in certain regulatory systems establish and maintain their mutual connections by recognizing certain intracellular or surface proteins. Further, we might hope that we could learn to stain (or otherwise detect) these proteins and decode the functional significance of the neurons. There is a general search for such methods in neurobiology, and we should be ready to use them when they are available.

Transection of the spinal cord causes profound changes, usually decreases, in sympathetic activity. Some of these changes are the result of "spinal shock" and are temporary, lasting from hours to months depending on the species (Head and Riddoch, 1914; Richter and Shaw, 1930; Brooks, 1933; Alexander, 1946; Polosa, 1968; Sato, 1973; Mannard and Polosa, 1973). Other effects of transection are permanent and consist of a generalized decrease in ongoing sympathetic activity, changes in the relative levels of sympathetic activity among sympathetic nerves, and alterations in the excitability and inhibitability of sympathetic activity in response to afferent stimuli.

Ongoing Sympathetic Activity after Spinal Transection

Total, ongoing sympathetic activity decreases after spinal transection. However, the magnitude of this reduction seems to vary from species to species and even from animal to animal within species. Nearly all investigators who have studied animals with chronic spinal cord transections (including man) have noted that resting arterial pressure and, where measured, sympathetic activity may vary from considerably below normal to only slightly below normal. This inter-animal variability in the response to spinal transection, especially when lesions are made in the same species under experimental control, is puzzling and, as yet, unexplained.

Measurement of arterial blood pressure does not properly represent the complexity of changes in sympathetic activity after transection. Actual recordings on sympathetic nerves indicate that ongoing activity is not reduced uniformly. For instance, Meckler and Weaver (1985) found that acute spinal transection in the cat reduced renal and cardiac nerve activity. However, the average level of splenic nerve activity was not altered. Schramm et al. (1985) observed a very different response of renal sympathetic activity to spinal transection in the rat. Although transection lowered arterial pressure slightly in their animals, indicating an overall decrease in ongoing sympathetic activity, renal sympathetic activity **almost doubled**. It is not yet clear whether this difference between the responses to spinal transection in cats and rats represents a true species difference or whether other factors are more important. For instance, to conveniently record from the renal nerves of the rat, it is ordinarily necessary to retract the kidney relatively further ventrad than in the cat. Perhaps the higher renal nerve activity after transection in the rat results from disinhibition of a spinal reflex which converts noxious input from the traction on the renal pelvis to renal sympathetic activity.

Spinal transection not only alters the average activity in many sympathetic nerves, it also changes discharge patterns of sympathetic neurons. For instance, after spinal transection, the overall variability of the interpulse intervals between discharges of SPNs decreases (Mannard and Polosa, 1973). This decrease in variability probably is the result of the loss of many regulatory signals which normally descend from supraspinal systems.

There is a wealth of data which confirm the importance of these descending pathways. In the intact animal, activity in many sympathetic nerves has a cardiac rhythmicity, imposed by the baroreceptor reflexes (Gebber and Barman, 1980), and transmitted from bulbar systems to spinal sympathetic systems. Of course, disruption of bulbospinal pathways by spinal transection abolishes this cardiac rhythmicity. Even in the absence of baroreceptor inputs, sympathetic activity in intact cats exhibits a 2 to 6 Hz periodicity (Gebber and Barman, 1980). This periodicity is present in most sympathetic nerves examined so far, and, in the cat with an intact nervous system, there appears to be strong coupling between the activity in various nerves. However, after spinal transection, this coupling disappears (Gootman and Cohen, 1981). Apparently the propriospinal pathways which provide a degree of intraspinal coupling for spinal sympathetic systems (Laskey et al., 1979) do not mediate synchronization of activity in different sympathetic nerves. Thus, the normal coordination of efferent sympathetic activity to different organs and tissues is most likely mediated by supraspinal systems.

How is ongoing sympathetic activity generated after spinal transection? One hypotheses is that the activity represents the response of spinal systems to the lowered arterial pressure and, perhaps, lowered spinal blood flow caused by the transection itself. Indeed, Alexander (1945) showed that reductions in spinal blood flow or spinal anoxia produced profound increases in sympathetic activity. Increases in arterial pressure, which presumably improved blood flow to the spinal cord in these preparations, decreased sympathetic activity. Recently, by recording directly from SPNs, Rohlicek and Polosa (1981) and Zhang,

Rohlicek and Polosa, (1982) have measured the thresholds for increased firing of SPNs in acute spinal cats exposed to hypoxic and hypercapnic stresses. All of these cats exhibited ongoing activity at normal blood levels of oxygen and carbon dioxide, and the thresholds for elevated sympathetic activity were well above levels which would have been likely in most experiments to date.

This question has been reinvestigated recently by Meckler and Weaver (1985). They reported only small effects of hypercapnia and elevated arterial pressure on integrated sympathetic activity in cardiac, splenic, and renal nerves in cats with acutely-transected spinal cords. Although the degree to which changes in blood gases and changes in spinal cord blood flow might alter sympathetic output may be controversial, results to date show that spinally-transected animals with normal arterial pressures, normal spinal cord blood flow, and normal blood gases have ongoing sympathetic activity.

A second hypothesis is that ongoing sympathetic activity after spinal transection is generated by inherent pacemaker potentials in SPNs. The first SPNs from which intracellular recordings were made did not exhibit spontaneous activity, and, of course, no pacemaker activity was seen (Fernandez de Molina et al., 1965). Subsequent intracellular investigations have demonstrated varying degrees of synaptic activity and ongoing discharges (see above). However, slow depolarizing potentials of a pacemaker type have never been observed. Apparently, then, the ongoing activity of SPNs after spinal transection is generated by intraspinal systems extrinsic to the SPNs. Dembowsky et al. (1985a) have stressed that excitatory input to SPNs in the intact cat is probably greater than had previously been appreciated. They also noted that spinal transection reduced, but did not eliminate, ongoing EPSPs. These EPSPs were still capable of bringing SPNs to threshold and sustaining ongoing discharge in some SPNs.

As SPNs are successively isolated, there seems to be a crude correlation between the complexity of the remaining systems and their discharge rate. Polosa noted that the discharge rate of SPNs was reduced by spinal cord section and then further reduced by isolation of a spinal cord segment (Polosa, 1968; Mannard and Polosa, 1973). McKenna and Schramm (1983) found that the maximum firing rates for SPNs in the completely isolated spinal cord preparation was much lower than normal. Finally, in the isolated spinal cord slice preparation, although ongoing excitatory postsynaptic potentials were observed, and SPNs could be caused to discharge by stimulation of afferents, there was apparently insufficient natural excitatory input to generate ongoing discharge of the SPNs (Yoshimura and Nishi, 1982).

Another source of ongoing sympathetic activity after spinal transection is spinal afferents. As discussed in the following section, electrical or natural stimulation of both visceral and somatic afferents may elicit either increases or decreases in sympathetic activity in both intact and spinally-transected animals. Long propriospinal pathways ensure that SPNs receive afferent information from many spinal levels (Mannard and Polosa, 1973).

If afferent activity may either increase or decrease ongoing sympathetic activity, what is the net effect of decreasing afferent sympathetic activity after spinal transection? The data pertaining to this question are in conflict. Polosa (1968) reported a very low discharge rate (0.3Hz) for one SPN recorded after both spinal cord transection and dorsal rhizotomy in the cat. Presumably, this was the only SPN which exhibited spontaneous activity under these conditions. Since spinal transection reduces the number of tonically active SPNs by only about half (Mannard and Polosa, 1973), it may be inferred that dorsal rhizotomy decreased even further the discharge rate of these rostral thoracic SPNs. Meckler and Weaver (1985) performed dorsal rhizotomies on their spinally-transected cats to determine whether ongoing afferent input was producing the sympathetic activity which remained after spinal transection. Rhizotomy in a previously isolated segment of spinal cord (T4 through L5) did not significantly alter splenic and renal sympathetic activity suggesting that this activity was entirely of intraspinal origin. On the other hand, Schramm et al. (1985) found that both baclofen and morphine, in doses which would be expected to reduce afferent input to the spinal cord, markedly reduced renal sympathetic activity after spinal transection.

One possible resolution to these disparate results is that spinal afferents may inhibit as well as excite sympathetic activity (Wyszogrodski and Polosa, 1973; Kirchner, Wyszogrodski and Polosa, 1975; Kirchner, Kirchner and Polosa, 1975). Differences in preparations, differences in techniques, even differences in the positions of the animals could result in considerable differences in the constellation of active afferents projecting information to spinal sympathetic systems, and this could lead to differences in the effects of subsequent decreases in afferent information.

Sympathetic Excitability after Spinal Transection

Spinal transection also alters sympathetic responses to afferent signals. Early experiments by Brooks (1933) indicated that stimulation of the sciatic nerve could elicit increases in blood sugar, arterial pressure, and heart rate as well as diminish clotting time and cause contraction of the denervated nictitating membrane after chronic spinal transection in cats. However, he noted that these indications of sympathoadrenal activation could be elicited only with stimulus intensities higher than those necessary in intact cats. Beacham and Perl, 1964a,b) conducted the first modern electrophysiological study of spinal sympathetic reflexes. They, too, noted that sympathetic responses to afferent stimulation were more difficult to elicit in cats with transected spinal cords. However, they also noted that afferent stimulation produced sympathoinhibition as well as sympathoexcitation and that the threshold for the inhibition was lower than that for the excitation. They suggested that this mixture of evoked inhibition and excitation could confound the measurement of sympathetic stimulus-response relationships, and that it might account for much of the confusion in the earlier literature on evoked sympathetic responses. As discussed above, the phenomenon of contemporaneous generation of sympathoinhibition and sympothoexcitation by ongoing afferent activity may have lead to confusion about the factors regulating ongoing sympathetic activity.

In a recent study, Araki et al. (1984) found that, in the intact rat, noxious stimulation of either the lower chest or the hindlimb elicited increases in adrenal nerve activity. However, after spinal transection, only noxious stimulation of the chest was able to elicit increases in adrenal nerve activity.

The traditional interpretation of these results is that, in intact animals, sympathetic responses to noxious stimuli are more generalized than in animals with transected spinal cords in which they tend to be more segmentally organized. An alternate hypothesis, suggested by the work of Polosa et al. (1982), is that, after spinal cord transection (but not before), some stimuli may have a sympathoinhibitory component which is greater than their sympathoexcitatory component.

There is good evidence that some of the inhibition of SPNs by sensory input impinges directly on the SPNs themselves, rather than on interneurons (Wyzogrodsky and Polosa, 1973; Dembowsky et al., 1985a). On the other hand, if ongoing sympathetic activity is being driven to a significant extent by sensory input from nearby dorsal roots, then afferent information from distant segments may inhibit this input at a site in the dorsal horn (Cadden et al., 1983; Villanueva, Cadden and LeBars, 1984), reducing the drive on the SPNs and, thus, decreasing ongoing activity or reducing the excitatory component of the same distal stimulus. The relative degree to which these mechanisms contribute to the overall sympathetic responses to afferent information after spinal transection needs to be determined.

An extensive series of electrophysiological experiments on spinal sympathetic reflexes was conducted by Sato (1973). These have been thoroughly reviewed by Sato and Schmidt (1973) and Koizumi and Brooks (1972), and they will not be discussed in detail here. However, these studies demonstrated that a spinal component of the somatosympathetic reflex exists in both the intact and in the spinally-transected cat. However, the spinal component of the reflex is sometimes difficult to demonstrate in the intact cat. In fact, the spinal component of some sympathetic reflexes appears to be under tonic inhibition by supraspinal systems. Thus, transection of the spinal cord or lesions of the spinal cord may accentuate the spinal component of a sympathetic reflex while abolishing the supraspinal component. For instance, Coote and Sato (1978) found that lesions of the ventral funiculus or total spinal cord transection "released" a short-latency somatosympathetic reflex of spinal origin in cats. Before transection, some of these cats had exhibited only a long-latency, (presumably) spino-bulbo-spinal reflex.

Spinal cord lesions or transection can also substantially change the spectrum of responses elicited by somatic stimuli, sometimes producing changes which result in increased sympathetic responses to afferent stimulation and sometimes producing changes which result in decreased responses. In intact, anesthetized rats, Araki et al. (1984) showed that gentle brushing of the lower chest decreased adrenal nerve activity and decreased adrenal catecholamine secretion. Noxious stimulation of this same region elicited increases in adrenal nerve activity and increases in adrenal catecholamine secretion. However, after spinal transection (resulting in an 80 to 90% decrease in ongoing adrenal nerve activity), both gentle brushing and noxious stimulation of the lower chest produced increases in adrenal nerve activity. Thus, a previously neutral and certainly innocuous stimulus produced an inappropriate response, and elevated catecholamine secretion after spinal transection.

After spinal cord transection, humans often exhibit varying degrees of resting hypotension and orthostatic hypotension. At the same time, abnormal sweating and hypertensive episodes are common responses to visceral afferent stimuli in these patients (Head and Riddoch, 1917; Guttman and Witteridge, 1947; Cunningham et al., 1953; Hodgson and Wood, 1958; Randall, Wurster and Lewin, 1966; Wurster and Randall, 1979; Naftchi et al., 1978; Mathius et al., 1979; Nieder, O'Higgins and Aldrete, 1979).

There is certainly a good neurophysiological basis for understanding these patients' hypotension. As reviewed above, spinal transection reduces excitatory drive to most SPNs, and this can be expected to decrease their rate of ongoing discharge. Because there is no longer any connection between the medullary baroreceptor systems and the SPNs, sympathetic activity (which is already low) cannot be raised to compensate for postural changes, leading to orthostatic hypotension. In addition, spinal transection may accentuate or prolong potentially hypotensive somatosympathetic reflexes (Kummel, 1983), leading to hypotensive episodes.

How does one account for the sympathetic hyper-reactivity sometimes observed after spinal cord transection? Recording from sympathetic nerves in humans, Wallin and Stjernberg (1984) have shown that sympathetic discharges in response to afferent stimuli are more prolonged in patients with spinal cord injuries than in intact subjects. The effects of such long after-discharges could contribute to episodes of hypertension, especially without the buffering action of the sympathoinhibitory limb of the baroreceptor reflex. The observation, described above, that spinal transection is capable of converting a previously sympathoinhibitory response to a nonnoxious stimulus into a sympathoexcitatory one might provide a parallel experimental observation (Araki et al., 1984). Another supporting observation is that spinal sympathoexcitatory reflexes, which are under such strong supraspinal inhibition that they are not observable in some intact animals, become evident after spinal transection (Coote and Sato, 1978). Finally, chronic spinal transection causes periods of somatic hyper-reflexia in both cat and rat which are reminiscent of sympathetic hyper-reflexia (Malmsten, 1983; Hultborn and Malmsten, 1983).

If sympathetic hyperreactivity is related to the abolition of tonic inhibition of spinal sympathetic reflexes, what is the normal site of this inhibition? Is it directly on the SPNs, on sympathetic interneurons, or upon the sensory-processing network in the dorsal horn? There are too few data to answer this question with any certainty. However, it seems unlikely from the effects of spinal transection on ongoing sympathetic activity that the inhibition would impinge directly on the SPN. The extensive literature on the descending modulation of both somatic and visceral afferent information at spinal levels (see Duggan, 1985 for review) suggests that sympathetic hyper-reactivity after spinal cord transection may be due as much to disinhibition of sensory pathways as to direct disinhibition of the sympathetic systems themselves.

Perspective on Spinal Transection

Spinal transection is a pathological state. It is unlikely that there has been much evolutionary pressure to survive massive spinal lesions, and much of the sympathetic behavior which follows such lesions is non-adaptive. We can hope to learn "something" about normal spinal regulatory function by studying the transected animal. However, this is likely to be even less rewarding than similar studies of spinal somatic processes because so many of the **receptors** necessary for adaptive metabolic regulation project only to the brain. After spinal transection, the spinal somatic nervous system at least maintains contact with proprioceptors and exteroceptors. The sympathetic nervous system maintains contact with some visceral and somatic receptors. However, this contact may be highly inappropriate since information from these receptors usually modulates sympathetic activity largely <u>via</u> spino-bulbo-spinal connections.

Sympathetic regulation organized by the acutely transected spinal cord may suffer from "spinal shock." In the chronically-transected animal, are we finally studying the spinal component of normal sympathetic regulation, or are we studying a new system, elaborated by regeneration and plasticity? It is at least of some interest to be able to observe the sympathetic behavior of the spinally-transected animal and reflect that only the spinal cord and ganglia (with or without plasticity) can generate that behavior.

I submit that a principal motivation for studying sympathetic processing in the "spinal animal" might be to understand the pathological processes exhibited by the thousands of patients who injure their spinal cords each year. When autonomic disfunction does not threaten their lives, it detracts from the quality of their lives. The efficacy and safety of future treatments will depend on careful analyses of the mechanisms which underlie these pathological processes.

ACKNOWLEDGEMENTS

I thank Mark Knuepfer, Renea Livingstone, and Diana Schramm for their careful reading of the manuscript and Kathy Hartzell for her editorial assistance. This work was supported by NIH grant HL16315.

REFERENCES

Adair, J. R., Hamilton, B. L., Scappaticci, K. A., Helke, C. J., and Gillis, R. A., 1977, Cardiovascular responses to electrical stimulation of the medullary raphe area of the cat, Brain Res., 128:141-145.

Alexander, R. S., 1945, The effect of blood flow and anoxia on spinal cardiovascular centers, Am. J. Physiol., 143:698-708.

Alexander, R. S., 1946, Tonic and reflex functions of medullary sympathetic cardiovascular centers, J. Neurophysiol., 9:205-217.

Altman, J., and Bayer, S. A., 1984, The development of the rat spinal cord, Springer-Verlag, Berlin.

Araki, T., Ito, K., Kurosawa, M., and Sato, A., 1984, Responses of adrenal sympathetic nerve activity and catecholamine secretion to cutaneous stimulation in anesthetized rats, Neuroscience, 12:289-299.

Backman, S., and Henry, J., 1983a, Effects of glutamate and aspartate on sympathetic preganglionic neurons in the upper thoracic intermediolateral nucleus of the cat, Brain Res., 277:370-374.

Backman, S. and Henry, J., 1983b, Effects of GABA and glycine on sympathetic preganglionic neurons in the upper thoracic intermediolateral nucleus of the cat, Brain Res., 277:365-369.

Backman, S., and Henry, J., 1984, Effects of substance P and thyrotopin-releasing hormone on sympathetic preganglionic neurones in the upper thoracic intermediolateral nucleus of the cat, Can. J. Physiol. Pharmacol., 62:248-251.

Barber, R. P., Phelps, P. E., Houser, C. R., Crawford, G. D., Salvaterra, P. M., and Vaughn, J. E., 1984, The morphology and distribution of neurons containing choline acetyltransferase in the adult rat spinal cord: an immunocytochemical study, J. Comp. Neurol., 229:329-346.

Barman, S. M., 1984, Spinal cord control of the cardiovascular system. in: "Nervous Control of Cardiovascular Function", W. C. Randall, ed., Oxford University Press, New York, pp. 321-345.

Barman, S. M., and Gebber, G. L., 1985, Anoxal projection patterns of ventrolateral medullospinal sympathoexcitatory neurons, J. Neurophysiol, 53:1551-1556.

Baron, R., Jänig, W., and McLachlan, E. M., 1985, The afferent and sympathetic components of the lumbar spinal outflow to the colon and pelvic organs in the cat. I. The hypogastric nerve, J. Comp. Neurol., 238:135-146.

Basbaum, A. I., Clanton, C. H., and Fields, H. L., 1978, Three bulbospinal pathways from the rostral medulla of the cat: autoradiographic study of pain modulating systems, J. Comp. Neurol., 178:209-224.

Basbaum, A. I., and Fields, H. L., 1979, The origin of descending pathways in the dorsolateral funiculus of the spinal cord of the cat and rat: further studies on the anatomy of pain modulation, J. Comp. Neurol., 187:513-531.

Baum, T. and Shropshire, A. T., 1977, Susceptibility of spontaneous sympathetic outflow and sympathetic reflexes to depression by clonidine, Eur. J. Pharmacol., 44:121-129.

Beacham, W. S., and Perl, E. R., 1964a, Background and reflex discharge of sympathetic preganglionic neurones in the spinal cat, J. Physiol., 172:400-416.

Beacham, W.S., and Perl, E.R., 1964b, Characteristics of a spinal sympathetic reflex, J. Physiol. (Lond) 173:431-448.

Beattie, J., Brow, G. R., and Long, C.N.H., 1930, Physiological and anatomical evidence for the existence of nerve tracts connecting the hypothalamus with spinal sympathetic centers, Proc. Royal Soc. (B), 106:253-275.

Bell, C., 1830, The Nervous System of the Human Body, Longman, Ress, Orme, Brown, Green, London.

Biedl, A., 1895, Uber die centra der splanchnici, Wien. Klin. Wochenschr., 8:915-919.

Bishop, G. H. and Heinbecker, P., 1932, A functional analysis of the cervical sympathetic nerve supply to the eye, Am. J. Physiol., 100:519-532.

Bodian, D., and Taylor, N., 1963, Synapse arising at central node of Ranvier, and note on fixation of the central nervous system, Science, 139:330-332.

Bok, S. T., 1928, Das Rückenmark. Handbuch der microscopischen Anatomie des Menschen., In: W. von Mollendorf (ed.), Springer Verlag, Berlin, 1928, pp. 478-578.

Brooks, C. M., 1933, Reflex activation of the sympathetic system in the spinal cat, Am. J. Physiol., 106:251-266.

Cabot, J. B., Bogan, N., and Reid, L. J., 1985, Somatic and dendritic morphology of intracellularly labelled sympathetic preganglionic neurons, Society for Neuroscience Abstracts, 11:34.

Cabot, J. B., Wild, J., and Cohen, D. H., 1979, Raphe inhibition of sympathetic preganglionic neurons, Science, 203:184-186.

Cadden, S. W., Villaneuva, L., Chitour, D., and Lebars, D., 1983, Depression of activities of dorsal horn convergent neurones by propriospinal mechanisms triggered by noxious inputs; Comparison with diffuse noxious inhibitory controls (DNIC), Brain Res., 275:1-11.

Carlson, A., Falck, B., Fuxe, K., and Hillarp, N.-.A., 1964, Cellular localization of monoamines in the spinal cord, Acta Physiol. Scand., 60:112-119.

Caverson, M., Ciriello, J., and Calaresu, F., 1983, Direct pathway from cardiovascular neurons in the ventrolateral medulla to the region of the intermediolateral nucleus of the upper thoracic cord; an anatomical and electrophysiological investigation in the cat, J. Auton, Nerv. Syst., 9:451-475.

Celler, B. G., and Schramm, L. P., 1981, Pre- and postganglionic sympathetic activity in splanchnic nerves of rats, Am. J. Physiol., 241:R55-R61.

Cervero, F., and Connell, L. A., 1984, Distribution of somatic and visceral primary afferents within the thoracic spinal cord of the cat, J. Comp. Neurol., 230:88-98.

Charlton, C. G., and Helke, C. J., 1985, Autoradiographic localization and characterization of spinal cord substance P binding sites: High densities in sensory, autonomic, phrenic, and Onuf's motor nuclei, J. Neurosci., 5:1653-1661.

Chiba, T., and Murata, Y., 1981, Architecture and synaptic relationships in the intermediolateral nucleus of the thoracic spinal cord of the rat: HRP labelling, catecholamine histochemistry and electron microscopic studies, J. Neurocytol., 10:315-329.

Chung, J. M., Chung, K., and Wurster, R. D., 1975, Sympathetic preganglionic neurons of the cat spinal cord: horseradish perixodase study, Brain Res., 91:126-131.

Chung, K., Chung, J. M., Lavelle, F. W., and Wurster, R. D., 1979, Sympathetic neurons in the cat spinal cord projecting to the stellate ganglion, J. Comp. Neurol., 185:23-30.

Chung, K., and Coggeshall, R. E., 1983, Propriospinal fibers in the rat, J. Comp. Neurol., 217:47-53.

Chung, K., Lavelle, F. W., and Wurster, R. D., 1980, Ultrastructure of HRP-identified sympathetic preganglionic neurons in cats, J. Comp. Neurol., 190:147-155.

Ciriello, J., and Calaresu, F. R., 1983, Central projections of afferent renal fibers in the rat: an anterograde transport study of horseradish peroxidase, J. Auton. Nerv. Syst., 8:273-285.

Coggeshall, R. E., and Galbraith, S. L., 1978, Categories of axons in rami communicantes II. J. Comp. Neurol. 181:349-360.

Coggeshall, R. E., Hancock, M. B., and Applebaum, M. L., 1976, Categories of axons in mammalian rami communicantes, J. Comp. Neurol., 167:105-124.

Coote, J. H., and Gilbey, M. P., 1980, Raphe-spinal sympathoinhibition antagonized by lysergic acid diethylamide, J. Physiol., 312:7-8.

Coote, J. H., and Macleod, V. H., 1974, Evidence for the involvement in the baroreceptor reflex of a descending inhibitory pathway, J. Physiol., 241:477-496.

Coote, J. H., and Macleod, V. H., 1975, The spinal route of sympatho-inhibitory pathways descending from the medulla oblongata, Pflugers Arch., 359:335-347.

Coote, J. H., Macleod, V. H., Fleetwood-Walker, S., and Gilbey, M. P., 1981a, The response of individual sympathetic preganglionic neurones to microelectrophoretically applied endogenous monoamines, Brain Res., 215:135-145.

Coote, J. H., Macleod, V. H., Fleetwood-Walker, S. M., and Gilbey, M. P., 1981b, Baroreceptor inhibition of sympathetic activity at a spinal site, Brain Res., 220:81-93.

Coote, J. H., and Sato, A., 1978, Supraspinal regulation of spinal reflex discharge into cardiac sympathetic nerves, Brain Res., 142:425-437.

Coote, J. H., and Westbury, D. R., 1979, Intracellular recordings from sympathetic preganglionic neurones, Neurosci. Lett., 15:171-175.

Cummings, J. F., 1969, Thoracolumbar preganglionic neurons and adrenal innervation in the dog, Acta Anat., 73:27-37.

Cunninghan, D.J.C., Guttman, L, Whitteridge, D., and Wyndham, C. H., 1953, Cardiovascular responses of bladder distension in paraplegic patients, J. Physiol., 121:581-592.

Dahlstrom, A., and Fuxe, K., 1965, Evidence for the existence of monoamine neurons in the central nervous system. II. Experimentally induced changes in the intraneuronal amine levels in bulbospinal neuron systems, Acta Physiol. Scand., 64:7-34, Suppl. 247.

Dalsgaard, C.-.J. and Elfin, L.-.G., 1981, The distribution of the sympathetic preganglionic neurons projecting onto the stellate ganglion of the Guinea pig. A horseradish peroxidase study, J. Auton. Nerv. Syst., 4:327-337.

Dashwood, M. R., Gilbey, M. P., and Spyer, K. M., 1985, The localization of adrenoceptors and opiate receptors in regions of the cat central nervous system involved in cardiovascular control, Neuroscience, 15:537-551.

Davis, B. M., Krause, J. E., McKelvy, J. F., and Cabot, J. B., 1984, Effects of spinal lesions on substance P levels in the rat: evidence for local spinal regulation, Neuroscience, 13:1311-1326.

DeGroat, W. C., and Ryall, R. W., 1967, An excitatory action of 5-hydroxytryptamine on sympathetic preganglioic neurons, Exp. Brain Res., 3:299-305.

Dembowsky, K., Czachurski, J., and Seller, H., 1985a, An intracellular study of the synaptic input to sympathetic preganglionic neurones of the third thoracic segment of the cat, J. Auton. Nerv. Syst., 13:201-244.

Dembowsky, K., Czachurski, J., and Seller, H., 1985b, Morphology of the sympathetic preganglionic neurones in the thoracic spinal cord of the cat: an intracellular horseradish peroxidase study, J. Comp. Neurol., 238:453-465.

Deuschl, G., and Illert, M., 1978, Location of lumbar sympathetic neurons in the cat, Neurosci. Lett., 10:49-54.

Dibona, G.F., 1982, The functions of the renal nerves, Rev. Physiol. Biochem. Pharmacol., 94:75-181.

Duggan, A. W., 1985, Pharmacology of descending control systems, Philos. Trans. Roy. Soc. Lond. [Biol.], 308(b):375-391.

Faden, A., Jacobs, T., and Woods, M., 1979, An intraspinal sympathetic preganglionic pathway: physiological evidence in the cat, Brain Res., 162:13-20.

Faden, A., Petras, J. M., and Woods, M., 1978, An intraspinal sympathetic preganglionic pathway: anatomic evidence in the dog, Brain Res., 144:358-362.

Fernandez De Molina, A., Kuno, M., and Perl, E. R., 1965, Antidromically evoked responses from sympathetic preganglionic neurones, J. Physiol. (Lond), 180:321-335.

Franz, D. N., Hare, B. D., and McCloskey, K. L., 1982, Spinal sympathetic neurons: possible sites of opiate-withdrawal suppression by clonidine, *Science*, 215:1643-1645.

Franz, D. N., and Madsen, P. W., 1982, Differential sensitivity of four central sympathetic pathways to depression by clonidine, *Eur. J. Pharmacol.*, 78:53-59.

Gagel, O., 1932, Zur Histologie and Topographie der vegetativen Zentrum im Ruckenmark, *Z. Anat. Entwickl-Gesch.*, 85:213-250.

Galabov, P., and Davidoff, M., 1976, On the vegetative network of Guinea pig thoracic spinal cord, *Histochemistry*, 47:247-255.

Gaskell, W. H., 1886, On the structure, distribution and function of the nerves which innervate the visceral and vascular systems, *J. Physiol.*, 7:1-80.

Gebber, G. L., and Barman, S. M., 1979, Inhibitory interaction between preganglionic sympathetic neurons, *in*: "Nervous System and Hypertension. Perspectives in Nephrology and Hypertension," P. Mayer and H. Schmitt (eds.), J. Wiley and Sons, New York, pp. 137-145.

Gebber, G. L., and Barman, S. M., 1980, Basis for 2-6 cycle/s rhythm in sympathetic nerve discharge, *Am. J. Physiol.*, 239:R48-R56.

Gebber, G. L., and McCall, R. B., 1976, Identification and discharge patterns of spinal sympathetic interneurons, *Am. J. Physiol.*, 231:722-733.

Gilbey, M. P., Coote, J. H., Fleetwood-Walker, S., and Peterson, D. F., 1982, The influence of the paraventriculo-spinal pathway, and oxytocin and vasopressin on sympathetic preganglionic neurones, *Brain Res.* 251:283-290.

Gilbey, M. P., Coote, J. H., Macleod, V. H., and Peterson, D. F., 1981, Inhibition of sympathetic activity by stimulating in the raphe nuclei and the role of 5-hydroxytryptamine in this effect, *Brain Res.*, 226:131-142.

Gilbey, M. P., McKenna, K., and Schramm, L. P., 1983, Effects of substance P on sympathetic preganglionic neurons, *Neurosci. Lett.*, 4:157-159.

Gilbey, M. P., Peterson, D. F., and Coote, J. H., 1982, Some characteristics of sympathetic preganglionic neurons in the rat, *Brain Res.*, 241:43-48.

Glazer, E. J., and Ross, L. L., 1980, Localization of noradrenergic terminals in sympathetic preganglionic nuclei of the rat: demonstration by immunocytochemical localization of dopamine-beta-hydroxylase, *Brain Res.*, 185:39-49.

Gootman, P. M., and Cohen, M. I., 1981, Sympathetic rhythms in spinal cats, *J. Auton. Nerv. Syst.*, 3:379-387.

Guttman, L., and Witteridge, D., 1947, Effects of bladder distension on autonomic mechanisms after spinal cord injuries, Brain, 70:361-404.

Guyenet, P. G., and Cabot, J. B., 1981, Inhibition of sympathetic preganglionic neurons by catecholamines and clonidine: mediation by an alpha-adrenergic receptor, J. Neurosci., 1:908-917.

Haase, P., Contestabile, A., and Flumerfelt, B.A., 1982, Preganglionic innervation of the adrenal gland of the rat using horseradish peroxidase, Exp. Neurol., 78:217-221.

Hadjiconstantinou, P., Panula, P., Lackovic, Z., and Neff, N. H., 1984, Spinal cord serotonin: a biochemical and immunohistochemical study following transection, Brain Res., 322:245-254.

Hancock, M. B., 1982, Leu-enkephalin, substance P, and somatostatin immunohistochemistry combined with the retrograde transport of horseradish peroxidase in sympathetic preganglionic neurons, J. Auton. Nerv. Syst. 6:263-272.

Hancock, M. B., and Peveto, C. A., 1979, A preganglionic autonomic nucleus in the dorsal gray commisure of the lumbar spinal cord of the rat, J. Comp. Neurol., 183:67-72.

Head, H., and Riddoch, G., 1917, The autonomic bladder, excessive sweating and some other reflex conditions, in gross injuries of the spinal cord, Brain, 40:188-263.

Helke, C. J., Neil, J. J., Massari, M. J., and Loewy, A. D., 1982, Substance P neurons project from the ventral medulla to the intermediolateral cell column and ventral horn in the rat, Brain Res., 243:147-152.

Hodgson, N. B., and Wood, J. A., 1958, Studies of the nature of paroxysmal hypertension in paraplegics, J. Urol., 79:719-721.

Hokfelt, T., Ljungdahl, A., Steinbusch, H., Verhofstad, A., Nilsson, G., Brodin, E., Pernow, B., and Goldstein, M., 1978, Immunohistochemical evidence of substance P-like immunoreactivity in some 5-hydroxytryptamine-containing neurons in the rat central nervous system, Neuroscience, 3:517-538.

Holets, V., and Elde, R., 1982, The differential distribution and relationship fo serotoninergic and peptidergic fibers to sympathoadrenal neurons in the intermediolateral cell column of the rat: a combined retrograde axonal transport and immunofluorescense study, Neuroscience, 7:1155-1174.

Hongo, T., and Ryall, R. W., 1966, Electrophysiological and microelectrophoretic studies on sympathetic preganglionic neurons in the spinal cord, Acta Physiol. Scand., 68:96-104.

Howe, P.R.C., Kuhn, D. M., Minson, J. B., Steadand, B. H., and Chalmers,

J. P., 1983, Evidence for a bulbospinal serotonergic pressor pathway in the rat brain, Brain Res., 270:29-36.

Hultborn, H., and Malmsten, J., 1983, Changes in the segmental reflexes following chronic spinal cord hemisection in the cat 1. Increased monosynaptic and polysynaptic ventral root discharges, Acta Physiol. Scand., 119:405-422.

Hwang, B. H., and Williams, T. H., 1982, Flourescence microscopy used in conjunction with horseradish peroxidase localization and electron microscopy for studying sympathetic nuclei of the rat spinal cord, Brain Res. Bull., 9:171-177.

Jänig, W., and Schmidt, R. F., 1970, Single unit responses in the cervical sympathetic trunk upon somatic nerve stimulation, Pflugers Arch., 314:199-216.

Jänig, W., and Szulczyk, P., 1981, The organization of lumbar preganglionic neurons, J. Autonomic Nervous System., 3:177-191.

Kadzielawa, K., 1983a, Antagonism of the excitatory effects of 5-hydroxytryptamine on sympathetic preganglionic neurones and neurones activated by visceral afferents, Neuropharmacology, 22:19-27.

Kadzielawa, K., 1983b, Inhibition of the activity of sympathetic preganglionic neurones and neurones activated by visceral afferents, by alpha-methylnoradrenaline and endogenous catecholamines, Neuropharmacology, 22:3-17.

Keeler, J. R., and Helke, C. J., 1985, Spinal cord substance P mediates bicuculline-induced activation of cardiovascular responses from the ventral medulla, J. Auton. Nerv. Syst., 13:19-33.

Kirchner, F., Kirchner, D., and Polosa, C., 1975, Spinal organization of sympathetic inhibiton by spinal afferent volleys, Brain Res., 87:161-170.

Kirchner, F., Wyszogrodski, I., and Polosa, C., 1975, Some properties of sympathetic neuron inhibition by depressor area and intraspinal stimulation, Pflugers Arch., 357:349-360.

Koizumi, K., and Brooks, C.McC., 1972, The integration of autonomic system reactions: a discussion of autonomic reflexes, their control and their association with somatic reactions, Ergebnisse der Physiologie, 67:1-68.

Krier, J., Thor, K. B., and DeGroat, W. C., 1979, Effects of clonidine on the lumbar sympathetic pathways to the large intestine and urinary bladder of the cat, Eur. J. Pharmacol., 59:47-53.

Kuhn, K. M., Wolf, W. A., and Lovenberg, W., 1980, Pressor effects of electrical stimulation of the dorsal and median raphe nuclei in anesthetized rats, J. Pharmacol. Exp. Therm., 214:403-409.

Kummel, H., 1983, Activity in sympathetic neurons supplying skin and skeletal muscle in spinal cats, J. Auton. Nerv. Syst., 7:319-327.

Kuo, D. C., Nadelhaft, I., Hisamitsu, T., and DeGroat, W. C., 1983, Segmental distribution and central projections of renal afferent fibers in the cat studied by transganglionic transport of horseradish peroxidase, J. Comp. Neurol., 216:162-174.

Kuo, D. C., Yamasaki, D. S., and Krauthamer, G. M., 1980, Segmental organization of sympathetic preganglionic neurons of the splanchnic nerve as revealed by retrograde transport of horseradish peroxidase, Neurosci. Lett., 17:11-16.

Kurosawa, K., Minami, M., Togashi, H., and Sato, H., 1985, The effect of clonidine on adrenal sympathetic nerve responses to mechanical, noxious and innocuous stimulation of the skin and rats, Neurosci. Lett., 53:253-258.

Kuypers, H.G.J.M., and Maiskey, V. A., 1975, Retrograde axonal transport of horseradish peroxidase from spinal cord to brainstem cell groups, Neurosci. Lett., 1:9-14.

Laruell, L, 1936, Contribution a letude du neuraxe vegetatif, Comp. Rend. Assoc. Anat., 31:210-229.

Laskey, W., Schondorf, R., and Polosa, C., 1979, Intersegmental connections and interactions of myelinated somatic and visceral afferents with sympathetic preganglionic neurons in the unanesthetized spinal cat, J. Auton. Nerv. Syst., 1:69-76.

Lebedev, V. P., Petrov, V. I., and Skobelev, V. A., 1980, Do sympathetic preganglionic neurones have a recurrent inhibitory mechanism?, Pflugers Arch., 383:91-97.

Lebedev, V. P., Petrov, V. I., and Skobelev, V. A., 1976, Antidromic discharges of sympathetic preganglionic neurons located outside of the spinal cord lateral horns, Neurosci. Lett., 2:325-329.

Light, A. R., and Metz, C. B., 1978, The morphology of the spinal efferent and afferent fibers contributing to the ventral root of the cat, J. Comp. Neurol., 179:501-522.

Loewy, A. D., 1981, Raphe pallidus and raphe obscuras projections to the intermediolateral cell column in the rat, Brain Res., 222:129-133.

Loewy, A. D., and McKellar, S., 1981, Serotonergic projections from the ventral medulla to the intermediolateral cell column in the rat, Brain Res., 211:146-152.

Loewy, A. D., McKellar, S., and Saper, C. B., 1979, Direct projections from the A5 catecholamine cell group to the intermediolateral cell column, Brain Res., 174:309-314.

Loewy, A. D., and Sawyer, W. B., 1982, Substance P antagonist inhibits vasomotor responses elicited from ventral medulla in rat, Brain Res., 245:379-383.

Loewy, A. D., Wallach, J. H., and McKellar, S., 1981, Efferent connections of the ventral medulla oblongata in the rat, Brain Research Reviews, 3:63-80.

Lorenz, R. G., Saper, C. B., Wong, D. L., Ciaranello, R. D., and Loewy, A. D., 1985, Co-localization of substance P- and phenylethanol-amine-N-methyltransferase-like immunoreactivity in neurons of the ventrolateral medulla that project to the spinal cord: Potential role in control of vasomotor tone, Neurosci. Lett., 55:255-260.

Malmsten, J., 1983, Time course of segmental reflex changes after chronic spinal cord hemisection in the rat, Acta Physiol. Scand., 119:435-443.

Mannard, A., Rajchgot, P, and Polosa, C., 1977, Effect of post-impulse depression on background firing of sympathetic preganglionic neurons, Brain Res., 126:243-261.

Mannard, A., and Polosa, C., 1973, Analysis of background firing of single sympathetic neurons of cat cervical nerve, J. Neurophysiol., 36:398-408.

Mathias, C. J., Reid, J. L., Wing, L.M.H., Frankel, H. L., and Christensen, N. J., 1979, Antihypertensive effects of clonidine in tetraplegic subjects devoid of central sympathetic control, Clin. Sci., 57:425S-428S.

McCall, R. B., 1983, Serotonergic excitation of sympathetic preganglionic neurons. A microiontophoretic study, Brain Res., 289:121-127.

McCall, R. B., 1984, Evidence for a serotonergically mediated sympathoexcitatory response to stimulation of medullary raphe nuclei, Brain Res., 311:131-139.

McCall, R. B., Gebber, G. L., and Barman, S. M., 1977, Spinal interneurons in the baroreceptor reflex arc, Am. J. Physiol., 232:H657-H665.

McKenna, K. E., and Schramm, L. P., 1983, Sympathetic preganglionic neurons in the isolated spinal cord of the neonatal rat, Brain Res., 269:201-210.

McKenna, K. E., and Schramm, L. P., 1985, Mechanisms mediating the sympathetic silent period. Studies in the isolated spinal cord of the neonatal rat, Brain Res., 329:233-240.

McLachlan, E. M., 1985, The components of the hypogastric nerve in male and female Guinea pigs, J. Auton. Nerv. Syst., 13(4):327-342.

McLachlan, E. M., Baron, R., and Janig, W., 1984, Functional neuroanatomy of preganglionic neurones in the lower lumbar spinal cord of the cat, Neurosci. Lett. Suppl., 15:S49.

McLachlan, E. M., and Hirst, G.D.S., 1980, Some properties of preganglionic neurons in upper thoracic spinal cord of the cat, J. Neurophysiol., 43:1251-1265.

Meckler, R. L., and Weaver, L. C., 1985, Splenic, renal, and cardiac nerves have uneven dependence on tonic supraspinal inputs, Brain Res., 338:123-135.

Morrison, S. F., and Gebber, G. L., 1985, Axonal branching patterns and funicular trajectories of raphespinal sympathoexcitatory neurons, J. Neurophysiol., 53:759-772.

Naftchi, N. E., Demeny, M., Lomman, E. W., and Tuckman, J., 1978, Hypertensive crises in quadriplegic patients, Circulation, 57:336-341.

Neil, J. J., and Loewy, A. D., 1982, Decreases in blood pressure in response to L-glutamate microinjections into the A5 catecholamine cell group, Brain Res., 241:271-278.

Neuhuber, W., 1982, The central projections of visceral primary afferent neurons of the inferior mesenteric plexus and hypogastric nerve and the location of the related sensory and preganglionic sympathetic cell bodies in the rat, Anat. Embryol., 164:413-425.

Neumayr, R. J., Hare, B. D., and Franz, D. N., 1974, Evidence for bulbospinal control of sympathetic preganglionic neurons by monoaminergic pathways, Life Sci., 14:793-806.

Nieder, R. M., O'Higgins, J. W., and Aldrete, J. A., 1979, Autonomic hyperreflexia in urologic surgery, J. Am. Med. Assoc., 213:867-869.

Oldfield, B. J., and McLachlan, E. M., 1980, The segmental origin of preganglionic axons in the upper thoracic rami of the cat, Neurosci. Lett., 18:11-17.

Oldfield, B. J., Sheppard, A., and Nilaver, G., 1985, A study of the substance P innervation of the intermediate zone of the thoracolumbar spinal cord, J. Comp. Neurol., 236:127-140.

Pardini, B. J., and Wurster, R. D., 1984, Identification of the sympathetic preganglionic pathway to the cat stellate ganglion, J. Auton. Nerv. Syst., 11:13-25.

Petras, J. M., and Cummings, J. F., 1972, Autonomic neurons in the spinal cord of the rhesus monkey: a correlation of the findings of cytoarchitectonics and sympathectomy with fiber degeneration following dorsal rhizotomy, J. Comp. Neurol., 146:189-218.

Petras, J. M., and Faden, A. I., 1978, The origin of sympathetic preganglionic neurons in the dog, Brain Res., 144:353-357.

Phelps, P. E., Barber, R. P., Houser, C. R., Crawford, G. D., Salvatera,

P. M., and Vaughn, J. E., 1984, Postnatal development of neurons containing choline acetyltransferase in rat spinal cord: An immunocytochemical study, J. Comp. Neurol., 229:347-361.

Pick, J., 1970, The Autonomic Nervous System, J. P. Lippencott Company, Philadelphia.

Polosa, C., 1968, Spontaneous activity of sympathetic preganglionic neurons, Can. J. Physiol. Pharmacol., 46:887-896.

Polosa, C., Schondorf, R., and Laskey, W., 1982, Stabilization of the discharge rate of sympathetic preganglionic neurons, J. Auton. Nerv. Syst., 5:45-54.

Randall, W. C., Wurster, R. D., and Lewin, R. J., 1966, Responses of patients with high spinal transection to high ambient temperatures, J. Appl. Physiol., 21:985-993.

Rando, T. A., Bowers, C. W., and Zigmond, R. E., 1981, Location of neurons in the rat spinal cord which project to the superior cervical ganglion, J. Comp. Neurol., 196:73-83.

Reis, D. J., Granata, A. R., Joh, T. H., Ross, C. A., Ruggiero, D. A., and Park, D. H., 1984, Brain stem catecholamine mechanisms in tonic and reflex control of blood pressure, Hypertension, 6 SupII:II-7--II-15.

Rethelyi, M., 1972, Cell and neuropil architecture of the intermediolateral (sympathetic) nucleus of cat spinal cord, Brain Res., 46:203-213.

Richter, C. P., and Shaw, M. B., 1930, Complete transections of the spinal cord at different levels, Arch. Neurol. Psychiat., 24:1107-1116.

Rohlicek, C. V., and Polosa, C., 1981, Hypoxic responses of sympathetic preganglionic neurons in the acute spinal cat, Am. J. Physiol., 241:H679-H683.

Ross, C. A., Armstrong, D. M., Ruggiero, D. A., Pickel, V. M., Joh, T. H., and Reis, D. J., 1981, Adrenaline neurons in the rostral ventrolateral medulla innervate thoracic spinal cord: a combined immunocytochemical and retrograde transport demonstration, Neurosci. Lett., 25:257-262.

Ross, C. A., Ruggiero, D. A., Joh, T. H., Park, D. H., and Reis, D. J., 1984a, Rostral ventrolateral medulla: selective projections to the thoracic autonomic cell column from the region containing C1 adrenaline neurons, J. Comp. Neurol., 228:168-185.

Ross, C. A., Ruggiero, D. A., Park, D. H., Joh, T. H., Sved, A. F., Fernandez-Pardal, J., Saavedra, J. M., and Reis, D. J., 1984b, Tonic vasomotor control by the rostral ventrolateral medulla: effect of electrical or chemical stimulation of the area containing

C1 adrenaline neurons on arterial pressure, heart rate, and plasma catecholamines and vasopressin, J. Neurosci., 4:474-494.

Rubin, E., and Purves, D., 1980, Segmental organization of sympathetic preganglionic neurons in the mammalian spinal cord, J. Comp. Neurol., 192:163-174.

Ryall, R. W., 1967, Effect of monoamines upon sympathetic preganglionic neurons, Circ. Res., 20 Sup3: 83-87.

Saper, C. B., Loewy, A. D., Swanson, L. W., and Cowan, W. M., 1976, Direct hypothalamo-autonomic connections, Brain Res., 117:305-312.

Sato, A., 1973, Spinal and medullary components of the somatosympathetic reflex discharges evoked by stimulation of group IV somatic afferents, Brain Res., 51:307-318.

Sato, A., and Schmidt, R. F., 1973, Somatosympathetic reflexes: afferent fibers, central pathways, discharge characteristics, Physiol. Rev., 53:916-947.

Schramm, L. P., Adair, J. R., and Stribling, J. M., 1975, Preganglionic innervation of the adrenal gland of the rat. A study using horseradish peroxidase, Exp. Neurol., 49:540-553.

Schramm, L. P., Livingstone, R. H., and Knuepfer, M. M., 1985, Spinal transection elevates renal nerve sympathetic activity in anesthetized rats, Society for Neuroscience Abstracts, 11:35.

Schramm, L. P., Stribling, J. M., and Adair, J. R., 1976, Developmental reorientation of sympathetic preganglionic neurons in the rat, Brain Res., 106:166-171.

Seller, H., 1973, The discharge pattern of single units in thoracic and lumbar white rami in relation to cardiovascular events, Pflugers Arch., 343:317-330.

Seybold, A. V., and Elde, R. P., 1984, Receptor autoradiography in thoracic spinal cord: correlation of neurotransmitter binding sites with sympathoadrenal neurons, J. Neurosci., 4:2533-2542.

Smith, O. A., 1965, Anatomy of central neural pathways mediating cardiovascular functions, in: "Nervous Control of the Heart," W. C. Randall , ed., Williams and Wilkins Company, Baltimore, pp. 34-53.

Swanson, L. W., and Sawchenko, P. E., 1983, Hypothalamic integration: organization of the paraventricular and supraoptic nuclei, Annu. Rev. Neurosci., 6:269-324.

Takahashi, T., and Otsuka, M., 1975, Regional distribution of substance P in the spinal cord and nerve roots of the cat and the effect of dorsal root section, Brain Res., 87:1-11.

Takano, Y., and Loewy, A. D., 1984, [3H] Substance P binding in the intermediolateral cell column and striatum of the rat, Brain Res., 311:144-147.

Tan, C. K., and Wong, W. C., 1975, An ultrastructural study of the synaptic glomeruli in the intermediolateral nucleus of the rat, Experientia, 31:201-203.

Torigoe, Y., Cernucan, D., Nishimoto, J.A.S., and Blanks, R.H.I., 1985, Sympathetic preganglionic efferent and afferent neurons mediated by the greater splanchnic nerve in rabbit, Exp. Neurol., 87:334-348.

Unnerstall, J. R., Kuhar, M. J., Grzanna, R., and Schramm, L. P., 1984, Alpha-2 binding sites in the thoracic spinal cord: Autoradiographic evidence for postsynaptic localization in the intermediolateral cell column, Society for Neuroscience Abstracts, 10:713.

Villanueva, L., Cadden, S. W., and Lebars, D., 1984, Evidence that diffuse noxious inhibitory controls (DNIC) are mediated by a final post-synaptic mechanism, Brain Res., 298:67-74.

Wallin, G., and Stjernberg, L, 1984, Sympathetic activity in man after spinal cord injury, Brain, 107:183-198.

Wong, W. C., and Tan, C. K., 1980, The fine structure of the intermediolateral nucleus of the spinal cord of the monkey (Macaca facicularis), J. Anat., 130:263-277.

Wurster, R. D., and Randall, W. C., 1979, Cardiovascular responses to bladder distention in patients with spinal transection, Am. J. Physiol., 228:1288-1292.

Wyszogrodski, I., and Polosa, C., 1973, The inhibition of sympathetic preganglionic neurons by somatic afferents, Can. J. Physiol. Pharmacol., 51:29-38.

Yoshimura, M., and Nishi, S., 1982, Intracellular recordings from lateral horn cells of the spinal cord in vitro, J. Auton. Nerv. Syst., 6:5-11.

Young, J. Z., 1939, Partial degeneration of the nerve supply of the adrenal. A study in autonomic innervation, J. Anat. (Lond), 73:540-550.

Zhang, T.-.X., Rohlicek, C. V., and Polosa, C., 1982, Responses of sympathetic preganglionic neurons to systemic hypercapnia in the acute spinal cat, J. Auton. Nerv. Syst., 6:381-389.

NEUROANATOMICAL SUBSTRATES OF CARDIOVASCULAR AND EMOTIONAL - AUTONOMIC REGULATION

James S. Schwaber

E. I. du Pont de Nemours and Company
Central Research and Development Department
Experimental Station
Wilmington, Delaware 19898

INTRODUCTION

During the past decade a specific group of brain nuclei has come to be seen as important in the central neuronal regulation of the circulation. These highly interconnected nuclei have gained prominence principally because of their relationship to the nucleus tractus solitarii (NTS) and autonomic preganglionic neuronal populations in the dorsal motor nucleus of the vagus nerve (DMN), nucleus ambiguus (NA) and spinal cord. Using modern tract-tracing methods, investigators have mapped the central connections of these visceral sensory and autonomic nuclei and delineated the network of cell groups illustrated in Fig. 1. In addition, immunochemical techniques have provided biochemical information indicating an abundance of peptides and other neurotransmitters within these cell groups. As a consequence of these findings, a picture of the cellular basis for central cardiovascular regulation is emerging.

The present review summarizes this recent information with emphasis on the circuitry and mechanisms involved in emotional influences, particularly stress, on cardiovascular and other autonomic/endocrine functions. It is clear that many of the cell groups and pathways described are also involved in the maintenance of vascular tone and in the integration of endocrine and visceral functions, and these functions are briefly discussed.

In describing the central circuitry for cardiovascular control it is important to first focus on central structures with a direct anatomical

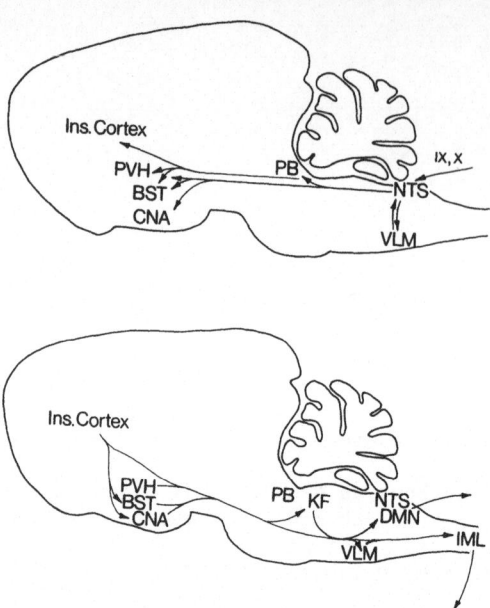

Fig. 1. Schematics of the rat brain in the sagittal plane. The
approximate positions are shown of neuronal groups, discussed in
the chapter, having a role in regulation of the cardiovascular
system. The top schematic shows connections among these
structures reflecting ascending, efferent projections of the NTS.
The bottom schematic shows descending projections afferent to the
preganglionic cell groups in the DMN, NA and spinal cord and to
the NTS.

Abbreviations: BST – bed nucleus of the stria terminalis; CNA –
central nucleus of the amygdala; DMN – dorsal motor nucleus of
the vagus nerve; Ins. Cortex – insular cortex; IML –
intermediolateral cell column; KF – Kolliker-Fuse nucleus; NTS –
nucleus tractus solitarii; PB – parabrachial nucleus of the pons;
PVH – paraventricular nucleus of the hypothalamus; VLM –
ventrolateral medulla; IX – ninth cranial (glossopharyngeal)
nerve afferent fibers; X – tenth cranial (vagus) nerve afferent
and cardiac efferent fibers.

relationship to the cardiovascular periphery. Consequently, this review
first concentrates on a description of the specifically cardiovascular
subdivision of the NTS and the cardiomotor preganglionic neurons of the
DMN, nucleus ambiguus and spinal cord. These structures, with their
cardiovascular connections and involvement in the baroreceptor reflexes

remain focal points for current research and for description of the cardiovascular relevance of other cell groups. The present review then describes nuclei directly connected to the NTS/DMN in three increasingly cephalic groupings:

(1) brainstem structures tied to the baroreflexes and the maintenance of vascular tone, including the parabrachial nucleus of the pons and the ventrolateral medulla;

(2) hypothalamic/preoptic structures involved in endocrine regulation including the paraventricular nucleus of the hypothalamus and the median preoptic nucleus, and;

(3) telencephalic structures involved in the translation of environmental stimuli and their emotional consequences into cardiovascular effects; namely the central nucleus of the amygdala and related basal forebrain structures and the insular cortex.

Since these structures were selected by their interconnectivity and relation to autonomic nuclei, some cell groups apparently important to cardiovascular regulation, such as the raphe nuclei, have not been discussed.

CELL GROUPS IN THE BARORECEPTOR REFLEXES: PRIMARY AFFERENTS TO THE NUCLEUS OF THE SOLITARY TRACT

Baroreceptors of the aortic arch and the carotid sinuses monitor the level of arterial blood pressure and contribute reflexively to its regulation (Heymans and Neil, 1958; Kirchheim, 1976). Baroreceptor afferents project centrally within the aortic depressor and carotid sinus branches of the IXth (glossopharyngeal) and Xth (vagus) cranial nerves respectively. The location of the first synapse of these nerves has been the subject of intense study. Cajal (1909) originally described afferents from the IXth and Xth cranial nerves as projecting to the NTS, including "circumferential", medial and commissural portions of the nucleus, in a variety of species. This finding was confirmed by studies of these nerves which localized degenerating myelinated fibers in the NTS following nerve section, although most studies reported vagal afferents to also project into the descending tract of the trigeminal nerve (Foley and DuBois, 1934; Ingram and Dawkins, 1945; Torvik, 1956; Rhoton et al., 1966).

Specific projections of the IXth and Xth nerves to the NTS were studied by Cottle (1964) in the cat by tracing degenerating fibers of the rootlets of each nerve. The rostral third of the NTS was shown to receive only glossopharyngeal afferents, the caudal regions only vagal fibers. Intermediate levels of the nucleus contained the most intense degeneration from both nerves; and Cottle concluded that this region contains the site of baroreceptor afferent termination. More recently Beckstead and Norgren (1979) examined the central projections of each nerve in the monkey, following injection of radiolabeled amino acids into each nerve's sensory ganglion. Autoradiographic examination confirmed that the pattern of innervation of both nerves closely resembles that found by Torvik (1956) and Cottle (1964), and provided more detail on the specific distribution of afferent fields of each nerve within the NTS.

These neuroanatomical studies of the whole IXth and Xth nerves do not provide data on the specific projections of the baroreceptor afferents. Rather, they define the limits within which these afferents must distribute. Many recent studies have sought to circumvent this problem by use of anterograde, transganglionic transport of HRP from the aortic or carotid sinus nerves. The first study of this kind was that of Katz and Karten (1979), who traced the aortic nerve of the pigeon specifically to the subnucleus lateralis sulcalis dorsalis of the NTS, located directly dorsal to the tractus solitarius.

Studies using this technique of tracing the central distribution of nerve branches arising from baroreceptors as well as from other specific viscera have provided compelling evidence for the presence of a visceral topographic map within the NTS. For example, gastric afferents project to parvocellular portions of the nucleus (Leslie et al., 1982) and pulmonary afferents end in magnocellular ventrolateral regions (Kalia and Mesulam, 1980). These results, taken together with results of physiological investigations of the projections of functionally specified afferents, particularly on the localization of baro- and chemo-receptor afferents, show the NTS to be divided into functionally and cytoarchitectonically distinct subnuclei (Donoghue et al., 1982; von Euler et al., 1973). Numerous attempts have been made to partition the NTS on the basis of alignment of these specific fields of peripheral sensory termination with cytoarchitectonic distinctions visible in Nissl stained sections. There is, however, little agreement on the number or nomenclature for these areas, which have been subdivided into as many as 26 distinct subnuclei

(Katz and Karten, 1983). Some agreement has begun to emerge on the most
obvious cell groups, and Fig. 2 shows our parcellation and nomenclature
for an intermediate level of the rabbit NTS (Higgins and Schwaber, 1981b).

Aortic Nerve Projections to the Dorsomedial Subnucleus of the NTS

Studies of the aortic nerve in the rabbit indicate that
cardiovascular afferents principally terminate dorsally and medially
within the nucleus (Wallach and Loewy, 1980; Higgins and Schwaber,
1981a,b). The aortic nerve in the rabbit provides an especially good
model system in this regard since it permits the separation of
baroreceptor afferents from the chemoreceptor afferents that are present
in the carotid sinus nerve and in the aortic nerve of most species. There
are no chemoreceptors in the aortic arch of the rabbit (Heymans and Neil,
1958; Gabriel and Seller, 1970). In addition, the aortic nerve in the
rabbit usually runs independent of the vagus nerve as far as the nodose
ganglion, making it easily accessible. Aortic nerve afferents enter the
ipsilateral medulla 1-2 mm rostral to the obex and travel caudally within
the lateral portions of the tractus solitarius. They distribute most
heavily to rostrocaudal levels of the NTS between 0.8 mm rostral and 1 mm
caudal to the obex. At these levels, the dorsomedial and medial subnuclei
receive the bulk of the aortic nerve afferents, with lesser inputs to
ventromedial and commissural subnuclei. Some aortic nerve afferents
appear to cross the midline in the commissural subnucleus and distribute
to the contralateral dorsomedial and medial subnuclei. Occasional fibers
are also present in the lateral subnucleus, the dorsomedial quadrant of
the area postrema and within the DMN.

In other species, the precise location of central baroreceptor
afferent projections again points up the importance of regions dorsal to
the tractus solitarius. In the Wistar rat, the aortic branch of the vagus
nerve, which may also contain only baroreceptor fibers (Sapru and Krieger,
1977; Sapru, Gonzalez and Kreiger, 1981), has a projection pattern similar
to that of the rabbit with the heaviest innervation reaching the region of
the ipsilateral NTS just dorsal and medial to the tractus solitarius at
rostrocaudal levels surrounding the obex (Higgins et al., 1984). In the
cat a number of neurophysiological studies suggest that the region capping
the tractus solitarius dorsally contains the first synapse in the
baroreceptor reflex (Seller and Illert, 1969; Gabriel and Seller, 1970;
Donoghue et al., 1982, 1984; Spyer et al., 1984). Transganglionic HRP
studies in the cat show that the aortic nerve, which contains a mixture of

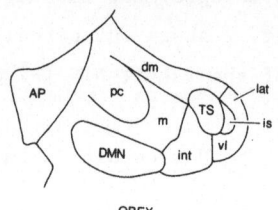

OBEX

Fig. 2. Schematic of the dorso-medial rabbit medulla in the transverse
plane of section at the level of the obex. The schematic shows
our subdivision of the NTS and nomenclature for subnuclei at
intermediate levels of the nucleus. The medial subnucleus (m)
occupies the NTS medial to the tractus solitarius (TS) and dorsal
to the dorsal motor nucleus (DMN) with the exceptions of the
dorsomedial subnucleus (dm) and the parvocellular subnucleus
(pc). The dorsomedial subnucleus is described more completely in
the text, and receives cardiovascular and limbic afferents as
well as a distinct input from the area postrema (Higgins and
Schwaber, 1981a; Shapiro and Miselis, 1985). The parvocellular
subnucleus is recipient to gastric afferents (Leslie et al.,
1982) and is distinguished as a domain of dendrites of DMN
preganglionic neurons projecting to the gut (Shapiro and Miselis,
pesonal communciation). Immediately lateral to the DMN is a
neuron-sparse intermediate subnucleus (int) which receives
descending projections from the amygdala and the bed nucleus of
the stria terminalis (Schwaber et al., 1982). This subnucleus
contains a heterogeneous population of neurons of various sizes
and shapes including large cells resembling those of the
subjacent reticular formation. Ventrally it contains vagal
efferent rootlets emanating from the DMN. The ventrolateral
subnucleus (vl) is distinguished by clusters of large,
darkly-stained cells situated among medium-sized cells, in which
the large neurons are involved in respiration and project to the
phrenic cell column. The lateral subnucleus (lat) contains
spherically-shaped neurons that are more darkly stained than
those of the dorsomedial subnucleus. The interstitial subnucleus
(is) is composed of a loose collection of neurons adjacent to the
tractus solitarius laterally and scattered cells within the
tractus proper. The commissural subnucleus (not illustrated) is
present from 0.5 to 2-3 mm caudal to the obex and is the result
of the fusion of the medial subnuclei of either side to form a
single midline structure.

chemo- and baroreceptor afferents, projects to dorsal portions of the NTS both medial and lateral to the tractus solitarius (Panneton and Loewy, 1980; Ciriello and Calaresu, 1981; Ciriello, 1983; Kalia and Welles, 1980). This may reflect a species difference in the subnuclear organization of the NTS such that the dorsomedial subnucleus is shifted laterally with respect to the tractus solitarius in the cat. Similarly, in the pigeon, the aortic nerve projects to the subnucleus sulcus dorsalis, which is located directly dorsal to the tractus solitarius (Katz and Karten, 1979, 1983).

The dorsomedial subnucleus in the rabbit (Fig. 2) is located dorsal and dorsomedial to the tractus solitarius, and ventral to the dorsal column nuclei, extending from the commissural subnucleus at caudal levels of the NTS to approximately 2 mm rostral to the obex. Morphologically, it is distinguished by the presence of numerous fibers oriented dorsomedially between the tractus solitarius and the area postrema or IVth ventricle. A computer assisted morphometric study of these neurons (Fig. 2) indicates the neurons in this subnucleus to be of medium size, elongate, and oriented in the plane of the fibers. This subnucleus is distinct from others in the NTS both by the high major to minor axis ratios of its neurons and in the angle of orientation of the major axis of the neurons. An exception, however, exists in the region directly medial to the tractus solitarius at and just rostral to the obex, where a distinct cluster of spherical cells are located. This region receives both limbic and cardiovascular inputs in both the rabbit and the rat (Higgins and Schwaber, 1981a,b; Higgins et al., 1984) and many of the cells are neurotensin and tyrosine hydroxidase positive in the rat (Higgins et al., 1984).

Several recent immunohistochemical surveys of the NTS have attempted to align the neuropeptide organization of the nucleus with the regions receiving baroreceptor inputs. Several neuropeptides show extensive fiber-terminal labeling within the dorsomedial subnucleus: in the cat met-enkephalin, substance P and somatostatin (Maley and Elde, 1982) and in the rat neurophysin II, substance P and met-enkephalin (Kalia, Fuxe and Hokfelt, 1983) and somatostatin, and neurotensin (unpublished personal observations) have been found in the dorsomedial subnucleus. Some of these peptides may function as transmitters in the baroreceptor reflex, since they have been identified within sensory neurons of the IXth and Xth nerves (Hokfelt et al., 1976).

In summary, abundant evidence has accumulated that the NTS and in particular its distinct dorsomedial subdivision is the location of the first synapse in cardiovascular afferent pathways. As such, efferent projections from this area define candidate nuclei for participation in the central circuitry for cardiovascular regulation, while afferents to this area may influence cardiovascular performance by modulating the flow of afferent information at an early synapse in regulatory pathways.

CELL GROUPS IN THE BARORECEPTOR REFLEXES; VAGAL AND SYMPATHETIC PREGANGLIONIC NEURONS

Nervous control of the heart is well known to be under the reciprocal control of its vagal and sympathetic innervations. Thus, the cardiac preganglionic populations form the final common path for central neuronal influences on the heart, and there has been intense interest in identifying structures which project to these populations. Working backwards in this way would seem to be the most straightforward way to identify the network of central structures participating in cardiovascular regulation. Unfortunately these cardiac neurons are a subset of the larger population of autonomic preganglionic neurons which mediate many other autonomic functions. Thus, not all inputs to autonomic preganglionic neurons are necessarily relevant to cardiovascular function. Considerable evidence has accumulated, however, on inputs to preganglionic populations, and recently there has been great progress on localizing cardiac preganglionics neurons.

Dorsal Motor Nucleus and Nucleus Ambiguus

The vagal cardiomotor neurons have been of special interest since the baroreceptor control of the heart is mediated primarily by modulation of the negative chronotropic influence of the vagal outflow (Heymans and Neil, 1958; Kirchheim, 1976; Levy, 1977). Vagus nerve efferents originate in two medullary nuclei, the dorsal motor nucleus (DMN) and the ventrolaterally positioned nucleus ambiguus (NA). The NA contains large alpha-motoneurons which innervate the pharyngeal-laryngeal musculature, while the neurons in the DMN are more typically autonomic preganglionic in appearance. In addition, there has been considerable anatomical and physiological evidence for the cardiovascular function of the DMN (Molhant, 1910; Malone, 1913; Getz and Sirnes, 1949; Mitchell and Warwick, 1955; Miller and Bowman, 1916). For these reasons, the DMN classically was taken to be sole source of all autonomic efferents in the vagus nerve (see Mitchell and Warwick, 1955; Spyer, 1981 for discussions).

More recently, however, by the use of retrogradely transported HRP from the central cut of the cervical vagus, it has been shown that there are smaller, autonomic-like vagal preganglionic neurons within the NA, as well as scattered between the DMN and NA (DeVito, Clausing and Smith, 1974; Miller, 1976; Robertson, et al., 1976; Geis and Wurster, 1978; Schwaber, Wray and Higgins, 1979). Furthermore as Cajal (1909) observed, the axons of NA cells exit by first coursing dorsomedially towards the DMN before turning laterally, and in the rabbit and cat these HRP labeled axons can be seen to pass very near the lateral border of the DMN (unpublished observations), making electrical stimulation or lesion of DMN neurons without NA involvement very difficult. It is now clear that the NA in fact contains cardiomotor neurons, since NA neurons can be antidromically activated from the cardiac branches in the cat (McAllen and Spyer, 1976; 1978a,b) and retrogradely labeled by HRP from injections in the epicardium and vagal cardiac branches (Geis and Wurster, 1978; Nosaka, Yamamoto and Yasunaga, 1979).

It now appears that both the DMN and NA contain vagal cardiac preganglionics in many mammalian species, although there appears to be some species variability in the proportion of neurons in each nucleus. Physiological studies have clearly shown cardiac neurons in the DMN of the rabbit (Schwaber and Schneiderman, 1975; Jordan et al., 1982) and the pigeon (Schwaber and Cohen, 1978a,b). Anterograde transport of radiolabeled amino acids from the DMN and NA has shown both nuclei to contribute to the innervation of the heart in the rabbit, rat and cat, although apparently with somewhat different cardiac targets (Schwaber, Wray and Higgins, 1979). Most of these vagal cardiac neurons appear to be in the NA in the cat (Geis and Wurster, 1978; McAllen and Spyer, 1976) while they all may be in the DMN in the pigeon (Cohen et al., 1970; Schwaber and Cohen, 1978b).

Vagal Cardiomotor Neurons

Some information has recently become available on the specific localization of cardiac neurons within the DMN and NA. In the NA of the cat and rabbit, these neurons are principally rostral to the obex and positioned ventrolaterally to the large NA neurons (Geis and Wurster, 1978; McAllen and Spyer, 1978b; Jordan et al., 1982). These NA neurons, however, are surrounded by the reticular formation and are very difficult to distinguish, thus making anatomical study of inputs to these neurons difficult.

In the DMN, the presence of a visceral topography has been described for some time. Malone (1913) initially termed the central portion of the DMN "nucleus cardiacus n. vagi", and Getz and Sirnes (1949) also concluded that the region of the nucleus surrounding the obex contains cardiac neurons. In the pigeon, cytoarchitectonically distinct subnuclei have been described that at least partially reflect a visceral topographic organization of the nucleus (Cohen et al., 1970; Schwaber and Cohen, 1978; Katz and Karten, Personal Communication. In both the pigeon and the rabbit, the vagal cardiac preganglionic neurons appear to be medium-sized neurons surrounding the level of the obex and principally located in a restricted lateral portion of the nucleus (Schwaber and Cohen, 1978; Schwaber and Schneiderman, 1975; Jordan et al., 1982). Since the DMN ventrally borders the NTS in the dorsomedial medulla, experiments on one structure often necessarily involve the other. Studies of the peripheral and central connectivity and chemoarchitecture of both structures (as a "vagal complex") are often conveniently performed together.

Spinal Cardiovascular Preganglionic Neurons

The sympathetic preganglionic neurons are largely located in the intermediolateral cell column (IML), which principally extends through the thoracic and upper lumbar spinal cord (Petras and Cummings, 1972). These neurons project to postganglionic neurons in the sympathetic ganglia and chain, where considerable mixing of functions and an extensive intermingling of projections occurs. This makes anatomical analysis of the specific distribution of the cardiovascular preganglionic neurons within the intermediolateral cell column difficult. It appears, however,that the greatest density of cardiac neurons is in the upper thoracic cord, from T1-4 (Wurster, 1977), while the vascular preganglionics distribute more widely, from T1 and L2 (Henry and Calaresu, 1972).

NTS Inputs to Preganglionic Neurons

NTS influence on cardiovascular preganglionic neurons may be by direct projections, forming disynaptic reflexes. NTS projections to the vicinity of the nucleus ambiguus have been reported (Loewy and Burton, 1978; Ricardo and Koh, 1978), but the proximity of the DMN has impeded a definitive analysis of its relationship to the NTS. NTS projections to the spinal cord appear to be largely related to respiratory function, with few neurons in cardiovascularly relevant parts of the nucleus involved (Loewy and Burton, 1978; Blessing et al., 1981). It seems more likely, however, that circuitry including other structures is involved since, for example, the "central delay" of the baroreceptor reflexes is calculated to be 20 msec. or more. In addition, the NTS does project heavily to structures which innervate the preganglionics neurons (Fig. 1).

BULBAR CARDIOVASCULAR STRUCTURES

In the past several years, considerable new information has accumulated identifying structures within the lower brain stem that are heavily connected to the NTS and to the autonomic preganglionic neurons, and are involved in the maintenance of vascular tone. This information promises to help clarify the large classical "vasomotor center" literature.

The medulla and pons have been recognized as essential for the regulation of cardiovascular reflexes and the normal maintenance of blood pressure since the later 1800´s when experiments by Dittmar and Owsjannikow in Ludwig´s laboratory were the first to localize vasomotor control. Dittmar (1870) found that pressor reflex responses to sciatic nerve stimulation could be produced in a pontine transected, "bulbar" animal, but disappeared when the medulla was destroyed. Owsjannikow (1871) then reported that, following successively lower transections of the brain stem in rabbits and cats, arterial blood pressure first fell at sections behind the caudal border of the inferior colliculi. More caudal sections produced increasingly greater falls in blood pressure until it fell to the low levels typical of a spinal animal with transections rostral to the obex. A more anatomically precise study by Dittmar (1873) concluded that, in the rabbit, the area primary for vasomotor control lies within ventral parts of the lateral reticular formation. A hugh subsequent literature (e.g., Ranson and Billingsly, 1916; Bayliss, 1923; Alexander, 1946; Chai and Wang, 1962) has attempted to localize these bulbar reflex and pressure-maintenance functions to single or multiple vasomotor "centers", but without great success in terms of cellular specificity. The recent literature, however, identifies two bulbar structures likely to play important roles in these functions: the parabrachial nucleus and the ventrolateral medulla (Fig. 1).

The parabrachial nucleus (PB) consists of an elongate collection of subnuclei surrounding the brachium conjunctivum (superior cerebeller peduncle) in the dorsolateral pons. In the cat and the rat, the nucleus is divided into connectionally and cytoarchitectonically distinct medial and lateral parts. The ventrolaterally located Kolliker-Fuse nucleus is included in the lateral part of the parabrachial complex. Attention was first directed to the PB by Norgren and his colleagues (see Norgren and Leonard, 1971, 1973). They initially described ascending projections from the rostral, gustatory part of the NTS to the medial part of the PB forming a relay for taste inputs to the forebrain. Subsequently the lateral part of the nucleus was described to receive inputs from the more caudal regions of the NTS involved with cardiovascular, respiratory and gastrointestinal autonomic functions (Beckstead, Morse and Norgren, 1980; Loewy and Burton, 1978; Ricardo and Koh, 1978; Hamilton et al., 1981). This generalized medial-lateral functional subdivision of the nucleus should not, however, be overdrawn. Gustatory and vagal afferent functions appear to be incompletely separated in the rat, with medial PB projections reaching nongustatory regions of insular cortex and the central nucleus of the amygdala. In addition, in the primate gustatory afferents may not relay in the PB at all (Beckstead et al., 1980; Norgren and Leonard, 1983).

Recent, more detailed anatomical studies by Voshart and Van Der Kooy (1981) and Fulwiler and Saper (1984) have demonstrated that, in the rat, cytoatchitectonically distinct subnuclei of the PB have connections to specific nuclear groups. These findings show that subnuclei of the lateral PB principally project to other structures implicated in autonomic regulation, both acting as a relay for ascending projections to forebrain nuclei as well as giving rise to decending projections to the brain stem and spinal cord. The descending projections appear to originate largely from the Kolliker-Fuse nucleus and to innervate 3 areas: the NTS (and DMN); the region of the intermediolateral cell column in the spinal cord, and; the ventrolateral medulla (Saper and Loewy, 1980; Fulwiler and Saper, 1984). Physiological studies in the cat (Mraovitch, Kumada and Reis, 1982) and the rabbit (Hamilton et al., 1981) also implicate the PB in cardiovascular regulation, describing strong PB influences on heart rate and blood pressure. Presumably the circuitry described above in part subserves these effects.

The Ventrolateral Medulla

Subsequent to the 19th Century studies of Dittmar (1873) outlined above, several lines of evidence have further implicated a region near the ventrolateral surface of the medulla and lower pons in cardiovascular system regulation. Attention initially was focused on this region by studies conducted by Feldberg and Guertzenstein (1972) in which application of drugs to restricted areas of the ventrolateral surface of the brainstem produced changes in blood pressure. Lesions in an area of the ventrolateral medulla close to the surface of the brain were found to cause a large fall in blood pressure, while electrical stimulation of the same area caused a pressor response and adrenal medullary catecholamine release in the rabbit (Dampney and Moon, 1980), rat and cat (Guertzenstein and Silver, 1974; Ross et al., 1983; Granata et al., 1983. This same region is a major terminal field for fibers arising from the general visceral part of the NTS (receiving cardiovascular inputs) and projects back to the NTS (Norgren, 1978; Ricardo and Koh, 1978). As noted above, it also receives inputs from PB. Numerous cells in this region of the ventrolateral medulla are retrogradely labeled from spinal injections of HRP or fluorescent tracers in cats and rabbits (Amendt et al., 1979; Blessing et al., 1981; Ross et al., 1984), and this region gives rise to highly specific projections to all levels of the intermediolateral column (Ross et al., 1984). Loewy, McKellar and Saper (1979) pointed out the potential importance of catecholamine neurons near this region to cardiovascular regulation, and subsequently many of these neurons have been shown to label with phenylethanolamine-N-methyl transferase (PNMT) as a part of the C1 adrenaline-synthesizing group (Ross et al., 1981, 1983; Hokfelt et al., 1974). Thus, this projection may be responsible for much of the abundant catecholamine innervation of the intermediolateral cell column.

The anatomical region defined by the intersection of the chemosensitive zone, connections and functional properties defined above lies in ventrolateral medulla (VLM) rostral to the level of the obex within a region classically part of the lateral reticular nucleus. Ross et al. (1984) have described the region in Nissl-stained material as being characterized by small to medium-sized neurons traversed by longitudinally-oriented fibers. The VLM is bounded by the facial nucleus rostrally, the spinal trigeminal tract and nucleus laterally, and by the nucleus ambiguus dorsally. It is distinguished from similar medullary regions caudal to the obex containing noradrenergic, A1 neurons and which do not give rise to spinal projections (Loewy, Wallach and McKellar, 1981; Blessing et al., 1981; Ross et al., 1984).

In summary, it is not unreasonable to believe that the brainstem circuitry described above including the NTS, the PB and the ventrolateral medulla functionally account for many of the cardiovascular reflex and tonic pressor functions first localized to bulbar levels over a century ago.

NEUROENDOCRINE-RELATED STRUCTURES

An alternate route by which the cardiovascular and other visceral periphery can interact with neuronal regulatory mechanisms is via humoral inputs, principally at the circumventricular organs, and via endocrine outputs, principally at the pituitary. It has been fascinating to note the development of evidence of neuronal circuitry relating these inputs and outputs to the NTS and autonomic preganglionic neurons. For example, it has already been noted that the area postrema, one of the circumventricular organs of the brain, projects to cardiovascular regions of the NTS and DMN and nucleus ambiguus, but it also projects rostrally to the PB (Shapiro and Miselis, in press). Two forebrain structures connected to the NTS, the paraventricular nucleus of the hypothalamus (PVH) and the median preoptic nucleus (MnPO), also receive circumventricular organ inputs (Fig. 1).

Paraventricular Nucleus of the Hypothalamus

The PVH was originally known as a compact group of magnocellular neurons projecting into the posterior pituitary and releasing oxytocin or vasopressin into the blood. Over the past decade, however, the nucleus has become the most intensively studied of several hypothalamic nuclei shown to have direct connections to the NTS, DMN and intermediolateral cell column (Kuypers and Maisky, 1975; Saper et al., 1976). On the basis of these connectional, cytoarchitectonic and immunocytochemical studies it has become clear that the nucleus has an additional, extensive parvocellular division and can be divided into several subnuclei with largely distinct connectional targets (Armstrong et al., 1980; Koh and Ricardo, 1980; Ono et al., 1978; Swanson and Kuypers, 1980; Swanson and

Sawchenko, 1983). The parvocellular divisions surround the magnocellular neurons. Medial and dorsally located periventricular neurons innervate the median eminence, while the longer descending projections principally arise in the other parvocellular subnuclei with very little involvement of the magnocellular division (Koh and Ricardo, 1980; Hosoya and Matsushita, 1979; Swanson and Kuypers, 1980). However, about 20 percent of the neurons projecting to brainstem and spinal cord do contain oxytocin and vasopressin and contribute predominately oxytocin fibers to the NTS, DMN and intermediolateral cell column (Swanson and Sawchenko, 1983).

Among parvocellular neurons innervating autonomic nuclei in the brain stem and spinal cord, those in the dorsal subnuclei appear to largely project to the spinal cord while more laterally and ventromedially located subnuclei project to both the spinal cord and the DMN and NTS, with a small number of neurons in these areas projecting to both (Koh and Ricardo, 1980; Swanson and Kuypers, 1980; Sawchenko and Swanson, 1981). Reciprocal connections to the PVH and particularly these subnuclei arise in the NTS, in part from the A2 noradrenergic cell group (Ricardo and Koh, 1978; McKeller and Loewy, 1981; Sawchenko and Swanson, 1981). The PVH is also reciprocally connected with the ventrolateral medulla (reciprocals are from only the more caudal, A1 levels) and the PB (Sawchenko and Swanson, 1981; Koh and Ricardo, 1980; Fulwiler and Saper, 1984).

In summary, as Swanson and Sawchenko (1983) have pointed out, largely distinct parts of the PVH participate in hormonal functions and in the circuitry for cardiovascular regulation. Nonetheless, the proximity of these parts may permit interactions and integration of these functions.

Median Preoptic Nucleus

The potential importance of the MnPO or nucleus medianus of the medial preoptic area was initially pointed out by Miselis and colleagues (Miselis, Shapiro and Hand, 1979; Miselis, 1981) who described neural efferents from the subfornical organ (SFO; a circumventricular organ), to the MnPO and to several other structures including the organum vasculosum of the lamina terminalis (OVLT; another circumventricular organ) and the entire, but particularly the magnocellular, PVH. Saper and Levisohn (1984) subsequently have indicated that the SFO projections to MnPO arise from its superficial part, which may bind and respond to peripheral angiotensin II (Buranarugsa and Hubbard, 1979), and that the MnPO appears to also project to the PVH.

Saper and Levisohn (1984) and several other investigators have shown the MnPO to receive inputs from many of the structures implicated above in cardiovascular regulation, including cardiovascular regions of the NTS, the lateral parabrachial nucleus (Saper and Loewy, 1980; Fulwiler and Saper, 1984; Lind, Van Hoesen and Johnson, 1982), the ventrolateral medulla (McKellar and Loewy, 1982) and the parvocellular PVH. Saper, Reis and Joh (1983) have shown about 70 percent of the NTS innervation to arise from A2 noradrenergic neurons and about 50 percent of the ventrolateral medulla innervation to be adrenergic.

The location of the MnPO within the tissue surrounding the anteroventral tip of the third ventricle (AV3V) makes it particularly interesting. A very large literature (reviewed recently by Brody and Johnson, 1981) indicates that this region plays some role in cardiovascular regulation and in the development of hypertension. The inclusion of the MnPO in this region, taken together with its afferent and efferent connections, suggest it plays a role in cardiovascular regulation.

In summary, the connectivity of the PVH and the MnPo relates both structures to the visceral periphery by both participation in the central circuitry related to cardiovascular regulation and by involvement in humoral and hormonal neuroregulation.

STRUCTURES MEDIATING EMOTIONAL-CARDIOVASCULAR INTERACTIONS

The powerful linkage between emotional state and visceral activity has spurred considerable controversy and a large body of research since the end of the last century, when James and Lange hypothesized that emotion in fact depends on visceral response (Fehr and Stern, 1970). As is well known, the work of Sherrington, Cannon, Bard, Hess, Penfield, Ranson, Magoun and Moruzzi established that cardiovascular and other visceral patterns of activity change moment to moment as appropriate to the requirements of changes in emotional status. Which neural structures integrate and organize these patterns has not been clear. The hypothalamus has been seen as a focal point for limbic-autonomic integration, since autonomic responses can be elicited from it (Karplus and Kreidl, 1910) and the elicitation of sham rage depends upon it (Bard, 1928). Subsequently, the work of Hilton and his colleagues on the "defense reaction" (reviewed in Hilton, 1966) and by Smith on conditioned emotional responding in the baboon (reviewed in Smith and DeVito, 1984) indicate a very restricted region of the hypothalamus, the perifornical region, to be specifically relevant. This region may partially overlap the lateral and posterior hypothalamic areas first described by Saper et al. (1976), which project to the NTS and both parasympathetic and sympathetic preganglionics.

Although the hypothalamus has classically been a focal point for theories of limbic-autonomic regulation, the central nucleus of the amygdala and the insular cortex have now come to appear to have important roles in mediating the influences of emotional status on cardiovascular function. With the introduction of sensitive neuroanatomical fiber tracing techniques these telecephalic regions, both long implicated in autonomic regulation, have been found to be directly connected with the NTS and preganglionic populations (Fig. 1). Since the descending projections of these structures traverse the region of the perifornical hypothalamus, the effects of electrical stimulation or lesion of the hypothalamus need to be interpreted carefully for involvement of the fibers of passage.

Central Nucleus of the Amygdala

Hopkins (1975) and Krettek and Price (1978) first reported long direct descending projections from the amygdala to the lower brain stem. Hopkins and Holstege (1978) described retrograde labeling confined to the amygdaloid central nucleus following large HRP injections into the medullary tegmentum as well as anterograde labeling in the NTS and DMN following large radiolabeled amino acid injections in the amygdala. Our experiments in the rabbit, following up these studies with much smaller injections confined to the NTS/DMN and to the amygloid central nucleus (central nucleus), pinpoint discrete regions of the NTS and DMN to receive projections from only the central nucleus within the amygdala (Schwaber, Kapp and Higgins, 1980; Schwaber et al., 1982; Schwaber et al., 1981).

Our data indicate that neurons throughout the body of the central nucleus, but not its capsular region, project to the NTS/DMN (Fig. 3). The great majority of projection neurons are in the large medial subnucleus, with the highest concentration in the rostral and dorsal parts of this subnucleus. The projections into the NTS/DMN are highly patterned, heavy to some regions and virtually absent to others. This pattern appears to at least partially respect the subnuclear organization of the NTS. Importantly, the central nucleus inputs include the

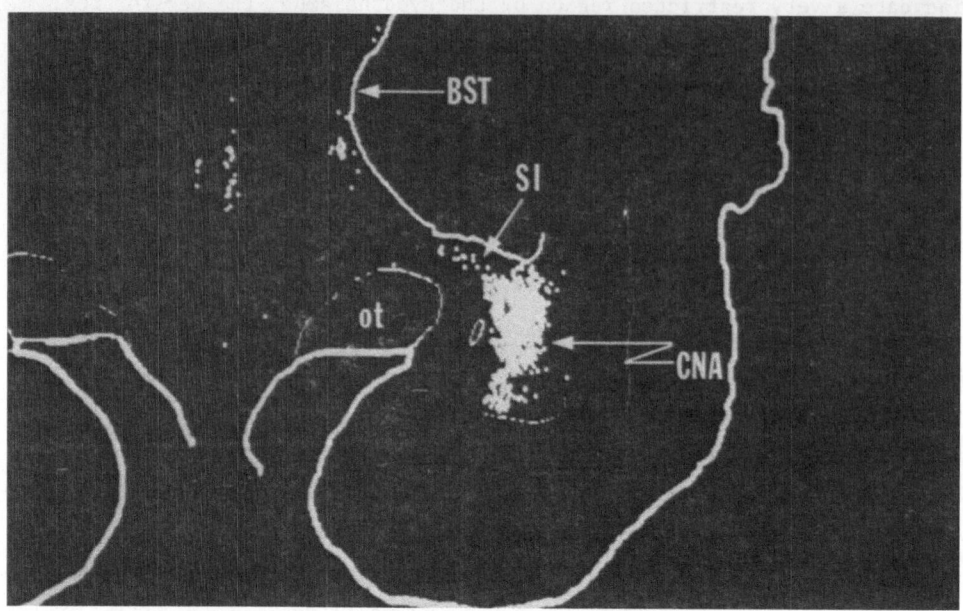

Fig. 3. Computer-generated graphic of the ventral part of the right side
of a transverse section through the rabbit forebrain. The dots
represent every cell in that section labeled with retrogradely
transported cholera toxin-HRP conjugate following injection of
the conjugate into the NTS and DMN. Neurons projecting to the
NTS and DMN can be seen to be distributed within the central
amygdaloid nucleus (CNA), with the most dense concentration of
neurons in dorsal and medial parts of the nucleus. Some labeled
neurons can be observed extending beyond the amygdala over the
optic tract (ot) into the substantia innominata (SI) and lateral
part of the bed nucleus of the stria terminalis (BST) at this
caudal level of these nuclei. Rostral to the amygdala labeled
neurons are abundant in these nuclei.

dorsomedial subucleus, thus regionally overlapping with cardiovascular
afferents.

Many neuropeptide containing cells and fiber/terminals are present
within the central nucleus and the distribution of these peptides appears
to partially respect the subnuclear structure of the nucleus (Schwaber et
al., 1981; Wray et al., 1981). Of peptide containing neurons in the
medial subdivision, some of the somatostatin and substance P positive
cells participate in the descending projections to the NTS/DMN (Higgins
and Schwaber, 1983).

The discovery of these direct descending projections has been of great interest because of the extensive evidence for amygaloid involvement in emotional behaviors, especially fear (e.g., Goddard, 1964; Ursin and Kaada, 1960; Fernandez deMolina and Hunsperger, 1962; Weingarten and White, 1978) and for profound influences on autonomic performance, for example in the defense reaction (e.g., Hilton and Zbrozyna, 1963; Bonvallet and Gary Bobo, 1972). Recent research in the rabbit by Kapp and his colleagues have made the central nucleus specifically a focal point for these amygdaloid functions. Thus, during Pavlovian fear conditioning of a bradycardic response, lesion of the central nucleus has been shown to greatly attenuate the conditioned response (Kapp et al., 1979), as have catecholamine and opioid pharmacological manipulations within the central nucleus (Gallagher et al., 1980, 1981). Electrical stimulation at sites localized specifically within the central nucleus elicits profound cardiovascular responses at low stimulation currents and not secondary to respiratory or somatic responses (Kapp et al., 1982; Applegate et al., 1983; Moruzzi et al., 1984). In addition, increases in multiunit activity in the central nucleus in response to a conditioned stimulus emerges over the course of Pavlovian conditioning (Applegate et al., 1982).

We have recently attempted to test the hypothesis that the direct central nucleus projections to the NTS and DMN mediate amygdaloid cardiovascular effects in the rabbit (Moruzzi et al., 1984). The physiological effects of activating central nucleus inputs were found to exert a profound influence on NTS neurons. In the dorsomedial NTS, where we have found overlapping innervations from aortic nerve baroreceptor afferents and central nucleus inputs (Higgins and Schwaber, 1981b), a convergence of physiological influences on the same neurons was observed. Within the DMN, vagal efferents were identified which were responsive to both aortic nerve and central nucleus stimulation. The physiological properties of the response (such as the strength and latency of the response), taken together with the existence of a heavy direct projection from the central nucleus, argues that these responses of DMN and NTS neurons are mediated by the direct amygdaloid inputs. Consequently, central nucleus cardiovascular effects may be mediated by direct inputs to DMN cardioinhibitory preganglionic neurons or by influencing the cardiovascular afferent pathway at an early synapse. The latter is consistent with abundant evidence for considerable integrative processing within the NTS (reviewed by Spyer, 1981).

Our anatomical studies of the central nucleus have shown that it is related to two structures, the substantia innominata (SI) and the bed nucleus of the stria terminalis (BST). The distribution of neurons projecting to the NTS and DMN extends dorsally, medially and rostrally beyond the central nucleus and through the sublenticular part of the SI, and then continues throughout the lateral part of the BST, forming a continuous diagonally-extending band through the basal forebrain (Schwaber et al., 1982). The anatomical similarity of these structures has been previously noted (Johnston, 1923; DeOlmos, 1972; Heimer, 1978) and their concentration of catecholamine inputs and peptides is well known (Lindvall and Bjorklund, 1974; Wray et al., 1981). Functionally, cardiovascular effects have been localized to the "anterior hypothalamic depressor area" (Hilton and Spyer, 1971) identified here as the lateral BST.

The central nucleus and also the SI and BST are reciprocally connected with the NTS (Ricardo and Koh, 1978) and the PB (Norgren, 1978). The projections arise from essentially separate populations of NTS neurons distributed within the dorsomedial and medial subnuclei (Kapp, Schwaber and Driscoll, 1985). Recently these nuclei have been found to also receive cortical inputs from the insular cortex (Saper, 1983; Kapp, Schwaber and Driscoll, 1985).

Insular Cortex

The insular cortex has been described in the rabbit and rat as lying and dorsal to the rhinal fissure overlying the claustrum (Rose, 1928; Saper, 1982a). This cell group has recently been identified as the cortical visceral/autonomic representation in the rat and mouse based on evidence for its direct projections to the NTS and autonomic preganglionic populations (Shipley, 1982; Saper, 1982a) and its receipt of NTS visceral inputs via relay in the PB, with which it is reciprocally connected (Saper, 1982b; Fulwiler and Saper, 1984; Yasui et al., 1985). Insular-amygdala projections have recently been described in several species (reviewed in Kapp, Schwaber and Driscoll, 1985), including a very heavy innervation of the regions of central nucleus, SI and BST in the rabbit containing neurons projecting to the NTS and DMN (Kapp et al., 1985). This is the only frontal cortical region to have such projections (Kapp et al., 1985). The regions of insular cortex projecting to the NTS/DMN and to the central nucleus in the rabbit are largely overlapping, although those projecting to the dorsal medulla extended more dorsally and were in layer V of the cortex while those projecting to the amygdala were

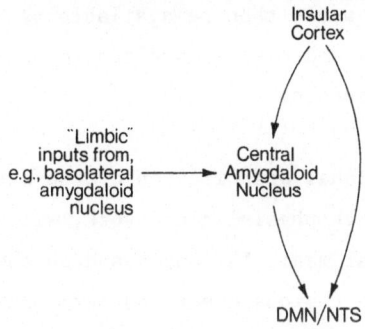

Fig. 4. Diagramatic representation of circuitry possibly subserving emotional influences, particularly stress, on cardiovascular function. These influences may, in part, be mediated by central amygdaloid neurons projecting to the NTS and DMN.

in layers III-V. Virtually no insular cortical neurons appear to project to both the amygdala and dorsal medulla (Kapp, Schwaber and Driscoll-Mendes, 1985). Cardiovascular regions of the NTS are innervated by the insular cortex in a pattern similar to that of the central nucleus in the rabbit (Schwaber et al., 1983).

The functional significance of the insular cortex projections to the central nucleus and the NTS/DMN is a matter for speculation at present. Turner, Mishkin and Knapp (1980) have shown insular cortex to have a role as a sensory association area, relaying relevant modality-specific information to the amygdala. Electrical stimulation of the insular cortex produces a wide range of cardiovascular and other autonomic responses (e.g., Kaada, 1951). These responses may be mediated via the amygdala or other structures, or via direct projections to autonomic nuclei.

DISCUSSION

Both the central nucleus and the insular cortex are connectionally well-positioned to integrate emotional behavior and autonomic response. The morphological substrate for cardiovascular regulation described here provides the basis for further experiments on the mechanisms by which these integrations are performed. One possible mechanism is pointed out by our work on central nucleus effects on NTS and DMN neurons: sensory inputs, including visceral inputs, may be integrated in the insular cortex and read out to the central nucleus where further integration with limbic inputs occurs (Fig. 4). A large repertoire of patterns of influence on cardiovascular function would then be available by the mix of projections from both insular cortex and central nucleus to both the NTS and the cardiac preganglionic neurons.

Study of the functional organization of the structures described above, taken together with physiological work, will test this hypothesis. In addition, experimental manipulations based on the unique chemoarchitecture in the regions described here may also provide insight into their functional organization.

REFERENCES

Alexander, R., 1946, Tonic and reflex functions of medullary sympathetic cardiovascular centers, J. Neurophysio., 9:205.

Applegate, C.D., Frysinger, R.C., Kapp, B.S., and Gallagher, M., 1982, Multiple unit activity recorded from amygdala central nucleus during Pavlovian heart rate conditioning in rabbit, Brain Res., 238:457-460.

Applegate, C.D., Kapp, B.S., Underwood, M.D., and McNall, C.L., 1983, Autonomic and somatomotor effects of amygdala central n. stimulation in awake rabbits, Physiol. Behav., 31:353-360.

Armstrong, W.E., Warach, S., Hatton, G.I., and McNeill, T.H., 1980, Subnuclei in the rat hypothalamic paraventricular nucleus: A cytoarchitectural, horseradish peroxidase and immunocytochemical analysis, Neuroscience 5:1931-1958.

Amendt, K., Czachurski, J., Dembowsky, K., and Seller, H., 1979, Bulbospinal projections to the intermediolateral cell column: A neuroanatomical study, J. Auton. Nerv. Syst., 1:103-117.

Bard, P., 1928, a diencephalic mechanism for the expression of rage with special reference to the sympathetic nervous system, Am. J. Physiol., 84:490-515.

Bayliss, W.M., 1923, The Vasomotor System. Longman, London.

Beckstead, R., and Norgren, R., 1979, An autoradiographic examination of the central distribution of the trigeminal, facial, glossopharyngeal, and vagal nerves in the monkey, J. Comp. Neurol., 184:455-472.

Beckstead, R.M., Morse, J., and Norgren, R., 1980, The nucleus of the solitary tract in monkey: Projections to thalamus and other brainstem nuclei, J. Comp. Neurol., 90:259-282.

Blessing, W.W., West, M.J., Chalmers, J., 1981, Hypertension, bradycardia, and pulmonary edema in the conscious rabbit after brainstem lesions coinciding with the A1 group of catecholamine neurones. Circ. Res., (49)4:949-958.

Bonvalet, M., and Gary Bobo, E., 1972, Changes in phrenic activity and
    heart rate elicited by localized stimulation of amygdala and adjacent
    structures, Electroencephalogr. Clin. Neurophysiol., 32:1016.

Brody, M.J. and Johnson, A.K., 1981, role of forebrain structures in
    models of experimental hypertension, in: "Disturbances in Neurogenic
    Control of Circulation", Clinical Physiology Series, F. M. Abboud, H.
    A. Fozzard, J. P. Gilmore, and D. J. Reis, eds., p. 105-117.

Buranarugsa, P. and Hubbard, J., 1979, The neuronal organization of the
    rat subfornical organ in vitro and test of the osmo- and
    morphine-receptor hypotheses, J. Physiol., 291:101-116.

Cajal, R.Y., 1909, Histologie due Systeme Nerveus de l´homme et des
    Vertebres, Volume 1. A. Maloine, Paris.

Chai, C.Y., Wang, S.C., 1962, Localisation of central cardiovascular
    control mechanism in lower brainstem of the cat, Am. J. Physiol.,
    202:25-30.

Ciriello, J. and Calaresu, F.R., 1981, Projections from buffer nerves to
    the nucleus of the solitary tract: An anatomical and
    electrophysilogical study in the cat, J. Auton. Nerv. Syst.,
    3:299-310.

Ciriello, J., 1983, Brainstem projections of aortic baroreceptor afferent
    fibers in the rat, Neurosci. Lett., 36:37-42.

Cohen, D.H., Schnall, A.M., MacDonald, R.L., and Pitts, L.H., 1970,
    Medullary cells or origin of vagal cardioinhibitory fibres in the
    pigeon. I. Anatomical structure of peripheral vagus nerves and the
    dorsal motor nucleus, J. Comp. Neurol., 140:299-320.

Cottle, M., 1964, Degeneration studies of primary afferents of IXth and
    Xth cranial nerves in the cat, J. Comp. Neurol., 122:329-343.

Dampney, R.A.L. and Moon, E.A., 1980, Role of ventrolateral medulla in
    vasomotor response to cerebral ischemia, Am. J. Physiol.,
    239:H349-H358.

DeOlmos, J.S., 1972, The amygdaloid projection field in the rat as studied
    with the cupric-silver method, in: "The Neurobiology of the
    Amygdala", B. E. Eleftheriou, ed., Plenum Press, New York, pp.
    145-204.

DeVito, J.L., Clausing, K.W., and Smith, O.A., 1974, Uptake and transport
    of horseradish peroxidase by cut ends of the vagus nerve, Brain Res.,
    82:269-271.

Dittmar, C., 1870, Ein Neurer Beweiss fur die Reizbarkeit de Centripetalen
    fasern des Ruckenmarks, K. Sachs. Ges. der Wiss.
    Mathematisch-Physisch Klasse. Ber., 22:18-45.

Dittmar, C., 1873, Uber die Loge des Sogenannten Gefasscentrums der

Medulla Oblongata, V. socks Ges. Wissenshaft Mathematisch—physiche Klass, 25:449.

Donoghue, S., Garcia, M., Jordan, D., Spyer, K.M., 1982, Identification and brain-stem projections of aortic baroreceptor afferent neurones in nodose ganglia of cats and rabbits, J. Physiol., 322:337–352.

Donoghue, S., Felder, R.B., Jordan, D., and Spyer, K.M., 1984, The central projections of baroreceptors and chemoreceptors in cat: A neurophysiological study, J. Physiol. 347:397–409.

Fehr, F.S. and Stern, J.A., 1970, Peripheral physiological variables and emotion: The James–Lange Theory revisited, Psychol. Bull. 74 No. 3, 411–424.

Feldberg, W. and Guertzenstein, P.G., 1972, A vasodepressor effect of pentobarbitone sodium, J. Physiol., 224:83–103.

Fernandez Demolina, A. and Hunsperger, R.W., 1962, Organization of the subcortical system governing defense and flight reactions in the cat, J. of Physiol., 200–213.

Foley, J.O. and DuBois, F.S., 1934, An experimental study of the rootlets of the vagus nerve in the cat, J. Comp. Neurol., 60:137–159.

Fulwiler, C.E. and Saper, C.B., 1984, Subnuclear organization of the efferent connections of the parabrachial nucleus in the rat, Brain Res. Rev., 7:229–259.

Gabriel, M. and Seller, H., 1970, Interactions of baroreceptor afferents from carotid sinus and aorta at the nucleus tractus solitarii, Pfluegers Arch., 318:7–20.

Gallagher, M., Kapp, B.S., Frysinger, R., and Rapp, R., 1980, β-Adrenergic manipulation in the amygdala central nucleus alters rabbit heart rate conditioning, Pharmacol. Biochem. Behav., 12:419–426.

Gallagher, M., Kapp, B.S., McNall, C.L., and Pascoe, J.P., 1981, Opiate effects in the amygdala central nucleus on heart reconditioning in rabbits, Pharmacol. Biochem. Behav., 14:497–505.

Geis, G.S. and Wurster, R.D., 1978, Localisation of cardiac vagal preganglionic soma, Neurosciences Abst., 4:20.

Getz, B. and Sirnes, R., 1949, The localization within the dorsal motor nucleus, J. Comp. Neurol., 90:95–110.

Goddard, G., 1964, Amygdaloid stimulation and learning in the rat, J. Comp. and Physiol. Psychol., 58:23–30.

Granata, A.R., Ruggiero, D.A., Park, D.H., Joh, T.H., and Reis, D.J., 1983a, Lesions of neurons in the rostral ventrolateral medulla abolish vasodepressor components of baroreflex and cardiopulmonary reflex, Hypertension 5(Suppl. V):V80–V84.

Guertzenstein, P.G. and Silver, A., 1974, Fall in blood pressure produced

from discrete region on the ventral surface of the medulla by glycine
and lesions, J. Physiol., 242:489-503.

Hamilton, R.B., Ellenberger, H.H., Liskowski, D., and Schneiderman, N.,
1981, Parabrachial area as mediator of bradycardia in rabbits, J.
Auto. Nerv, System, 4:261-281.

Heimer, L., 1978, The olfactory cortex and the central striatum, in:
"Limbic Mechanisms", K. Livingstone and O. Hornykiewics, eds., Plenum
Press, New York, pp. 95-187.

Henry, J.L. and Calaresu, F.R., 1972, Topography and numerical
distribution of neurons in the thoraco-lumbar intermediolateral
nucleus in the cat, J. Comp. Neurol., 144:205-214.

Heymans, C. and Neil, E., 1958, Reflexogenic Areas of the Cardiovascular
System, Churchill, London.

Higgins, G.A. and Schwaber, J.S., 1981a, Afferent organization for the
nucleus tractus solitarius of the rabbit: Evidence for overlap of
forebrain and primary afferent inputs, Anat. Rec. 199:144a.

Higgins, G.A. and Schwaber, J.S., 1981b, Subnuclear organization and
topography of forebrain inputs to the nucleus tractus solitarius in
the rabbit, Neurosci. Abstr., 7:116.

Higgins, G.A., Hoffman, G.E., Wray, S., and Schwaber, J.S., 1984,
Distribution and neurotensin immunoreactivity within baroreceptive
portions of the nucleus of the tractus solitarius and the dorsal
vagal nucleus of the rat, J. Comp. Neurol., 266:155-164.

Higgins, G.A. and Schwaber, J.S., 1983, Somatostatinergic projections from
the central nucleus of the amygdala to the vagal nuclei, Peptides,
4:1-6.

Hilton, S.M., 1966, Hypothalamic regulation of the cardiovascular system,
Br. Med. Bull., 22:243-248.

Hilton, S.M. and Zbrozyna, A.W., 1963, Amygdaloid region for defense
reactions and its efferent pathway to the brain stem, J. Physiol.,
165:160-173.

Hilton, S.M. and Spyer, K.M., 1971, Participation of the anterior
hypothalamus in the baroreceptor reflex, J. Physiol., 218:271-293.

Hokfelt, T., Edle, R., Johansson, O., Luft, R., Nilsson, G., and Arimura,
A., 1976, Immunohistochemical evidence for separate populations of
somatostatin-containing and substance P-containing primary afferent
neurons in the rat, Neuroscience, 1:131-136.

Hokfelt, T., Fuxe, K., Goldstein, M., and Johansson, O., 1974,
Immunohistochemical evidence for the existence of adrenaline neurones
in the rat brain, Brain Res., 66:235-251.

Hopkins, D.A., 1975, Amygdalotegmental projections in the rat, cat, and rhesus monkey, Neurosci. Lett., 1:263-270.

Hopkins, D.A. and Holstege, G., 1978, Amygdaloid projections of the mesencephalon, pons, and medulla oblongata in the cat, Exp. Brain Res., 32:329-347.

Hosoya, Y. and Matsushita, M., 1979, Identification and distribution of the spinal and hypophyseal projection neurons in the paraventricular nucleus of the rat. A light and electron microscopic study with the horseradish peroxidase method, Exp. Brain Res., 35:315-331.

Ingram, W.R. and Dawkins, E.A., 1945, The intramedullary course of afferent fibres of the vagus in the cat, J. Comp. Neurol., 82:157-168.

Johnston, J.B., 1923, Further contributions to the study of the evolution of the forebrain, J. Comp. Neurol., 35:337-481.

Jordan, D., Khalid, M.E.M., Schneiderman, N., and Spyer, K.M., 1982, The location and properties of preganglionic vagal cardiomotor neurones in the rabbit, Pflugers Arch. Ges. Physiol., 395:244-250.

Kaada, B.R., 1951, Somato-motor, autonomic and electrocorticographic responses to electrical stimulation of "rhinencephalic" and other structures in primates, cat, and dog. A study of repsonses from the limbic, subcallosal, orbito-insular, piriform and temporal cortex, hippocampus-fornix and amygdala, Acta Physiol. Scand., 24, Suppl. 83:1-285.

Kalia, M. and Mesulam, M.M., 1980, Brain stem projections of sensory and motor components of the vagus complex in the rat. II. Laryngeal, tracheobronchial, pulmonary, cardiac, and gastrointestinal branches, J. Comp. Neurol., 193:465-471.

Kalia, M. and Welles, R.V., 1980, Brain stem projections of the aortic nerve in the cat: A study using tetramethyl benzidine as the substrate for horseradish peroxidse, Brain Res., 188:523-553.

Kalia, M., Fuxe, K. and Hokfelt, T., 1983, Distribution of neuropeptide nerve terminals within the subnuclei of the nucleus of the tractus solitarius of the rat, Soc. Neurosci. Abstr., 9:772.

Kapp, B.S., Gallagher, M., Underwood, M.D., McNall, C.L., and Whitehorn, D., 1982, Cardiovascular responses elicited by electrical stimulation of the amygdala central nucleus in the rabbit, Brain Res., 234:251-262.

Kapp, B.S., Frysinger, R.C., Gallagher, M., and Haselton, J., 1979, Anygdala central nucleus lesions: Effects on heart rate conditioning in the rabbit, Physiol. Behav., 23:1109-1117.

Kapp, B.S., Schwaber, J.S., and Driscoll, P.A., 1985, Frontal cortex projections to the amygdalois central nucleus in rabbit,

Neuroscience, 15:327-346.

Karplus, J.P. and Kreidl, A., 1910, Gehirn und Sympathicus. II Mitteilung. Ein Sympathicuszentrum in Zwischenburn, _Pflugers Arch. Physiol._, 135:401-416.

Katz, D.M. and Karten, H.J., 1979, The discrete anatomical localization of vagal aortic afferents within a catecholamine-containing cell group on the nucleus solitarius, _Brain Res._, 171:187-195.

Katz, D.M. and Karten, H.J., 1983, Visceral representation within the nucleus of the tractus solitarius in the pigeon, _Columbia Livia_, _J. Comp. Neurol._, 218:42-73.

Krettek, J.E. and Price, J.L., 1978, Amygdaloid projections to subcortical structures within the basal forebrain and brainstem in the rat and cat, _J. Comp. Neurol._, 178:225-254.

Kirchheim, H.R., 1976, Systemic arterial baroreceptor reflexes, _Physiol. Rev._, 56:100-176.

Koh, E.T. and Ricardo, J.A., 1980, Paraventricular nucleus of the hypothalamus: Anatomical evidence of ten functionally discrete subdivisions, _Neurosci. Abstr._, 6:521.

Kuypers, H.G.J.M. and Maisky, V.A., 1975, Retrograde axonal transport of horseradish peroxidase from spinal cord to brain stem cell groups in the cat, _Neurosci. Lett._, 1:9-14.

Leslie, R.A., Gwyn, D.G., and Hopkins, D.A., 1982, The central distribution of the cervical vagus nerve and gastric afferent and efferent projections in the rat, _Brain Res. Bull._, 8:37-43.

Levy, M.N., 1977, Parasympathetic control of the heart, _in_: "Neural Regulation of the Heart", W. C. Randal, ed., Oxford Univerity Press, New York, pp. 95-130.

Lind, R.W., Van Hoesen, G.W., and Johnson, A.K., 1982, An HRP study of the connections of the subfornical organ of the rat, _J. Comp. Neurol._, 210:265-277.

Lindvall, O. and Bjorklund, A., 1974, The organization of the ascending catecholamine neuron system in the rat brain as revealed by the glyoxylic acid fluorescence method, _Acta Physiol. Scand. Suppl._, 412:1-48.

Loewy, A.D. and Burton, H., 1978, Nuclei of the solitary tract: Efferent projections to the lower brain stem and spinal cord of the cat, _J. Comp. Neurol._, 181:421-450.

Loewy, A.D., McKellar, S., and Saper, C.B., 1979, Direct projections from the A5 catecholamine cell group to the intermediolateral cell column, _Brain Res._, 174:309-314.

Loewy, A.D., Wallach, J.H., and McKellar, S., 1981, Efferent connections of the ventral medulla oblongata in the rat, Brain Res. Rev., 3:63-80.

Maley, B. and Elde, R., 1982, Immunohistochemical localization of putative neurotransmitters within the feline nucleus tractus solitarius, Neuroscience, 7:2469-2490.

Malone, E., 1913, The nucleus cardiacus nervi vagi and the three distinct types of muscle, Am. J. Anat., 15:121-129.

McAllen, R.M. and Spyer, K.M., 1976, The location of cardiac vagal pregalglionic motorneurons in the medulla of the cat, J. Physiol., 258:187-204.

McAllen, R.M. and Spyer, K.M., 1978a, Two types of vagal preganglionic motorneurones projecting to the heart and lungs, J. Physiol., 282:353-364.

McAllen, R.M. and Spyer, K.M., 1978b, The baroreceptor input to cardiac vagal motorneurones, J. Physiol., 282:365-374.

McKellar, S. and Loewy, A.D., 1981, Organization of some brain stem afferents to the paraventricular nucleus of the hypothalamus in the rat, Brain Res., 217:351-357.

McKellar, S. and Loewy, A.D., 1982, Efferent projections of the Al catecholamine cell group in the rat: An autoradiographic study, Brain Res., 241:11-29.

Miller, R., 1976, Identification of vagal efferent neurones; a horseradish peroxidase study, J. Physiol., 256:69-70.

Miller, R. and Bowman, J., 1916, The cardioinhibitory center, Am. J. Physiol., 39:149-153.

Miselis, R.R., 1981, The efferent projections of the subfornical organ of the rat: A circumventricular organ within a neural network subserving water balance, Brain Res., 230:1-23.

Miselis, R.R., Shapiro, R.E., and Hand, P.J., 1979, Subfornical organ efferents to neural systems for control of body water, Science, 205:1022-1025.

Mitchell, G. and Warwick, R., 1955, The dorsal vagal nucleus, Acta Anatomica., 25:371-395.

Molhant, M., 1910, Les connexions anatomiques et la valeur functionelle du noyau dorsal du vague, Nervaxe, 12:221-236.

Moruzzi, P., Schwaber, J.S., Spyer, K.M., and Turner, S.A., 1984, Amygdloid influences on brainstem neurones in the rabbit, J. Physiol., 353:43P.

Mraovitch, S., Kumada, M., and Reis, D.J., 1982, Role of the nucleus parabrachialiss in cardiovascular regulation in cat, Brain Res., 232:57-76.

Norgren, R., 1978, Projections from the nucleus of the solitary tract in the rat, Neuroscience, 3:207-218.

Norgren, R. and Leonard, C.M., 1971, Taste pathways in the rat brainstem, Science, 173:1136-1139.

Norgren, R. and Leonard, C.M., 1973, Ascending central gustatory pathways, J. Comp. Neurol., 150:217-238.

Nosaka, S., Yamamoto, T., and Yasunga, K., 1979, Localization of vagal preganglionic neurons with rat brainstem, J. Comp. Neurol., 186:79-92.

Ono, T., Nishino, H., Sasaka, K., Muramoto, K., Yano, I., and Simpson, A., 1978, Paraventricular nucleus connections to spinal cord and pituitary, Neurosco. Lett., 10:141-147.

Owsjannikow, P., 1871, Die tonischen und reflektorischen centrer der gefassnerven, V. Sachs Ges. Wissenshaft Mathematischphysische Klasse, 23:135.

Panneton, W.M. and Loewy, A.D., 1980, Projections of the carotid sinus nerve to the nucleus of the solitary tract in the cat, Brain Res., 191:239-224.

Petras, J.M. and Cummings, J.F., 1972, Autonomic neurons in the spinal cord of the rhesus monkey: A correlation of the findings of cytoarchitectonics and sympathectomy with fiber degeneration following dorsal rhizotomy, J. Comp. Neurol., 146:189-218.

Ranson, S. and Billingsley, P., 1916, Vasomotor reactions from stimulation of the floor of the fourth ventricle, Am. J. Physiol., 41:85.

Rhoton, A.L., O'Leary, L., and Furguson, J.P., 1966, The trigeminal facial, vagal, and glossopharyngeal nerves in the monkey, Arch. Neurol., 14:530-541.

Ricardo, J.A. and Koh, E.T., 1978, Anatomical evidence of direct projections from the nucleus of the solitary tract to the hypothalamus, amygdala, and other forebrain structures in the rat, Brain Res., 153:1-26.

Robertson, T.W., Wallach, J., Schneiderman, N., and Neumann, P., 1976, Fiber origins of cervical vagus nerve in rabbits, cats and rhesus monkeys investigated by injection of horseradish peroxidase, Neurosci. Abstr., 2:175.

Rose, M., 1928, Die Inselrinde des Menschen und der Tiere, J. Phychol. Neurol., 37:467-624.

Ross, C.A., Armstrong, D.M., Ruggiero, D.A., Pickel, V.M., Joh, T.H., and Reis, D.J., 1981, Adrenaline neurons in the rostral ventrolateral medulla innervate thoracic spinal cord: A combined immunocytochemical and retrograde transport demonstration, Neurosci. Lett., 25:257-262.

Ross, C.A., Ruggiero, D.A., Joh, T.H., Park, D.H., and Reis, D.J., 1983, Adrenaline synthesizing neurons of the rostral ventrolateral medulla: A possible role in vasomotor control, Brain Res., 273:356-361.

Ross, C.A., Ruggiero, D.A., Park, D.H., Joh, T.H., Sved, A.F., Fernandez-Pardal, J., Saavedra, J.N., and Reis, D.J., 1984, Tonic vasomotor control by the rostral ventrolateral medulla: Effect of electrical or chemical stimulation of Cl adrenaline containing neurons on arterial pressure, heart rate, and plasma catecholamines and vasopressin, J. Neurosci., 4:474-494.

Saper, C.B., Loewy, A.D., Swanson, L.W., and Cowan, W.M., 1976, Direct hypothalamo-autonomic connections, Brain Res., 117:305-312.

Saper, C.B. and Loewy, A.D., 1980, Efferent connections of the parabrachial nucleus in the rat, Brain Res., 197:291-317.

Saper, C.B., 1982a, Reciprocal parabrachial-cortical connections in the rat, Brain Res., 242:33-40.

Saper, C.B., 1982b, Convergence of autonomic and limbic connections in the insular cortex of the rat, J. Comp. Neurol., 210:163-173.

Saper, C.B., Reis, D.J., and Jon, T., 1983, Medullary catecholamine inputs to the anterventral third ventricular cardiovascular regulatory region in the rat, Neurosci. Lett., 42:285-291.

Saper, C.B. and Levisohn, D., 1984, Afferent connections of the median preoptic nucleus in the rat: Anatomical evidence for a cardiovascular integrative mechanism in the anteroventral third ventricular (AV3V) region, Brain Res., 288:21-31.

Sapru, H.N., Gonzalez, E., and Krieger, A.J., 1981, Aortic nerve stimulation in the rat: Cardiovascular and respiratory responses, Brain Res. Bull., 6:393-398.

Sapru, H.N. and Krieger, A.J., 1977, Carotid and aortic chemoreceptor function in the rat, J. Appl. Physiol., 42:344-348.

Sawchenko, P.E. and Swanson, L.W., 1981, A method for tracing biochemical defined pathways in the central nervous system using combined fluorescence retrograde transport and immunohistochemical techniques, Brain Res., 210:31-51.

Schwaber, J.S. and Cohen, D.H., 1978a, Electrophysiological and electron microscopic analysis of the vagus nerve of the pigeon, with particular reference to the cardiac innervation, Brain Res., 147:65-78.

Schwaber, J.S. and Cohen, D.H., 1978b, Field potential and single unit analyses of the avian dorsal motor nucleus of the vagus and criteria for identifying vagal cardiac cells of origin, Brain Res., 147, 79-90.

Schwaber, J.S. and Schneiderman, N., 1975, Aortic nerve-activated cardioinhibitory neurons and interneurons, Am. J. Physiol., 229:783-785.

Schwaber, J.S., Wray, S., and Higgins, G., 1979, Vagal cardiac innervation: Comparative contributions of the dorsalmotor nucleus and the nucleus ambiguus determined by liquid scintillation counting, Neurosci. Abst., 5:172.

Schwaber, J.S., Kapp, B.S., and Higgins, G.A., 1980, The origin and extent of direct amygdala projections to the region of the dorsal motornucleus of the vagus and the nucleus of the solitary tract, Neurosci. Lett., 20:15-20.

Schwaber, J.S., Kapp, B.S., Higgins, G.A., and Rapp, P.R., 1982, Amygdloid and basal forebrain direct connections with the nucleus of the solitary tract and the dorsal motor nucleus, J. Neurosci., 2:1424-1438.

Schwaber, J.S., Wray, S., Higgins, G.A., and Hoffman, G., 1981, The central nucleus of the amygdala: Decending autonomic connections and neuropeptide systems in the rat, Anat. Rec., 199:228A.

Seller, H. and Illert, M., 1969, The localization of the first synapse in the carotid sinus baroreceptor reflex pathway and its alteration of the afferent input, Pfluegers Arch., 306:1-19.

Shapiro, R.E. and Miselis, R.R., The neural connections of the area postrem of the rat, J. Comp. Neurol., in press.

Shapiro, R.E. and Miselis, R.R., 1985, The central organization of the vagus nerve innervating the stomach of the rat, J. Comp. Neurol., 238:473-488.

Shipley, M.T., 1982, Insular cortex projections to the nucleus of the solitary tract and brainstem visceromotor nuclei in the mouse, Brain Res. Bull., 8:139-148.

Smith, O.A. and DeVito, J.L., 1984, Central neural integration for the control of autonomic responses associated with emotion, Ann. Rev. Neurosci., 4:43-65.

Spyer, K.M., 1981, Neural organization and control of the baroreceptor reflex, Rev. Physiol. Biochem. Pharmacol., 88:23-124.

Spyer, K.M., 1984, Central control of the cardiovascular system, Recent Adv. Physiol., 10:163-200.

Spyer, K.M., Donoghue, S., Felder, R.B. and Jordan, D., 1984, Processing of afferent inputs in cardiovascular control, Clin. Exp. Hyperten., 6:173-184.

Swanson, L.W. and Kuypers, H.G.J.M., 1980, The paraventricular nucleus of the hypothalamus: Cytoarchitectonic subdivisions and organization of projections to the pituitary, dorsal vagal complex, and spinal cord as demonstrated by retrograde fluorescence double-labeling methods, J. Comp. Neurol., 194:555-570.

Swanson, L.W. and Sawchencko, P.E., 1983, Hypothalamic integration: Organization of the paraventricular and supraoptic nuclei, Ann. Rev. Neurosci., 6:269-324.

Torvik, A., 1956, Afferent connections to the sensory trigeminal nuclei, the nucleus of the solitary tract, and adjacent structures, J. Comp. Neurol., 106:51-141.

Turner, B.H., Mishkin, M., and Knapp, M., 1980, Organization of the amygdalopetal projections from modality-specific cortical association areas in the monkey, J. Comp. Neurol., 191:515-543.

Ursin, H. and Kaada, B.R., 1960, Functional localization within the amygdaloid complex in the cat, Electroenceph. Clin. Neurophysiol., 12:1-20.

Voshart, K. and Van Der Kooy, D., 1981, The organization of the efferent projections of the parabrachial nucleus to the forebrain in the rat: A retrograde fluorescent double labeling study, Brain Res., 212:271-286.

Von Euler, C., Hayward, J.N., Martilla, I., and Wyman, R.J., 1973, Respiratory neurons of the ventrolateral solitary tract to the cat: Vagal inputs, spinal connections and morphological identification, Brain Res., 61:1-22.

Wallach, J.H. and Loewy, A.D., 1980, Projections of the aortic nerve to the nucleus tractus solitarius in the rabbit, Brain Res., 188:247-251.

Weingarten, M. and White, N., 1978, Exploration evoked by electrical stimulation of the amygdala in rats, J. Physiol. Psych., 6:229-235.

Wray, S., Schwaber, J., and Hoffman, G., 1981, Neuropeptide localization in the rat central amygdaloid nucleus, Anat. Rec., 199:282A.

Wurster, R.D., 1977, Spinal sympathetic control of the heart, in: "Neural Regulation of the Heart", W.C. Randall, ed., Oxford University Press, New York, pp. 211-246.

Yasui, Y., Itoh, K., Takada, M., Mitani, A., Kaneko, T., and Mizuno, N., 1985, Direct cortical projections to the parabrachial nucleus in the cat, J. Comp. Neurol., 234:77-86.

CENTRAL AND PERIPHERAL ACTIONS OF ANGIOTENSIN II

M. Ian Phillips, Elaine M. Richards and Ahnke Van Eekelen

Department of Physiology
Box J-274 J.H.M.H.C.
University of Florida
Gainesville, Florida 32610

INTRODUCTION

The renin-angiotensin system is involved in fluid balance and cardiovascular function in several ways. Renin, the essential enzyme that converts angiotensinogen to angiotensin I, is released from the kidney to produce angiotensin under a variety of conditions, most notably low pressure in the renal artery. The increase of angiotensin I caused by renin leads to an increase of angiotensin II caused mostly by a conversion in the lungs through converting enzymes. Angiotensin II is a very powerful vasoconstrictor which acts directly on blood vessels to cause an increase in peripheral resistance and thereby increase blood pressure. At the same time, receptors in the adrenal gland are stimulated by angiotensin II to trigger the release of aldosterone. Aldosterone acts on the kidneys to retain sodium and by these two methods, vasoconstriction and aldosterone release, angiotensin II can effectively counteract low volume in the blood. So much has been known for many years and more recently, experiments with angiotensin II on the brain have shown two more effects that could be significant in counteracting decreased fluid volume in the body. When angiotensin II is injected into the brain ventricles, it produces thirst in rats and other species which compel the animals to drink and thereby, increase their fluid volume. At the same time, low doses of angiotensin directly injected into the brain cause release of vasopressin. Vasopressin has two effects. One is water retention by a direct action on the kidney and the second is vasoconstriction. the vasoconstriction by vasopressin has been debated as a mechanism for increasing blood pressure because of the apparent lack of effect of

vasopressin when injected directly into the peripheral circulation. The
reason for this seems to be the rapid compensation for vasoconstriction by
the baroreceptor reflex. Thus, wherever the baroreceptor reflex is
impaired, vasopressin causes vasoconstriction which is not compensated by
a decrease in heart rate. A fifth action of angiotensin II is to cause
vasoconstriction by an increase in sympathetic nerve activity. It can do
this in the periphery by a direct action on the catecholamine terminals
surrounding blood vessels where it increases release of norepinephrine and
decreases uptake. In the brain, angiotensin appears to increase
sympathetic output directly. Finally, a sixth method by which the
octapeptide can alter blood pressure is through an action directly on the
baroreceptor reflex. As we shall show later, angiotensin II injected
directly into the nucleus tractus solitarius (NTS), which is the first
synapse of the baroreceptor reflex, causes an increase in blood pressure.
This implies that angiotensin is released from neurons in the brain to
alter the baroreflex. Thus, with all of its effects including vasopressin
release, angiotensin could be a powerful agent for raising blood pressure
and a possible source of hypertension. In addition to these fluid balance
related effects, Ang II has been shown to have multiple interactions with
other hormonal systems such as estrogens, luteinizing hormone (LH),
corticotropin releasing factor (CRF) and with neurotransmitters such as
norepinephrine and serotonin. What we need to understand is the relative
importance of peripheral versus central effects of Ang II and whether they
interact and under what conditions. In this review, we have described
some of the actions of Ang II and tried to separate the central from
peripheral Ang II.

2. BRAIN ANGIOTENSIN

In considering the separate influences of peripheral versus central
angiotensin, we first have to separate the two sources of angiotensin. If
we look back historically at the story of the renin-angiotensin system, it
began 100 years ago with Tiegerstedt who extracted from the kidneys a
substance that could cause vasoconstriction when injected into the blood.
Fuller understanding of this finding was not clear until the late 1950´s
when various groups around the world proposed that renin was part of a
cascade which produced a vasoconstrictive substance that also stimulated
aldosterone release. During the 1960´s with the advent of
radioimmunoassays, the presence of angiotensin and its role became widely
accepted and the basis for new therapies in hypertensive disease. In the

late 1960´s, questions began to rise whether the kidney was the only source of renin. In 1971, two groups, Ganten et al. in Germany and Fisher-Ferraro et al. in Argentina, published papers on the renin-like activity that they found in the brain. However, these results were not accepted without question. Day and Reid, in 1976, made a strong case for the renin-like activity that had been discovered earlier being cathepsin D which could cleave angiotensinogen to angiotensin at a low pH. In 1972, Yang and Neff discovered converting enzyme in the brain. In 1975, Ganten et al. and Phillips et al. independently showed that the blood pressure of spontaneously hypertensive rats could be lowered by central injections of the angiotensin II antagonist, Sar^1ala^8-angiotensin II. Cathepsin D and brain renin were finally separated by Hirose et al. in 1978 using the correct pH studies and assays. The levels of renin in the brain were fairly low. Angiotensinogen was indicated in the brain by physiological and isoelectric focusing experiments which established its presence independent of peripheral or blood-borne angiotensinogen (Printz et al., 1982). Immunocytochemistry demonstrated the presence of fibers and cells staining for angiotensin II-like material in the brain (Phillips et al., 1979) (Fig. 1). The cell bodies were located in the supraoptic nucleus (SON) and paraventricular nucleus (PVN). Of all the antibodies made for

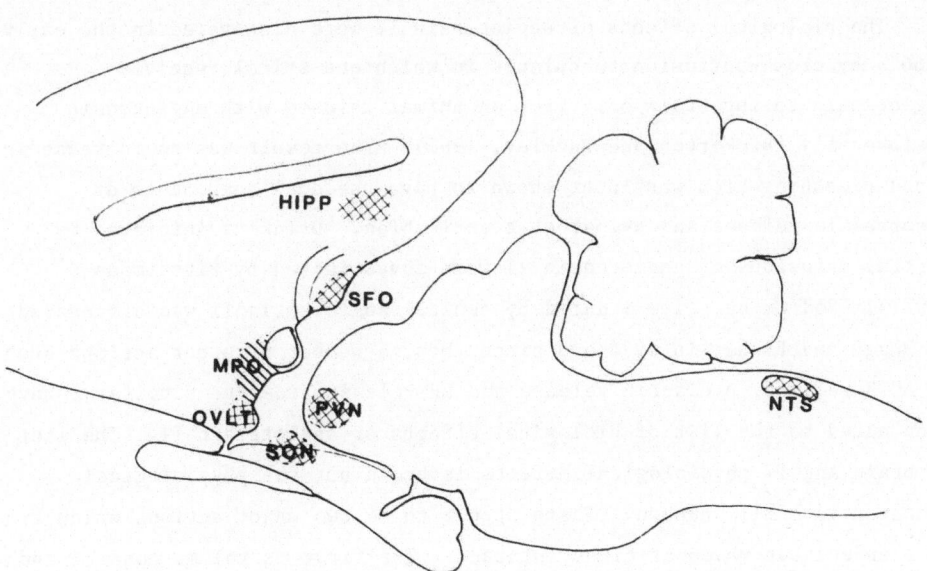

Fig 1. Distribution of Ang II-like immunoreactivity in brain cells based on many studies. Fiber tracts are not shown. (For abbreviations, see Fig. 2).

use in radioimmunoassay, however, only two antibodies seemed to be successful for immunocytochemical staining (Ganten et al., 1978; Fuxe et al., 1976b; Brownfield et al., 1982; Kilcoyne et al., 1980). The difficult problem for establishing angiotensin in the brain independent of peripheral or blood-borne angiotensin was overcome by looking for angiotensin in cells grown in culture which would have no peripheral angiotensin present. Angiotensin-like material was found in cultured neurons of 21-day-old rat brains with the use of immunocytochemistry (Weyhenmeyer et al., 1980). This was later proven to be Ang II when we characterized the immunoreactivity (Raizada et al., 1984b). Surprisingly, all the components of the renin-angiotensin system were found in neuroblastoma cells (Fishman et al., 1981; Inagami et al., 1982). Finally, angiotensin II in brain tissue has now been quantified and characterized using high performance liquid chromatography (HPLC) (Hermann et al., 1984; Phillips and Stenstrom, 1985).

In addition to establishing the presence of angiotensin in the brain, angiotensin receptors have been localized in several locations including the septal region, the superior colliculus, the nucleus tractus solitarius (NTS) and the anterior wall of the third ventricle including two important circumventricular organs, the subfornical organ (SFO) and the organum vasculosum lamina terminalis (OVLT) (Fig. 2).

The biological effects of angiotensin II were discovered in the early 1960's by cross-perfusion techniques in which one animal received angiotensin to the brain only from an animal infused with angiotensin peripherally (Bickerton and Buckley, 1961). The result was an increase in blood pressure which was later shown to have the dual components of vasopressin release and sympathetic activation. Drinking initiated by central infusions of angiotensin II were investigated by Fitzsimons (1971). Sodium appetite induced by central angiotensin II was discovered by Buggy and Fisher in 1974 and since then, a number of other actions such as ACTH release, prolactin release and LH release from the pituitary, have been added to the list of biological effects of angiotensin II. The list of brain Ang II physiological effects is broad but the physiological function is still unknown. There appear to be two major actions which are independent but which at times interact. The first is volume control and the second is reproduction. All the effects on thirst, AVP release, sodium appetite and blood pressure can be related to homeostatic responses for preserving blood volume. The interaction with estrogen and the release of LH point to a role in control of the reproductive cycle.

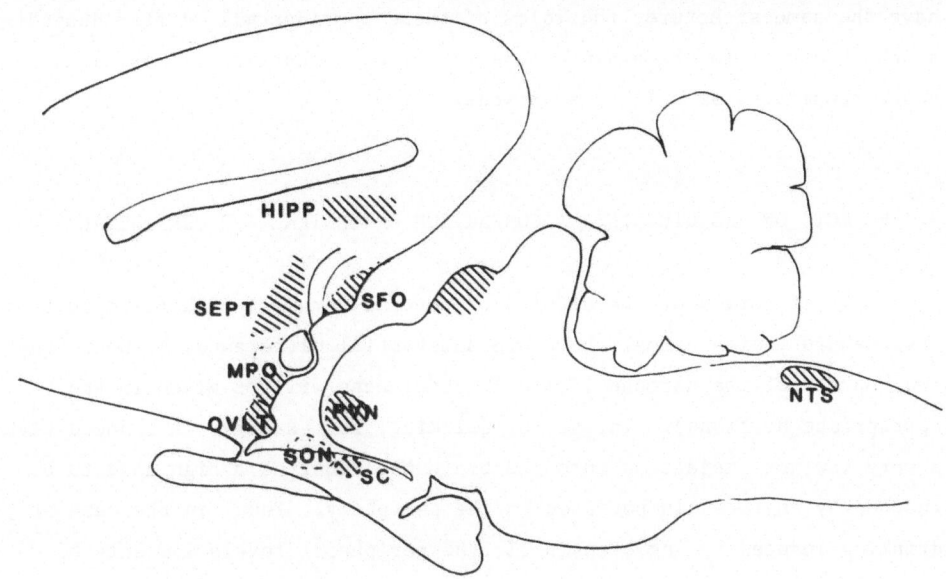

Fig 2. Distribution of Ang II receptors in the rat brain, based on
studies using homogenates of excised tissue and autoradiography.
(The presence of receptors in SON is minor or absent.)

Hipp: Hippocampus

Sept.: Septum

SFO: Subfornical organ

OVLT: Organum vasculosum lamina terminalis

MPO: Median preoptic

PVN: Paraventricular nucleus

SC: Suprachiasmatic nucleus

NTS: Nucleus tractus solitarius

There still remain many problems to be resolved such as the synthesis
of angiotensin and the function of brain angiotensin. However, the main
task of establishing that brain angiotensin exists independently of
peripheral angiotensin is now done. The thrust of future studies will be
in the molecular genetic approach to clone the DNA and understand the
conditions for the expression of mRNA for angiotensin II. In addition to
brain angiotensin, there is very good indication that angiotensin is
independently synthesized in the kidneys, uterus, testicles and vascular
endothelium cells as well as other tissues. Thus, the brain angiotensin
is but one of several locally produced angiotensins. Although they may

have the same structure, the roles of these peptides will differ depending on the tissue. In the brain, there is evidence that Ang II serves as a neurotransmitter as well as a hormone.

3. EFFECTS OF ANGIOTENSIN: A COMPARISON OF CENTRAL AND PERIPHERAL

A brief survey of the effects of central versus peripheral effects of angiotensin indicate that there are substantial differences between these two sources of the hormone (Table I, references will be given in the appropriate sections). Thirst, or drinking, for example, is induced with a very low dose injection into the brain but requires a high dose to be reached by angiotensin perfused in the periphery. Thus, in the case of drinking induced by angiotensin II, the peripheral levels may only be experienced in conditions of severe dehydration. The central actions of angiotensin on thirst, however, may be involved under more physiological conditions. Blood pressure responses to injections into the brain show a slower increase than with peripheral injections and a longer-lasting effect. This difference probably reflects the difference of mechanisms. In the periphery, angiotensin has a local action on the receptors of the blood vessels. In the brain, the blood pressure response requires the combined actions of vasopressin release, sympathetic activation and inhibition of the baroreceptor reflex. Vasopressin release to central injection of angiotensin has been controversial. All investigators agree that central injections whether in the brain of dogs or rats produce a prompt release of vasopressin. In the periphery, infusions of angiotensin have produced variable results, some finding vasopressin released and others finding it is not. Sympathetic nerve activation is increased by central injections as indicated by the rise in blood pressure which can be inhibited by hexamethonium blockade. Also, injections of angiotensin into the brain lead to a rise in norepinephrine. In the periphery, angiotensin interacts with norepinephrine postganglionic terminals on blood vessels, increasing the action of sympathetic activity when present. Sodium appetite is increased by central injections of Ang II. It is still debated whether peripheral infusions of Ang II stimulate sodium appetite or are due to mineralocorticoid interactions which may have their effects on the brain. Heart rate in response to central injections of angiotensin generally increases. This indicates a special action of angiotensin since the normal response to an increased blood pressure, would lead to a decrease in heart rate. Peripheral infusions of angiotensin on the other hand are reported to decrease heart rate. Prolactin release has been

TABLE 1

Summary of Effects of Angiotensin by Different Routes of Administration

Actions	Central	Peripheral
Drinking	Low dose	High dose required
Blood pressure increased	Slow rise/long duration	Fast rise/short duration
	Complex mechanisms	Local vasoconstriction
Heart rate	Up or no change	Down
AVP release	Prompt	Questionable
Symp. nerv activation	Increase	Local NE changes
Catecholamines	Levels increased in plasma/brain	Prolongs action in periphery
Na$^+$ appetite	Slow onset/Long lasting	Questionable
Prolactin release	Positive	Questionable
LH release	Positive	Questionable
ACTH release	Positive	Not released without CRF being present
Natriuresis	Stimulated	Inhibited
Aldosterone secretion	Inhibited	Stimulated

shown with angiotensin added to the media of pituitary cells in culture.
So far, prolactin release with peripheral infusions of angiotensin has not
been established. ACTH is also released by central injections of
angiotensin II. These injections were aimed at the anterior pituitary and

are presumed to be by direct action on the corticotropes of the anterior pituitary. Peripheral angiotensin has been shown to release ACTH but the doses are very high. Finally, CRF, while it is not known to be released by central angiotensin II, interacts in a way that increases the release of ACTH. Again, peripheral injections or infusions of angiotensin II have to be used in large doses to interact with CRF for ACTH release.

In summary, there are a number of differences between central and peripheral injections of Ang II. Some of these are a matter of degree. For example, the dose difference with thirst, but this may reflect different mechanisms. It is hard to assess the actual dose equivalency between blood-borne angiotensin and angiotensin injected into the brain. Clearly, in the case of vasopressin and thirst, the response is easy to obtain centrally but a difficult response to achieve with peripheral infusions. One of the reasons for the difference is that direct injection into the brain allows angiotensin to reach all receptor sites. Peripheral infusion of angiotensin can only reach specific limited sites which are predominantly the circumventricular organs (CVO). The types of receptors could be different and certainly the connections from the binding sites are different.

4. MECHANISMS OF PERIPHERAL-CENTRAL INTERACTIONS

In order for angiotensin to get into the brain from the blood it must cross the blood brain barrier (BBB). Careful studies of Ang II in cerebrospinal fluid of rats and dogs showed that none crossed the BBB from plasma (Schelling et al., 1977; Suzuki et al., 1983). Normally, the BBB would exclude large molecules from brain tissue. This can be seen when an injection of a protein such as horseradish peroxidase is made i.v. (Brightman, 1977). Horseradish peroxidase (HRP) does not cross into brain tissue although it reaches the choroid plexus. The one exception to this is that the tracer spreads into the tissues of circumventricular organs (CVO). These organs are dotted around the ventricular system and have no blood brain barrier due to the presence of fenestrated capillaries and lack of tight junctions. The spread outside the circumventricular organs into brain parenchyma, however, appears to be very limited even when we injected high doses of HRP. The mechanism for this limitation is not entirely clear and several mechanisms are possible. Protein diffusion is limited by molecular size. The intricate tangle of arterioles, venules and capillaries allows re-uptake by adjacent leaky vessels. In this case, the CVOs would act like a push-pull apparatus. The escaping protein attracts water by oncotic forces which slow diffusion. Finally, we have noted that the CVOs are rich in glial cells located in close proximity to the capillary walls. The presence of these glial cells could restrict diffusion by taking up proteins or having on their surfaces enzymes that break them down. Obviously, more research is needed to establish the roles of CVO glial cells. For whatever reason, under physiological conditions access to the brain via the circumventricular organs is confined to the tissue immediately surrounding the microvascular bed. A consequence of this finding is that in hypertension where the arterial pressure is high, substances such as angiotensin II from plasma could be forced further into the brain beyond the capillary-glial tangle and not taken back up, creating a vicious cycle of increasing spread of angiotensin leading to maintained hypertension.

Over the surface of the organum vasculosum laminae terminalis (OVLT) and subfornical organ (SFO), there are tight junctions which exclude passive diffusion across the surface from CSF to the brain tissue (Phillips et al., 1978). The only uptake of protein from CSF is via active tanycytes. Around the organs the ependyma have gap junctions and protein material can diffuse into the surrounding tissue through these gaps. However, again there seems to be a barrier - this time keeping proteins out of the CVOs. Thus, the physiology of the blood brain barrier produces a compartmentalization between the ventricular system and the vascular system. The result is that a peptide in the CSF may have different effects from its effects via the blood. Bradykinin when injected into the blood causes vasodilation and when injected into the brain causes vasoconstriction (Lewis and Phillips, 1984). With Ang II, the many differences reported between the effects of central injections and peripheral infusions (see Table 1) may result from the different compartments and their receptors and connections (Fig. 3). Only when excessive doses are given does the compartmentalization break down. This might be due to forcing the tight junctions open or by overflowing diffusion in CVOs. Therefore, to understand the true physiology of Ang II, it is essential to maintain an intact BBB.

Angiotensin receptors in the brain. In the rat, studies of the site of action of angiotensin II have focused on the SFO and the OVLT. The evidence for their role as central angiotensin II receptors are very similar. Lesioning one or the other results in a loss of response to angiotensin II (Simpson and Routtenberg, 1973; Buggy et al., 1975). Very

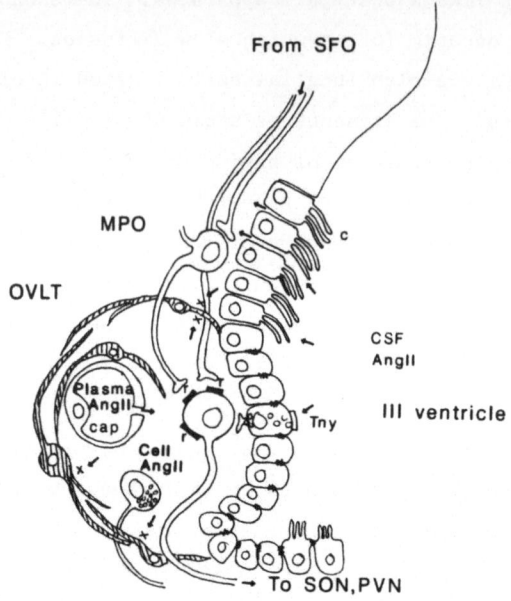

Fig 3. Glial barrier to Ang II diffusion in a circumventricular organ
(CVO). We have found staining to glia fibrillary protein (GFP)
antibodies in the area postrema, subfornical organ and OVLT. the
role of these glia is hypothesized to limit the spread of peptides
that leak from the blood into the CVO while preventing infusion of
the endogenous brain peptide from brain parenchyma into the CVO.
The ependyma surface of the CVO has tight junctions so that the
only way for Ang II in the third ventricle to stimulate receptors
in the OVLT is through surface receptors (r) or transport in
tanycytes (tny) on the ependymal wall. Where cilia (c) are
present, the ependyma have gap junctions through which Ang II can
pass but because of the glia, it does not penetrate the CVO. In
this way, Ang II from blood is limited in its spread and brain Ang
II has access to many more receptor sites.

low doses of angiotensin II into these regions produced thirst,
vasopressin release and sympathetic activation (Phillips, 1978). Both
areas have been shown by electrophysiological studies to contain neurons
which respond to angiotensin II with excitatory effects (Phillips and
Felix, 1976; Knowles and Phillips, 1980). In both organs, membrane tissue
has angiotensin II binding sites (Healy and Printz, 1984b; Mendelsohn et
al., 1984). The main difference between the two sites is in their

responsiveness to intraventricular and intravenous injections of angiotensin II (Phillips, 1978; Simpson, 1981). Earlier plugging studies revealed that when the surface of the OVLT was blocked there was no response to Ang II i.v.t., but still a response to Ang II i.v., even though the surface of the subfornical organ was not covered (Hoffman and Phillips, 1976a; Buggy et al., 1975). On the other hand, lesioning the SFO reduced drinking and blood pressure responses to Ang II i.v. (Simpson and Mangiapane, 1980; Simpson, 1981) In this case, however, the response to Ang II i.v.t. was not affected by the lesioning. To abolish the effects of i.v.t. Ang II, a lesion has to be produced which involves the OVLT. In goats, a lesion below the SFO but including the OVLT causes hypodipsia (Andersson and Eriksson, 1971) and in rats, smaller, more limited lesions which include the OVLT prevent Ang II i.v.t. effects and also result in adipsia producing the AV3V syndrome (see Table 2). Finally, when fluorescent labeled angiotensin is injected into the brain ventricle, the label if found only over the ventral portion of the OVLT

TABLE 2

Physiological Effects of AV3V Lesions

Acute Phase

1. Adipsia
2. Inappropriate diuresis
3. Inappropriate hypernatremia
4. Hypovolemia

Chronic Phase

1. Absence of drinking response to injections of:
   a) angiotensin II peripherally or centrally
   b) hypertonic saline peripherally or centrally
2. Absence of i.v.t. angiotensin II pressor response
3. Absence of i.v.t. angiotensin II vasopressin release
4. Absence of hypertension in several models
5. Hypernataremia
6. Hyperreninemia

(Landas et al., 1980). This was done in rats with implanted cannulas and the labeled Ang II was tested for its activity by observing the animals drink. The results implied that binding occurred on the surface of the OVLT but not the SFO. Autoradiographic studies of Ang II given i.v. have shown the Ang II in the blood has binding sites in both the SFO and OVLT (VanHouten et al., 1980). Therefore, we can hypothesize that angiotensin II in the blood can reach the SFO and OVLT on the blood side to produce its various effects. Angiotensin II given i.v.t., however, reaches the brain via receptors on the OVLT surface. The receptors are not necessarily the same because of the limited diffusion effect on the blood side.

The other Ang II receptor site in a CVO is the area postrema. This is apparently of great importance for the blood pressure response in the dog (Ferrario et al., 1970) but not in the rat (Haywood et al., 1980). This may represent a difference in the evolution of the brains of these species and ultimately the significance of these structures will have to be clarified in the human brain. The close proximity of the AP to the nucleus tractus solitarius places it in a key position to modify baroreceptor reflexes. It is also important that lesion studies of the AP do not encroach on the NTS. We have found a barrier of glial cells around the AP separating it from the NTS (see Fig. 3). Angiotensin rupture in this glial "barrier" may alter the functional relationship between the AP and the NTS. However, the baro-reflexes are associated with short-term changes in blood pressure and yet long-term infusions of angiotensin II produce continuous hypertension (Brunner et al., 1985). Clearly, the mechanism of action after the Ang II reception on these receptor sites has yet to be better understood before any conclusions can be drawn. In addition to these circumventricular organs, there are areas with concentrated binding sites for Ang II in the septum, superior colliculus and lateral hypothalamus (Sirett et al., 1977). Further work of the type described here should show if these are receptor sites that lead to a physiological response to Ang II.

Angiotensin binding. Central angiotensin II binding has been characterized by a number of methods and various laboratories, yielding similar results. Initially, binding of iodinated Ang II to brain membranes was studied. The binding was saturable, reversible, time-, temperature-, and pH-dependent and of high affinity, suggesting a physiological receptor. In the calf brain, binding was highest in the cerebellum. In the rat brain, highest binding was observed in the

thalamus-hypothalamus, midbrain and brainstem; the Kd of the receptor was
0.2 nM at 37°C in both calf and rat. Analogs of Ang II and fragments of
Ang II competed with the binding in a fashion which would be expected from
their biological activity (Bennett and Snyder, 1976). Other studies
followed revealing very similar Kd values and giving a binding capacity of
approximately 11 fmoles/mg protein. The septum was also found to contain
a high number of receptors, especially the lateral septum (Sirett et al.,
1977; Mann et al., 1981; Hawkins and Printz, 1983; Tonnaer et al., 1983;
Cole et al., 1981; Mendelsohn et al., 1984). Since then analogs of Ang II
such as Sar^{1}-Ile^{8}-Ang II or Sar^{1}-Ala^{8}-Ang II have been used as the
radioactive ligand, giving similar results (Bennett and Snyder, 1980).

Thus, there is a receptor for Ang II in the brain which has all the
characteristics of a physiological receptor, and is located in highest
concentrations in areas of the brain associated with Ang II's biological
activities. An elegant study by Mann et al. (1981) showed that the
binding characteristics of Ang II and its analogs correlate exceptionally
well with Ang II's known effects. The correlation coefficient was 0.975
for the effects of analogs on blood pressure and 0.9 for drinking,
suggesting very strongly that the receptor investigated by in vitro
methods is the same one involved in the physiological effects of Ang II.
It has been argued that in species other than rat such as primates and
hamsters, Ang III is the active metabolite of Ang II in drinking and
brains of those species contain high affinity binding sites for Ang III
(Harding et al., 1981).

Recently, the use of autoradiography to study Ang II binding has
allowed a much finer dissection of the areas of the brain containing Ang
II binding than was possible by the anatomically crude membrane method.
It has been possible to show that there is very high binding of Ang II to
the circumventricular organs, especially the SFO and the OVLT (Israel et
al., 1984). A complete autoradiographic study by Mendelsohn et al. (1984)
showed high binding also in the nucleus of the solitary tract, median
eminence, area postrema, and much of the hypothalamus and limbic system.
High densities of receptors were revealed in the anterior olfactory
nucleus and the lateral olfactory tract nucleus, the periventricular and
suprachiasmatic nuclei of the hypothalamus, the subthalamic nucleus, locus
coeruleus and the inferior olivary nuclei. Moderate densities occur in
the median preoptic nucleus, median eminence, medial habenular nucleus,
superior colliculus, lateral septum, medial geniculate nucleus and the
nucleus of the spinal trigeminal tract. Low concentrations of Ang II

binding sites were seen in the caudate-putamen, nucleus accumbens, amygdala and grey matter of the spinal cord. Cortical areas, the cerebellum and white matter such as the corpus callosum showed very little binding. Also of interest is the finding that Ang II receptors are closely associated with Ang II-like immunoreactivity in the lateral septal nucleus (Healy and Printz, 1984a) and angiotensinogen and Ang II receptors are similarly associated in the thalamus, hypothalamus, midbrain and brainstem, suggesting again a physiological role for the Ang II receptors (Chen, Hawkins, and Printz, 1982). The pituitary also shows binding sites for angiotensin II by autoradiography (Paglin et al., 1984).

Further evidence to support the physiological importance of these receptors can be revealed by studies which elucidate the regulation of these receptors. For example, during dehydration the binding in the SFO is increased by 50%, due to an increase in the number of binding sites, while the OVLT, area postrema and cortex were unaffected (Mendelsohn et al., 1984). In normal rats, the SFO had higher binding levels than the OVLT but in spontaneously hypertensive rats the binding for Ang II increased over 120%, whereas there was no change in the SFO (Stamler et al., 1980).

Angiotensin receptors and their connections. In order for an area to contain receptors one must show not only binding but also a biological response. In both the OVLT and the SFO, there is evidence of a biological response when Ang II is applied directly to the tissue. This comes from studies on microiontophoresis (Phillips and Felix, 1976; Knowles and Phillips, 1980; Felix and Phillips, 1979). In studies on anesthetized cat, rat and unanesthetized rat brain slices, angiotensin II was shown to produce a marked excitation of cells within the SFO and OVLT. These excitatory responses were specifically inhibited by the specific antagonist for angiotensin II, saralasin. Some of the cells identified were also sensitive to acetylcholine while other cells were exclusively sensitive to angiotensin II. Having established that there are receptors for Ang II in the circumventricular organs, the next step has been to trace the neural connections from these structures. In two studies, the techniques of horseradish peroxidase tracing and radioactive labeled amines tracing have been used. Horseradish peroxidase injected discreetly into the OVLT has revealed connections to cell bodies in the SFO (Camacho and Phillips, 1981b). Injections of labeled amines into the SFO demonstrated terminal fields in the OVLT (Miselis, 1981). Both organs have connections to the preoptic area and possibly the supraoptic nucleus.

It is probable, however, that the connections to the supraoptic nucleus from these organs are not direct but involve another synapse before arriving at the supraoptic. Other areas connected to the OVLT include the hippocampus, septum and sites as far away as the locus coeruleus and dorsal vagal motor nucleus. Which of these connections are involved in the effects of Ang II has still to be established. The effects which Ang II produces of vasopressin release, sympathetic activation and water and sodium drinking require a widespread involvement of various brain sites. For example, the connection to the hippocampus from the OVLT was at first surprising. When one considers, however, that drinking behavior must involve a component of memory in seeking out the source of water even within the small confines of a cage for our experimental subjects, a memory function in the hippocampus would not be inappropriate. Finally, there have been several investigations into the pharmacology of Ang II-induced effects. We separated the blood pressure response from the drinking response pharmacologically (Camacho and Phillips, 1980a). Phentolamine, an alpha adrenergic blocking agent, inhibits the Ang II pressor effect substantially without significant effects on the drinking response. Phillips et al., 1981) have shown that dopamine blockers significantly inhibit the drinking response to Ang II. Thus, we are beginning to see a separation of the neurocircuits involved the various responses to Ang II (see Table 1). So far, no separation of sodium drinking versus water drinking has been attempted.

In cell culture, where the Ang II receptor has been shown to be similar to the brain receptor (Raizada et al., 1981), it has been demonstrated that catecholamines may regulate the Ang II receptors. Here, decreasing norepinephrine (NE) and dopamine (DA) levels with α-methyl-p-tyrosine (α-MPT) leads to an increase in Ang II receptors, which returned to normal following removal of α-MPT and recovery of catecholamine levels. Pargyline, while increasing catecholamines, decreased Ang II binding. Both effects were time- and dose-dependent and due to changes in receptor affinity and not receptor number (Sumners and Raizada, 1984; Raizada et al., 1984a).

Regulation of Receptors. Unfortunately, other studies on regulation of Ang II receptors are not as clear cut as those cited above. For example, as Ang II infused centrally affects sodium balance by increasing sodium loss in the urine and increasing water intake, the effects of sodium on the Ang II receptor would be interesting to understand. Three studies have indeed looked at this and found three different responses, an increase, a decrease and no change. Each study has its own merits, and the differing factors may be the time course of the changes on sodium level and the different strains of rat used. Mann et al. (1980) placed animals on a low sodium diet for seven days. After this time, the blood pressure increase to i.v.t. Ang II was blunted along with the drinking in both SH and normotensive rats. The response to hyperosmotic sodium was increased, but all carbachol responses were unaltered. Ang II binding was decreased by 30% due to a change in receptor number. The K_d was also altered from 0.16 nM to 0.24 nM in sodium replete versus depleted rats. The positive side of this study is the agreement of the binding data and the physiological responses (Mann et al., 1981).

A second study compared spontaneously hypertensive rats (SHR) and Wistar Kyoto rats (WKY) in their responses to a change in sodium intake. Here the WKY showed a decreased binding capacity on a high sodium diet compared to the low, while the SHR exhibited no difference. The SHR had a higher binding capacity than the WKY on a high sodium diet, however (Ashida et al., 1982). It has also been shown that SHR cell cultures exhibit higher Ang II receptor binding than WKY cultures (Raizada et al., 1984).

The third study had two merits. Firstly, the alteration in diet was a long-term change of five weeks, and the changes of peripheral receptors were compared to those of central ones. No change in central receptors was seen in the hypothalamus-thalamus-septal (HTS) region, but the adrenal gland showed an increased Kd and binding capacity, while the bladder smooth muscle had a decreased binding capacity (Speth et al., 1984). Thus, the possible regulation of central Ang II receptors by sodium still needs clarifying.

The possible regulation of central Ang II receptors by Ang II itself has also been studied. The point of this exercise was that peripherally, Ang II can influence its own receptors on vascular smooth muscle (Gunther et al., 1980), adrenal gland (Aguilera and Ratt, 1983; Douglas and Brown, 1982), and uterus (Devynck et al., 1976). Central infusions of a single dose of Ang II for six days had no effect on central angiotensin receptors of the hypothalamus-thalamus septum-midbrain (HTSM) region of the rat brain. Adrenal binding was also unaffected (Singh et al., 1984).

The most recent area of regulation which has been studied is the difference between the spontaneously hypertensive (SHR) rat and its

400

normotensive control (WKY). In all situations studied, the SHR appears to have a greater binding capacity than the WKY. It was shown in a preliminary study that the OVLT of the SHR binds nearly twice as much Ang II as the OVLT of the WKY, but there was no difference between the SFO of either WKY or SHR (Stamler et al., 1980). On a high sodium diet, the SHR has greater binding capacity than the WKY for Ang II (Ashida et al., 1982), and in neuronal cell culture, the SHR also show an increased number of binding sites compared to WKY. Interestingly, the regulation of Ang II binding sites by catecholamines (described in section 16) was apparent in WKY cultures, but was not seen in the SHR (Raizada et al., 1984). These results as a group suggest that brain Ang II receptors may play a role in the hypertension of the SHR.

In summary, there are receptors in the brain for Ang II which show all the characteristics of a physiological receptor, and which interact in vitro with analogs of Ang II in a fashion predicted by their biological activity. They appear to be regulated by catecholamines and steroid hormones, especially estrogen (see section 12), and possibly the sodium status of the animal. In the SHR, there is an increased ability to bind Ang II which may contribute to the hypertension demonstrated by these animals.

5. ANGIOTENSIN II AND THIRST

One of the most compelling biological effects produced by a peptide is the drinking induced by injections of angiotensin II (Ang II) into the brain (Fitzsimons, 1980). This appears to be a direct effect and is an absolutely reliable phenomenon which totally alters an animal's behavior to seek water and drink. For this reason alone, angiotensin II in the brain is attractive to study, but, in addition, it raises blood pressure and releases hormones. The rise in blood pressure appears to be indirect since it is mediated by two processes which are triggered simultaneously by Ang II. These are the release of vasopressin and the increase in sympathetic activation. The question of how truly physiological and thirst effect of Ang II is has not been fully answered. Although very low doses injected into the third ventricle have been shown to produce the Ang II effects (Phillips, 1978), the concentrations are still in the upper limits of normal concentrations found in cerebrospinal fluid (CSF). If, however, the CSF levels merely represent a spillover from brain tissue, then the doses used may be considered to be physiological. When one reflects on the concentration of neurotransmitter released into the synaptic cleft of neurons, the levels must be extremely high. the synaptic cleft is 200 A wide, offering only a minute volume of extracellular fluid into which the neurotransmitter of hormone can be diluted. Therefore, in order to achieve similar concentrations at the receptor sites for Ang II, much higher concentrations have to be applied to the CSF.

Several attempts have been made to eliminate the angiotensin II stimulus from the other two by inhibition with a specific antagonist. The dose to produce thirst by intravenous infusion is orders of magnitude higher than that required by the intraventricular (i.v.t.) brain route. It has been argued that the rise in plasma angiotensin II with dehydration never reaches the levels which have to be injected to produce thirst. Ang II is the end product of the cascade of the renin angiotensin system and, therefore, the various components of this system should also produce the same physiological effects. Based on measurements of renin levels in the blood, Stricker (1978) calculated that the amount of renin which has to be injected i.v. to elicit thirst is more than can be reached physiologically.

The amount of renin, however, is only an indirect measure of Ang II. Mann et al. (1981) infused i.v., different amounts of Ang II and measured the plasma Ang II levels after infusion. Their data is incomplete because no drinking was demonstrated when they made these infusions. They only compared the infusion rates to another study by Epstein (1978). Epstein showed very little water intake with an infusion of 8 ng Ang II/min i.v. We can calculate this amount to yield about 500 pg Ang II/ml (based on 8 ng being diluted in 16 ml plasma equaling .5 ng/ml). At this dose, the average intake was only 0.86 ml and this occurred in as few as 10 out of 18 subjects tested. Based on that, 500 pg Ang II/ml would seem to be threshold. However, we have reported that 48 hour water deprivation produces 399 pg Ang II/ml plasma and all the subjected showed thirst and drank large amounts of water (Hoffman et al., 1978). Although i.v. Ang II can induce thirst and cause animals to drink, the physiological levels at which this occurs have not been fully established. An attempt by Mann et al. (1980) to measure the levels of Ang II in the blood required for drinking measured everything but drinking. They measured the Ang II levels after infusing i.v., after hours of water deprivation, but they did

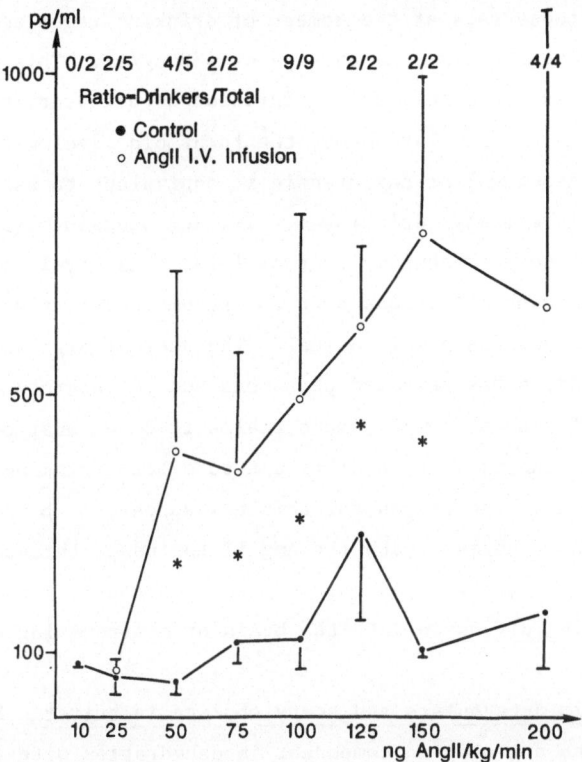

Fig 4. Plasma Ang II levels at the time when drinking occurs to i.v.
infusion of Ang II. Note that at each infusion dose the level is
about the same. This is because the rats do not drink until the
plasma levels hve reached a threshold level of about 400 pg/ml.
In untreated rats, such a plasma concentration is found only after
48 hours of water deprivation. Therefore, drinking to peripheral
Ang II is probably in response to extreme situations only (data
from Van Eekelen and Phillips, unpublished).

not measure the levels at which animals drank. We have repeated this
experiment and taken the blood of rats at the moment of drinking (Van
Eekelen and Phillips, Fig. 4). Rats were given infusions of 10, 25, 50,
75, 100, 125, 150 or 200 ng Ang II/kg/min. At the lowest dose, none of
the animals drank. At 25 ng Ang II, two out of five animals drank but
only after a long latency. At 50 ng Ang II, four out of five rats drank
but again, the latency to drink was 40-50 min. At 100 ng Ang II, all
animals tested drank at a latency of 10-12 min. At the highest dose, all
animals drank at a latency of 3-5 min. Measuring the levels of Ang II in

the plasma of these rats at the moment of drinking revealed that the amount was always the same, an average of 458 pg/ml. This means that in the animals given low doses, a long infusion is necessary to build up to the dose level. At the high doses, the threshold level was met sooner. In any case, this level of angiotensin is equivalent to water-deprivation of 48 hours. In summary, drinking to i.v. angiotensin requires a high threshold level to be reached. In real life, this level is probably only met during emergency situations such as extreme dehydration. One cannot describe such levels as physiological. The uses of high doses of Ang II in numerous studies has revealed phenomena but to understand the significance of such phenomena, more realistic doses must be used. The fact is that in the rat, it is difficult to elicit drinking with peripheral Ang II. The reason for this now appears to be because the periphery is not a normal route for Ang II to induce thirst. Only with the high doses can the natural barriers to this abnormal route be broken down - by massive diffusion into the brain or overpowering the BBB.

Angiotensin antagonists and acetylcholine in thirst. Efforts to separate out the angiotensin component in dehydration with the antagonist sar^1ala^8-angiotensin II (saralasin) have had mixed success. Malvin et al. (1977) continuously infused saralasin at 66 ng/3.3 µl/minute intraventricularly to rats deprived of water for 30 hours. When they infused this dose 75 minutes before access to water but not for a lesser period, the animals showed a delay in the onset of drinking. The long infusion period was considered to overcome the agonistic effects and reveal antagonistic effects of saralasin. However, saralasin (10 µg i.v.t.) injected immediately prior to drinking by rats deprived of water for 12, 24 or 48 hours, did not alter the water intake over a 30 minute period (Phillips et al., 1982). In another study (Hoffman et al., 1978), saralasin was infused to a total dose of 42 µg i.v.t. but no decrease in thirst was found, even in water deprived rats where plasma Ang II levels were raised significantly. In these experiments, we had expected to be able to diminish the water intake with saralasin pre-treatment by eliminating the percentage of water intake contributed by plasma Ang II.

A clue to the reason for the negative results comes from electrophysiological studies. We had noted that in the brain there are neurons which respond to angiotensin II alone and others that respond to acetylcholine alone. In two areas there were neurons which responded to both acetylcholine and angiotensin (Phillips and Felix, 1976). Carbachol, a cholinergic agonist, evokes thirst when injected into the rat brains

(Grossman, 1962). The effects are strikingly similar to angiotensin in dose, time order and also in producing a pressor response (Hoffman and Phillips, 1976a). The difference between the stimuli is seen in sodium appetite. Angiotensin II i.v.t. elicits a preference for hypertonic saline solutions but carbachol does not (Buggy and Fisher, 1974). Thus, the cholinergic stimulus may mediate osmotic dehydration which does not require sodium intake while angiotensin mediates hypovolemic dehydration, which does. Water alone passes into both the intracellular and extracellular compartments. Sodium does not easily get into cells and, therefore, stays in the extracellular compartment. As the sodium levels increase, there is an osmotic effect drawing water from the cells to fill up the extracellular space. Therefore, two different drinking circuits may act in parallel and when one is inhibited the other can take over. To reveal the role of Ang II, we had to use a combination of atropine, the cholinergic antagonist, and saralasin, the angiotensin antagonist. Combinations of these antagonists on water deprived rats produced a significant decrease in the amount of water that rats drank after 48 hours of deprivation (Phillips et al., 1982). We hypothesize that there must exist parallel circuits for thirst involving on the one hand angiotensin and on the other acetylcholine. Presumably, to abolish drinking completely would require the double blockade of these transmitter pathways plus effective denervation of the volume receptors.

6. CIRCUMVENTRICULAR ORGANS AND THIRST

The subfornical organ (SFO), organum vasculosum laminae terminalis (OVLT) and the area postrema (AP) have all been implicated as receptor sites for the actions of angiotensin II. What they have in common is that they contain capillaries with gap junctions as opposed to the tight junctions found in capillaries elsewhere in the brain and this endows them with the lack of a blood-brain-barrier even though they are brain tissue. The evidence for angiotensin from the periphery having an action of the brain via the circumventricular organs is quite voluminous. Simpson and Routtenberg (1973) were the first to show that lesioning the SFO could abolish drinking to i.v. angiotensin. Later we found that injections into the brain directly would continue to cause drinking after the SFO had been lesioned (Buggy et al., 1975). One explanation for this was that the area involved in drinking was another CVO not damaged by the SFO lesions. The evidence for the OVLT being the site was that when we blocked access to the OVLT by the use of cream plugs, then injections of angiotensin into the ventricles were no longer effective, even though angiotensin could reach the SFO (Hoffman and Phillips, 1976a). After many debates, the picture which emerged was of the SFO being important for i.v. Ang II and the OVLT being important for intraventricular or i.v.t. Ang II. More recent studies on receptor binding have confirmed that in both organs there are numerous binding sites for Ang II. The importance of the OVLT area (AV3V) for Ang II mechanisms in thirst and blood pressure is indicated after AV3V lesions (Table 2).

7. BLOOD PRESSURE AND CENTRAL ANGIOTENSIN II

There are major differences in the blood pressure responses to peripheral and central angiotensin II. Firstly, there is the topology of the response. Peripheral Ang II produces a sharp rise in blood pressure which is short-lasting and declines rapidly. Central Ang II injections cause a slow increase in blood pressure which reaches a maximum and stays at that level for some minutes, declining back to baseline by about 20 minutes. Heart rate is unchanged with the i.v. injections and decreased with the i.v. injections. Secondly, the mechanisms for these blood pressure increases are entirely different. Peripheral injection of infusion of Ang II causes a local vasoconstriction in peripheral beds due to angiotensin receptors on the endothelial wall. The mechanisms of blood pressure increase by i.v.t. Ang II is complex and involves several different systems. These include the release of vasopressin and the increase of sympathetic nerve activity. Recently, we have discovered that the nucleus tractus solitarius (NTS) is also involved, possibly by dampening the baroreflex (Casto and Phillips, 1985).

All these systems appear to act in parallel although the total interaction has not yet been worked out. The parallel organization, however, means that if one system is abolished, the other systems can take over and contribute to the maintenance of the blood pressure response. This explanation clarifies some of the earlier literature where it was debated whether for example, vasopressin was involved in the response or not. The role of vasopressin has been demonstrated by hypophysectomy which, of course, prevents the release of vasopressin. In hypophysectomized rats, i.c.v. Ang II increased blood pressure about 50% less than the controls (Severs et al., 1970). Intravenous pretreatment with cyclo-dVP AVP partially reduces the blood pressure responses to i.c.v. Ang II (Unger et al., 1980). Since this synthetic analog has been demonstrated to be a specific antagonist for the vaso-depressor response of AVP, this finding supports the role of plasma AVP contributing to the central Ang II pressor response. In Brattleboro rats there was no blood pressure increase to i.v.t. Ang II until a very high dose level (500 ng) was reached (Hutchinson et al., 1976). It now seems likely that in those animals, which are congenitally unable to produce vasopressin, there was not only a lack of hormonal mediation of the blood pressure response but also a dysfunction of the sympathetic nervous system.

The increase in sympathetic nerve activity has been assumed, based on pharmacological experiments and a few studies directly measuring the levels of norepinephrine in blood. Hexamethonium or 6-hydroxydopamine alter the response to Ang II centrally. Falcon et al. (1978) showed that the lack of sympathetic activation caused by 6-OHDA treatment did not prevent the rise in blood pressure but only delayed the response. Again, this indicated that there was a parallel system acting and that another system was able to compensate for the lack of sympathetic nerve activity. Within the brain, blocking alpha receptors of the noradrenergic system by phentolamine significantly lowered the pressor response to Ang II i.v.t. but did not alter the drinking response. This differentiation demonstrated a pharmacological distinction between the mediation of the central pressor response and the drinking response to Ang II. Beta blockade in the brain by injections of l-propranolol i.v.t. prevented either the pressor or thirst response to i.v.t. Ang II (Simon et al., 1981).

The nucleus tractus solitarius (NTS) and Ang II in the baroreflex. Recently, we have found that micro-injections of Ang II into the NTS increase blood pressure (Casto and Phillips, 1985). This response, which has also been found with micro-injections of AVP (Matsuguchi et al., 1982), suggests that these peptides control the blood pressure to some extent by dampening the first synapse of the baroreceptor reflex. In the experiments with Ang II, the heart rate was not changed. Inhibiting the reflex overcomes the decrease in heart rate that normally occurs when blood pressure is raised. Without the HR decrease, blood pressure rises even further. The increase in blood pressure was not due to vehicle injections and a study of the mechanism of the response showed that sympathetic activation was involved. Hypophysectomized rats did not show any reduction in their response to Ang II in the NTS. Hexamethonium-treated animals, however, showed almost complete blockade to the Ang II injections (Casto and Phillips, 1985). This has led us to conclude that the NTS controls to some extent the sympathetic outflow system and can be modified by central Ang II. The source for this central Ang II may be from the cells containing Ang II in the PVN. Swanson and Sawchenko (1983) have shown that there are extensive projections from the PVN to the brainstem with collateral branches terminating in the NTS. Healy and Printz (1984a) demonstrated that in the NTS there are regions that stain immunocytochemically with Ang II antibodies. Thus, it seems possible that the brain angiotensin system with its major manufacturing sites in the hypothalamus (Phillips et al., 1979; Phillips and Stenstrom, 1985), projects to the NTS and release endogenous Ang II at terminals to alter blood pressure by inhibiting the baroreflex. This central release of Ang II is mimicked when we inject Ang II directly into the NTS by micropipettes.

A blunting of the baroreflex has been found in spontaneously hypertensive rats. This finding fits in with a general picture of the SHR

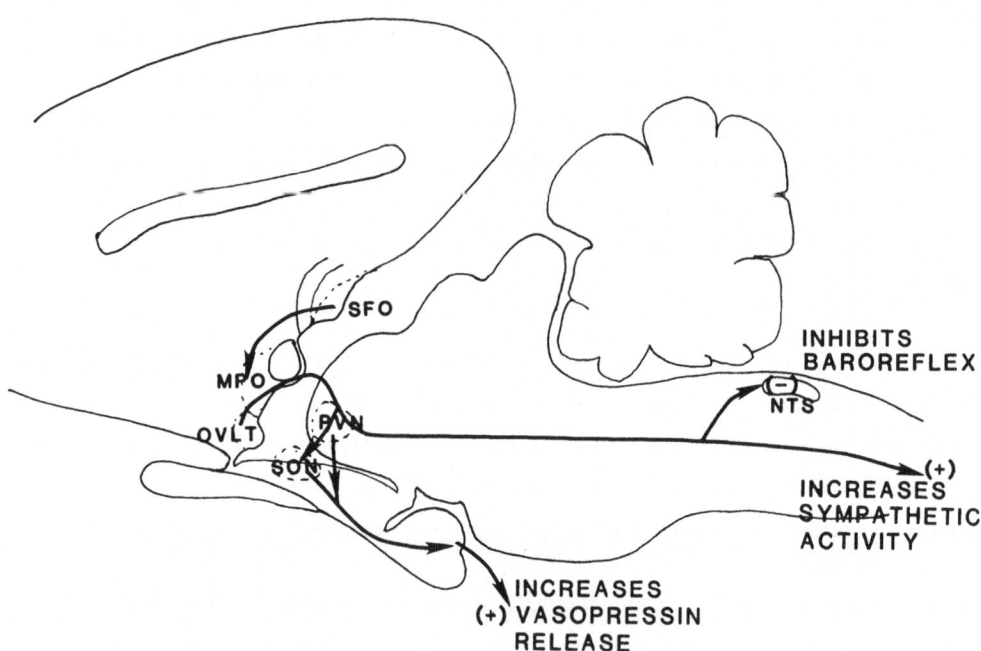

Fig 5.   Summary of the pathways from Ang II receptor sites in the AV3V region to the three major sites of action: 1) the magnocellular cells from which vasopressin is released, 2) the brainstem-hypothalamic path through which sympathetic activation occurs and 3) the NTS in which the baroreflex is inhibited by Ang II. These three actions based on Ang II release, receptors and pathways are the major sources by which brain Ang II increases blood pressure.

having central Ang II as a controlling hormone in their blood pressure
(Phillips et al., 1977). Central injections of saralasin or infusions of
Captopril reduce blood pressure in SHR. These central effects indicate
that endogenous Ang II is produced in the brain and contributes to their
blood pressure.

In summary, i.v.t. Ang II induces a blood pressure response which is
distinctly different from that produced by peripheral injections or
infusions of Ang II. The mechanisms controlling i.v.t. Ang II involve
vasopressin release, sympathetic activation and NTS synaptic inhibition of
the baroreflex by Ang II. Each contributing factor is probably
independent and the multiple mechanisms permit over-compensation when one
is removed or under-compensation when one is increased to maintain
constant blood pressure. Vasopressin is a powerful vasoconstrictor but
its actions are not obvious until the baroreflex is removed. In
hypertension, however, there is a new setpoint for blood pressure and all
the mechanisms are more activated. Brain Ang II may play a role in
hypertension because of its central control of three major mechanisms of
blood pressure elevation: Vasopressin, sympathetic activation and the
baroreflex (Fig. 5).

8. SODIUM APPETITE

One of the effects of central angiotensin is on sodium appetite.
Buggy and Fisher (1974) were the first to show that rats could be induced
to shift their preference from water to isotonic saline after intracranial
angiotensin administration. At high doses or after long infusions, the
rats showed a preference for concentrated saline and went into positive
sodium balance. A criticism of this study was the high dose used and the
natural preference of a population of rats for isotonic saline.
Nevertheless, they had discovered a basic phenomenon and one which has
been repeated and confirmed, that central angiotensin induces sodium
appetite. Angiotensin II induced at low rates of 1 pmole per hour into
the third ventricle causes rats to drink large quantities of 2.7% sodium.
This increase in sodium appetite was not secondary to sodium excretion
because animals went into positive balance and secondly, anuric animals
also demonstrated sodium appetite to central angiotensin (Avrith and
Fitzsimons, 1980). Compared to the water intake response to i.v.t.
infusions of Ang II, the response to sodium is slower. Fluharty and
Manaker (1983) demonstrated that there are two phases in the response - an

early phase with low intake and a later phase (8-12 hrs. later) with large Na intake. The later phase was in response to natriuresis and could be considered a compensatory behavioral action. It was hypothesized by Fitzsimons (1980) that angiotensin was the hormone of thirst whenever there was volume reduction. The slow onset of sodium drinking to angiotensin is in keeping with this hypothesis. Sodium appetite can be induced by various methods of volume reduction but the process is slow. In experiments by Leavitt (1975), sodium appetite was induced by a single injection of polyethyleneglycol (PE). PE draws extracellular fluids as an oncotic stimulus and sodium appetite appears six hours or so after the injection of PE (Stricker, 1966). In Leavitt's data, however, we were surprised to find that the effect was extremely long-lasting. Rats given a single exposure to polyethyleneglycol and the induction of sodium appetite, showed a permanent preference for concentrated sodium solutions whenever tested from one week to six weeks. This could not be due to a compensatory behavior as the effect was not related to Na^+ loss or hyperaldosteronism. We were not able to find a mechanism for this effect but at that time we were unable to measure brain Ang II and the possibility of its involvement needs to be considered. The evidence for peripheral renin angiotensin being involved in sodium appetite is weak. Infusions of angiotensin peripherally do not induce sodium appetite until extremely high doses have been reached and the results are variable.

The latest studies in this area have focused on the relationship of sodium appetite to the renin-angiotensin system and its connection to aldosterone. In the periphery, one of the principal roles of angiotensin is to excite receptors on the adrenal cortex and release aldosterone into the circulation. Aldosterone can cross the blood brain barrier and influence the brain. It is known that in rats receiving a high dose of aldosterone or deoxycorticosterone (DOCA), NaCl appetite is induced. At the opposite extreme, adrenalectomized rats in the absence of mineralcorticoid hormones also have a profound NaCl preference. Thus, there is a U-shaped relationship between NaCl intake and mineralcorticoid levels with the highest levels being at either end and the lowest levels being in the normal range where aldosterone secretion is present. It is interesting to note that i.v.t. administration of Ang II suppresses aldosterone secretion (Brooks and Malvin, 1980). Therefore, the central Ang II induction of sodium appetite appears to resemble the situation seen in adrenalectomized rats. On the other hand, i.v.t. Ang II produces an even greater NaCl appetite in rats given DOCA than with either substance alone (Fluharty and Epstein, 1983). This interaction suggests that there are two processes sustaining Na^+ appetite which are independent of one another. One resemble the situation where an animal may suffer an extreme lack of sodium in the diet which would induce more mineralocorticoid release. The other is a situation where no mineralocorticoids are present. In both cases, the addition of central actions of angiotensin would be of survival value to the animal.

In summary, sodium appetite is a complex function, important for survival. Although there are many peripheral factors involved such as the role of the salivary glands and the adrenal glands, there is a distinction between the peripheral renin-angiotensin system and centrally applied angiotensin. Central Ang II induces sodium appetite reliably and predictably, in contrast to infusions of Ang II in the blood which do not produce reliable Na^+ intake except in the presence of mineralocorticoids. An interesting feature of sodium appetite is its slow onset compared to thirst and secondly, its apparent long-lasting duration. In the human population, there is a clear link between sodium intake and hypertension in sodium-sensitive individuals. Whether this sodium intake is related to brain angiotensin still remains to be investigated.

Brain receptor binding and sodium. When Ang II is injected into the brain ventricles of normotensive, sodium deplete rats, there is a significant difference in the blood pressure response compared to the same injections into sodium replete rats (Mann et al., 1980). At 100 ng Ang II i.v.t., blood pressure increase in sodium replete rats was 16.3 ± 1.0 mm Hg but in sodium deplete rats, the pressor effect was 12.2 ± 0.8 mm Hg ($p < 0.05$). Carbachol is also known to raise blood pressure (Hoffman and Phillips, 1976b) but on a test of sodium deplete versus replete rats there was no difference in their response to this cholinergic agent. These data imply that the receptors in the brain for Ang II are altered by giving the rats a low sodium diet. It has earlier been suggested by Andersson and Eriksson (1971) that angiotensin II requires sodium at the receptor level in order for it to have biological effects. Buggy et al. (1979), however, were not able to find an interaction between sodium injections and angiotensin injections which would support this view. Receptor binding studies in our laboratory on brain tissue of sodium replete rats did not reveal any alterations in the level of binding in the presence of high sodium or low sodium concentrations in the incubation medium. One possibility for failure to find an interaction is that the effect of altering sodium levels in the brain may take time to manifest itself. In a study by Mann et al. (1980), rats were fed on a normal or low salt diet.

The brains were dissected to produce a block of mid-brain tissue and the binding method of Bennett and Snyder (1976), using ^{125}I-Ang II, was applied. The saturation curves indicate that with increasing concentrations of angiotensin II the low sodium rats have less binding than the sodium-replete rats. A Scatchard analysis of these data revealed that sodium-depleted rat brains have fewer binding sites for Ang II than sodium replete rat brains. The K_d calculated from the slope of the regression line changed from 0.16 to 0.24 nM. This study implied that lowering the sodium in the diet reduces the responsiveness to angiotensin II in the brain because there are fewer receptors for angiotensin II to bind to in the sodium deplete state.

More recently, however Speth et al. (1984), could not confirm this finding. Such experiments are not easy because a slight procedural change in the binding assay could obliterate the dynamic changes of prolonged diet. The newer techniques of peptide binding which do not require homogenization of tissue may be useful in resolving this question.

Central angiotensin II and sodium excretion. So far, angiotensin in the periphery or in the brain appears to have a single function, that of conserving water and sodium. However, Kapsha et al (1979) have demonstrated that with both acute and chronic injections of angiotensin II i.v.t. there is a natriuresis. This occurs even when the smaller urine volumes are taken into account because of the vasopressin that angiotensin II releases. This central action of angiotensin II is in contrast to the peripheral action of aldosterone release by angiotensin at adrenal gland receptors. Aldosterone acts on the kidneys, primarily to conserve sodium and produce an antinatriuresis. Thus, the blood angiotensin II and the CSF-borne angiotensin II have opposite actions on sodium balance. As we have seen, angiotensin II clearly increases sodium appetite when injected into the brain which would seem to counteract the sodium loss. Mouw and Vander (1970) showed that the natriuresis after i.v.t. sodium chloride (NaCl) is unrelated to changes in blood pressure and in view of the similarity between injection effects of i.v.t. NaCl and Ang II, it suggests that the natriuresis is not indirect. Another effect of central Ang II injections is a decrease in plasma renin activity (Malayan et al., 1979). This decrease, whatever the cause, would also reduce the possibility of compensation by peripheral angiotensin II to conserve sodium. Severs has suggested the two systems, brain and periphery, appear to have a reciprocal effect on absolute sodium excretion. Therefore, angiotensin II may play a role centrally during hypernatremia to reduce its effect. It is interesting that in the AV3V rats, one of the characteristics is loss of responsiveness to central angiotensin II and a state of hypernatremia (see Table 2). It would be hypothesized that brain angiotensin II is invoked when brain levels of sodium are raised so that natriuresis occurs and normal levels are again achieved. In the absence of angiotensin II action by removal of the principal receptors in the AV3V region, this circuit is destroyed and persistent hypernatremia developed.

9. ANGIOTENSIN II AND VASOPRESSIN RELEASE

Antidiuretic hormone or arginine vasopressin (AVP) is part of the triangle of hormones involved in fluid balance which include vasopressin, angiotensin and aldosterone. On the face of it, it would seem logical that peripheral angiotensin should release vasopressin. This would give the renin angiotensin system two arrows to its goal - one to fire off aldosterone release and conserve sodium and the other to release vasopressin and conserve water. The evidence for peripheral angiotensin releasing vasopressin, however, is debatable. Some investigators have established increases in ADH when Ang II was infused i.v. in dogs (Bonjour and Malvin, 1970; Ramsay et al., 1978). Many others, have been unable to stimulate vasopressin release in either anesthetized dogs with i.v. or intracarotid infusions of Ang II (Cadnapaphornchai et al., 1975; Claybaugh et al., 1972; Share, 1979) or in conscious dog (Cowley et al., 1981; Reid et al., 1982). The key difference between these studies appears to be in the concentration of Ang II in the plasma. In a study by Reid et al. a dose of 20 ng/kg/min increased plasma vasopressin release slightly (1.7 pg/ml) but this was only achieved at a very high level of Ang II in the blood (449 pg/ml).

In the rat, i.v. Ang II does not appear to release ADH except at unphysiological doses. In our experiments using an ADH bioassay, we were unable to show antidiuretic responses to infusions of Ang II i.v. in conscious rats. Knepel and Meyer (1980) could only elicit vasopressin release in the plasma with enormous Ang II concentrations of 1600 pg/ml. In the human, the story is much the same. In each case where Ang II injected i.v. has been shown to increase plasma vasopressin concentration in normal human subjects, the doses given or the levels of Ang II in the plasma had to be very high in order to produce quite small vasopressin increases (Uhlich et al., 1975; Padfield and Morton, 1977). Thus, peripheral Ang II at physiological levels does not release vasopressin.

Vasopressin secretion can be induced by high concentrations of Ang II in the plasma or by an interaction with plasma osmolality (Shimizu et al., 1973). The interaction of sodium and angiotensin may be on receptors or may be by opening the blood brain barrier.

On the other hand, it is undisputed that central Ang II releases AVP. In conscious rats, using the bioassay (Hoffman et al., 1977) or vasopressin radioimmunoassay (Keil et al., 1975) there was a consistent and prompt release of vasopressin with Ang II i.v.t. It is always difficult to state the dose of central injections of Ang II and impossible to claim whether these are physiological or not until one knows the exact amount that arrives at the receptors. In the studies of vasopressin release, very small quantities of Ang II i.v.t. have not been studied. However, water intake can be induced by i.v.t. injections of Ang II in concentrations compatible with the K_d of brain Ang II receptors.

The site of action of central Ang II for vasopressin secretion is still unknown. Several possibilities exist based on preliminary data. One possibility is that Ang II which has been localized by immunocytochemistry in the supraoptic nucleus (Phillips et al., 1979) could be released into the nucleus and have a direct action on the neurons containing AVP. Electrophysiological evidence indicates that Ang II directly stimulates SON cells (Nicoll and Barker, 1971; Sakai et al., 1974). Also, Ang II binding sites have been demonstrated in the PVN (Mendelsohn et al., 1984) and, therefore, a direct mechanism is possible in that nucleus. Studies by Sladek and Joynt (1979) on isolated hypothalamic tissue with the pituitary intact indicate that Ang II can directly act on the tissue to release AVP. This block of tissue contains the OVLT and the median eminence, as well as the PVN, SON and neurohypophysis. There has been some debate whether the action of Ang II is on any or all of these potential receptor sites. In addition, we must consider that Ang II could act via an interneuron and other transmitters. The isolated hypothalamic block is devoid of noradrenergic inputs from the brainstem which may be crucial in AVP release (Blessing et al., 1982).

On the other hand, the circumventricular organs may play a role in the release. For example, lesions to the SFO or to the OVLT (or AV3V) inhibit or abolish the vasopressin release response to i.v.t. Ang II (Mangiapane et al., 1982; Bealer et al., 1979). In the isolated hypothalamic block taken from AV3V lesioned rats, in which the OVLT is destroyed, however, there is still release of AVP to Ang II (Sladek and

Johnson, 1983). This finding shows that the CVOs may not be the only site controlling AVP release and that the hypothalamic block is lacking circuitry that is active in vivo. The connection from the OVLT and the SFO to the supraoptic nucleus has been demonstrated by HRP tracer studies from the SFO (Miselis, 1981) and by aminoacid transport studies from the OVLT (Camacho and Phillips, 1981b). The role of the circumventricular organs in vasopressin release does not necessarily require blood-borne angiotensin II to be present. Recent immunocytochemical studies by Lind et al. (1985) and by Healy and Printz (1984a) have demonstrated small spherical cells in the SFO and OVLT which contain Ang II-like immunoreactivity. If these cells are the ones that are connected to the supraoptic nucleus or PVN along the axons demonstrated by HRP and aminoacid transport, then central injections of Ang II would be mimicking the release of endogenous brain Ang II from these types of cells.

10. CRF AND ANGIOTENSIN II INTERACTION

That angiotensin II (Ang II), administered either centrally or peripherally, can induce ACTH and corticosteroid release has been demonstrated many times since the initial experiments by Maran and Yates (1977) and DeWeid et al. (1961). The mechanism of the effect is still unclear. Several possibilities can be inferred from the available experimental data. Firstly, Ang II may cause the release of CRF. Secondly, Ang II may act to potentiate the action of CRF, and thirdly, Ang II causes the release of AVP. AVP itself is able to stimulate ACTH release, either by acting directly at the anterior pituitary or by releasing CRF.

The third possibility seems unlikely to be a total explanation as Ang II causes the same increases in ACTH in vasopressin deficient homozygous Brattleboro rats as in their Long-Evans controls (Spinedi and Negro-Vilar, 1983). However, Ganong, Shinsako and Reid (1982) showed a partially blunted ACTH response to i.v.t. Ang II administration in the Brattleboro rat which suggests that the Ang II effect may be mediated partly by AVP, and partly by another mechanism.

The definitive experiments to distinguish between the first two possible mechanisms of Ang II mediated ACTH release have not yet been performed. That is, the ability of Ang II to release CRF has not been tested. However, the experimental data do show a strong interaction between CRF and Ang II. If isolated pituitary cells are treated with CRF, ACTH is released. If Ang II is used, instead of CRF, ACTH is released also but to a much lower extent, findings which are comparable with the in vivo studies of Spinedi and Negro-Vilar (1983). However, if CRF and Ang II are concurrently presented to the cells, the ACTH release is larger than that to either CRF or Ang II alone. The interaction has been determined to be additive and not synergistic (Vale et al., 1983).

The evidence which suggests that Ang II exerts its effects via CRF release come from Spinedi and Negro-Vilar (1983). Here, rats were centrally blocked by pretreatment with chloropromazine-morphine sulphate and nembutol. If CRF or AVP was then infused i.v., ACTH release occurred, but the ACTH response to i.v. administration of Ang II was blocked. This suggests that Ang II acts, not in the pituitary, but at a higher center, possibly to cause CRF release. However, recent studies in trained unstressed dogs infused with Ang II and CRF have shown that CRF may be permissive to Ang II's action on ACTH release (Keller-Wood et al., 1986). Therefore, if endogenous CRF release was inhibited in the centrally-blocked rats, Ang II would have no effect. However, as stated above, the definitive studies to sort out these possibilities have not been performed. Thus, Ang II causes ACTH release partially indirectly through AVP release and partially via another mechanism.

Although the mechanisms are not completely understood, the release of ACTH can occur by central or peripheral Ang II infusions. The peripheral route is less effective than the central route but both are less effective than intrapituitary infusion (Maran and Yates, 1977), and the amounts released by circulating levels of Ang II are small and probably due to a central nervous system and pituitary mechanism (Spinedi and Negro-Vilar, 1983).

11. ALDOSTERONE AND ANGIOTENSIN II

In sodium replete sheep, intraventricular infusion of Ang II results in increased plasma aldosterone levels. This response is probably mediated by ACTH, as it can be blocked by dexamethasone, and because the infusion did not alter plasma Ang II, norepinephrine or epinephrine levels (Nichols et al., 1983).

In anesthetized, sodium depleted dogs, i.v.t. Ang II infusion has been shown to decrease plasma aldosterone levels (Brooks and Malvin, 1980). The mechanism for this decrease is unknown.

Eguchi and Bravo (1984) and Brooks and Malvin (1980) have performed studies to determine how the sodium status of dogs affects the response of aldosterone to central Ang II administration. In Eguchi et al.´s studies of anesthetized, sodium depleted dogs, intracerebroventricular administration of Ang II resulted in decreased plasma aldosterone concentrations, as shown by Brooks and Malvin (1980). However, in salt-replete dogs, the infusion had no effect on plasma aldosterone levels, until ACTH release was suppressed by dexamethasone. Dexamethasone pretreatment had no further depressant effect on aldosterone secretion in the salt-depleted dogs. Thus, both the sodium status and the species of the animal has a definite effect on the aldosterone response to i.v.t. Ang II. However, in all situations central Ang II administration have very different actions from peripheral infusion of Ang II, which is stimulatory to aldosterone secretion.

There is another important central action of Ang II (Brooks and Malvin, 1982). Anesthetized sodium depleted dogs received i.v.t. infusions of either SQ20881 or P-113, inhibitors of Ang II, which caused plasma aldosterone concentrations to rise. This implies that central Ang II tonically inhibits aldosterone secretion. This becomes particularly interesting when it is considered that peripheral infusions of the inhibitors did not alter aldosterone levels.

A further interaction is suggested by the evidence that adrenal steroids may regulate angiotensingoen levels centrally (Printz et al., 1983). Reserpine treatment, which depletes catecholamines, leads to an increase in brain angiotensinogen levels. This increase is mediated by adrenal corticosterone, which increases following reserpine treatment, as adrenalectomy prevented the change. Furthermore, adrenalectomy decreases brain angiotensinogen levels in certain brain areas, e.g., the anterior hypothalamus, arcuate nucleus and periaquiductal grey, changes which can be brought back to normal by replacement of corticosterone. However, the interaction may not be direct, as there are indications that adrenal steroids alter norepinephrine levels, which in turn alter angiotensinogen levels (Printz et al., 1979). Schelling et al. (1983) showed that adrenalectomy decreased CSF levels of angiotensinogen, confirming these results, although in a different "area" of the brain.

During the estrus cycle, at estrus, spontaneous drinking is reduced and there is a tendency for the receptors in the preoptic area, lateral hypothalamus, septum and pituitary to be reduced. There is also a drop in angiotensinogen levels at estrus in the olfactory bulbs and area postrema. This is in contrast to the peripheral effects of estrus where plasma angiotensinogen is raised and uterine angiotensin II receptors are much increased (Chen, Hawkins and Printz, 1982; Chen and Printz, 1983).

The ability of estrus to alter binding is probably due to estrogen, as several recent studies have shown that estrogen administration decreased Ang II binding centrally at the same time as reducing the amount of water ingested following central angiotensin II administration (Fregly, 1980; Jonklaas and Buggy, 1984).

Jonklaas and Buggy (1984) showed that drinking to Ang II i.v.t. was depressed in ovariectomized female rats one day after estradiol benzoate i.v.t. The interaction of estrogen and central Ang II is specific because drinking to carbachol or polyethyleneglycol in the same rats was not altered and no change in drinking to Ang II i.v.t. occurred in males. The pressor response to Ang II i.v.t. was also reduced by estrogen pretreatment.

In ovariectomized rats pretreated with estrogen, i.v.t. Ang II increased plasma levels of luteinizing hormone (LH) (Steele et al., 1982)). In ovariectomized rats that did not have estrogen replacement, Ang II lowered the level of LH in blood, the pulse frequency of release and the amplitude (Steele et al., 1985). In rats pretreated with both estradiol and progesterone, Ang II i.v.t. increased LH in plasma. Blocking brain Ang II with saralasin or enalapril i.v.t. in the presence of progesterone and estrogen prevented the LH surge on proestrus (Steele et al., 1985). In our laboratory, we have measured brain angiotensin throughout the estrus cycle. The data shown in Figure 6 indicate an increase in brain Ang II during proestrus. The results, however, is preliminary.

These experiments and the report of Ang II releasing prolactin (Aguilera et al., 1982), point to a significant role for brain Ang II in the control of reproduction. The mechanism for LH release is unknown but electrophysiological data suggest that Ang II influences LH releasing

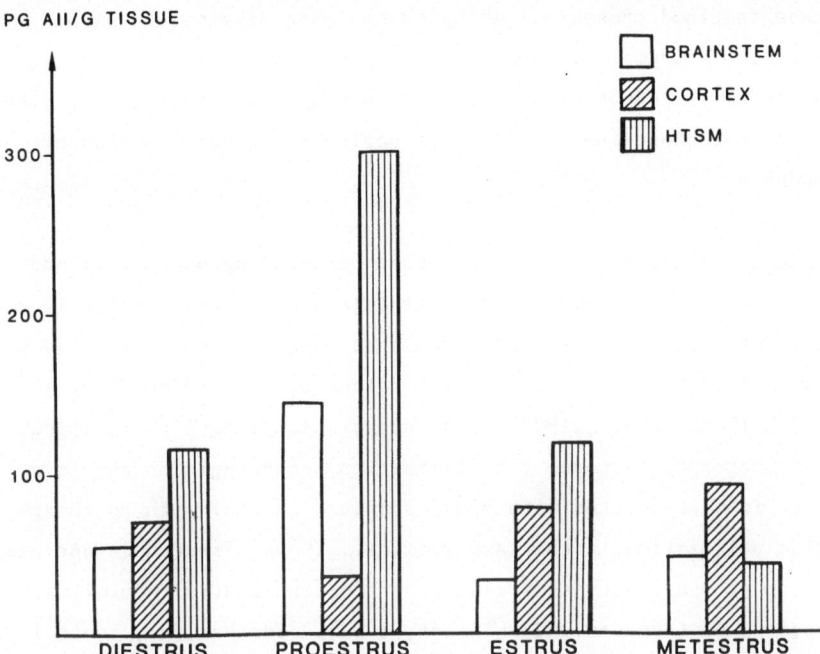

Fig 6.   Brain Ang II extracted from hypothalamus and measured by HPLC
         during different phases of the estrus cycle.   There appears to be
         a rise in brain Ang II coincident with the proestrus phase during
         which time the LH surge occurs.   These are, however, prelimanry
         results (N = 5).

hormone (Felix and Phillips, 1979).   The recording study was carried out
in the OVLT because of the previous work on Ang II and the OVLT and
because it is a site of numerous cells staining to LHRH antibodies.

    In summary, there is increasing evidence that central angiotensin has
actions related to reproduction in addition to fluid balance.   At times
these two major survival functions are interrelated in terms of thirst
during the estrus cycle and water retention during pregnancy.

13.   CATECHOLAMINES AND ANGIOTENSIN II

    The interaction between angiotensin II and catecholamines is
phylogenetically very old, at least in the periphery.   In seven classes of
non-mammalian vertebrates, intravenously injected angiotensin II caused an

419

increase in blood pressure. In all cases phentolamine, an alpha adrenoceptor antagonist, could either reduce or totally abolish the angiotensin II pressor effect (Carroll and Opdyke, 1982). This shows how important catecholamines are for the peripheral pressor action of angiotensin II.

There is also a strong interaction between angiotensin II and catecholamines in the brain. Central injection of angiotensin II causes a pressor response which is partly mediated by activation of the sympathetic nervous system (Severs et al., 1970; Falcon et al., 1978; Unger et al., 1981; Scholkens et al., 1982). Angiotensin II given intraventricularly induces increased turnover (or utilization) of norepinephrine in specific brain regions associated with blood pressure control with no change in dopamine utilization in the same regions. These effects are not caused by the blood pressure increases resulting from the central administration of angiotensin II (Sumners and Phillips, 1983). Furthermore, central injection of renin, the enzyme which cleaves angiotensin I from angiotensinogen, can increase DA turnover in the external layer of the median eminence, while decreasing NE turnover in the magnocellular paraventricular nucleus and the dorsomedial hypothalamic nucleus. Renin administration also increased epinephrine turnover in the DCMO, a block of brainstem tissue containing the nucleus tractus solitarius, nucleus commisuralis and the dorsal motor nucleus of the vagus. the effects are assured to be specifically mediated by angiotensin II as they can be blocked by SQ 14225, an inhibitor of the conversion of angiotensin I to angiotensin II (Fuxe et al., 1980).

Angiotensin II can be shown to alter the catecholamine content of neuronal brain cells in culture. this is important because it circumvents many of the confounding variables of the in vivo situation, such as increased blood pressure, increased water and salt intake, which can complicate interactions between angiotensin II and catecholamines. Here angiotensin II treatment resulted in increased catecholamines in both the cells and the growth media, an effect blocked by saralasin, an angiotensin II receptor antagonist (Sumners et al., 1983).

More evidence to support the interrelationship of angiotensin II and catecholamines is the finding that catecholamines and angiotensinogen or catecholamines and angiotensin II receptors are often located in the same brain regions. If angiotensinogen concentrations are measured in various brain regions there is a good correlation with norepinephrine levels, but not with dopamine levels (Printz et al., 1979). Angiotensin II receptors are localized in high concentrations on the circumventricular organs (Mendelsohn et al., 1984) where catecholamines can be located (Saavedra et al., 1976) and especially on the locus coeruleus, a major source of norepinephrine pathways throughout the brain.

Similarly, the nucleus tractus solitarius containing a high concentration of both catecholamines and angiotensin II receptors, as do many parts of the hypothalamus, which also contains angiotensin II. Thus, components of the angiotensin II system and catecholamines are closely related anatomically in areas of the brain associated with the control of blood pressure.

There is a functional as well as an anatomical relationship between Ang II and catecholamines. In neuronal cell culture, decreasing catecholamine levels lead to an increase in Ang II receptors, and increasing catecholamines causes decreased Ang II receptors, suggesting that catecholamines and angiotensin II may interact in a negative feedback fashion (Sumners and Raizada, 1984).

Of interest in this regard is the action of 6-OHDA, an agent which destroys catecholaminergic neurons, on angiotensinogen content of the brain. 6-OHDA caused no change in blood pressure, but it lowered the catecholamine content of the brain while raising the levels of angiotensinogen in the brainstem, CSF and cerebral cortex (Basso et al., 1984), thus suggesting a possible mechanism for the Ang II receptor-catecholamine interaction. Also, treatment with 6-OHDA intraventricularly decreases both the drinking and the pressor response to centrally injected angiotensin II (Gordon et al., 1979).

At what level of catecholamine action is the interaction with angiotensin II occurring? There is evidence to suggest that angiotensin II affects both the release and re-uptake of catecholamines in the brain.

Ang I can evoke release of $[^3H]$-DA from slices of striatum exposed to $[^3H]$-tyrosine without affecting synthesis of $[^3H]$-DA. This could be blocked by either removing calcium from the buffer or by Sar^1-Ile^8-Ang II, an angiotensin II receptor antagonist (Simmonet and Giorguieff-Chesselet, 1979).

Ang II does not seem to be able to evoke spontaneous NE release from brain tissue, although it affects electrically stimulated NE release. At low doses (0.1 and 1 nM), angiotensin II enhances electrically stimulated NE release, whereas at high doses (1 μM) it decreases it. Captopril and HOE-498 diacid at 1 μM were able to decrease the stimulated outflow. Especially interesting in this study was the finding that HOE-498 diacid had the same effect on release as clonidine, an α_2 adrenoceptor agonist, and that HOE-498 diacid could partially inhibit the effects of rauwolscine, an α_2 adrenoceptor antagonist (Schacht, 1984), strengthening even further the interaction between angiotensin II and catecholamines. Similarly, in a dose-dependent fashion, Ang II can increase the potassium evoked release of NE from the rabbit hypothalamus, an effect blocked by saralasin (Garcia-Sevilla, 1979).

This kind of approach has been extended in the studies of Meldrum et al. (1984). They found no effect of angiotensin II or a low sodium diet on NE release from the hypothalamus, but Ang II increased NE overflow from the A_2 region of rats. This stimulation of NE overflow by Ang II could be blunted by placing the animals on a low sodium diet. Furthermore, in Wistar compared to Sprague-Dawley rats, the effect of angiotensin II on overflow from the A_2 region was more marked, but the stimulatory effect could be completely blocked by administration of a low sodium diet. Thus, an area of the brain known to contain Ang II receptors and catecholamines and to be involved in blood pressure control demonstrates interactions between angiotensin II and catecholamines which can be modulated by the sodium status of the animal, another variable important in the control of blood pressure. A study by Chevillard et al., (1979) shows that the in vitro studies mentioned above are also applicable to the in vivo situation. Here, ventriculocisternal infusion of angiotensin II in rabbits caused an increase in the CSF concentration of NE, which was highly correlated with the increase in blood pressure due to angiotensin II, thus strengthening the evidence for a central peptide-CA interaction in blood pressure regulation.

The effects of angiotensin II on NE uptake is a little more complex. Palaic and Khairallah (1967) showed that angiotensin partially inhibits NE uptake by brain, with no effect on release of NE after one hour of exposure to Ang II. However, recent studies in neuronal cell culture have demonstrated a biphasic effect. At short times after Ang II exposure, Ang II enhances NE uptake, while after longer treatment inhibition is seen (Sumners and Raizada, 1985). Possibly, this effect was overlooked by

Palaic and Khairallah as they only looked at one time point, or it may represent a difference between cell culture and brain tissue.

The effects on release and re-uptake are seen when peripheral tissues are examined (Reit, 1972; Felderg and Lewis, 1964; Starke, 1971; Palaic and Khairallah, 1967). However, more peripheral studies have been done than central and, therefore, less information is available on central effects. Firstly, angiotensin II appears to increase the amount of NE released spontaneously (Zimmerman et al., 1972) as well as increasing release to sympathetic stimulation (Zimmerman and Gisslen, 1968; Hughes and Roth, 1971) and secondly, an acceleration of catecholamine biosynthesis is seen to angiotensin II treatment in sympathetically innervated tissues (Boadle et al., 1969).

It is possible that these responses are all interrelated, for example, increased release of NE would be expected to increase biosynthesis of NE (Davila and Khairallah, 1971). It remains to be seen whether all of these responses can be seen centrally as well as peripherally as the studies have not yet been performed.

14. SEROTONIN AND ANGIOTENSIN II

Although Camacho and Phillips (1981a) were unable to show an interaction of serotonin with the angiotensin II induced drinking or pressor response, several other studies suggest that an interaction may exist. If rats were treated with para-chlorophenylalamine (pCPA), an inhibitor of serotonin (5-HT) synthesis, and then given intraventricular Ang II, a depressor rather than a pressor response was seen. Similarly, methylsergide, a 5-HT antagonist, caused the same effect. Increasing 5-HT levels with morphine produced an exaggerated pressor response to Ang II. Concomitantly, hypothalamic turnover of 5-HT was decreased while brainstem 5-HT turnover was increased. In synaptosomal preparations from hypothalamus and brainstem, Ang II increased 5-HT synthesis, which was not mediated by increased ^{3}H-L-tryptophan incorporation, as both Ang II and angiotensin III decreased this. And finally, Ang II was shown to affect the rate-limiting enzyme of serotonin synthesis, tryptophan hydroxylase, in a dose-dependent fashion. At low concentrations, Ang II decreased the enzyme's activity and at high concentrations, it increased it, as did angiotensin III (Nahmod et al., 1978). Davila and Davila (1975) showed that Ang II decreased uptake of exogenously applied ^{14}C-tryptophan in rats, supporting these findings. Simon et al. (1981) agreed that Ang II stimulated the turnover of 5-HT and showed that α-adrenoceptor blockers blunt the pressor and drinking responses to Ang II, probably through the inhibition of serotonin binding.

The site of 5-HT and Ang II interaction has been more closely studied by Benarroch and colleagues (1981). Ang II causes a pressor response when injected into the anterior hypothalamus/preoptic area (AV3V) region of rat brain. This response could be potentiated by increasing the 5-HT content of the forebrain bundle and brainstem, and abolished by decreasing the medial forebrain bundle concentration of 5-HT. (In fact, Ang II caused a depressor effect and a decrease in heart rate under these conditions). Thus, the interaction appears to be at the level of the medial forebrain bundle input to the anterior hypothalamic preoptic area region. However, this interaction is in turn deponent on an intact brain norepinephrine system as 6-OHDA abolished it.

Peripheral administration of Ang II also leads to increases in serotonin content of the hypothalamus, brainstem and pineal gland of dogs (Haulica et al., 1980). These workers have also shown that a renin-like substance has a diurnal rhythm in the pineal similar to those of norepinephrine and melatonin. Furthermore, if 5-HT synthesis is blocked with pCPA, the activity of the pineal renin-like activity is increased (Haulica et al., 1980). Thus, it is possible that Ang II is involved with serotonin in control of diurnal rhythm. This would appear attractive as drinking shows a definite diurnal rhythm in rats.

15. PROSTAGLANDINS AND ANG II

Many of the peripheral effects of angiotensin II are opposed by the biosynthesis and release of prostaglandins, for example, in the kidney (McGiff and Wong, 1979). However, the central interaction of these two substances is less well understood. If prostaglandins of the E series, especially E_1 and E_2, are administered centrally, they can inhibit the dipsogenic effect of angiotensin administered by the same route (Fluharty, Kenney and Epstein, 1981). Arachadonic acid administered centrally has the same effect, which can be blocked by indomethacin, suggesting that it acts through central increases in prostaglandin E_2. However, if it is suggested that endogenous PGE_2 is produced to "turn off" centrally induced Ang II drinking, then it should be possible to use prostaglandin synthetase inhibitors to enhance angiotensin II-induced drinking. In the study of Phillips and Hoffman (1976), meclofenamate caused this response, but indomethacin, used by Fluharty, did not. This apparent anomaly could be clarified by studying the effects of these inhibitors on the levels of PGEs in the brain, as it is possible that one could more easily attain access to the sites involved with prostaglandin synthesis than the other. The solution of this anomaly is potentially very important to the understanding of control of angiotensin II-induced drinking.

While it is established that in the periphery Ang II releases prostaglandins, this has not been established in the brain by direct measures. Indirect evidence points to an interaction of prostaglandins and central Ang II but the nature of that interaction and whether Ang II acts independently or interdependently remains to be seen.

16. OPIOID-ANGIOTENSIN II INTERACTION

The interrelationship of opioids and angiotensin II in the brain is poorly understood because very few studies have been directed towards that end. At first sight, the interaction seems simple. Enkephalins, β-endorphins and morphine can inhibit the centrally administered Ang II-stimulated increase in plasma-vasopressin and oxytocin, drinking behavior and pressor response in a dose-dependent fashion (Keil et al., 1984; Summy-Long et al., 1981,1983). Naloxone potentiates the pressor effect of intraventricularly administered Ang II, and naloxone can disinhibit the opioid inhibition of Ang II-induced effects. The site of the interaction appears to be a site which can be reached from the lateral and third ventricle (Summy-Long, 1983).

However, when it is shown that naloxone and naltrexone also decrease Ang II induced drinking, the situation becomes more complex (Summy-Long, 1983). Several points may aid in understanding this situation. Firstly, there are indications that a full dose response curve of morphine would be biphasic, that is, low doses potentiate, while high doses inhibit angiotensin-induced effects. Thus, these studies may have missed this effect of dose. Furthermore, naloxone and naltrexone act through inhibition of endogenous opioids, which may have very different effects than exogenous administration of opioids. Studies of Ang II's effects on opioid turnover and vice versa along with studies with specific inhibitors of both systems would help unravel this, but these techniques are currently unavailable.

Two other studies of opioids are interesting also. Appel and VanLoon (1983) showed that the effects of intracisternally-administered β-endorphin to increase plasma catecholamines could be enhanced by Ang II and blunted by saralasin. Secondly, naloxone inhibits while morphine enhances the pressor effects of intravertebrally administered Ang II, without affecting the pressor response to intravenous Ang II (Szilagyi and Ferrario, 1981).

We found that naloxone inhibits the actions of Ang II. In this case, naloxone significantly reduced pressor and drinking actions of central Ang II in unanesthetized rats (Fig. 6). This implies that angiotensin expresses its effects via an enkephalin component at some stage in the neural circuitry. Enkephalin alone increases blood pressure when injected into the brain but inhibits vasopressin release (Summy-Long et al., 1981,1983). Therefore, the pressor effect of the opiate would seem to be via sympathetic activation. Whether this enkephalin component comes before, after or at the same time as the alpha adrenergic component in the circuit has not yet been established. Other evidence that the opioid peptides are involved when angiotensin activates brain receptors is the observation of Haulica et al. (1982), that Ang II induces analgesia. Ang II induced analgesia was measured by the tailflick method and could be antagonized by naloxone. One of the actions of the opioid peptides is to produce immobility. This may account for a reduction in drinking reported when enkephalins were injected i.v.t. together with Ang II. A common link between angiotensin and enkephalins is the possibility that converting enzyme is also an enkephalinase (Swerts et al., 1979). Captopril, which is a converting enzyme inhibitor and would, therefore, prevent he hydrolysis of enkephalins by the enkephalinase activity, also produces analgesia (Swerts et al., 1979). Converting enzyme is also an important link between bradykinin and angiotensin II because in its role as kininase, it brings about the degradation of bradykinin.

17. SUMMARY AND CONCLUSIONS

In this review, we have described a number of different actions of angiotensin II by central injections into the brain or the brain ventricles and by peripheral injections or infusions into the blood. The number of different actions are summarized in Table 1. The major effects can be classified into three groups. The first and largest group are actions all in one way or another associated with the regulation of body

426

fluid volume. This includes thirst, blood pressure increase, vasopressin release, sodium appetite and aldosterone release. This function alone has important implications for the control of blood pressure and the disease of hypertension. Another area which is emerging is that angiotensin plays a role in the timing of hormone release during the reproductive cycle. This may be linked to the volume control function since the menstrual cycle involves changes in volume homeostasis and during pregnancy blood pressure responses to the renin-angiotensin system are altered. A third group of functions is a miscellaneous set of findings of different actions of angiotensin on opiates, prostaglandins and other peptides, not all of which have been reviewed here. This interaction leads to alterations in pain, memory and possibly learning. The amount of data available, however, is so limited that to claim angiotensin plays any major role in such functions would be premature. Finally, the theme of this review has been to analyze central versus peripheral effects. Many of the effects can be elicited by both routes of administration of angiotensin. However, in reviewing this literature it is clear that supraphysiological doses of angiotensin have to be sued peripherally to achieve many of the effects. The central route produces reliable responses. From receptor binding, we can quantify the effective dose by central injections. The central effects are physiological and mimic the release of brain angiotensin. We are suggesting that where central and peripheral effects are the same, this may be due to a breakdown in the blood brain barrier caused by the excessive peripheral doses. At least that possibility should be carefully analyzed in the future. There are distinct differences in the responses to the two routes of administration. In the domain of peripheral angiotensin is the increase of blood pressure by smooth muscle vasoconstriction and the release of aldosterone, leading to conservation of sodium. In the domain of central angiotensin are the numerous effects reviewed here. Given this division of actions by the peptide, we may consider all central injections of angiotensin to mimic the actions of brain angiotensin produced endogenously. Only by precise study of the brain angiotensin system under different physiological conditions will we be able to define which actions are truly physiological.

ACKNOWLEDGMENTS

We are very grateful for Mrs. Carolyn Gabbard for typing this review. Unpublished data shown in some of the figures is from work supported by NIH Grant 2R01-HL27334.

Aguilera, G., and Catt, K.J., 1983, Regulation of aldosterone secretion during altered sodium intake, J. Steroid Biochem., 19:525-530.

Aguilera, G., Hyde, C.L., and Catt, K.J., 1982, Angiotensin II receptors and prolactin release in pituitary lactotrophs, Endocrinology 111:1045-1050.

Andersson, B., and Eriksson, L., 1971, Conjoint action of sodium and angiotensin on brain mechanisms controlling water and salt balances, Acta Physiol. Scand., 81:18-29.

Appel, N.M., and Van Loon, G.R., 1983, Activation of angiotensin II receptors in brain potentiates the stimulating effect of endogenous opioid neurons on central sympathetic outflow, Peptides 4:59-62.

Ashida, T., Ohuchi, Y., Saito, T., and Yazaki, Y., 1982, Effects of dietary sodium on brain angiotensin II receptors in spontaneously hypertensive rats, Jpn. Circ. J., 46:1328-1336.

Avrith, D.B., and Fitzsimons, J.T., 1980, Increases sodium appetite in the rat induced by intracranial administration of components of the renin-angiotensin system, J. Physiol. (Lond)., 301:349-364.

Basso, N., Kurnjek, M.L., Mikulic, L., Cannata, M.A., and Taguini, A.C., 1984, Effect of intraventricular 6-hydroxydopamine (6OHDA) on the angiotensinogen concentration of the brain, Clin. and Exp. Theory and Practice, A6(10&11):1773-1776.

Bealer, S.L., Phillips, M.I., Johnson, A.K., and Schmid, P.G., 1979, Effect of anteroventral third ventricle lesions on antidiuretic responses to central angiotensin II, Am. J. Physiol., 236:E610-E615.

Benarroch, E.E., Pirola, C.J., Alvarez, A.L., and Nahmod, V.E., 1981, Serotonergic and noradrenergic mechanisms involved in the cardiovascular effects of angiotensin II injected into the anterior hypothalamic preoptic region of rats, Neuropharmacology 29:9-13.

Bennett, J.P., Jr., and Snyder, S.H., 1976, Angiotensin II binding to mammalian brain membranes, J. Biol. Chem., 251:7423-7430.

Bennett, J.P., Jr., and Snyder, S.H., 1980, Receptor binding interactions of angiotensin II antagonist, ^{125}I-[sarcosine^{1}, Leucine^{8}]angiotensin II, with mammalian brain and peripheral tissues, Eur. J. Pharmacol., 67:11-25.

Bickerton, R.K., and Buckley, J.P., 1961, Evidence for a central mechanism in angiotensin-induced hypertension, Proc. Soc. Exp. Biol. Med., 106:834-836.

Blessing, W.W., Sved, A.F., and Reis, D.J., 1982, Destruction of noradrenergic neurons in rabbit brainstem elevating plasma vasopressin, causing hypertension, Science 217:661-663.

Boadle, M.C., Hughes, J., and Roth, R.H., 1969, Acceleration of angiotensin accelerates catecholamine biosynthesis in sympathetically innervated tissues, Nature (Lond.), 222:987-988.

Bonjour, J.P., and Malvin, R.L., 1970, Stimulation of ADH release by the renin-angiotensin system, Am. J. Physiol., 218:1555-1559.

Brightman, M.W., 1977, Morphology of blood-brain interfaces, Exp. Eye Res., Suppl:1-25.

Brooks, V.L., and Malvin, R.L., 1980, Intracerebroventricular infusion of angiotensin II inhibits aldosterone secretion, Am. J. Physiol., 239:E447-E453.

Brooks, V.L., and Malvin, R.L., 1982, Intracerebroventricular infusions of angiotensin II increases sodium excretion (41385), Proc. Soc. Exp. Biol. Med., 169:532-537.

Brownfield, M.W., Reid, I.A., Ganten, D., and Ganong, W.F., 1982, Differential distribution of immunoreactive angiotensin and angiotensin converting enzyme in rat brain, Neuroscience 7:1759-1769.

Bruner, C.A., Weaver, J.M., and Fink, G.D., 1985, Sodium-dependent hypertension produced by chronic central angiotensin II infusion, Am. J. Physiol., 249:H321-H327.

Buggy, J., Fink, G.D., Johnson, A.K., and Brody, M.J., 1977, Prevention of the development of renal hypertension by anteroventral third ventricular tissue lesions, Cir. Res., 40(1):10-17.

Buggy, J., and Fisher, A.E., 1974, Evidence for a dual central role for angiotensin in water and sodium intake, Nature 250:733-735.

Buggy, J., Fisher, A.E., Hoffman, WE., Johnson, A.K., and Phillips, M.I., 1975, Ventricular obstruction: Effect on drinking induced by intracranial injection of angiotensin, Science 190:72-74.

Buggy, J., Hoffman, W.E., Phillips, M.I., Fisher, A.E., and Johnson, A.K., 1979, Osmosensitivity of rat third ventricle and interactions with angiotensin, Am. J. Physiol., 236:R75-R82.

Cadnapaphornchai, P., Boykin, J., Harbottle, J., McDonald, K., and Schrier, R., 1975, Effect of angiotensin II on renal water excretion, Am. J. Physiol., 228:155-159.

Camacho, A., and Phillips, M.I., 1981a, Separation of drinking and pressor responses to central angiotensin by monoamines, Am. J. Physiol., 240:R106-R113.

Camacho, A., and Phillips, M.I., 1981b, Horseradish peroxidase study in rat of the neural connections of the organum vasculosum of the lamina terminalis, Neurosci. Lett., 25:201-204.

Carroll, R.G., and Opdyke, D.F., 1982, Evolution of angiotensin II-induced catecholamine release, Am. J. Physiol., 243:R65-R69.

Casto, R., and Phillips, M.I., 1985, Neuropeptide action in nucleus tractus solitarius: Angiotensin specificity in hypertensive rats, Am. J. Physiol., 249:R341-R347.

Chen, F.M., Hawkins, R., and Printz, M.P., 1982, Evidence for a functional, independent brain-angiotensin system: correlation between regional distribution of brain angiotensin receptors, brain angiotensinogen and drinking during the estrus cycle of rats, Exp. Brain Res., Suppl. 4:157-168.

Chen, F.M., and Printz, M.P., 1983, Chronic estrogen treatment reduces angiotensin II receptors in the anterior pituitary, Endocrinology 113:1503-1510.

Chevillard, C., Duchene, N., Pasquier, R., and Alexander, J.M., 1979, Relation of the centrally evoked pressor effect of angiotensin II to central noradrenaline in the rabbit, Eur. J. Pharmacol., 58:203-206.

Claybaugh, J.R., Share, L., and Shimizu, K., 1972, The inability of infusions of angiotensin to elevate the plasma vasopressin concentration in the anesthetized dog, Endocrinology 90:1647-1652.

Cole, F.E., Blakesley, H.L., Graci, K.A., Frohlich, E.D., and MacPhee, A.A., 1981, the specificity of angiotensin II receptor binding in rat brain, Peptides 2:441-444.

Cowley, A.W., Jr., Switzer, S.J., and Skelton, M.M., 1981, Vasopressin, Fluid, and electrolyte response to chronic angiotensin II infusion, Am. J. Physiol., 240:R130-R138.

Davila, D., and Davila,T., 1975, Effect of angiotensin II and its analogs on uptake and release of ^{14}C-5-hydroxyteyptamine by rat brain, Eur. J. Pharmacol., 34:329-335.

Davila, D., and Khairallah, P.A., 1971, Angiotensin and the biosynthesis of norepinephrine, Arch. Int. Pharmacodyn., 1983:307-314.

Day, R.P., and Reid, I.A., 1976, Renin activity in dog brain: Enzymological similarity to cathepsin D., Endocrinology 99:93-100.

Devynck, M.A., Rouzaire-Dubois, B., Chevillotte, E., and Meyer, P., 1976, Variations in the number of uterine angiotensin receptors following changes in plasma angiotensin levels, Eur. J. Pharmacol., 40:27-37.

Deweid, D., Sideirus, P., and Mirsky, I.A., 1961, The antidiuretic- and ACTH-releasing effect of various octapeptides, Arch Intern. Pharmacodyn., 133:50-59.

Douglas, J.G., and Brown, G.P., 1982, Effect of prolonged low dose infusions of angiotensin II and aldosterone on rat smooth muscle and adrenal angiotensin II receptors, Endocrinology 111:988-992.

Eguchi, T., and Bravo, E.L., 1984, Humoral responses to intracerebroventricularly administered angiotensin II in dogs, Am. J. Physiol., 247:E336-E342.

Epstein, A.N., 1976, The physiology of thirst, Canad. J. Physiol. and Pharmacol., 54:639-649.

Falcon, J.E., Phillips, M.I., Hoffman, W.E., and Brody, M.J., 1978, Effects of intraventricular angiotensin II mediated by the sympathetic nervous system, Am. J. Physiol., 235(4):H392-H399.

Feldberg, W., and Lewis, G.P., 1964, The actions of peptides on the adrenal medulla. Release of adrenaline by bradykinin and angiotensin, J. Physiol., 171:98.

Felix, D., and Phillips, M.I., 1979, Inhibitory effects of luteinizing hormone releasing hormone (LHRH) on neurons in the organum vasculosum lamina terminalis (OVLT), Brain Res., 169:204-208.

Ferrario, C.M., Dickinson, C.J., and McCubbin, J.W., 1970, Central vasomotor stimulation by angiotensin, Clin. Sci., 39:239-245.

Fisher-Ferraro, C., Nahmod, V.E., Goldstein, D.J., and Finkielman, S., 1971, Angiotensin and renin in rat and dog brain, J. Exp.. Med., 133:353-361.

Fishman, M.C., Zimmerman, E.A., and Slater, E.E., 1981,, Renin and angiotensin II: The coupler system within the neuroblastoma and glioma, Science 214:922-923.

Fitzsimons, J.T., 1980, Angiotensin stimulation of the central nervous system, Rev Physiol. Biochem. Pharmacol., 87:117-167.

Fluharty, S.J., and Epstein, A.N., 1983, Sodium appetite elicited by intracerebroventricular infusion of angiotensin II in the rat. II Synergistic interaction with systemic mineralcorticoids, Behav. Neurosci., 97:746-758.

Fluharty, S.J., Kenney, N., and Epstein, A.N., 1981, Cerebral prostaglandin biosynthesis and angiotensin-induced drinking in rats, J. Comp. Physiol Psychol., 95:915-923.

Fluharty, S.J., and Manaker, S., 1983, Sodium appetite elicited by intracerebroventricular infusion of angiotensin II in the rat: I. Relation to urinary sodium extraction, Behav. Neurosci., 97:738-745.

Fregly, M.J., 1980, Effect of chronic treatment with estrogen on the dipsogenic response of rats to angiotensin, Pharm. Biochem. Behav., 12:131-136.

Fregly, M.J., Rowland, N.E., Sumners, C., and Gordon, D.B., 1985, Reduced dipsogenic responsiveness to intracerebroventricularly administered angiotensin II in estrogen-treated rats, Brain Res., 338:1151-121.

Fuxe, K., Andersson, K., Locatelli, V., Mutt, V., Lundberg, J., Hokfelt, T., Agnati, L.F., Eneroth, P., and Bolme, P., 1980, Neuropeptides and central catecholamine systems: Interactions in neuroendocrine and central cardiovascular regulation, in: "Neural Peptides and Neuronal Communication," E. Costa, and M. Trabucchi, eds., pp. 37-50, Raven Press, New York.

Fuxe, K., Ganten, D., Bolme, P., Agnati, L.F., Hokfelt, T., Andersson, K., Goldstein, M., Harfstrand, A., Unger, T., and Rascher, W., 1976a, The role of central catecholamine pathways in spontaneous and renal hypertension in rats, in: "Central Adrenaline Neurons, Wenner-Gren Center International Symposium Series," K. Fuxe et al., eds., Vol. 33, pp. 259-276, Pergamon Press, Oxford and New York.

Fuxe, K., Ganten, D., Hokfelt, T., and Bolme, P., 1976b, Immunohisto-chemical evidence for the existence of angiotensin II-containing nerve terminals in the brain and spinal cord in the rat, Neurosci. Lett., 2:229-234.

Ganong, W.F., Shinsako, J., and Reid, I.A., 1982, Role of vasopressin in the renin and ACTH responses to intraventricular angiotensin II, in: "The Brattleboro Rat," H.W. Sokol, and H. Valtin, ed., Vol. 394, pp. 619-624, Ann. N.Y. Acad. Sci., New York.

Ganten, D., Fuxe, J., Phillips, M.I., Mann, J.F.E., and Ganten, U., 1978, The brain iso-renin angiotensin system: Biochemistry localization and possible role in drinking and blood pressure regulation, in: "Frontiers in Neuroendocrinology", W.F. Ganong, and L. Martini, eds., pp. 61-69,Raven Press, New York.

Ganten, D., Hutchinson, J.S., and Schelling, P., 1975,The intrinsic brain iso-renin-angiotensin system in the rat: Its possible role in central mechanism of blood pressure regulation, Clin. Sci. Mol. Med., 48:265s-268s.

Ganten, D., Minnich, J.L., Granger, P., Hayduk, K., Brecht, H.M., Barbeau, A., Boucher, R., and Genest, J, 1971, Angiotensin-forming enzyme in brain tissue, Science 173:64-65.

Garcia-Sevilla, J.A., Dubocovich, M.L., and Langer, S.Z., 1979, Angiotensin II facilitates the potassium-evoked release of ^{3}H-noradrenaline from the rabbit hypothalamus, Eur. J. Pharmacol., 56:173-176.

Gordon, F.J., Brody, M.J., Fink, G.D., Buggy, J., and Johnson, A.K., 1979, Role of central catecholamines in the control of blood pressure and drinking behavior, Brain Res., 178:161-173.

Grossman, S.P., 1962, Direct adrenergic and cholinergic stimulation of hypothalamic mechanisms, Am. J. Physiol., 202:872-882.

Gunther, S., Gimbrone, M.A., Jr., and Alexander, R.W., 1980, Regulation by angiotensin II of its receptors in resistance blood vessels, Nature 287:230-232.

Harding, J.W., Stone, L.P., and Wright, J.W., 1981, The distribution of angiotensin II binding sites in rodent brain, Brain Res., 205:265-274.

Haulica, I., Petrescu, G., Stratone, A., Branisteanu, D., Rosca, V., Neamtu, C., and Slatineanu, S., 1982, Possible functions of brain renin, in: "The Renin Angiotensin System in the Brain," D. Ganten, M. Printz, M.I. Phillips, and B.A. Scholkens, eds., pp. 335-352, Springer-Verlag, Heidelberg.

Haulica, I., Petrescu, G., Uluitu,M., Rosca, V., and Slatineanu, S., 1980, Influence of angiotensin II on dog pineal serotonin content, Neurosci. Lett., 18:329-332.

Hawkins, R.L., and Printz, M.P., 1983, Distribution of angiotensinogen in Brattleboro rat brain, Brain Res. Bull., 10:163-166.

Haywood, J.R., Buggy, J., Phillips, M.I., and Brody, M.J., 1980, The area postrema plays no role in the pressor action of angiotensin in the rat, Am. J. Physiol., 239:H108-H113.

Healy, D.P., and Printz, M.P., 1984a, Distribution of immunoreactive angiotensin II, angiotensin I, angiotensinogen and renin in the CNS of intact and nephrectomized rats, Hypertension 6:I130-I136.

Healy, D.P., and Printz, M.P., 1984b, Localization of angiotensin II binding sites in rat septum by autoradiography, Neurosci. Lett., 44:167-172.

Hermann, K., Lang, R.E., Unger, T., Bayer, C., and Ganten, D., 1984, Combined high-performance liquid chromatography radioimmunoassay for the characterization and quantitative measurement of neuropeptides, J. Chromatogr., 312:273-284.

Hirose, S., Yokosawa, H., and Inagami, T., 1978, Immunochemcial identification of renin in rat brain and distinction from acid proteases, Nature 274:392-393.

Hoffman, W.E., Ganten, U., Phillips, M.I., Schmid, P.G.,Schelling, P., and Ganten, D., 1978, Inhibition of drinking on water-deprived rats by combined central angiotensin II and cholinergic receptor blockade. Am. J. Physiol., 234:F41-F47.

Hoffman, W.E., and Phillips, M.I., 1976a, Regional study of cerebral ventricular sensitive sites to angiotensin II, Brain Res., 110:313-330.

Hoffman, W.E., and Phillips, M.I., 1976b, A pressor response to intraventricular injections of carbachol, Brain Res., 105:157-162.

Hoffman, W.E., Phillips, M.I., and Schmid, P.G., 1977, The role of catecholamines in central antidiuretic and pressor mechanisms, Neuropharmacology 16:563-569.

Hughes, J., and Roth, R.H., 1971, Evidence that angiotensin enhances transmitter release during sympathetic nerve stimulation, Br. J. Pharmacol., 41:239-255.

Hutchinson, J.S., Schelling, P., Mohring, J., and Ganten, D., 1976, Pressor action of centrally perfused angiotensin II in rats with hereditary hypothalamic diabetes insipidus, Endocrinology 99:819-823.

Inagami, T., Okamura, T., Hirose, S., Clemens, D., Celio, M.R., Naruse, K., Takii, Y., and Yokosawa, H., 1982, Identification characterization and evidence for interneuronal function of renin in the brain and neuroblastoma cells, in: "The Renin Angiotensin System in the Brain," D. Ganten, M. Printz, M.I. Phillips, and B.A. Scholkens, eds., pp. 64-75, Springer-Verlag, Heidelberg.

Israel, A., Correa, F.M.A., Niwa, M., and Saavedra, J.M., 1984, Quantitative determination of angiotensin II binding sites in rat brain and pituitary gland by autoradiography, Brain Res., 322:341-345.

Jonklaas, J., and Buggy, J., 1984, Angiotensin-estrogen interaction in female brain reduces drinking and pressor responses, Am. J. Physiol., 247:R167-R172.

Kapsha, J.M., Keil, L.C., Klase, P.A., and Severs, W.B., 1979, Centrally mediated hydration effects of angiotensin in various states of sodium balance, Pharmacology 81:25-33.

Keil, L.C., Rosella-Dampman, L.M., Emmert,S., Chee, O., and Summy-Long, J.Y., 1984, Enkephalin inhibition of angiotensin-stimulated release of oxytocin and vasopressin, Brain Res., 297:329-336.

Keil, L.C., Summy-Long, J.,and Severs,W.B., 1975, Release of vasopressin by angiotensin II, Endocrinology 96:1063-1065.

Keller-Wood, M., Stenstrom, B., Shinsako, J., and Phillips, M.I., 1986, Interaction between CRF and Ang II in control of ACTH and adrenal steroids, Am. J. Physiol., In Press.

Kenney, J., and Epstein, A.N., 1978, Antidipsogenic role of the E-prostaglandins, J. Comp. Physiol. Psychol., 92:204-219.

Kilcoyne, M.M., Hoffman, D.L., and Zimmerman, E.A., 1980, Immunocytochemical localization of angiotensin II and vasopressin in rat hypothalamus: Evidence for production in the same neuron,

Clin. Sci., 59:57s–60s.

Knepel, W., and Meyer, D.K., 1980, Role of the renin-angiotensin system in isoprenaline-induced vasopressin release, J. Cardiovas. Pharmacol., 2:815–824.

Knowles, W.D., and Phillips, M.I., 1980, Angiotensin II responsive cells in the organum vasculosum lamina terminalis (OVLT) recorded in hypothalamic brain slices, Brain Res., 195:256–259.

Landas, S., Phillips, M.I., Stamler, J.F., and Raizada, M.K., 1980, Visualization of specific angiotensin II binding sites in the brain by fluorescent microscopy, Science 210:791–793.

Leavitt, M.L., 1975, The phenomenon of permanent Na$^+$ appetite in rats, Ph.D. Thesis, University of Iowa, Iowa City, Iowa.

Lewis, R.E., and Phillips, M.I., 1984, Localization of the central pressor action of bradykinin to the cerebral third ventricle, Am. J. Physiol., 16:R63–R68.

Lind, R.W., Swanson, R.W., and Ganten, D., 1985, Organization of angiotensin II immunoreactive cells and fibers in the rat central nervous system, Neuroendocrinology 40:2–24.

Malayan, S.A., Keil, L.C., Ramsay, D.J., and Reid, I.A., 1979, Mechanism of suppression of plasma renin activity by centrally administered angiotensin II, Endocrinology 104:672–675.

Malvin, R.L., Mouw, D., and Vander, A.J., 1977, Angiotensin: Physiological role in water deprivation-induced thirst of rats, Science 197:171–173.

Mangiapane, M.L., and Simpson, J.B., 1980, Subfornical organ lesions reduce pressor effects of intravenous angiotensin, Neuroendocrinology 31:380–384.

Mangiapane, M.L., Thrasher, T.N., Keil, L.C., Simpson, J.B., and Ganong, W.F., 1982, Subfornical organ lesions impair the vasopressin (AVP) response to hyperosmolality or angiotensin II (AII), Fed. Proc.., 41:1105.

Mann, J.F.E., Johnson, A.K., and Ganten, D.. 1980, Plasma angiotensin II: Dipsogenic levels and angiotensin generating capacity of renin, Am. J. Physiol., 238:R372–R377.

Mann, J.F.E., Schiller, P.W., Schiffrin, E.L., Boucher, R., and Genest, J., 1981, Brain receptor binding and central actions of angiotensin analogs in rats, Am. J. Physiol., 241:R124–R129.

Maran, J.W., and Yates, F.E., 1977, Cortisol secretion during intrapituitary infusion of angiotensin II in conscious dogs, Am. J. Physiol., 233:E273–E285.

Matsuguchi, H., Sharabi, F.M., Gordon, F.J., Johnson, A.K., and Schmid,

P.G., 1982, Blood pressure and heart rate responses to vasopressin microinjection into the nucleus tractus solitarius region of the rat, Neuropharmacol., 21:687-693.

McGiff, J.C., and Wong, P.Y.K., 1979, Compartmentalization of prostaglandins and prostacyclins within the kidney: Implications for renal function, Fed. Proc., 38:89-93.

Meldrum, M.J., Xne, C.S., Badino, L., and Westfall, T.C., 1984, Angiotensin facilitation of noradrenergic neurotransmission in central tissues of the rat: Effects of sodium restriction, J. Cardiovas. Pharmacol., 6:989-995.

Mendelsohn, F.A.O., Quirion, R., Saavedra, J.M., Aguilera, G., and Catt, K.J., 1984, Autoradiographic localization of angiotensin II receptors in rat brain, Proc. Natl. Acad. Sci., USA, 81:1575-1579.

Miselis, R.R., 1981, The efferent projections of the subfornical organ of the rat: A circumventricular organ within a neural network subserving water balance, Brain Res., 230:1-23.

Mouw, D.R., and Vander, A.J., 1970, Evidence for brain Na receptors controlling renal Na excretion and plasma renin activity, Am. J. Physiol., 219:822-832.

Nahmod, V.E., Finkielman,S., Benarroch, E.E., and Pirola, C.J., 1978, Angiotensin regulates release and synthesis of serotonin in brain, Science 202:1091-1093.

Nicholls, M.G., Malvin, R.L., Espiner, E.A., Lun, S., and Miles, K.D., 1983, Adrenal hormone responses to cerebroventricular angiotensin II in the sheep (41565), Proc. Soc. Exp. Biol. Med., 172:330-333.

Nicoll, R.A., and Barker, J.L., 1971, Excitation of supraoptic neurosecretory cells by angiotensin II, Nature New Biol., 233:172-174.

Padfield, P.L., and Morton, J.J., 1977, Effects of angiotensin II on arginine-vasopressin in physiological and pathological situations in man, J Endorcin., 74:251-259.

Paglin, S., Stukenbrok, H., and Jamieson, J.F., 1984, Interaction of angiotensin II with dispersed cells from the anterior pituitary of the male rat, Endocrinology 114:2284-2292.

Palaic, D., and Khairallah, P.A., 1967,Effect of angiotensin on uptake and release of norepinephrine by brain, Biochem. Pharmacol., 16:2291-2298.

Phillips, M.I., 1978, Angiotensin in the brain, Neuroendocrinology 25:354-377.

Phillips, M.I., 1981,Central effects of angiotensin II on hypertension and sodium intake, in: "Role of Sodium in Cardiovascular

Hypertension," M. Fregly and M. Kare,, eds., pp. 127-143, Academic Press

Phillips, M.I., Camacho, A., and Sumners, C., 1981, Thirst induced by angiotensin II analyzed pharmacologically, Psychopharm. Bull., 17(3):53-56.

Phillips, M.I., Deshmukh, P., and Larsen, W., 1978, Morphological comparisons of the ventricular wall of subfornical organ and organum vasculosum of the lamina terminalis, Scan. Elec. Microsc., 2:349-356.

Phillips, M.I., and Felix, D., 1976, Specific angiotensin II receptive neurons in the cat subfornical organ, Brain Res., 109:531-540.

Phillips, M.I., and Hoffman, W.E., 1976, Sensitive sites in the brain for the blood pressure and drinking responses to angiotensin II, in: "Central Actions of Angiotensin II and Related Hormones," J.P. Buckley, and C. Ferrario, eds., pp. 325-356, Pergamon Press, Oxford.

Phillips, M.I., Hoffman, W.E.,and Bealer, S.L., 1982, Dehydration and fluid balance: Central effects of angiotensin, Fed. Proc.., 41:2520-2527.

Phillips, M.I., Mann, J.F.E., Haebara, H., Hoffman, W.E, Dietz, R., Schelling, P., and Ganten, D., 1977, Lowering of hypertension by central saralasin in the absence of plasma renin, Nature 270:445-447.

Phillips, M.I., Phipps, J., Hoffman, W.E., and Leavitt, M., 1975, Reduction in blood pressure by intracranial injection of angiotensin blocker (P113) in spontaneoulsy hypertensive rats (SHR), Physiologist 18:350.

Phillips, M.I., and Stenstrom, B., 1985, Angiotensin II in rat brain comigrates with authentic angiotensin II in HPLC, Circ. Res., 56:212-219.

Phillips, M.I., Weyhenmeyer, J.A., Felix, D., and Ganten, D., 1979, Evidence for an endogenous brain renin angiotensin system, Fed. Proc., 38:2260-2266.

Printz, M.P., Ganten, D., Unger, T.,and Phillips, M.I., 1982, Minireview: The brain renin angiotensin system, in-: "The Brain Renin Angiotensin System," D. Ganten, M. Printz, M.I. Phillips, and B.A. Scholkens, eds., pp. 30-52, Springer Verlag, Heidelberg.

Printz, M.P., Hawkins, R.L., Wallis, C.J., and Chen, F.M.,1983, Steroid hormones as feedback regulators of brain angiotensinogen and catecholamines, Chest 83(2):308-311.

Printz, M.P., Wallis, C.J., Lewicki, J.A.,and Fallon, J.H., 1979,

Correlation between CNS angiotensinogen and central catecholamine pathways, in: "Catecholamines: Basic and Clinical Frontiers," E. Usdin, I.J., Kopin, and J. Barchas, eds., Vol. 2, pp. 1419-1421, Pergamon Press, New York and Oxford.

Raizada, M.K., Muther, T.F., and Sumners, C., 1984a, Increased angiotensin II receptors in neuronal cultures from hypertensive rat brain, Am. J. Physiol., 247:C364-C372.

Raizada, M.K., Stenstrom, B., Phillips, M.I., and Sumners, C., 1984b, Angiotensin II in neuronal cultures from the brains of normotensive and hypertensive rats, Am. J. Physiol., 247:C115-C119.

Raizada, M.K., Yang, J.W., Phillips, M.I., and Fellows,R.E., 1981, Rat brain cells in primary culture: Characterization of angiotensin II binding sites, Brain Res., 207:343-355.

Ramsay, D.J., Keil, L.C., Sharpe, M.C., and Shinsako, J., 1978, Angiotensin II infusion increases vasopressin, ACTH, and 11-hydroxycorticosteroid secretion, Am. J. Physiol., 234:R66-R71.

Reid, I.A., Brooks, V.L., Rudolph, C.D., and Keil, L.C., 1982, Analysis of the actions of angiotensin on the central nervous system of conscious dogs, Am. J Physiol., 243:R82-R91.

Reit, E., 1972, Actions of angiotensin on the adrenal medulla and autonomic ganglia, Fed. Proc., 31:1338-1343.

Saavedra, J.M., Brownstein, M.J., Kizer, J.S.,and Palkovits, M., 1976, Biogenic amines and related enzymes in the circumventricular organs of the rat, Brain Res., 107:412-417.

Sakai, K.K., Marks, B.H., George, J., and Koostner, A., 1974, Specific angiotensin II receptors in organ cultured canine supra-optic nucleus cells, Life Sci., 14:1337-1344.

Schacht, U., 1984, Effects of angiotensin II and ACE inhibitors on electrically stimulated noradrenaline release from superfused rat brain slices, Clin. and Exper. Hyper., Theory and Practice A6(10&11):1847-1851.

Schelling, P., Ganten, D., Heckel, R., Hayduk, K., Hutchinson, J.S., Sponer, G., and Ganten, U., 1977, On the origin of angiotensin-like peptide in cerebrospinal fluid, in: "Central Actions of Angiotensin and Related Hormones," J.P. Buckley, and C.M. Ferrario, eds., pp. 519-526, Pergamon Press, New York.

Schelling, P., Clauser, E., and Felix, D., 1983, Regulation of angiotensinogen in the central nervous system, Clin. and Exper. Hyper., theory and Practice A5(7&8):1047-1061.

Scholkens, B.A., Jung, W., Rascher, W., Dietz, R., and Ganten, D., 1982, Intracerebroventricular angiotensin II increases arterial blood pressure in rhesus monkeys by stimulation of pituitary hormones and sympathetic nervous system, Experentia 38:469-470.

Severs, W.B., and Daniels-Severs,A.E., 1973,Effects of angiotensin on the central nervous sytem, Pharmacol Rev., 25:415-449.

Severs, W.B., Summy-Long, J., Taylor, J.A., and Connor, J.D., 1970, A central effect of angiotensin: Release of pituitary pressor material, J. Pharmacol. Exp. Ther., 174:27-34.

Share, L., 1974, Blood pressure, blood volume, and the release of vaso-pressin, in: "Handbook of Physiology, Sect. 7: Endocrinology," R.O. Greep, and E.B. Astwood, eds., Vol. IV, Part 1,pp. 243-255, American Physiological Society, Washington, D.C.

Shimizu, K., Share, L., and Claybaugh, J.R., 1973, Potentiation by angiotensin II of the vasopressin response to an increasing plasma osmolality, Endocrinology 93:42-50.

Simmonet, G., and Giorguieff-Chesselet, M.F., 1979, Stimulating effect of angiotensin II on the spontaneous release of newly synthesized [^{3}H]-dopamine in rat striatal slices, Neurosci. Lett., 15:153-158.

Simon, W.,Schaz, K., Mann, J.F.E., Ganten, U., Johnson, A.K., Unger, T., Rascher, W., and Ganten, D., 1981, The effects of beta-adrenoceptor blockers on blood pressure responses to central angiotensin II, Neuropharmacology 20:719-726.

Simpson, J.B., 1981, The circumventricular organs and the central actions of angiotensin, Prog. Neuroendocr., 32:248-256.

Simpson, J.B., and Routtenberg,A., 1973, Subfornical organ:Site of drinking elicitation by angiotensin II, Science 181:1172-1175.

Singh, R., Husain, A., Ferrario,C.M., and Speth, R.C., 1984, Rat brain angiotensin II receptors: Effects of intraventricular angiotensin II infusion, Brain Res., 303:133-140.

Sirett, N.E., McLean, A.S., Bray, J.J., and Hubbard, J.I., 1977, Distribution of angiotensin II receptors in rat brain, Brain Res., 122:299-312.

Sladek, C.D., and Johnson, A.K., 1983, Effect of anteroventral third ventricle lesions on vasopressin release by organ cultured hypothalamo-neuropophyseal explants, Neuroendocrinology 37:78-84.

Sladek, C.D., and Joynt, R.J., 1979, Angiotensin stimulation of vasopressin release from the rat hypothalamo-neurohypophyseal system in organ culture, Endocrinology 104:148-153.

Speth, R.C., Singh, R., Smeby, R.R., Ferrario, C.M., and Husain, A., 1984, Restricted dietary sodium intake alters peripheral but not central angiotensin II receptors, Neuroendocrinology 38:387-392.

Spinedi, E., and Negro-Vilar, A., 1983,angiotensin II and ACTH release:

Site of action and potency relative to corticotropin releasing factor and vasopressin, Neuroendocrinology 37:446-453.

Stamler, J.F., Raizada, M.K., Fellows,R.E., and Phillips, M.I., 1980, Increased specific binding of angiotensin II in the organum vasculosum of the lamina terminalis area of the spontaneously hypertensive rat brain, Neurosci Lett., 17:173-177.

Starke, K., 1971, Action of angiotensin on uptake, release and metabolism of ^{14}C-noradrenaline by isolated rabbit hearts, Eur. J. Pharmacol., 14:112-123.

Steele, M.K., Gallo, R.V.,and Ganong, W.F., 1985, Stimulatory or inhibitory effects of angiotensin II upon LH secretion in ovariectoized rats: A function of gonadal steroids, Neuroendocrinology 40:210-216.

Steele, M.K., Negro-Vilar, A., and McCann, S.M., 1982, Modulation by dopamine and estradiol of the central effects of angiotensin II on anterior pituitary hormone release, Endocrinology 111:722-729.

Stricker, E.M., 1978,The renin angiotensin system and thirst: Some unanswered questions, Fed. Proc., 37:2704-2710.

Summy-Long, J.Y., Keil, L.C., Deen, K., Rosella, L., and Severs, W.B., 1981, Endogenous opioid peptide inhibition of the central actions of angiotensin, J. Pharmacol. Exp. Therap., 217:619-623.

Summy-Long, J.Y., Keil, L.C., Sells, G., Kirby, A., Ohee, O., and Severs, W.B., 1983, Cerebroventricular sites for enkephalin inhibition of the central actions of angiotensin, Am. J. Physiol., 244:R522-R529.

Sumners, C., and Phillips, M.I., 1983, Central injection of angiotensin II alters catecholamine activity in rat brain, Am. J. Physiol., 244:R257-R263.

Sumners, C., Phillips, M.I., and Raizada, M.K., 1983, Angiotensin II stimulates changes in the norepinephrine content of primary cultures of rat brain, Neurosci Lett., 36:305-309.

Sumners, C., and Raizada, M.K., 1984, Catecholamine-angiotensin II receptor interaction in primary cultures of rat brain, Am. J. Physiol., 246:C502-C509.

Sumners, C., and Raizada, M.K., 1985, Angiotensin II stimulates norepinephrine uptake in hypothalamus/brainstem neuronal cultures, Am. J. Physiol., In Press.

Suzuki, H., Ferrario, C.M., Speth, R.C., Brosnihan, K.B., Smeby, R., and deSilva, P., 1983, Alterations in plasma and cerebrospinal fluid, norepinephrine and angiotensin II during the development of renal hypertension in conscious dogs, Hypertension 5 (Suppl. I):I139-I148.

Swanson, L.W., and Swachenko, P.E., 1983, Hypothalamic integration:

Organization of the paraventricular and supraoptic nuclei, Ann. Rev. Neurosci., 6:269-324.

Swerts, J.P., Perdrisot, R., Malfroy, B., and Schwartz, J.C., 1979, Is "enkephalinase" identical with "angiotensin converting enzyme?", Eur. J. Pharmacol., 53:209-210.

Szilagyi, J.E., and Ferrario, C.M., 1981, Central opiate modulation of the area postrema pathway, Hypertension 3:313-317.

Tigerstedt, R., and Bergman, P.G., (1898), Niere and Kreislauf, Skand. Arch Physiol., 8:223-271.

Tonnaer, J.A.D.M., Engels, G.M.H., Voshart, K., Wiegant, V.M.,a nd Dejong, W., 1983, Binding of angiotensins to rat brain tisue: Structure activity relationship, Brain Res. Bull., 10:295-300.

Uhlich, E.P., Weber, P., Eigler, J., and Groschel-Stewart, U., 1975, Angiotensin stimulated AVP-release in humans, Klin. Wschr., 53:177-180.

Unger, T., Rascher, W., Shuster, C., Pavlovitch, R., Schonig, A., Dietz, R., and Ganten, D., 1981, Central blood pressure. Effects of substnce P and angotensin II: Role of the sympathetic nervous system and vasopressin, Eur. J. Pharmacol., 71:33-42.

Vale, W., Vaughan, J., Smith, M., Yamamoto, G., Rivier, J., and Rivier, C., 1983, Effects of synthetic ovine corticotropin-releasing factor, glucocorticoids, catecholamines, neurohypophysial peptides, and other substances on cultured corticotropic cells, Endocrinology 113:1121-1131.

Van Houten, M., Schriffrin, E.L., Mann, J.F.E., Posner, B.I., and Boucher, R., 1980, Radioautographic localization of specific binding sites for blood borne angiotensin II in the rat brain, Brain Res., 86:480-485.

Weyhenmeyer, J.A., Raizada, M.K., Phillips, M.I., and Fellows, R.E., 1980, Presence of angiotensin II in neurons cultured from fetal rat brain, Neurosci. Let., 16:41-46.

Yang, H.Y.T., and Neff, N.H., 1972, Distribution and properties of angiotensin converting enzyme of rat brain, J. Neurochem., 19:2443-2450.

Zimmerman, B.G., and Gisslen, J., 1968, Pattern of renal vasoconstriction and transmitter release during sympathetic stimulation in the presence of angiotensin and cocaine, J. Pharmacol. Exp. Therap., 163:320-329.

Zimmerman, B.G., Gomez, S.K., and Liao, J.C., 1972, Action of angiotensin on vascular adrenergic nerve endings: Facilitation of noreinephrine release, Fed. Proc., 31:1344-1348.

ROLE OF PARAVENTRICULAR AND ARCUATE NUCLEI IN CARDIOVASCULAR REGULATION

Michael J. Brody*, Timothy P. O'Neill** and James P. Porter***

*Department of Pharmacology and Cardiovascular Center, University of Iowa College of Medicine, Iowa City, IO; **Procter and Gamble Company, Miami Valley Laboratories, P.O. Box 39175, Cincinnati, OH 45247; ***Department of Physiology, University of Louisville, A-1115 Health Science Center, Louisville, KY 40292

INTRODUCTION

There has been an increasing awareness in recent years of the potential for structures in forebrain to influence the cardiovascular system. Much attention has been focused on the paraventricular nucleus, which has been defined neuroanatomically to contain both magnocellular units that project to the neurohypophysis and parvocellular neurons that descend monosynaptically to the intermediolateral column of the spinal cord. Interest in the paraventricular nucleus has also extended to its role as a source of both humorally derived vasopressin and of descending projections from parvocellular regions of the nucleus that contain separate vasopressin and oxytocin-containing fibers. Portions of the studies reviewed in this chapter describe experiments designed to evaluate the role of the parvocellular paraventricular nucleus in cardiovascular regulation and to determine whether vasopressinergic projections to the spinal cord contribute to cardiovascular effects produced by excitation of this region.

The arcuate nucleus of the hypothalamus has been demonstrated neuroanatomically to be the bed nucleus for the pro-opiomelanocortin system of the brain. We found that virtually nothing was known about interactions between the arcuate nucleus and the cardiovascular system despite the fact that the nucleus projects to many important sites with the capability to influence cardiovascular function. Included among the projections from arcuate nucleus are direct connections with the

paraventricular nucleus. The second portion of the chapter describes
studies that demonstrate a role of the arcuate nucleus in cardiovascular
regulation that is mediated by release of pressor quantities of
vasopressin.

The Paraventricular Nucleus

Recent advances in techniques for identification of neuronal
connectivity have identified the paraventricular nucleus of the
hypothalamus (PVN) as a potentially important site involved in the
integration of cardiovascular functions. Before 1975, it was generally
thought that the hypothalamus exerted its effects on the autonomic nervous
system through polysynaptic pathways which terminated in the pons and
medulla (Pitts, Larrabee and Bronk, 1941; Nauta and Haymaker, 1969).
However, in 1975, Kuypers and Maisky (1975) employed the relatively new
technique of using retrograde transport tracers to show that the dorsal
hypothalamus sent long projections which extended without synapses to the
spinal cord. Saper et al. (1976) confirmed this finding and identified
the PVN as the site where the majority of these descending fibers
originated. Furthermore, anterograde transport of tritiated amino acids
which were injected into the PVN showed that these projections terminated
in the nucleus of the solitary tract (NTS) and the intermediolateral
column of the spinal cord. This raised the possibility that the PVN could
influence autonomic functions, including cardiovascular regulation,
through direct connections in the medulla and spinal cord.

Since these original observations, many neuroanatomical studies have
been undertaken to determine more precisely the pathways by which the PVN
could exert a regulatory influence on the cardiovascular system. The PVN
can be divided, based on potential function, into three different
anatomical subdivisions. Magnocellular cell bodies which produce oxytocin
and vasopressin, send axons primarily to the posterior pituitary
(Armstrong et al., 1980). This neurosecretory system has long been known
to exist and is involved in homeostatic mechanisms related to fluid and
electrolyte balance, cardiovascular regulation, suckling reflex and
partuition (Valiquette, 1980). A second group of neurons is located in
parvocellular PVN and projects to the median eminence. These neurons are
presumably involved in the neuroendocrine control of hypophyseal hormone
secretion (Swanson and Kuypers, 1980). A third group of neurons in the
PVN, also parvocellular, send long descending projections to the medulla

and spinal cord (Armstrong et al., 1980). It is these neurons which presumably play a role in regulation of sympathetic outflow. Approximately 15% of these neurons send collateral axons to both the dorsal medulla and the spinal cord (Swanson and Kuypers, 1980). The remaining 85% appear to project solely to one site or the other. Immunohistochemistry has been used to demonstrate that a portion of the long descending neurons contain vasopressin or oxytocin and neurophysin (Swanson, 1977; Nilaver et al., 1980; Sofroniew, 1983). However, the majority of the fibers have not been characterized as to transmitter type.

Anatomical studies have also shown that many of the cardiovascular sites in the medulla which receive afferent input from the PVN also send reciprocal efferent projections which terminate in this nucleus. The PVN has long been known to be a major termination site for adrenergic neurons (Carlsson, Falck and Hillarp, 1962). Techniques employing combined retrograde labeling and catecholamine cell body markers have shown that the A1 catecholamine area in the ventrolateral medulla (VLM), the A2 region in the dorsomedial medulla, and the A6 area (locus coeruleus) all send adrenergic projections to parvocellular regions of PVN. Only the A1 area sent catecholaminergic projections to magnocellular PVN (Sawchenko and Swanson, 1982). However, one recent report showed that some of the catecholaminergic terminals in parvocellular PVN actually made contact with dendrites from the magnocellular region (Silverman et al., 1985). This raises the possibility that axons from the A2 and A6 areas also make connections with magnocellular PVN.

Electrophysiological techniques employing antidromic and orthodromic neuronal activation have been used to confirm the existence of the neuroanatomical connections mentioned above. Connections from the PVN to the NTS, VLM, and spinal cord have all been confirmed (Kannan and Yamashita, 1983; Caverson, Ciriello and Calaresu, 1984; Yamashita, Inenaga and Koizumi, 1984; Rogers and Nelson, 1984). Cell bodies have been identified in the VLM and NTS which are influenced by both baroreceptor inputs and efferent fibers from the PVN (Caverson, Ciriello and Calaresu, 1983; Kannan and Yamashita, 1985). Some baroreceptor-sensitive neurons in the NTS and VLM have also been shown to orthodromically alter the firing rate of PVN cell bodies (Ciriello and Caverson, 1984; Calaresu and Ciriello, 1980). Furthermore, direct activation of baroreceptor afferent fibers resulted in a change in the firing of PVN neurons (Calaresu and Ciriello, 1980). Taken together, these data strengthen the possibility that the PVN is involved in modulation of baroreflex function.

Thus the PVN is situated, anatomically, to exert a major integrative influence on the cardiovascular system. It can receive afferent input from medullary centers which are influenced by baroreceptor activity and exert regulatory effects on the cardiovascular system through release of vasopressin into the peripheral circulation and through efferent signals to the medulla (baroreflex and vasomotor centers) and to the spinal cord (origin of sympathetic outflow).

Anatomical and electrophysiological data can only suggest a functional role for a given central nervous site. There are only a few studies which have directly investigated the functional contribution of the PVN to cardiovascular regulation. Ciriello and Calaresu (1980) reported that electrical stimulation of the PVN in anesthetized cats produced increases in heart rate and arterial pressure. In addition, stimulation of PVN inhibited the reflex bradycardia evoked by stimulating the carotid sinus nerve. These data suggested that the PVN can indeed modulate cardiovascular functions including baroreflex regulation. Lesions in the PVN have been reported to attenuate the development of hypertension in spontaneously hypertensive rats and prevent or reverse the elevation in pressure produced by denervation of the aortic baroreceptors (Ciriello et al., 1984; Zhang and Ciriello, 1982). This suggests that this nucleus may contribute to the pathogenesis of certain forms of hypertension. The PVN may also be an important relay site for cardiovascular outputs from the subfornical organ (SFO). The SFO is a circumventricular organ without a blood brain barrier and contains receptors for angiotensin II which mediate in part the central pressor effect of this peptide. Several investigators have shown that lesions in the PVN significantly reduced the increase in pressure produced by electrical activation of the SFO (Ferguson and Renaud, 1984; Mangiapane and Brody, 1984).

For the past few years, studies in our laboratory have been aimed at determining more precisely the functional role of the PVN in regulating cardiovascular functions. We have used knife cuts, chemical lesions, local anesthetics and peptide antagonists within the medulla and spinal cord to determine the course taken and function subserved by fibers descending from the PVN.

In initial studies we stimulated the PVN in anesthetized rats instrumented with Doppler flow probes on the renal and mesenteric arteries and the lower abdominal aorta to determine the pattern of change in

regional vascular resistance. We found that unilateral activation of the
PVN with relatively large stimulating electrodes produced an integrated
response similar to the defense reaction, i.e., vasoconstriction in the
gut and kidney and vasodilation in the hindquarters (skeletal muscle
vascular beds) (Porter and Brody, 1985a). This response was mediated by
increased sympathetic outflow since prior ganglionic blockade prevented
the increase in arterial pressure and changes in vascular resistance (Fig.
1). This was somewhat surprising since the PVN contains many vasopressin
neurosecretory neurons which project to the posterior pituitary (Swanson
and Kuypers, 1980; Armstrong et al., 1980). Acute bilateral adrenalectomy
reduced the hindquarter vasodilation produced by stimulation of PVN but
the vasoconstriction was not affected (Porter and Brody, 1985a). These
data are in agreement with other work in our laboratory which suggested
the circulating adrenal catecholamines are required for centrally mediated
skeletal muscle vasodilation (Berecek and Brody, 1982).

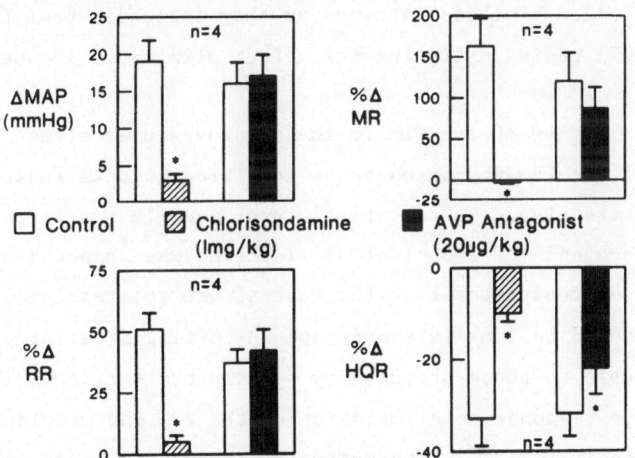

Fig. 1. Effect of chlorisondamine (1 mg/kg) and vasopressin (AVP)
         antagonist (20 μg/kg)on response to right paraventricular nucleus
         stimulation. Each drug was given to a different group of 4 rats.
         Values in this and all subsequent figures are means and SE. MAP:
         mean arterial pressure; MR: mesenteric resistance; RR: renal
         resistance; HQR: hindquarter resistance. *P <0.05. From Porter
         and Brody, 1985a.

Once the hemodynamic response to stimulation of the PVN was characterized, we sought to functionally determine which pathways mediated the changes in vascular resistance. Anatomical data suggested the long descending projections from PVN separated at the level of the rostral medulla (Swanson, 1977; Luiten et al., 1985). Fibers destined to terminate in the spinal cord coursed down the ventrolateral medulla. Fibers destined to terminate in the area of the NTS swept dorsomedially away from the ventrolateral medulla just caudal to the facial nucleus. We sought to determine separately the contribution of each of these major neuronal pathways to the hemodynamic response produced by activation of the PVN. The contribution of the pathway to the NTS region was assessed by stimulating the PVN before and after making a knife cut which was designed to interrupt the PVN-NTS connections. This cut was made by undercutting the NTS region unilaterally with an L-shaped knife. Following this cut, the vasoconstrictor responses produced by stimulation of PVN were unaffected, however, the skeletal muscle vasodilation was significantly reduced (Porter and Brody, 1985a). This suggested that a component of the vasodilator effect was mediated by fibers projecting to or through the NTS area. Other work in our laboratory had previously identified a vasodilator pathway from the AV3V region which also passed through the dorsomedial medulla (Knuepfer, Johnson and Brody, 1984). It is not clear whether the vasodilation with PVN stimulation was due to indirect activation of these pathways as they descended past the PVN or whether the cell bodies mediating the effect originated in the nucleus.

The contribution of the VLM to the cardiovascular effect of stimulation of the PVN was assessed by injecting a local anesthetic directly into the tissue. The ventrolateral medulla was exposed using the ventral approach and 100-200 nl of 5% lidocaine was injected bilaterally through glass micropipettes into the rostral ventrolateral medulla. Lidocaine produced an immediate decrease in resting arterial pressure to levels comparable to those produced by spinal cord section. The vasoconstrictor response to stimulation of the PVN was completely blocked by the lidocaine (Fig. 2). Interestingly, the vasodilation in the hindquarters was significantly enhanced. Thus vasoconstrictor outflow from the PVN appears to travel through the ventrolateral medulla.

Anatomical and electrophysiological evidence have suggested that the ventrolateral medulla is an important termination site for neurons from the PVN. In our experiment with lidocaine, we could not determine if the

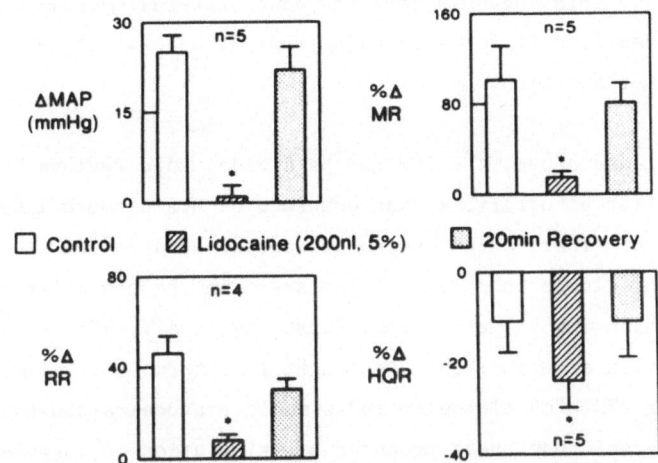

Fig. 2. Effects of bilateral injections of 5% lidocaine (200 nl) into
ventrolateral medulla on response to right paraventricular
nucleus stimulation. MAP: mean arterial pressure; MR: mesenteric
resistance; RR: renal resistance; HQR hindquarter resistance. *P
<0.05. From Porter and Brody, 1985a.

anesthetic blocked the vasoconstrictor response by inhibiting axons from
the PVN which were passing though the area or by acting to block
transmission in axons terminating in the area. In a separate experiment
we investigated the effects of bilateral ventrolateral medullary
injections of kainic acid on the response to stimulation of PVN. Kainic
acid is reported to act on cell bodies, but not fibers of passage, to
produce excitation followed by a prolonged depolarization block (McGeer
and McGeer, 1982). We reasoned that if the vasoconstrictor response to
stimulation of the PVN was mediated by axons passing through the
ventrolateral medulla without synapsing then the kainic acid would have no
effect on the response. On the other hand if the important fibers
terminated in the ventrolateral medulla, kainic acid should inhibit the
firing of the postsynaptic cell body and prevent the vasoconstriction. As
a control, we also determined the effects of kainic acid in the
ventrolateral medulla on the reflex decrease in vascular resistance
produced by electrical stimulation of aortic baroreceptor afferent fibers
(superior laryngeal nerve). The secondary neurons in the baroreflex arc
probably terminate on cell bodies in the ventrolateral medulla (Ross et
al., 1984). We found that the kainic acid attenuated, in parallel, the
vasoconstriction produced by stimulation of the PVN and the vasodilation
produced by baroreflex-induced withdrawal of sympathetic tone (Porter and

Brody, 1985a). This suggests that the ventrolateral medulla may be an important synaptic site for descending PVN pathways which subserve a vasoconstrictor function.

As mentioned above, the PVN can be divided into regions that project to the posterior pituitary, median eminence or brainstem and spinal cord. The majority of fibers which descend to autonomic areas in the medulla and spinal cord originate in parvocellular areas in the posterior part of the nucleus (Swanson and Kuypers, 1980; Armstrong et al., 1980). In a different series of experiments we sought to determine if stimulation of parvocellular PVN with microelectrodes would produce cardiovascular effects different than those produced by stimulation of magnocellular areas. Stimulations were performed in urethane anesthetized rats using small (O.D. 100 μm) monopolar electrodes and currents ranging from 25 to 200 μA. In initial experiments in rats with intact baroreflexes, we found that parvocellular or magnocellular stimulation produced only modest differences. Magnocellular stimulation tended to produce small depressor effects and parvocellular stimulation produced small pressor responses (Porter and Brody, 1985b). In an attempt to make the animals more responsive to electrical stimulation we performed acute sinoaortic denervation to eliminate the buffering effect of the baroreflex. Following sinoaortic denervation a clear difference between parvocellular and magnocellular stimulation was observed (Fig. 3). Parvocellular stimulation produced a marked pressor effect accompanied by generalized vasoconstriction (including the hindquarter vascular beds). On the other hand, magnocellular stimulation produced small depressor effects with a rather selective hindquarter vasodilation. We had, in effect, separated the vasoconstrictor and vasodilator components at the level of the PVN. Since the long descending projections from PVN originate in parvocellular areas we suggest that this pathway (which courses through the synapses in VLM) subserves a selective vasoconstrictor function. The vasodilation produced by magnocellular stimulation is difficult to explain. The magnocellular PVN projects almost exclusively to the posterior pituitary. One might expect that magnocellular stimulation would release vasopressin into the peripheral circulation and produce an increase in arterial pressure. However, the effects of magnocellular activation are opposite to what would be expected if vasopressin mediated the response. Furthermore, peripheral blockade of vascular vasopressin receptors did not abolish the vasodilator response produced by magnocellular stimulation. It may be that the vasodilation resulted from indirect activation of a vasodilator pathway originating anterior to the PVN.

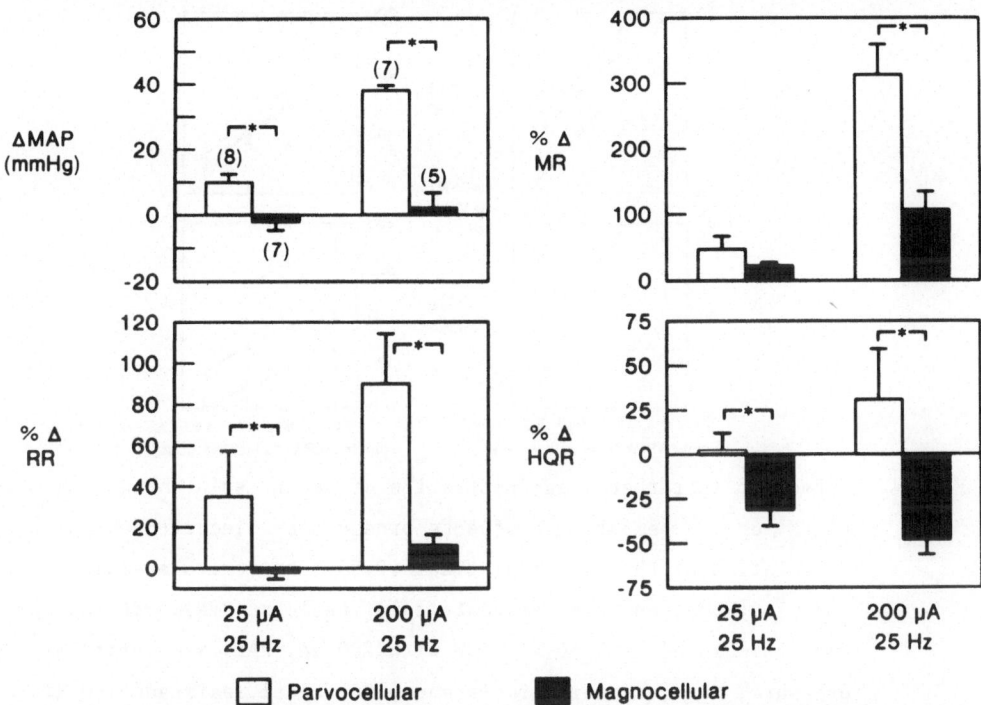

Fig. 3. Comparison of hemodynamic effects produced by stimulating parvocellular or magnocellular PVN in rats with SAD. Numbers in parentheses represent n. MAP: mean arterial pressure; MR: mesenteric resistance; RR: renal resistance; HQR: hindquarter resistance. *P <0.05. Porter and Brody, 1985b.

Given the recent evidence which suggests that vasopressin of neural origin may be involved in central nervous regulation of the cardiovascular systems (Berecek, Webb and Brody, 1983; Berecek et al., 1984; Matsuguchi et al., 1982; Pittman, Lawrence and McLean, 1982) we hypothesized that the vasopressinergic fibers which extend from the PVN to the spinal cord might be involved in mediating the hemodynamic effects produced by stimulation of the PVN. To test this hypothesis we used the straight forward approach of stimulating the PVN before and after injecting a specific vasopressin antagonist into the spinal subarachnoid space. In initial experiments we found that intrathecal injection of 2.6 nmoles of $d(CH_2)_5Me(Tyr)AVP$ significantly attenuated the hemodynamic effects produced by stimulation of the PVN (Porter and Brody, 1984). However, in subsequent experiments aimed at determining the specificity of the intrathecal antagonist, we found that pressor and vasoconstrictor responses produced by stimulating other brain sites, which were devoid of vasopressin-containing cell

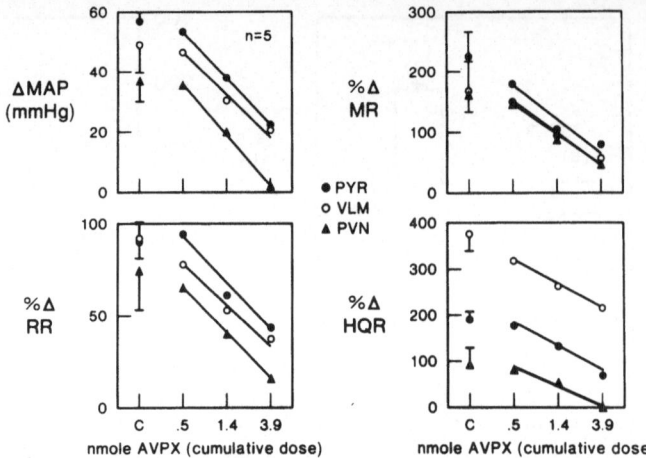

Fig. 4. Effect of intrathecal administration of vasopressin antagonist (AVPX) on the hemodynamic effects produced by electrical stimulation of the pyramidal tract (PYR), the ventrolateral medulla (VLM), and the paraventricular nucleus (PVN). PYR and VLM were stimulated with 50 µA, PVN with 200 µA. MAP: mean arterial pressure; MR: mesenteric resistance; RR: renal resistance; HQR: hindquarter resistance; C: control response. Porter and Brody, 1985c.

bodies, were also blocked by the antagonist (Fig. 4). The effect of the antagonist appeared to be due to a nonspecific neurodepression in the spinal cord. This hypothesis was confirmed when we tested the antagonist in Brattleboro rats which lack hypothalamic and spinal vasopressin (Porter and Brody, 1985c). In these rats, intrathecally administered vasopressin antagonist significantly attenuated the response produced by stimulation of the PVN. Since these rats do not produce vasopressin, the antagonist could not have exerted its effects by blocking the effects of vasopressin released in the spinal cord during the stimulation. We were surprised to find that Brattleboro rats responded with normal increases in arterial pressure and vascular resistance during the hypothalamic stimulation. Nelson et al. (1984) reported that in Brattleboro rats, the increase in blood pressure during stimulation of the PVN required higher threshold currents and was less sensitive than in Long Evans control rats. Since acute sinoaortic denervation was performed on our Brattleboro rats before stimulation of the PVN we thought that might explain the discrepancy. However, in separate experiments when we compared the effects of PVN stimulation in baroreflex-intact Brattleboro and Long Evans rats, we again found no significant difference in the hemodynamic response (Porter and Brody, 1985d).

Since Brattleboro rats responded normally to stimulation of the PVN, spinal vasopressinergic mechanisms may not be very important for cardiovascular regulation.

As illustrated in Fig. 4, low doses of the intrathecal vasopressin antagonist (0.5 nmole) did not produce any effect on the response to brain stimulation. We wondered if these lower doses were still sufficient to block spinal vasopressin receptors. Riphagen and Pittman (1984) reported that vasopressin given intrathecally produced an increase in arterial pressure in anesthetized rats. We wanted to determine if 0.5 nmoles of the vasopressin antagonist given intrathecally could block the hemodynamic effects produced by subsequent intrathecal injection of vasopressin. These experiments were performed in conscious, freely moving rats with indwelling intrathecal catheters. We found that doses of vasopressin as low as 1 pmole could produce increases in arterial pressure when given intrathecally. This increase in blood pressure was accompanied by marked vasoconstriction in renal, mesenteric and hindquarter vascular beds. The effect appeared to be due to an action of vasopressin within the spinal cord since prior blockade of peripheral vasopressin receptors had no effect on the response to intrathecal vasopressin. In contrast, prior intrathecal injection of the antagonist at 0.5 nmole completely prevented the rise in arterial pressure expected with subsequent intrathecal injection of vasopressin (Fig. 5). The effect appeared to be specific for vasopressin receptors since pretreatment with intrathecal saralasin, an angiotensin II receptor antagonist did not prevent the increase in pressure with subsequent intrathecal vasopressin. In our initial experiments, using anesthetized rats, 0.5 nmoles of the vasopressin antagonist did not appear to have any nonspecific neurodepressant effects in the spinal cord. We needed to confirm this observation in conscious rats. Some of the rats used in the experiment outlined above also had electrodes placed in the PVN. We found that although the antagonist prevented the actions of intrathecal vasopressin it had no effect on the hemodynamic response produced by stimulating the PVN (Porter and Brody, 1985d). These data strongly suggest that the cardiovascular effects produced by stimulation of the PVN do not depend on the release of vasopressin from spinal nerve terminals. However, we cannot rule out the possibility that that intrathecally administered vasopressin antagonist had access to a different group of receptors than those reached by neurally released vasopressin.

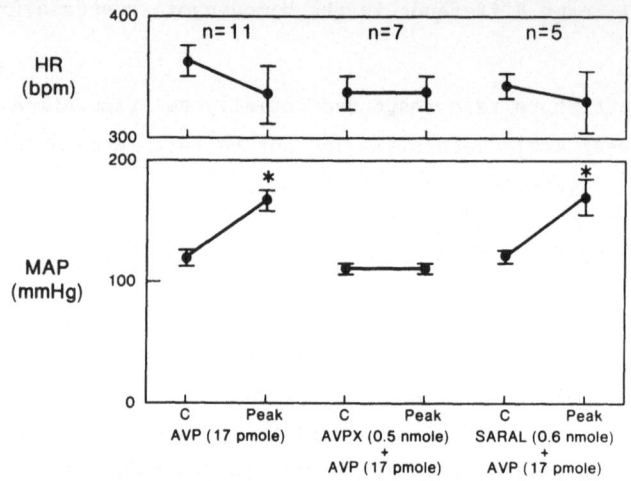

Fig. 5. Effect of intrathecal administration of vasopressin antagonist (AVPX) on hemodynamic responses produced by electrical stimulation of the paraventricular nucleus in Brattleboro rats. For 50 μA, n = 3 except for % ΔR where n = 2. For 100 μA, n = 4 for all comparisons. MAP: mean arterial pressure; MR: mesenteric resistance; RR: renal resistance; HQR: hindquarter resistance; C: control response. Porter and Brody, 1985c.

SUMMARY

Studies directly investigating the role of the PVN in subserving cardiovascular functions are beginning to emerge which appear to agree with the hypothesis generated from neuroanatomical and electrophysiological experiments. The PVN has now been shown to exert a possible regulatory influence on baroreflex control of heart rate and has been implicated in the pathogenesis of experimental hypertension. Cardiovsacular effects originating in the SFO may also be mediated, in part, by projections from this circumventricular organ to the PVN. Work in our laboratory has suggested that the long descending pathways which extend from parvocellular PVN to the ventrolateral medulla or spinal cord mediate vasoconstrictor responses. A vasodilator pathway also originates in or passes through magnocellular PVN and travels to the area of the dorsomedial medulla. Our experiments suggest that the vasopressinergic neurons which extend from the PVN to the spinal cord do not play an important role in mediating the hemodynamic effects produced by stimulation of the PVN. Nevertheless, the PVN appears to be a major integrative site within the central nervous system involved in regulation of the cardiovascular system.

The Arcuate Nucleus

The hypothalamic arcuate nucleus is a paired, medial structure that forms the boundary between the median eminence and more lateral basal hypothalamic regions that has been extensively studied for its role in neuroendocrine regulation. Inputs to this region include projections from preoptic, limbic and midbrain regions (Anschel et al., 1978, 1982; Poulain, 1977; Renaud, 1977).

It has recently been recognized that the arcuate nucleus also serves as a "bed nucleus" from which pro-opiomelanocortin-containing projections emanate to numerous brain regions that participate in cardiovascular regulation. These regions include the median preoptic nucleus of the lamina terminalis, the paraventricular nucleus, and the nucleus of the tractus solitarius (Knigge and Joseph, 1984). Since these connections are consistent with a possible role of the arcuate nucleus in cardiovascular regulation we determined if electrical or chemical activation of this region would alter arterial pressure or regional vascular resistance. Since our preliminary studies indicated that activation of the arcuate nucleus increased arterial pressure primarily by the release of vasopressin (O'Neill and Brody, 1984), we went on to determine which pool of magnocellular neurons participate in the pressor response (O'Neill and Brody, 1985).

Electrical Activation of Arcuate Nucleus

Anesthesized animals. Electrical activation of the arcuate nucleus in urethane anesthesized rats with intact baroreflexes produced only small, frequency dependent increases in arterial pressure and regional vascular resistances. The response had an onset latency of 10-20 seconds and a duration of 3-5 minutes. To examine the possibility that baroreflex activation decreased the amplitude of the response the experiments were repeated in rats with baroreflexes removed by acute sinoaortic deafferentation (SAD). Baroreceptor denervation produced a marked enhancement of both the pressor and regional vasoconstrictor responses to arcuate stimulation. Thus, arcuate stimulation produced a frequency dependent increase in mean arterial pressure and in renal, mesenteric, and hindquarter vascular resistances that were significantly enhanced following acute SAD.

Since the pressor response to arcuate stimulation persisted for several minutes after stimulation, and arcuate neurons project to magnocellular neurons in the paraventricular nucleus, we hypothesized that the response was mediated by the release of pressor quantities of vasopressin. To test this hypothesis, we determined the effect of 2 μg/kg of the vascular V_1 vasopressin receptor antagonist PMP-1-o-methyl, Tyr^2, Arg^8-vasopressin, administered intravenously, on the response evoked by electrical activation of the arcuate nucleus. Following vasopressin receptor blockade, the pressor and regional vasoconstrictor responses to arcuate stimulation were abolished. Thus, in urethane-anesthetized rats, electrical activation of the arcuate nucleus appears to increase arterial pressure and regional vascular resistances primarily by causing the release of pressor quantities of vasopressin.

Conscious animals. In conscious animals electrical activation of the arcuate nucleus produced frequency-dependent increases in arterial pressure and regional vascular resistances similar to those seen in anesthetized animals both in terms of the magnitude and in the duration of the response. However, in rats with intact baroreflexes, bradycardia was observed. In conscious rats, SAD also significantly enhanced the effects of arcuate stimulation on renal and mesenteric vascular resistance and arterial pressure, whereas the bradycardia was abolished. Finally, as was the case in anesthetized animals, vasopressin receptor blockade abolished the effect of arcuate stimulation in both baroreflex-intact and SAD animals.

To summarize, stimulation of the arcuate nucleus increased arterial pressure, renal, mesenteric, and hindquarter vascular resistances, and decreased heart rate in baroreflex-intact, conscious animals. These effects were blocked by peripheral vasopressin receptor blockade. Removal of baroreflexes produced a significant enhancement of the pressor and of the renal and mesenteric constrictor responses, and prevented the bradycardia, suggesting that the heart rate changes were of reflex origin. Finally, the effects of arcuate stimulation in SAD rats were also blocked by vasopressin receptor blockade. Thus, effects evoked by arcuate stimulation in urethane anesthetized and conscious animals both appear to be mediated by vasopressin release and differ primarily in that bradycardia is seen in baroreflex-intact, conscious animals, and that the hindquarter constriction is not significantly enhanced following SAD in conscious animals.

456

Chemical Activation

Fibers projecting to the posterior pituitary from magnocellular neurons in the supraoptic and paraventricular nuclei project through the median eminence, which lies immediately ventral and medial to the arcuate nucleus. To assure ourselves that the response evoked by arcuate stimulation was not due to activation of fibers of passage in or near the structure, two studies were performed. In the first, stimulation of the median eminence per se at parameters equal to those producing large pressor and vasoconstrictor effects in response to stimulation of the arcuate itself, elicited much smaller effects. In the second study we attempted to selectively activate cell bodies in the arcuate nucleus by microinjecting the excitotoxin kainic acid. Injections of 50 pmole of kainic acid into the arcuate nucleus produced significant increases in arterial pressure and regional vascular resistances, without producing any change in heart rate. Additional injections of kainic acid into the same sites failed to evoke any further changes in arterial pressure or vascular resistances, however, injection into the contralateral arcuate nucleus evoked similar changes. To determine if the response evoked by kainic acid was evoked by release of vasopressin, the vasopressin receptor antagonist, was again employed. The response by kainic acid activation, using separate injections on both sides was prevented by treatment with the vasopressin receptor antagonist administered between kainic acid injections.

In summary, microinjection of kainic acid into the arcuate nucleus produced a significant increase in arterial pressure, and in renal, mesenteric, and hindquarter vascular resistances of urethane-anesthetized animals without producing significant changes in heart rate. These changes were prevented by pretreatment with a vasopressin receptor antagonist. We can conclude that activation of cell bodies in arcuate nucleus by kainic acid injection, like electrical stimulation, evokes the release of pressor quantities of vasopressin.

Source of Vasopressin

Since arcuate neurons project heavily to magnocellular neurons in the paraventricular nucleus, we examined whether these neurons participate in the pressor response evoked by arcuate stimulation. Our approach was to make microinjections of the local anesthetic lidocaine into the ipsi- and contralateral paraventricular nucleus. We reasoned that if paraventricular neurons participate in this response, inactivation of these neurons with lidocaine should reduce the magnitude of the response.

In urethane anesthetized animals microinjections of 500 nl of 4% lidocaine solution into either ipsi- or contralateral paraventricular nucleus (relative to the arcuate electrode) produced a 15-20 mm fall in arterial pressure accompanied by a decrease in regional vascular resistances. However, injection of lidocaine into the contralateral paraventricular nucleus had no effect on the pressor and regional vasoconstrictor responses to arcuate stimulation, whereas injection into the ipsilateral paraventricular nucleus produced a 30% reduction in these responses. Thus, it appears that an uncrossed projection to the paraventricuar nucleus participates in the pressor response evoked by electrical activation of the arcuate nucleus. Furthermore, these data suggest that ongoing neuronal activity in paraventricular nucleus is involved in the maintenance of arterial pressure in urethane-anesthetized animals, presumably through tonic release of vasopressin.

Although stimulus spread from the arcuate region to the nearby median eminence might have evoked the release of vasopressin, since it contains the vasopressinergic projection from the supraoptic and paraventricular nuclei to the posterior pituitary, several arguments suggest that this was not the case. First, the injection of lidocaine into the ipsi-, but not into the contralateral paraventricular nucleus, attenuated the response arcuate stimulation. If axons of passage were being activated by stimulation in the arcuate region, it is not clear why injections of lidocaine into relatively distant regions would have different effects on the pressor and regional vasoconstrictor response. Second, the effects evoked by electrical stimulation of the arcuate were qualitatively very similar to those produced by kainic acid injection, which presumably would activate only cell bodies in the arcuate region.

Additional evidence that stimulation of structures originating in the arcuate nucleus is responsible for the vasopressin-mediated response was obtained from preliminary studies showing that stimulation of the peri-arcuate region in rats that were treated neonatally with monosodium glutamate (a treatment that results in a marked loss of neurons in the arcuate nucleus), failed to exhibit pressor responses. Overall, these data suggest that the hemodynamic effects evoked by either electrical or chemical activation of the arcuate nucleus result from activation of cell bodies in this region, and not of fibers of passage in the median eminence.

The finding that the inactivation of the ipsilateral paraventricular nucleus by the microinjection of lidocaine inhibits the pressor response evoked from the arcuate nucleus suggests that an uncrossed projection participates in this response. Since the response is mediated primarily by the release of vasopressin, and arcuate neurons project magnocellular paraventricular neurons, it appears likely that the inactivation of magnocellular neurons underlies the attenuation of the response.

If it is assumed that the neuronal blockade achieved by lidocaine was substantial (as is suggested by the depressor effect produced by lidocaine), then projections to the paraventricular nucleus do not account for all of the pressor response evoked by arcuate stimulation. Injection of lidocaine produced only a 30% reduction of this response. A recent finding using tracer dyes demonstrates arcuate neurons also project to supraoptic nuclei (Wilkins and Mitchell, personal communication). This second projection to the magnocellular neuronal pool could account for the response that remains after inactivation of the paraventricular nucleus.

The arcuate nucleus contains many pro-opiomelanocortin cell bodies that project to a number of cardiovascular regulatory regions in the brain. It was interesting to note that, despite projections to the nucleus of the solitary tract and to the anteroventral third ventricular region, the response to acute activation of the arcuate was mediated almost entirely by the release of vasopressin, and not through any effects on sympathetic outflow. The potential for these and other projections from the arcuate to influence neural regulation of arterial pressure appears to be low, however, alternate studies that employ other types of activation of the nucleus, may be required.

Removal of baroreflexes by SAD increases the pressor activity of intravenously administered vasopressin approximately 10-fold (Webb, Osborn and Cowley, 1985). However, SAD produced only a three-fold potentiation of the vasopressin-mediated response to arcuate stimulation. These data suggest that projections to the nucleus of the solitary tract may serve to decrease the ability of baroreflexes to buffer the increase in pressure evoked by arcuate stimulation in baroreflex-intact animals.

In summary, electrical or chemical activation of cell bodies in the arcuate nucleus increases arterial pressure by stimulating the release of vasopressin. This response appears to be mediated, at least in part, by an uncrossed projection from the arcuate to the paraventricular nucleus.

And finally, neurons in the paraventricular nucleus appear to participate in the maintenance of basal arterial pressure under urethane anesthesia.

CONCLUSIONS

The results summarized in these experiments provide evidence for interactions between the arcuate nucleus and the paraventricular nucleus in arterial pressure regulation mediated through the release of vasopressin. The studies also demonstrate a significant sympathoexcitatory role for parvocellular regions of the paraventricular nucleus that results in pressor and regional vasoconstrictor responses. Although vasopressin is a putative neurotransmitter for these sympathoexcitatory effects, our results while finding no evidence for vasopressin-mediated effects in the response to paraventricular stimulation, have identified the potential for vasopressin to exert receptor mediated pressor activity originating at the cord level. Future studies on cardiovascular control regulated through both arcuate and paraventricular nuclei should focus on the specific physiological and pathophysiological circumstances that may be capable of activating these forebrain structures.

REFERENCES

Anschel, S., Alexander, M. and Perachio, A.A., 1978, Convergence of inputs from limbic and brainstem areas upon antidromically identified neurons in the hypothalamus, Soc. Neurosci., 4:215.

Anschel, S., Alexander, M. and Perachio, A.A., 1982, Multiple connections of medial hypothalamic neurons in the rat. Exp. Brain Res., 46:383-392.

Armstrong, W.E., Warach, S., Hatton, G.I. and McNeill, T.H., 1980, Subnuclei in the rat hypothalamic paraventricular nucleus: A cytoarchitectural, horseradish peroxidase and immunocytochemical analysis, Neuroscience, 5:1931-1958.

Berecek, K.H. and Brody, M.J., 1982, Evidence for a neurotransmitter role for epinephrine derived from the adrenal medulla, Am. J. Physiol., 242:H593-H601.

Berecek, K.H., Webb, R.L. and Brody, M.J., 1983, Evidence for a central role for vasopressin in cardiovascular regulation, Am. J. Physiol., 244:H852-H859.

Berecek, K.H., Olpe, H.R., Johes, R.S.G. and Hofbauer, K.G., 1984,

Microinjections of vasopressin into the locus coeruleus of conscious rats, Am. J. Physiol., 247:H675–H681.

Calaresu, F.R. and Ciriello, J., 1980, Projections to the hypothalamus from buffer nerves and nucleus tractus solitarius in the cat, Am. J. Physiol., 239:R130–R136.

Carlsson, A., Falck, B. and Hillarp, N.-A., 1962, Cellular localization of brain monoamines, Acta. Physiol. Scand., 56 (Suppl. 196):1–25.

Caverson, M.M., Ciriello, J. and Calaresu, F.R., 1984, Paraventricular nucleus of the hypothalamus: An electrophysiological investigation of neurons projecting directly to intermediolateral nucleus in the cat, Brain Res., 305:380–383.

Caverson, M.M., Ciriello, J. and Calaresu, F.R., 1983, Cardiovascular afferent inputs to neurons in the ventrolateral medulla projecting directly to the central autonomic area of the thoracic cord in the cat, Brain Res., 274:354–358.

Ciriello, J. and Calaresu, F.R., 1980, Role of paraventricular and supraoptic nuclei in central cardiovascular regulation in the cat, Am. J. Physiol., 239:R137–R142.

Ciriello, J. and Caverson, M.M., 1984, Ventrolateral medullary neurons relay cardiovascular inputs to the paraventricular nucleus, Am. J. Physiol., 246:R968–R978.

Ciriello, J., Kline, R.L., Zhang, T.-X. and Caverson, M.M., 1984, Lesions of the paraventricular nucleus alter the development of spontaneous hypertension in the rat, Brain Res., 310:355–359.

Ferguson, A.V. and Renaud, L.P., 1984, Hypothalamic paraventricular nucleus lesions decrease the pressor response to subfornical organ stimulation, Brain Res, 305:360–364.

Kannan, H. and Yamashita, H., 1983, Electrophysiological study of paraventricular nucleus neurons projecting to the dorsomedial medulla and their response to baroreceptor-stimulation in rats, Brain Res.,279:31–40.

Kannan, H. and Yamashita, H., 1985, Connections of neurons in the region of the nucleus tractus solitarius with the hypothalamic paraventricular nucleus: Their possible involvement in neural control of the cardiovascular system, Brain Res., 329:205–212.

Knigge, K.M. and Joseph, S.A., 1984, Anatomy of the opioid-systems of the brain, Can. J. Neurol. Sci., 11:14–23.

Knuepfer, M.M., Johnson, A.K. and Brody, M.J., 1984, Identification of brainstem projections mediating hemodynamic responses to stimultion of the anteroventral third ventricle (AV3V) region, Brain Res., 305:305–314.

Kuypers, H.G.J.M. and Maisky, V.A., 1975, Retrograde axonal transport of horseradish peroxidase from spinal cord to brain stem cell groups in the cat, Neurosci. Let., 1:9-14.

Luiten, P.G.M., ter Horst, G.J., Karst, H. and Steffens, A.B., 1985, The course of paraventricular hypothalamic efferents to autonomic structures in medulla and spinal cord, Brain Res., 329:374-378.

Mangiapane, M.L. and Brody, M.J., 1984, Attenuation of hemodynamic responses to electrical stimulation of subfornical organ (SFO) by anteroentral third ventricle (AV3V) region or paraventricular nucleus (PVN) lesions, Fed. Proc., 43:311.

Matsuguchi, H., Sharabi, F.M., Gordon, F.J., Johnson, A.K. and Schmid, P.G., 1982, Blood pressure and heart rate responses to microinjection of vasopressin into the nucleus tractus solitarius of the rat, Neuropharmacol., 21:687-693.

McGeer, P.L. and McGeer, E.G., 1982, Kainic acid: The neurotoxic breakthrough, CRC Crit. Rev. Toxicol., March, 1-26.

Nauta, W.J.H. and Haymaker, W., 1969, Hypothalamic nuclei and fiber connections, in: "The Hypothalamus", W. Haymaker, E. Anderson, W.J.H. Nauta, Ed., Thomas, Springfield, IL, pp. 136-209.

Nelson, D.O., Graham, C. and Ernsberger, P., 1984, Role of descending vasopressin pathways in arterial pressure control by the paraventricular nucleus (PVN), Fed. Proc., 43:1067.

Nilaver, G., Zimmerman, E.A., Wilkins, J., Michaels, J., Hoffman, D. and Silverman, A.-J., 1980, Magnocellular hypothalamic projections to the lower brain stem and spinal cord of the rat, Neuroendocrinology, 30:150-158.

O'Neill, T.P. and Brody, M.J., 1984, Hemodynamic effects evoked by electrical stimulation of the arcuate nucleus (AN), Soc. Neurosci., 10:32.

O'Neill, T.P. and Brody, M.J., 1985, Hemodynamic effects produced by arcuate stimulation are mediated in part by a projection to paraventricular nucleus, Fed. Proc., 44:1553.

Pittman, Q.J., Lawrence, D. and McLean, L., 1982, Central effects of arginine vasopressin on blood pressure in rats, Endocrinology, 110:1058-1060.

Pitts, R.F., Larrabee, M.G. and Bronk, D.W., 1941, An analysis of hypothalamic cardiovascular control, Am. J. Physiol., 134:359-383.

Porter, J.P. and Brody, M.J., 1984, Do vasopressin-containing neurons in the spinal cord influence vasomotor outflow? The Physiologist, 27:224.

Porter, J.P. and Brody, M.J., 1985a, Neural projections from paraventricular nucleus that subserve vasomotor functions, Am. J. Physiol., 248:R271-R281.

Porter, J.P. and Brody, M.J., 1985b, A comparison of the hemodynamic effects produced by electrical stimulation of subnuclei of the paraventricular nucleus, Brain Res., in press.

Porter, J.P. and Brody, M.J., 1985c, A V1 vasopressin receptor antagonist has nonspecific neurodepressant actions in the spinal cord, Neuroendocrinology, in press.

Porter, J.P. and Brody, M.J., 1985d, Role of spinal vasopressinergic mechanisms in the cardiovascular effects produced by stimulation of the paraventricular nucleus, Soc. for Neurosci., 11:36.

Poulain, P., 1977, Septal afferents to the arcuate-median eminence region in the guinea pig: Correlative electrophysiological and horseradish peroxidase studies, Brain Res., 137:150-153.

Renaud, L. P., 1977, Influence of medial preoptic-anterior hypothalamic area stimulation on the excitability of mediobasal hypothalamic neurons in the rat, J. Physiol. (Lond.), 264:541-564.

Riphagen, C.L. and Pittman, Q.J., 1984, Blood pressure responses following spinal application of arginine vasopressin in the rat, Can. J. Physiol. Pharmacol., 62:Axxiii.

Rogers, R.C. and Nelson, D.O., 1984, Neurons of the vagal division of the solitary nucleus activated by the paraventricular nucleus of the hypothalamus, J. Autonom. Nerv. Sys., 10:193-197.

Ross, C.A., Ruggiero, D.A., Park, D.H., Joh, T.H., Sved, A.F., Fernandez-Pardal, J., Saavedra, J.M. and Reis, D.J., 1984, Tonic vasomotor control by the rostral ventrolateral medulla: Effect of electrical or chemical stimulation of the area containing C1 adrenaline neurons on arterial pressure, heart rate, and plasma catecholamines and vasopressin, J. Neurosci., 4:474-494.

Saper, C.B., Loewy, A.D., Swanson, L.W. and Cowan, W.M., 1976, Direct hypothalamic-autonomic connections, Brain Res., 117:305-312.

Sawchenko, P.E. and Swanson, L.W., 1982, The organization of noradrenergic pathways from the brainstemm to the paraventricular and supraoptic nuclei in the rat, Brain Res., 4:275-325.

Silverman, A.-J., Oldfield, B., Hou-Yu, A. and Zimmerman, E.A., 1985, The noradrenergic innervation of vasopressin neurons in the paraventricular nucleus of the hypothalamus: An ultrastructural study using radioautography and immunocytochemistry, Brain Res., 325:215-229.

Sofroniew, M.V., 1983, Vasopressinn and oxytocin in the mammalian brain
and spinal cord, TINS, November, 467–472.

Swanson, L.W., 1977, Immunohistochemical evidence for a
neurophysin–containing autonomic pathway arising in the
paraventricular nucleus of the hypothalamus, Brain Res.,
128:346–353.

Swanson, L.W. and Kuypers, H.G.J.M., 1980, The paraventricular nucleus of
the hypothalamus: Cytoarchitectonic subdivisions and organization
of projections to the pituitary dorsal vagal complex, and spinal
cord as demonstrated by retrograde fluorescence double–labeling
methods, J.Comp. Neurol., 194:555–570.

Valiquette, G., 1980, Posterior pituitary hormones and neurophysins, in:
"The Endocrine Functions of the Brain", M. Motta, ed., Raven Press,
New York, pp. 385–417.

Webb, R.L., Osborn, Jr., J.W. and Cowley, A.W., 1985, The effects of
sinoaortic denervation (SAD) on the hemodynamic responses to
arginine vasopressin (AVP), angiotensin II (AII) and phenylephrine
(PE) in conscious rats, Fed. Proc., 44:816.

Yamashita, H., Inenaga, K. and Koizumi, K., 1984, Possible projections
from regions of paraventricular and supraoptic nuclei to the spinal
cord: Electrophysiological studies, Brain Res., 296:373–378.

Zhang, T.-X. and Ciriello, J., 1982, Lesions of paraventricular nucleus
reverse the elevated arterial pressure after aortic baroreceptor
denervation in the rat, Neurosci. Abst., 8:434.

THE CENTRAL NERVOUS CONTRIBUTION TO VASOMOTOR TONE

Sidney Hilton

Department of Physiology
The Medical School
Vincent Drive
Birmingham B15 2TJ, U.K.

INTRODUCTION

Vasomotor tone is to some extent a manifestation of the myogenic response of vascular smooth muscle. In resistance vessels on the arterial side of the circulation of key organs such as the brain and the heart, this is a major factor underlying their basal tone; but in the splanchnic region, which even between periods of food digestion receives some 25% of the cardiac output, maintained vascular tone requires continuing activity of the vasoconstrictor nerve supply. Even in skeletal muscle, where the myogenic tone of resistance vessels is high, resting tone requires a certain level of vasoconstrictor nerve discharge, and this is important in a tissue which constitutes 40-50% of the tissue mass in an average individual. It has long been known that when this tonic vasoconstrictor nerve activity is suddenly interrupted, as after high spinal transection, there is a profound fall of arterial blood pressure, because of the great reduction in total peripheral resistance which immediately ensues. A point to emphasize, however, is that this fall of blood pressure, which can be quite prolonged, must be due mainly to vasodilation in the splanchnic bed, because this, more than any other, is the vascular bed which dilates - and remains dilated - when vasconstrictor tone is eliminated.

LIMITED CAPABILITIES OF REFLEX CONTROL

The source and causation of the tonic discharge in sympathetic preganglionic neurones have often been discussed (Polosa, Mannard &

465

Laskey, 1979). Of the inputs which could be involved, afferents from skeletal muscle, skin and the viscera have been considered, in addition to inputs from central and peripheral chemoreceptors and central temperature sensors, and even vestibular and visual afferents. Indeed, the regulation of vasomotor tone by the central nervous system is usually discussed as though it were entirely a matter of reflex control, with the baroreceptors playing a major part and the peripheral chemoreceptors also exerting a significant influence, particularly after haemorrhage.

To give the baroreceptors their due, there can be little doubt that they do help to stabilise arterial blood pressure; and they are important in man at least, certainly in the special case of a change of posture, and particularly from the prone to the upright position. But, apart from this, how important are the baroreceptors for moment-to-moment control; and can they be responsible for setting the general level of blood pressure? It has long been known that the baroreceptor reflex has too small an open-loop gain to function really effectively as a pressure control system (Scher & Young, 1963) and, indeed, blood pressure does fluctuate widely about the equilibrium level in ordinary everyday life with almost everything we do. As shown some ten years ago by Guyton and his coworkers in conscious dogs (Cowley, Laird & Guyton, 1973), though chronic baroreceptor denervation leads to significantly greater swings of blood pressure, after a few days the equilibrium level is unchanged.

SOURCES OF TONIC DISCHARGE

Many workers have demonstrated that the baroreceptors themselves are readily reset: even in hypertension they appear to adapt themselves to the general level of arterial blood pressure by altering the range over which they work (e.g. Mancia et al., 1980), and this range must therefore be set by the nervous system independently of the baroreceptor input. On the other hand, the nucleus of the tractus solitarius (NTS) in which the baroreceptor afferent fibres relay, clearly does influence the equilibrium level of blood pressure; for bilateral lesions of this nucleus can lead to severe hypertension in experimental animals: this has been demonstrated in rats (Doba & Reis, 1973), cats (Nathan & Reis, 1977) and dogs (Carey et al, 1979). Reis and his co-workers (Talman, Perrone & Reis, 1981) have also shown that microinjection of kainic acid bilaterally into this nucleus in the rat has the same effect, so there are neurones within the nucleus which exert a tonic inhibitory influence on some part of the neural network which produces the ordinary level of on-going activity in sympthetic vasoconstrictor nerve fibres. There is no information as yet, however, to indicate whether these neurones in NTS may be spontaneously active or whether they require an input from elsewhere in the nervous system.

Before finishing this introduction, reference must be made to the peripheral chemoreceptors for, as mentioned already, they can have an important effect on arterial blood pressure. It was pointed out by Kenney & Neil (1951) many years ago in experiments on anaesthetised cats that, after a haemorrhage sufficient to reduce blood pressure, the carotid chemoreceptors are responsible for much of the vasoconstrictor nerve activity remaining. As would be expected, therefore, stimulation of the carotid bodies, with ventilation controlled, causes an overall increase in peripheral vascular resistance (see Daly, 1984, for review). Perhaps these chemoreceptors in the carotid and aortic bodies, which incidentally also relay in the NTS, play an important part in generating the tonic drive in vasoconstrictor nerve fibres; for they are never silent, even under conditions of normoxia (Biscoe, 1971).

IS THERE A VASOMOTOR CENTRE?

These considerations lead us directly to the nub of the problem. Whereabouts in the central nervous system is this tonic drive generated? To be sure, the chronic spinal animal or man does regain, after a longish period of time which is different for each species, a level of blood pressure little different from normal. So spinal mechanisms can take over, but they are normally dominated by bulbospinal inputs, and the chief of these has long been thought to arise in a separate vasomotor centre, located in the medulla. This is an idea about which I have been somewhat scornful for some years and, as will be clear, I will have to eat some of my words. Before that, however, I should outline the main reasons for my scepticism: the first reason was that large lesions within the so-called pressor region of the dorsal medulla were found to cause only small falls of blood pressure (e.g. Manning, 1965; Chai & Wang, 1968). This common result had led to the conclusion that neurones exerting tonic vasomotor effects were quite diffusely distributed: indeed, when searches were made for medullary neurones with a firing pattern correlated to the discharge in sympathetic efferent nerves (e.g. Hukuhara & Takeda, 1975), they were seen to occupy almost all of the medullary reticular formation. This leads me to my second reason, which was that no area, or neuronal group, within the medulla could be shown to have an exclusively cardiovascular function: apart perhaps from the special case of the NTS, they all appeared to have complex and non-specific inputs and outputs (see Hilton (1980) for review).

As a way out of the dilemma presented by findings like these, I proposed that the search for a separate vasomotor centre had led nowhere because the concept itself was mistaken, and that the main influence on vasomotor tone is actually exerted by regions longitudinally arrayed from the hypothalamus through the brainstem, which initiate and integrate those patterns of physiological response which include significant cardiovasular components. I took as my paradigm the defence-alerting system; for not only is a lot known about its anatomical organisation and the details of the pattern of cardiovascular changes which it imposes, but also - and this is the main point - this pattern includes a significant increase in mean arterial blood pressure.

DEFENSE REACTION AS A PARADIGM FOR CARDIOVASCULAR CONTROL

The regions of the amygdala, hypothalamus and mid-brain in which this pattern of response is integrated have been located precisely in a number of mammalian species, including monkeys and baboons (Hilton, 1980; Smith et al., 1980), and more recent studies have been revealing the details of the anatomical connections between them. As an important part of the overall pattern of autonomic response, the cardiovascular changes comprise an increase in cardiac output with a strong and maintained vasoconstriction in the splanchnic area, accompanied by a more transient vasoconstriction in kidney and skin, and vasodilation in skeletal muscle, which in the cat and dog is largely caused by cholinergic vasodilator nerve fibres. This complex pattern of changes which we have called the visceral alerting response, can lead to a large increase in mean arterial pressure and pulse pressure when the system is activated reflexly as well as by the artificial means of electrical stimulation of the integrating areas in the brain (Abrahams, Hilton & Zbrozyna, 1960; 1964; Hilton & Marshall, 1982). Indeed, it is evoked by any sudden or novel stimulus, and, in the conscious animal, it is graded with the stimulus strength (Caraffa-Braga, Granata & Pinotti, 1973). During sleep, particularly in the paradoxical phase, blood pressure falls, and the pattern of cardiovascular change is the mirror image of the alterting response, i.e. vasodilatation occurs in the gut and kidneys, and vasoconstriction in skeletal muscles (Mancia et al., 1971). Awakening from sleep, not surprisingly therefore, is accompanied by the same alerting pattern of cardiovascular change (Caraffa-Braga et al., 1973). So it looks as though the general level of blood pressure in the normal awake animal could be determined mainly by that level of activity in the defense-altering system, which is necessary to the state of arousal in the awake state.

Ironically perhaps, attempts to test this idea have led us back to the medulla, specifically to the ventrolateral medulla very near the surface (Hilton, 1980; 1982); for evidence is now accumulating as to the importance of a circumscribed group of neurones in this location, which is the site of the final relays in the efferent pathway from the defense-alterting areas of the amygdala, hypothalamus and mid-brain to the spinal cord. In addition, and more germane to the main topic of this talk, continuing activity of these same neurones (or others in the same pool) seems to be essential for the maintenance of blood pressure at its normal level. The nerve cell bodies concerned are all found close to the surface of the medulla, ventral and caudal to the facial nucleus and lateral to the rostral pole of the inferior olive. This corresponds to a superficial portion of the nucleus paragigantocellularis lateralis (PGL) as defined in the cat by Taber (1961) and more recently in the rat (Andrezik, Chan-Palay & Palay, 1981).

NUCLEUS PARAGIGANTOCELLULARIS LATERALIS (PGL) A CELL GROUP OF SPECIAL SIGNIFICANCE FOR VASCULAR CONTROL

It is now clear that several cell groups in the ventrolateral medulla can exert significant effects on the heart and circulation. We must include here cells in the region of the A1 and A5 groups of noradrenaline neurones, which exert mainly inhibitory effects (Coote and MacLeod, 1974; Neil and Loewy, 1982). But PGL lies between these (in the rostro-caudal line, that is) and there are a number of reasons for picking it out as being of special importance. First and foremost is the evidence that this particular region, so superficially and precisely situated, plays a crucial part in cardiovascular control. The earliest evidence for this conclusion came from a series of studies by Feldberg and Guertzenstein in the early 1970s. They showed, in experiments on cats, that bilateral application of several substances that would be expected to block synaptic transmission caused a profound fall of blood pressure, whereas excitatory substances had the opposite effect (Feldberg, 1976). In an important paper, Guertzenstein & Silver (1974) defined what they called a glycine-sensitive area of the ventral surface, just caudal to the trapezoid body, from which the vasodepressor effect was specifically

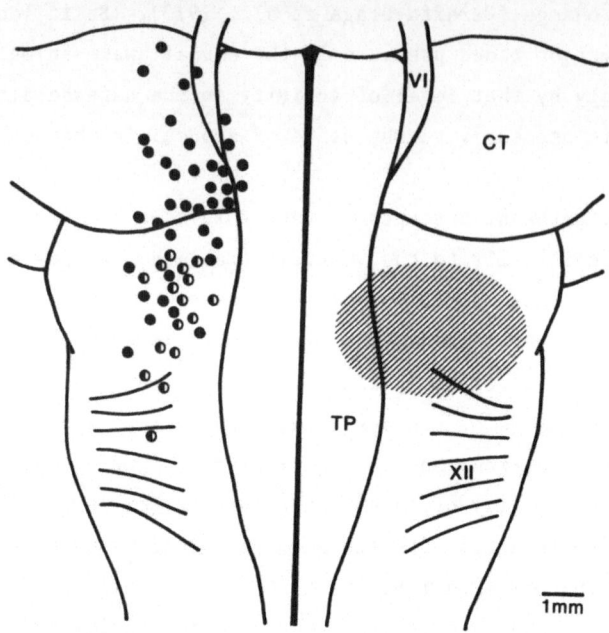

Fig. 1. Diagrammatic plan of the ventral medulla of the cat showing sites
at which stimulation evoked the full pattern of visceral
alterting response (indicated by filled circles), and similar
responses with one or another component absent (indicated by
half-filled circles). For clarity, all symbols are shown on the
left-hand side: the shaded area on the left-hand side indicates
the area sensitive to glycine. Abreviations: CT, corpus
trapezoideum; TP, tractus pyramidalis; VI, nervus abducens; XII,
nervus hypoglossus. (Reproduced from Hilton et al., 1983, with
permission).

evoked. Like the application of glycine, bilateral electrolytic lesions
of this region reduced the level of arterial blood pressure to that of the
spinal animal.

Earlier work had shown that the pathway descending from the
defense-alerting areas of the hypothalamus and mid-brain, for the
cardiovascular components of the alerting response, runs ventrally in the
medulla (Lindgren & Uvnas, 1953; Abrahams et al., 1960), and that it
becomes extremely superficial in the glycine-sensitive region (Schramm &
Bignall, 1971). This led us to ask the question whether this pathway
relays within this particular group of ventral medullary neurones. We

470

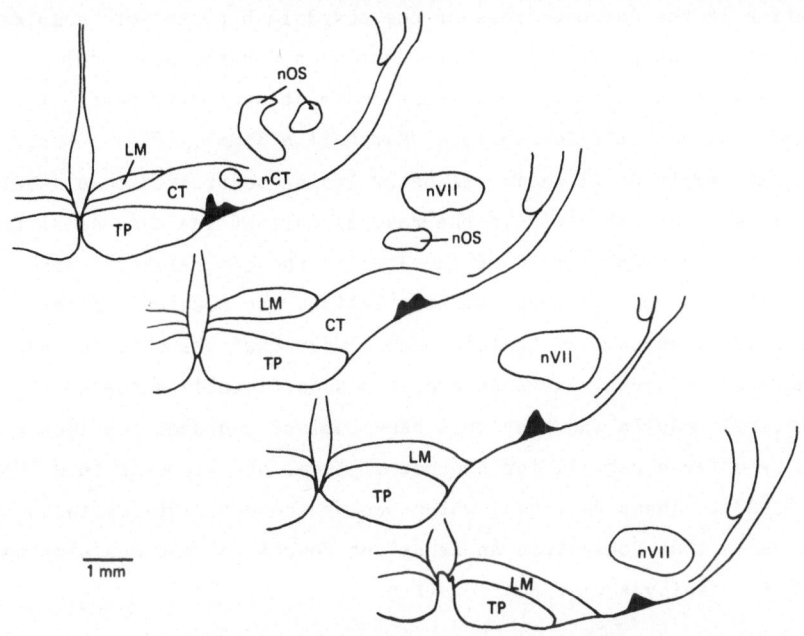

Fig. 2. Drawings from stained coronal sections through the medulla
oblongata from each of four cats, showing the extent of the
visible tissue destruction (filled black) resulting from the
passage of radio-frequency current via the stimulating electrode.
Each lesion was roughly cone-shaped, and a section through the
central (i.e. largest) part of the lesion is illustrated. The
sections are arranged from rostral (uppermost section) to caudal
(lowermost section): they represent roughly levels P 6.5-7.5 on
the atlas of Snider & Niemer (1961). Abbreviations: CT, corpus
trapezoideum; LM, lemniscus medialis; nCT, nucleus corporis
trapezoidei; nOS, nucleus olivaris superior; nVII, nucleus nervus
facialis; TP, tractus pyramidalis. Reproduced from Hilton et al.
(1983) with permission.

therefore set about mapping the pathway by electrical stimulation as far
down the length of the medulla as we could. We found it to run as a
narrow strip that becomes extremely superficial within the glycine-
sensitive area (Fig. 1). At this level also, the response to electrical
stimulation became fragmented, one or another feature of whole pattern of
response dropping out. Bilateral application of glycine not only caused a
large fall of blood pressure within 5-10 minutes, but also, and in
parallel with this fall, the visceral alerting response evoked by

stimulation in the defense areas of the amygdala-hypothalamic complex, or
the mid-brain, was progressively attenuated: the vasoconstrictor
components were those reduced earliest and most strongly, particularly
splanchnic vasoconstriction (Hilton, Marshall & Timms, 1983). Moreover,
after an extremely small radio-frequency lesion restricted to a section of
the previously located strip in the rostral part of the area sensitive to
glycine (Fig. 2), application of the drug to the contralateral side
produced all the changes that, without lesion, are seen only after
bilateral application. We therefore concluded that the efferent pathway
from the degense areas relays on synapses superficially situated in the
ventrolateral medulla and that this same pool of neurones provides a tonic
excitatory drive essential for the normal level of vasomotor tone (Hilton
et al., 1983). These neurones, which occupy the more superficial portion
of PGL, would thus constitute an important functional nucleus playing a
key role in cardiovascular control (Fig. 3).

SOME FUNCTIONAL ANATOMY

In discussing just how far these conclusions or suggestions seem
justified or are supported by other evidence, the first point to emphasize
is that there really are monosynaptic connections linking the hypothalamus
and mid-brain to PGL: Lovick (1985a) has recently established this point
in an anatomical study in the cat, using horseradish peroxidase injected
into PGL, and similar findings have been made in the rat (Andrezik,
Chan-Palay & Palay, 1981; Marson, 1984). More particularly to the point,
there is also electrophysiological evidence for a convergence of inputs
from the hypothalamic and mid-brain defense areas onto units recorded in
the ventral portion of PGL at the level of the rostral pole of the
inferior olive (Hilton & Smith, 1984; Lovick, Smith & Hilton, 1984; Li &
Lovick, 1985).

In Lovick's (1985a) study, cell bodies of projecting neurones were
found to be quite concentrated in the mid-brain defense area, in the
dorsal half of the central grey matter and the tegmentum adjacent to it.
In the hypothalamus, however, cells were labelled in the tuberal region,
in the dorsomedial nucleus and lateral to it. This does not correspond
exactly to the defense area as defined by electrical stimulation, and the
same is true in the rat. Moreover, we have now found that, in the rat,
the hypothalamic defense area as located by microinjection of nanomolar
quantities of the cell excitant, DL-homocysteic acid (Hilton & Redfern,
1983; Redfern & Hilton, 1985), overlaps but does not correspond exactly

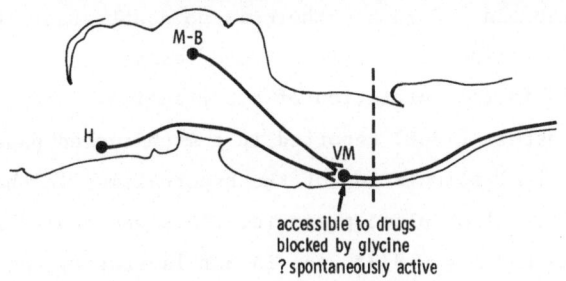

accessible to drugs
blocked by glycine
?spontaneously active

Fig. 3. Diagrammatic paramedian sagittal section of cat brainstem
indicating location of representative neurones in hypothalamic
(H) and mid-brain (M-B) defense areas giving rise to axons
hypothetically converging on cardiovascular neurones in nucleus
paragigantocellularis lateralis near the ventral surface of the
medulla (VM) which generate the alerting pattern of
cardiovascular response, and vasomotor tone. Interrupted line
indicates plane of section which, as shown originally by Dittmar
(1873), eliminates vasomotor tone of central nervous origin.
Reproduced from Hilton (1982) with permission.

with that located on the basis of electrical stimulation (Yardley, 1985;
Yardley & Hilton, 1986). Antidromic activation of cells in the
hypothalamus on electrical stimulation in PGL has also shown direct
projections to originate mainly in the dorsal part of the lateral
hypothalamus and the zona incerta (Li & Lovick, 1985) in agreement with
the anatomical findings of Andrezik et al. (1981). Of course, the
defense-alerting areas may project to PGL by polysynaptic as well as
monosynaptic pathways; but our findings with chemical and electrical
stimulation in the hypothalamus contrast with those in mid-brain where the
correspondence is exact.

When stimulating the mid-brain defense area which lies in the
dorsomedial portion of the periaqueductal grey matter, chemically as well
as electrically, the full pattern of cardiovascular change is regularly
obtained. So far, however, we have not been able to reproduce the full
pattern of autonomic changes seen in response to electrical stimulation in
the hypothalamic defense area when we use selective chemical activation of
the cell bodies in this region (Hilton & Redfern, 1983; Redfern & Hilton,
1985). In particular, the vasoconstrictor component has been lacking, so
we have to think again about the precise neuronal organisation within this

part of the forebrain. Even so, there is no doubt about its involvement in the defense reaction, and there is good reason to suggest that it could also be involved in the initiation of hypertension. Many years ago, Folkow and Rubinstein (1966) reported in a much quoted paper that long-term electrical stimulation of the hypothalamus in the rat could raise the basal level of blood pressure. This was denied more recently in a study by Bunag & Riley (1979), so, in our laboratory, once we had determined the exact location of the hypothalamic defense area, we started out to look for such an effect again, by long-term stimulation within the defense area and comparing the results with those of electrical stimulation dorsal to it. After daily, intermittent stimulation over a period of 6-10 weeks, it was found that there was a small trend towards hypertension. This trend was statistically significant, though far from dramatic (Yardley, 1985), but it was only obtained when stimulating within the defense area itself.

To return to PGL (Fig. 4), there is a good electrophysiological evidence of a bulbospinal projection from cells in this nucleus, in the cat and rat (Lebedev, Krasiukov & Nikitin, 1984; Brown & Guyenet, 1984; Lovick, 1985b; McAllen, 1985) backed up by anatomical evidence of a direct projection to the region of the intermediolateral column in the spinal cord (IML), from which the sympathetic preganglionic neurones arise (Amendt et al., 1979; Blessing et al., 1981; Loewy, Wallach & McKellar, 1981; Martin et al., 1979; Miura, Onai & Takayama, 1983; Ross et al., 1983).

SELECTIVE LOCALIZATION OF CARDIOVASCULAR CONTROL NEURONES IN PGL

Microinjection of excitant amino acids into PGL in the rat has recently been shown in several laboratories to evoke large vasopressor responses (Howe et al., 1983; Ross et al., 1983), and the same is true of the rabbit (Dampney et al., 1982) and the cat (McAllen, 1984; Lovick, 1985b; Lovick & Hilton, 1985). In our studies on the cat using microinjection of DL-homocysteic acid, in addition to blood pressure and heart-rate, we have also registered blood flows to hind-limb muscle, kidney and splanchnic area separately and the results show an interesting separation of effects (Lovick, 1985b and unpublished observations; Lovick & Hilton, 1985). Pressor effects were readily obtained on injection into the more caudal part of PGL, and these could be large and prolonged; but the relative contribution of each vascular bed to the underlying rise in

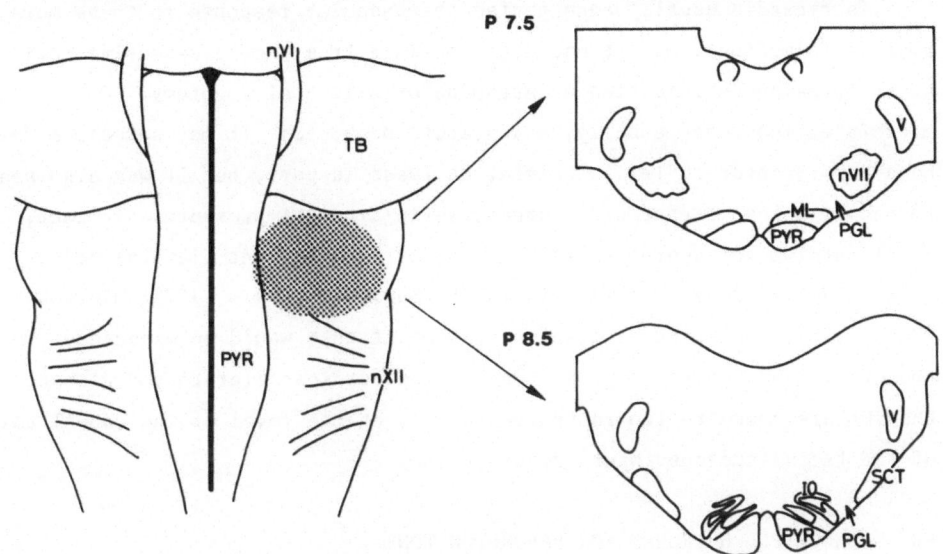

Fig. 4.   On left, diagrammatic plan of the ventral surface of the cat
          medulla showing location of area sensitive to glycine (shaded
          region). On right, diagrammatic coronal sections of the medulla
          at rostral and caudal extents of this area (approximately at the
          levels indicated, according to the atlas of Snider & Niemer,
          1961). Neuronal cell bodies lie in close proximity to the
          ventral surface in nucleus paragigantocellularis lateralis (PGL).
          Other abbreviations as before, and V, spinal trigeminal tract;
          SCT, spinocerebellar tract. Reproduced from Lovick (1986) with
          permission.

peripheral resistance varied according to the exact location
rostrocaudally of the site of injection. On the other hand, injection
into the most rostral part of PGL, superficial to the caudal pole of the
facial nucleus, did not cause a large pressor effect, although the flow
recordings often revealed a large vasodilation in skeletal muscle balanced
by renal vasoconstriction, sometimes with a moderate vasoconstriction in
the splanchnic bed also. This muscle vasodilation was atropine-resistant,
so there may not be a relay in PGL on the pathway activating the
cholinergic vasodilator nerve fibres which are known to be present in the
cat. On the other hand, intravenous injection of β-receptor antagonists
reduced it significantly, so it seems due in part to an increase in
circulating adrenaline released from the adrenal medulla (Ross et al.,
1984; McAllen et al., 1985), and in part to inhibition of background
vasoconstrictor nerve activity.

Tachycardia usually accompanied the vascular response to these more rostral injections, and it could be obtained from a more extensive region after intravenous injection of atropine or bilateral vagotomy. On microinjection more caudally, bradycardia occurred: it may sometimes have been baroreceptor reflex in origin, at least in part, but it was also seen to occur in the absence of a large rise in arterial pressure. In fact, vagal cardiomotor neurones are known to be located ventrolateral to nucleus ambiguus, as well as within it (Sugimoto et al., 1979; Miura & Okada, 1981) so direct excitation of some of them would be expected. On the other hand, neurones projecting to the cardioacceleratory region of the IML are known to lie rostrally in PGL, at the level of the caudal pole of the facial nucleus (Miura et al., 1983).

KEY ROLE OF PGL NEURONES FOR VASOMOTOR TONE

It looks very much, from these results, as though the neurones which have been excited really are cardiovascular control neurones, which are arranged in separate, though overlapping pools, and each controlling an individual vascular bed or the heart (Fig. 5). From the evidence dealt with already, it seems safe to conclude that their activity is modified by the inputs to them from the defense areas, to produce the pattern of cardiovascular changes characteristic of the alerting response : and it is now reasonable to ask to what extent they are the mediators of other patterns of cardiovascular change also, in temperature regulation, for example, or during muscular exercise. But the main point to emphasize here is that when they are inactivated, as described earlier, blood pressure is not maintained.

There are other sympatho-excitatory pathways. In this connection, mention must be made of the neuronal mechanism which generates central respiratory activity and which also modulates sympathetic nerve activity, such that it is most conspicuous during inspiration. This rhythm survives in vagotomised, thoracotomised preparations under artificial ventilation, so bulbospinal inspiratory neurones may well be involved (Lipski, Coote & Trzebski, 1977). However, a powerful case has been presented by Gebber and his colleagues that sympathetic tone is generated independently (Gebber, 1984), and work from several, widely separated laboratories, including our own, has demonstrated the key role in this respect of neurones in the region of PGL (Guertzenstein & Silver, 1974; Hilton, 1980; Dampney et al., 1982; Hilton, Marshall & Timms, 1983; Ross et al., 1983).

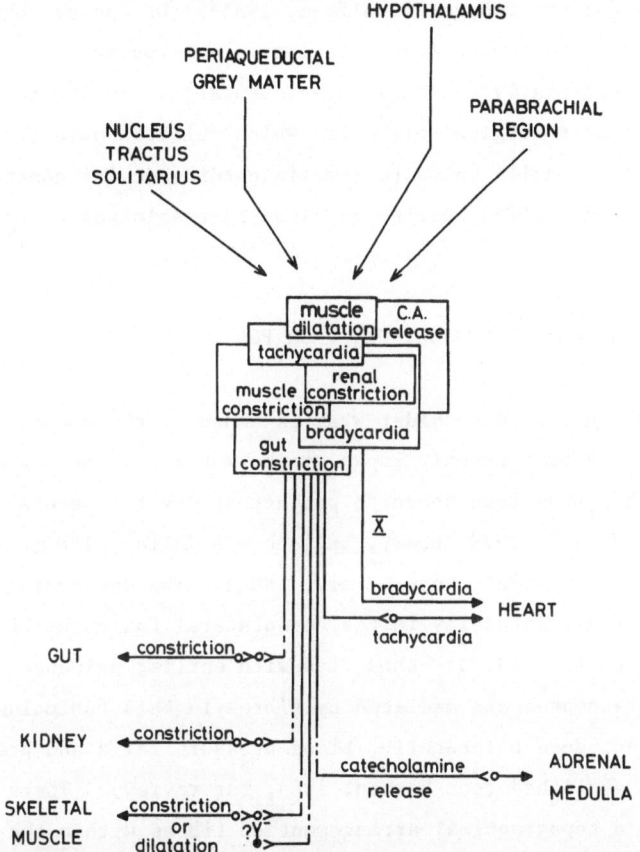

Fig. 5. Schematic diagram to illustrate the rostro-caudal location of cell-bodies in the region of PGL of particular cardiac or vascular dedication. Distinct but overlapping neuronal pools within the area control the level of activity of sympathetic preganglionic neurones supplying the resistance vessels in the splanchnic region, kidney and muscle, and innervating the heart and adrenal medulla (CA release). Some well-established inputs to PGL are indicated by arrows. One possible mechanism of muscle vasodilatation by inhibition of vasoconstrictor nerve activity is also included. Reproduced from Lovick (1986) with permission.

In the light of this conclusion, it is fascinating that neurosurgeons have recently made the discovery that hypertension in man can be caused by vascular abnormalities developing in middle life that impinge upon, and actually indent, the surface of the ventrolateral medulla in the region of the inferior olive (Janetta & Gendell, 1979). In a number of such cases, vascular decompression has been followed by complete resolution of the hypertension (Janetta, Segal & Wolfson, 1985). Of course, these cases had come to their attention because of symptoms attributable to cranial nerve dysfunction, especially glosspharyngeal neuralgia, but it must be significant that a surgical procedure which relieves pulsating pressure on a region of the medulla known to contain cardiovascular control neurones in a number of mammalian species can lead to resolution of hypertension in man.

BULBOSPINAL PATHWAY OF PROJECTIONS FROM PGL

Finally, we should consider what is known of the bulbospinal pathway originating in PGL. Its only known projection is to the region of the IML, to which it has been shown to project at several levels of the spinal cord (Amendt et al., 1979; Loewy, Wallach & McKellar, 1981; Martin et al., 1981; Miura et al., 1983; Ross et al., 1984). The descending axons have been seen to run bilaterally in the dorsolateral funiculus (Loewy et al., 1981; Martin et al., 1979): this fits with earlier evidence that vasopressor responses are mediated by fibres in this funiculus and that, when it is sectioned bilaterally, blood pressure falls and pressor reflexes are abolished (see Barman, 1984, for review). There is even evidence for a topographical arrangement of fibres within the funiculus supplying the heart and different vascular beds (Barman & Wurster, 1975). Antidromic stimulation of the spinal cord in the cat evokes responses in neurones in PGL, when stimulating the contralateral dorsolateral white matter or the ipsilateral ventrolateral white matter (Lovick, 1985b), and excitatory orthodromic projections onto such neurones from the defense areas of the hypothalamus and mid-brain have been demonstrated (Lovick, Smith & Hilton, 1984). In the rat also, neurones in PGL have been activated by electrical stimulation in the region of the IML: these neurones were tonically active and their activity was found to be inversely related to the level of arterial blood pressure when this was altered by vasoactive drugs (Brown & Guyenet, 1984).

OUTSTANDING QUESTIONS

Some of these neurones could well be identical with the cardiovascular control neurones which we have identified, and the question arises whether these neurones are themselves capable of spontaneous activity, or whether they require an excitatory input. If the latter is the case, where do the inputs originate, and how important in this respect are the defense-alerting areas of the amygdalo-hypothalamic complex and the mid-brain? In the experiments of Li & Lovick (1985), most of the cells in these areas projecting to PGL were tonically active, but results from different studies on the effect on blood pressure of removal of these inputs by mid-collicular decerebration have given differing results (see Peiss, 1965, for review), varying from no effect to considerable falls in arterial pressure. However, there is evidence of a tonic inhibitory input onto these neurones; for application of bicuculline to the neuropil superficial to them causes arterial pressure to rise, while GABA has the opposite effect (Yamada et al., 1982; Yamada, McAllen & Loewy, 1984; Ross et al., 1984). This permits the speculation that the nerve fibres known to run to PGL from the NTS (Loewy & Burton, 1978; Dampney et al., 1982; Lovick, unpublished observations) are inhibitory and that this is the source of the baroreceptor-mediated inhibition of their activity (Brown & Guyenet, 1984; McAllen, 1985); but this still remains to be demonstrated.

Finally, what is the transmitter (or transmitters) used by these bulbospinal neurones? Lovick & Hunt (1983) have demonstrated neurones containing substance P-like material and serotonin in this location in the cat, while some show met-enkephalin and β-lipotropin-like immunoreactivity (Hunt & Lovick, 1982). In the rat, there are known to be neurones in this region containing substance P-like material and 5-HT (Helke et al., 1982; Loewy & McKellar, 1981), neuropeptide-Y (Hokfelt et al., 1983), and phenyl-ethanolamine N-methyltransferase (and, therefore, presumably adrenaline) (Ross et al., 1984). So there are already a number of candidates, and more than one may be involved. Microiontophoresis of some of these substances onto preganglionic sympathetic neurones has been carried out: so far, 5-HT (de Groat & Ryall, 1967; McCall, 1983) and substance P (Gilbey, McKenna & Schramm, 1983) have been shown to be excitatory, while adrenaline has only inhibitory actions (Coote et al., 1981; Guyenet & Cabot, 1981; Sangdee & Franz, 1983).

CONCLUSIONS

Altogether we are left with a number of questions which cannot be definitely answered at present, but this is one of those exciting times in the development of a field of work when important discoveries seem just around the corner. At least it is certain that neurones in PGL provide one of the most important sources of the sympathoexcitatory drive originating from the medulla and that their activity is modulated by other systems which generate biologically important patterns of cardiovascular change. Neurones in this group are essential for vasomotor tone of central nervous origin, and they are surely those which were predicted by Dittmar (187?) on the basis of his original work on vasomotor tone so many years ago.

REFERENCES

Abrahams, V.C., Hilton, S.M., and Zbrozyna, A.W., 1960, Active muscle vasodilation produced by stimulation of the brainstem: its significance for the defense reaction, J. Physiol. (Lond.), 154:491-513.

Arahams, V.C., Hilton, S.M., and Zbrozyna, A.W., 1964, The role of active muscle vasodilation in the alerting stage of the defense reaction, J. Physiol. (Lond)., 171:189-202.

Amendt, K., Czachurski, J., Dembowsky, K., and Seller, H., 1979, Bulbospinal projections to the intermediolateral cell column: a neuroanatomical study, J. Auton. Nerv. Syst., 1:103-117.

Andrezik, J.A., Chan-Palay, V., and Palay, S.L., 1981, The nucleus paragigantocellularis lateralis in the rat. Demonstration of afferents by retrograde transport of horseradish peroxidase, Anat. Embryol., 161:373-390.

Barman, S.M., 1984, Spinal cord control of the cardiovascular system, in: "Nervous control of cardiovascualr function," W.C. Randall (ed.), pp. 321-345, Oxford University Press.

Barman, S.M., and Wurster, R.D., 1975, Visceromotor organisation within descending sympathetic pathways in the dog, Circ. Res., 37:209-214.

Biscoe, T.J., 1971, Carotid body: structure and function. Physiol. Rev., 51:437-495.

Blessing, W.W., Goodchild, A.K., Dampney, R.A.L., and Chalmers, J.P., 1981, Cell groups in the lower brain stem of the rabbit projecting to the spinal cord, with special reference to catecholamine-containing neurons, Brain Res., 221:35-55.

Brown, D.L., and Guyenet, P,.G., 1984, Cardiovascular neurons of brain stem with projections to spinal cord, Amer. J. Physiol., 247:R1009-1016.

Bunag, R.D., and Riley, E., 1979, Chronic hypothalamic stimulation in awake rats fails to induce hypertension, Hypertension 1:498-507.

Caraffa-Braga, E., Granata, L., and Pinotti, O., 1973, Changes in blood flow distribution during acute emotional stress in dogs, Pflugers Arch., 339:203-216.

Carey, W.E., Dacey, R.G., Jane, J.A., Winn, H.R., Ayers, C.R., and Tyson, G.W., 1979, Production of sustained hypertension by lesion in the

nucleus tractus solitarii of the American foxhound, Hypertension
1:246-254.

Chai, C.Y., and Wang, S.C., 1968, Integration of sympathetic
cardiovascular mechanisms in medulla oblongata of the cat, Amer. J.
Physiol., 215:1310-1315.

Coote, J.H., and MacLeod, V.H., 1974, Evidence for the involvement in the
baroreceptor reflex of a descending inhibitory pathway, J. Physiol.
(Lond)., 241:477-496.

Cowley, Jr., A.W., Laird, J.F., and Guyton, A.C., 1973, Role of the
baroreceptor reflex in daily control of arterial blood pressure and
other variables in dogs, Circ. Res., 32:564-576.

Daly, M. de Burgh, 1984, Breath-hold diving: mechanisms of cardiovascular
adjustments in the mammal, in: "Recent Advances in Physiology 10,"
P.F. Baker (ed.), pp. 201-245, Churchill Livingstone.

Dampney, R.A.L., Goodchild, A.K., Robertson, L.G., and Montgomery, W.,
1982, Role of ventrolateral medulla in vasomotor regulation: a
correlative anatomical and physiological study, Brain Res.,
249:223-235.

deGroat, W.C., and Ryall, R.W., 1967, An excitatory action of 5-ht on
sympathetic preganglionic neurones, Exp. Brain. Res., 3:299-303.

Dittmar, C., 1873, Uber die Lage des sogenannten Gefasscentrums der
Medulla Oblongata, Ber verh sachs Ges der Wiss Math, Phys. Kl,
25:449-469.

Doba, N., and Reis, D.J., 1973, Acute fulminating neurogenic hypertension
produced by brainstem lesions in the rat, Circ. Res., 32:584-593.

Feldberg, W.S., 1976, The ventral surface of the brainstem: a scarcely
explored region of pharmacological sensitivity, Neuroscience
1:427-441.

Folkow, B., and Rubinstein, E.H., 1966, Cardiovascular effects of acute
and chronic stimulations of the hypothalamic defense area in the
rat, Acta Physiol. Scand., 68:48-57.

Gebber, G.L., 1984, Brainstem systems involved in cardiovascular
regulation, in: "Nervous control of cardiovascular function," W.C.
Randall (ed.), pp. 346-368, Oxford University Press.

Gilbey, M.P., McKenna, K.E., and Schramm, L.P., 1983, Effects of substance
P on sympathetic preganglionic neurones, Neurosci. Lett.,
41:157-159.

Guertzenstein, P.G., and Silver, A., 1974, Fall in blood pressure produced
from discrete regions of the ventral surface of the medulla by
glycine and lesions, J. Physiol. (Lond)., 242:489-503.

Guyenet, P.G., and Cabot, J.B., 1981, Inhibition of sympathetic

preganglionic neurons by catecholamines and clonidine: mediation by
an alpha-adrenergic receptor, J. Neurosci. 1:908-917.

Helke, C.J., Neil, J.J., Massari, V.J., and Loewy, A.D., 1982, Substance P
neurons project from the ventral medulla to the intermediolateral
cell column and ventral horn in the rat, Brain Res., 243:147-152.

Hilton, S.M., 1980, Central nervous origin of vasomotor tone, Adv.
Physiol. Sci., 8:1-12.

Hilton, S.M., 1982, The defense-arousal system and its relevance for
circulatory and respiratory control, J. Exp. Biol., 100:159-174.

Hilton, S.M., and Marshall, J.M., 1982, The pattern of cardiovascular
response to carotid chemoreceptor stimulation in the cat, J.
Physiol. (Lond)., 326:495-513.

Hilton, S.M., Marshall, J.M., and Timms, R.J., 1983, Ventral medullary
relay neurones in the pathway from the defense areas of the cat and
their effect on blood pressure, J. Physiol. (Lond)., 345:149-166.

Hilton, S.M., and Redfern, W.S., 1983, Exploration of the brainstem
defense areas with a synaptic excitant in the rat, J. Physiol.
(Lond)., 345:134P.

Hilton, S.M., and Smith, P.R., 1984, Ventral medullary neurones excited
from the hypothalamic and mid-brain defense areas, J. Auton. Nerv.
Syst., 11:35-42.

Hokfelt, T., Lundberg, J.M., Tatemoto, K., Mutt, V., Terenius, L., Polak,
J., Bloom, S., Sasek, C., Elde, R., and Goldstein, M., 1983,
Neuropeptide-Y (NPY)- and FMRFamide neuropeptide-like
immunoreactivities in catecholamine neurons of the rat medulla
oblongata, Acta Physiol. Scand., 117:315-318.

Howe, P.R.C., Kohn, D.M., Minson, J.B., Stead, B.H., and Chalmers, J.P.,
1983, Evidence for a bulbospinal pressor pathway in the rat brain,
Brain Res., 270:29-36.

Hukuhara, T., and Takeda, R., 1975, Neuronal organization of central
vasomotor mechanisms in the brain stem of the cat, Brain Res.,
87:419-429.

Hunt, S.P., and Lovick, T.A., 1982, The distribution of serotonin,
met-enkephalin and β-lipotropin-like immunoreactivity in neuronal
perikarya of the cat brainstem, Neurosci. Lett., 30:139-145.

Jannetta, P.J., and Gendell, H.M., 1979, Clinical observations on etiology
of essential hypertension, Surg. Forum., 30:431-432.

Jannetta, P.J., Segal, R., and Wolfson, S.K., 1985, Neurogenic
hypertension: etiology and surgical treatment - 1. Observations in
53 patients, Ann. Surg., 201:391-398.

Kenney, R.A., and Neil, E., 1951, The contribution of aortic chemoreceptor mechanisms to the maintenance of arterial blood pressure of cats and dogs after haemorrhage, J. Physiol. (Lond)., 112:223–228.

Lebedev, V.P., Krasiukov, A.V., and Nikitin, S.A., 1984, Neuronal organization of the sympathoactivating structures of the bulbar ventrolateral surface, Fiziol. Zh, 70:761–772.

Li, P., and Lovick, T.A., 1985, Excitatory projections from hypothalamic and midbrain defense areas to nucleus paragigantocellularis lateralis in the rat, Exp. Neurol., in press.

Lindgren, P., and Uvnas, B, 1953, Activation of sympathetic vasodilator and vasoconstrictor neurones by electrical stimulation in the medulla of the cat, Circ. Res. 1:479–485.

Lipski, J., Coote, J.H., and Trzebski, A., 1977, Temporal patterns of antidromic invasion latencies of sympathetic preganglionic neurons related to central inspiratory activity and pulmonary stretch receptor reflex, Brain Res., 155:162–166.

Loewy, A.D., and Burton, H., 1978, Nuclei of the solitary tract: Efferent projections to the lower brain stem and spinal cord of the cat, J. Comp. Neurol., 181:421–450.

Loewy, A.D., and McKellar, S., 1981, Serotonergic projections from the ventral medulla to the intermediolateral cell column in the rat, Brain Res. 211:146–152.

Loewy, A.D., Wallach, J.H., and McKellar, S., 1981, Efferent connections of the ventral medulla oblongata in the rat, Brain Res. Rev., 3:63–80.

Lovick, T.A., 1985a, Projections from the diencephalon and mesencephalon to nucleus paragigantocellularis lateralis in the cat, Neuroscience 14:853–861.

Lovick, T.A., 1985b, Descending projections from the ventrolateral medulla and cardiovascular control, Pflugers Arch, 404:197–202.

Lovick, T.A., 1986, Cardiovascular control from neurones in the ventrolateral medulla, in: "Neurobiology of the Cardiorespiratory System," E.W. Taylor (ed.), Manchester University Press, in press.

Lovick, T.A., and Hilton, S.M., 1985, Vasodilator and vasoconstrictor neurones in the ventrolateral medulla in the cat, Brain Res., 331:353–357.

Lovick, T.A., and Hunt, S.P., 1983, Substance P-immunoreactive and serotonin-containing neurones in the ventral brainstem of the cat, Neurosci. Lett., 36:223–228.

Lovick, T.A., Smith, P.R., and Hilton, S.M., 1984, Spinally projecting neurones near the ventral surface of the medulla in the cat, J. Auton. Nerv. Syst., 11:27–33.

McAllen, R.M., 1984, Do ventral medullary surface neurones integrate the "visceral alerting response" in the cat?, J. Physiol., 353:120P.

McAllen, R.M., 1985, Bulbospinal neurones of the 'glycine-sensitive' area in the cat, J. Physiol. (Lond)., 361:48P.

McAllen, R.M., Kilpatrick, I.C., Jones, M.W., and Woollard, S., 1985, Adrenal catecholamine release following chemical activation of ventrolateral brainstem neurones in cats, Neurosci. Lett. Suppl., 21:S57.

McCall, R.B., 1983, Serotonergic excitation of sympathetic preganglionic neurons: a microiontophoretic study, Brain Res., 289:121–128.

Mancia, G., Baccelli, G., Adams, D.B., and Zanchetti, A., 1971, Vasomotor regulation during sleep in the cat, Amer. J. Physiol., 220:1085–1093.

Mancia, G., Ferrari, A., Ludbrook, J., and Zanchetti, A., 1980, Carotid baroreceptor influences on blood pressure in normotensive and hypertensive subjects, in: "Arterial baroreceptors and hypertension," P. Sleight (ed.), pp. 484–491, Oxford University Press.

Manning, J.W., 1965, Cardiovascular reflexes following lesions in medullary reticular formation, Amer. J. Physiol., 208:283–288.

Marson, L., 1984, Neuronal pathways involved in the defense reaction, Ph.D. thesis, University of Birmingham.

Martin, G.F., Humbertson, A.O., Laxson, C., and Panneton, W.M., 1979, Evidence for direct bulbospinal projections to laminae IX, X and the intermediolateral cell column. Studies using axonal transport techniques in the North American oposuum, Brain Res., 170:165–171.

Miura, M., and Okada, J., 1981, Cardiac and non-cardiac preganglionic neurons of the thoracic vagus nerve: an HRP study in the cat, Jap. J. Physiol., 31:53–66.

Miura, M., Onai, T., and Takayama, K., 1983, Projections of upper structure to the spinal cardioacceleratory center in cats: an HRP study using a new microinjection method, J. Auton. Nerv. Sys., 7:119–139.

Nathan, M.A., and Reis, D.J., 1977, Chronic labile hypertension produced by lesions of the nucleus tractus solitarii in the cat, Circ. Res., 40:72–81.

Neil, J.J., and Loewy, A.D., 1982, Decreases in blood pressure in response to L-glutamate microinjections into the A5 catecholamine cell group, Brain Res., 241:271–278.

Peiss, C.N., 1965, Concepts of cardiovascular regulation: past, present and future, in: "Nervous control of the heart," W.C. Randall (ed.), pp. 154-197, Baltimore: Williams & Williams.

Polosa, C., Mannard, A., and Laskey, W., 1979, Tonic activity of the autonomic nervous system: functions, properties, origins, in: "Integrative functions of the autonomic nervous system," pp. 342-354, University of Tokyo Press.

Redfern, W.S., and Hilton, S.M., 1985, Location of brain-stem cell groups controlling the behavioural, respiratory and autonomic components of the defense reaction in the rat, Neurosci. Lett. Suppl., 21:S57.

Ross, C.A., Ruggiero, D.A., Park, D.H., Joh, A., Fernandez-Pardal, J., Saavedra, J.M., and Reis, D.J., 1984, Tonic vasomotor control by the rostral ventrolateral medulla: effect of electrical and chemical stimulation of the area containing Cl adrenaline neurons on arterial pressure, heart rate and plasma catecholamines and vasopressin, J. Neuroscience., 4:474-494.

Ross, C.A., Ruggiero, D.A., Joh, T.H., Park, D.H., and Reis, D.J., 1983, Adrenaline synthesising neurones in the rostral ventrolateral medulla: a possible role in tonic vasomotor control, Brain Res., 273:356-361.

Sangdee, C., and Franz, D.N., 1983, Evidence for inhibition of sympathetic preganglionic neurons by bulbospinal epinephrine pathways, Neurosci. Lett., 37:167-173.

Scher, A.M., and Young, A.C., 1963, Servoanalysis of carotid sinus reflex effects on peripheral resistance, Circ. Res., 12:152:162.

Schramm, L.P., and Bignall, K.E., 1971, Central neural pathways mediating active sympathetic muscle vasodilatation in cats, Amer. J. Physiol., 221:754-767.

Smith, O.A., Astley, A.C., DeVito, J.L., Stein, J.M., and Walsh, K.E., 1980, Functional analysis of hypothalamic control of the cardiovascular responses accompanying emotional behaviour, Fed. Proc., 39:2487-2494.

Snider, R.S., and Niemer, W.T., 1961, A stereotaxic atlas of the cat brain, Chicago: University of Chicago Press.

Sugimoto, T., Itoh, K., Misuno, N., Nomura, S., and Konishi, A., 1979, The site of origin of cardiac preganglionic fibers in the vagus nerve: an HRP study in the cat, Neurosci. Lett., 12:53-58.

Taber, E., 1961, The cytoarchitecture of the brain stem of the cat - 1. Brain stem nuclei of cat, J. Comp. Neurol., 116:27-70.

Talman, W.T., Perrone, M.H., and Reis, D.J., 1981, Acute hypertension after the local injection of kainic acid into the nucleus tractus solitarii of rats, Circ. Res., 48:292-298.

Yamada, K.A., McAllen, R.A., and Loewy, A.D., 1984, GABA antagonists

applied to the ventral surface of the medulla oblongata block the
baroreceptor reflex, Brain Res., 297:175-180.

Yamada, K.A., Norman, W.P., Hamosh, P., and Gillis, R.A., 1982, Medullary
ventral surface GABA receptors affect respiratory and
cardiovascular function, Brain Res., 248:71-78.

Yardley, C.P., 1985, Cardiovascular and behavioural responses to acute and
chronic stimulation of the defense area in the rat, Ph.D. thesis,
University of Birmingham.

Yardley, C.P., and Hilton, S.M., 1986, The hypothalamic and brainstem
areas from which the cardiovascular and behavioural components of
the defense reaction are elicited in the rat, J. Auton. Nerv. Syst.
In press.

THE C1 AREA OF ROSTRAL VENTROLATERAL MEDULLA: ROLE IN TONIC AND REFLEX REGULATION OF ARTERIAL PRESSURE

Donald J. Reis

Laboratory of Neurobiology, Department of Neurology
Cornell University Medical College
New York, New York   10021

INTRODUCTION

Although it has been known from the middle of the last century that
the medulla oblongata is required for maintaining normal levels of
arterial pressure (AP) (see Alexander, 1946), the identification of the
specific neuronal groups within that structure which serve as sympathetic
promoter neurons was not established for almost a century. Indeed, the
failure of investigators to reproduce the effects of spinal cord
transection on AP by local lesions within the medulla, led to the concept
that the "tonic vasomotor centers" represented the action of neurons
distributed along a large extent of the neuraxis rather than the activity
of a collection of nerve cells within a restricted zone (Hilton, 1975).

The possibility that a highly localized neuronal network within the
medulla could in fact serve this vasomotor function came from two lines of
investigation. The first were studies directed at examining the role of
structures lying in apposition to the ventral surface of the medulla
oblongata in the control of the circulation and respiration. Schläfke and
Löschke (1967) demonstrated that a cold probe applied to a specific zone
on the ventral surface of the cat's medulla resulted in the collapse in AP
to levels comparable to those produced by spinal cord transection.
Pharmacological studies by Feldberg & Guertzenstein (1972, 1976)
demonstrated that the application of pharmacological agents to a
homologous region would also produce a collapse of AP of a comparable
magnitude. These findings, and the observations that small localized
lesions of the ventral surface effected a comparable collapse of AP

(Guertzenstein & Silver, 1974; Guertzenstein et al. 1978), pointed to the probability that neurons lying somewhere in apposition to the ventral surface of the medulla were critical in maintaining background drive to sympathetic neuronal discharge.

The second line of investigation derived from studies on the cerebral ischemic response, a potent vasopressor reflex elicited by rendering the brainstem ischemic. Kumada, Dampney & Reis (1979) first demonstrated that selective lesions of a restricted zone of the dorsal medullary reticular formation (DMRF) would not only abolish the cerebral ischemic response but also lower blood pressure to spinal levels. Subsequent investigations by Dampney and his associates (Dampney & Moon, 1980; Dampney et al., 1982) demonstrated that comparable effects could be elicited by lesions of neuronal groups in the rostral ventrolateral medulla. Dampney et al. (1982) proposed indirect evidence that the DMRF contained axons of neurons within the ventral medulla which projected down to innervate autonomic centers in the spinal cord. While collectively the pharmacological and physiological studies demonstrated that neurons localized in the ventrolateral medulla oblongata probably represented the tonic vasomotor center of classic physiology, the precise identity of this neuronal group, however, was not clearly established.

C1 AREA NEURONS OF RVL: TONIC VASOMOTOR FUNCTION

It was of interest to our group that the localization of the purported neurons in rostral medulla coincided with those described by histochemists as corresponding to catecholamine networks in the lower brainstem (Dahlström & Fuxe, 1964). In particular, our attention was focused on a group of neurons designated as the C1 group (Hökfelt et al., 1974) which contained adrenalin. It was the objective of our group to establish whether or not such neurons played a critical role in maintaining sympathetic tone and hence normal resting levels of AP, and whether or not they were involved in the reflex control of the circulation as well.

In an initial series of studies, Armstrong et al. (1982) carefully mapped the medulla for the immunocytochemical localization of tyrosine hydroxylase (TH), dopamine beta-hydroxylase (DBH) and phenylethanolamine

N-methyltransferase (PNMT) within catecholamine-containing groups lying in a column of cells within the ventrolateral medulla. In the more caudal portions of the catecholamine grouping, all of the catecholamine neurons were noradrenergic and hence consisted of A1 neurons. Rostrally, PNMT-containing neurons began to appear; here they were admixed with neurons of the A1 group. In its most rostral extent only C1-adrenalin neurons were seen. C1 neurons resided within a nucleus termed the nucleus reticularis rostroventrolateralis (RVL) (Ross et al., 1984a). The region has also been referred to as the C1 area (see e.g. Reis, 1984).

Neurons in the RVL project to the spinal cord where they exclusively innervate the autonomic neurons of the intermediolateral column (Ross et al., 1981; Ross et al., 1984b). Although quantitative studies remain to be done, it appears that many, if not all, of the neurons of the RVL which project to the spinal cord are of the C1 group.

Axons of the C1 neurons leave the nucleus projecting dorsomedially through radially arranged axon bundles (Ross et al., 1984a; Granata et al., 1985a; Benarroch et al., 1985; Ruggiero et al., 1985b). They form into a pathway in the dorsomedial medulla, the principal tegmental track, and descend from there down to the intermediolateral columns of the spinal cord. This fiber bundle is easily identified by staining with PNMT (Fig. 1).

Evidence that neurons of the C1 area of RVL participate in autonomic control has come from physiological studies (Ross et al., 1984a). Electrical stimulation of the C1 area with microelectrodes results in sympathoexcitation with elevation of AP and heart rate. Catecholamines are released from adrenal medulla and arginine vasopressin (AVP) from the posterior pituitary. The fact that the local microinjection of excitatory amino acids replicates the cardiovascular responses and that with doses eliciting maximum rises of AP the elevations of AP are comparable in magnitude to those produced with supermaximal stimulus currents indicates first, that the response is elicited from intrinsic neurons and not fibers of passage, and second, that the preponderance of the sympathetic effects elicited by electrical stimulation of the C1 area are evoked from intrinsic neurons. When the medulla is mapped with electrical stimulation of low intensity, the distribution of pressor sites almost exactly overlaps the distribution of C1 neurons ventrally and the PNMT fiber bundles dorsally (Fig. 2).

Fig. 1.   Distribution of phenylethanolamine N-methyltransferase (PNMT, right side) and tyrosine hydroxylase (TH, left side) immunoreactive neurons and processes in rat medulla. Cells in rostral ventrolateral medulla (C1 area) synthesize epinephrine (A-C); those in the caudal ventrolateral medulla synthesize norepinephrine (F-G) and at intermediate levels both cell types are admixed (D-E). Axons of C1 neurons arch dorsally to form a longitudinal fiber tract, principal adrenergic tegmental bundle (PT). PT constitutes the dorsal-most segment of the more extensive and heterogeneous TH-labeled fiber system. Fibers of PT ascend and descend. At caudal levels (D-G) descending limb of PT (PT_d), also referred to as spinal limb (PT_s), shifts medially and sprays ventrolaterally and caudally (E-G) en route to the lateral funiculus of the spinal cord. Note that A_1 fibers do not contribute to PT_d and rather pass lateral to it toward NTS (D-G). PNMT-stained perikarya (open circles); TH-stained perikarya (filled circles); PNMT- and TH-labeled axons (lines)1; PNMT-and TH-labeled terminals (stipple). A_1-Tt, norepinephrine transtegmental tract; A_1, ventrolateral norepinephrine cell area; AP, area postreme; C_1, ventrolateral epinephrine cell area; C_2, dorsomedial epinephrine cell area; CNV, commissural nucleus of vagus, CST, corticospinal tract; ECN, external cuneate nucleus, IVN, inferior vestibular nucleus, LRN_1, precerebellar lateral reticular nucleus pars medialis; MAO, medial accessory olive; MLF, medial longitudinal fasciculus; MVN, medial vestibular nucleus; NC, nucleus cuneatus; NG, nucleus gracilis; NTS_r, nucleus tractus solitarii pars rostralis; PION, principal nucleus of inferior olive; PP, nucleus prepositus; RP_a, nucleus raphe pallidus; STN, spinal trigeminal nucleus; STN_c, caudal subdivision of STN; VS, ventral medullary surface; X, dorsal vagal motor nucleus; XII, hypoglossal nucleus (from Granata et al., 1985a).

The neurons of the C1 area, moreover, appear to be the same ones whose activity has been modified by physiologists working along the ventral surface of the medulla. Benarroch et al. (1986) have recently demonstrated that: a) there is a direct topographic correspondence between the localization of the C1 area and the area of the ventral surface of the medulla defined by pharmacologists as being sensitive to topical

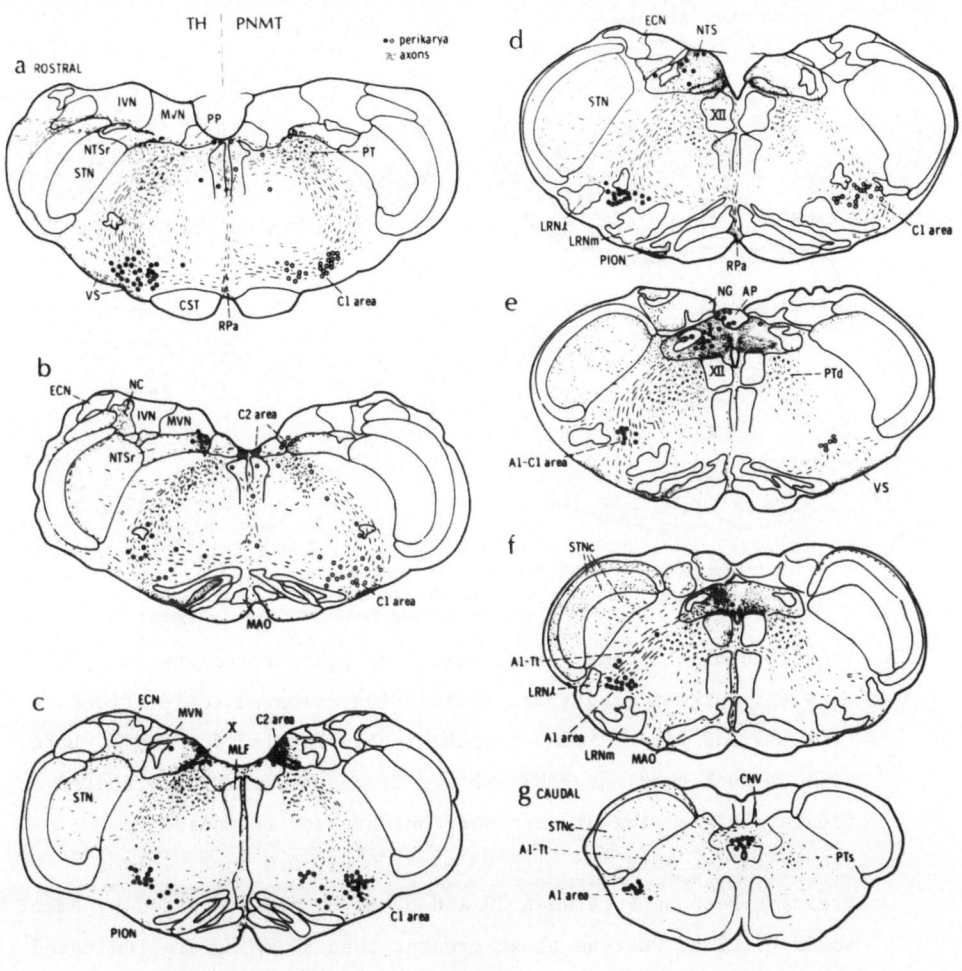

Fig. 1.

stimulation; b) the area of the ventral surface which is the most
sensitive to pressor actions of glutamate or to pressor responses to GABA
directly overlies the Cl area as demonstrated in the same animal by
combined physiological and immunocytochemical techniques; c) the responses
elicited by application of drugs to the ventral surface of the medulla are
identical to those produced by microinjection of the agents into the
medulla but with less sensitivity; and finally d) lesions of the Cl area
or its projection pathway dorsally abolish responses to chemical
stimulation of the ventral surface. The anatomical substrate for the
response appears related to the fact that processes of Cl neurons extend
to lie directly in apposition to the ventral surface of the medulla.

Interference of the function of Cl area neurons by the local
injection of tetrodotoxin (Ross et al., 1984a) or by placement of

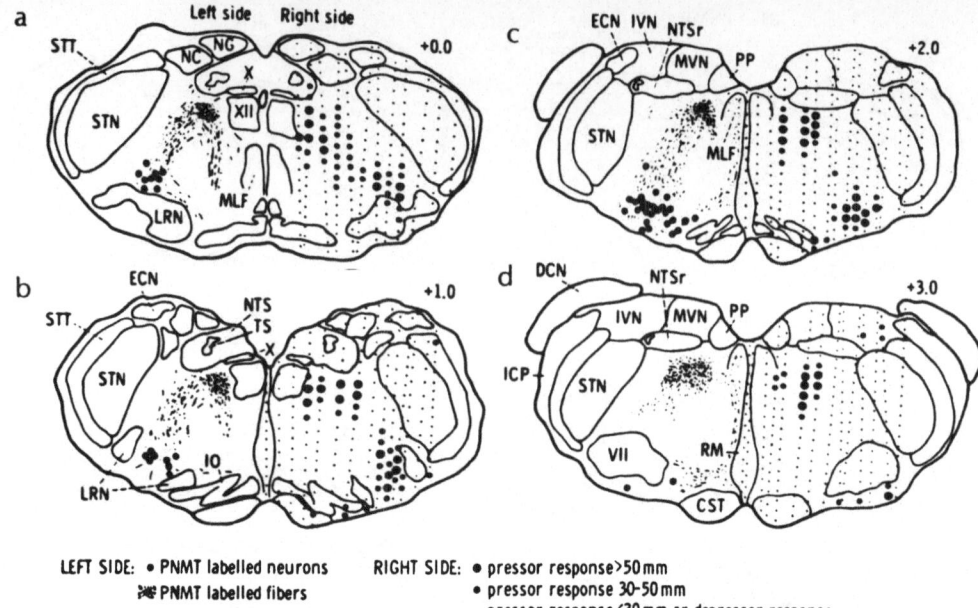

LEFT SIDE: • PNMT labelled neurons   RIGHT SIDE: • pressor response>50mm
          PNMT labelled fibers   • pressor response 30-50mm
                                       · pressor response <30mm or depressor response

Fig. 2.   Locations of the most active medullary pressor regions to
constant current electrical stimulation compared to locations of
C1 cells and fibers immunocytochemically labeled for PNMT. Left
side of each section, PNMT-labeled C1 neurons and PNMT-labeled
fibers. Right side of each section, pressor responses to
electrical stimulation with 25 µA (100 Hz, 0.5 msec, 10-sec
train). Responses between 30 and 55 mm Hg are indicated by small
solid circles, whereas those greater than 50 mm Hg are indicated
by larger solid circles. Responses less than 30 mm Hg or
depressor responses are indicated by small dots. The location of
the pressor region in the ventrolateral medulla corresponds to
the location of the C1 neurons, and the location of the pressor
region in the dorsomedial medulla corresponds to the location of
the PNMT-labeled fiber bundle. CST, corticospinal tract; DCN,
dorsal cochlear nucleus; ECN, external cuneate nucleus; ICP,
inferior cerebellar peduncle; IO, inferior olive; IVN, inferior
vestibular nucleus, LRN, lateral reticular nucleus; MLF, medial
longitudinal fasciculus; MVN, medial vestibular nucleus; NC,
nucleus cuneatus; NG, nucleus gracilis; NTS, nucleus tractus
solitarius; NTSr, nucleus tractus solitarius pars rostralis; PP,
nucleus prepositus; RM, raphe magnus; STN, spinal trigeminal
nucleus; STT, spinal trigeminal tract; TS, tractus solitarius;
VII, facial nucleus; X, dorsal motor nucleus of vagus; XII,
hypoglossal nucleus. (From Ross et al., 1984b.)

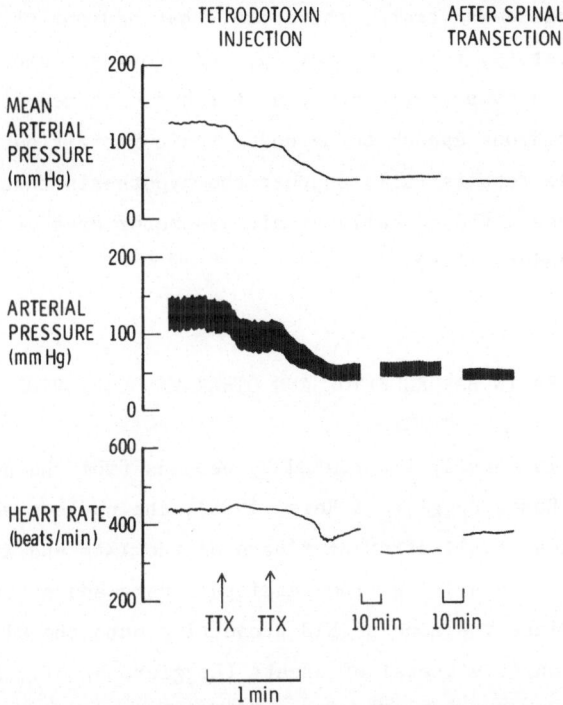

Fig. 3. Cardiovascular responses to bilateral injections of tetrodotoxin
(10 pmole/100 nl saline) into the Cl area. The first injection
(TTX), indicated by an arrow on the bottom of chart recording,
produced a partial reduction of AP, while injection into the
contralateral side (second arrow) resulted in a collapse of AP to
levels comparable to that produced by subsequent spinal cord
transection. CST = corticospinal tract; IO = inferior olive; IVN
= inferior vestibular nucleus; MVN = medial vestibular nucleus;
NA = nucleus abiguus; NTSr = nucleus tractus solitarius pars
rostralis; PP = nucleus prepositus; RPa = raphe pallidus; STN =
spinal trigeminal nucleus; STT = spinal trigeminal tract; TS =
tractus solitarius. (From Ross et al., 1984b.)

bilateral electrolytic lesions of the Cl area (Granata et al., 1985a)
reduces AP to levels comparable to those produced by spinal cord
transection (Fig. 3). The bilateral microinjection of GABA within the
same area at maximal doses also lowers AP to near spinal levels (Ross et
al, 1984a; Benarroch, 1986). In contrast, the local microinjection of the
GABA antagonist bicucculine will elevate AP and heart rate substantially
(Ross et al., 1984a; Benarroch et al., 1986).

These findings demonstrate, therefore, that neurons of the C1 area are sympathoexcitatory, tonically active, and provide basically all of the background drive to sympathetic neurons attributed to medullary networks. Moreover, these neurons appear to be under tonic inhibition by GABAergic mechanisms. These results fully support the hypothesis that neurons of the C1 area represent the so-called tonic vasomotor area of classical physiology (Alexander, 1946).

ROLE OF THE C1 AREA IN BARORECEPTOR AND OTHER VASODEPRESSOR REFLEXES

The C1 area is heavily innervated by neurons from the nucleus tractus solitarii (NTS) (Ross, Ruggiero & Reis, 1985), the nucleus which is the site of termination of all afferent fibers of the IXth and Xth cranial nerves including those arising from cardiopulmonary and arterial baroreceptors. Since the zone of NTS projecting onto the C1 area is the site of termination of visceral efferents (Ruggiero et al., 1985a), the question is raised: Does the C1 area mediate the vasodepressor responses elicited from arterial baro- and other cardiopulmonary receptors?

In a series of experiments, Granata et al. (1985a) observed that the vasodepressor responses elicited by natural stimulation of the carotid sinus or by electrical stimulation of the vagus nerve in rat were abolished by bilateral electrolytic lesions of the C1 area (Fig. 4a,b). However, since the lesions themselves reduced AP to spinal levels it was difficult to establish whether or not the reflex was abolished. A model was then devised whereby AP was maintained in the normal range even after interruption of baroreceptor reflexes (Fig. 4c). The model was based on the anatomical facts that inputs from efferents of the IXth and Xth to the NTS are bilateral but that the projections from the NTS to the C1 area are principally unilateral. With this model, unilateral lesion of the left NTS would disconnect the left C1 area from inputs from NTS on that side yet leave the C1 spinal pathway intact, presumably maintaining AP. A lesion of the C1 area on the other side, however, would, if the C1 area was involved in vasodepressor responses, abolish them.

The combined lesion of the left NTS and right C1 maintained AP at normal levels but completely abolished the baroreflexes (Fig. 4d). Lesions placed outside of the C1 area had no effect on the reflex. Comparable effects were obtained by chemical blockade of the C1 area after the NTS lesion. The neurons within the C1 area mediated the vasodepressor

494

response to stimulation of arterial, baro- and other cardiopulmonary receptors innervated by the IXth and Xth nerves.

THE ROLE OF NEURONS OF THE A1 AREA OF THE CAUDAL VENTROLATERAL MEDULLA (CVL) IN CIRCULATORY CONTROL

Further dramatic demonstration of the unique role of C1 neurons of the RVL in producing sympathoexcitation has come from studies on the physiological function of neurons more caudally situated in and around the A1 area. This zone of the caudal ventrolateral medulla has been termed by us the A1 area with respect to its relationship to the A1 noradrenergic group (e.g. Imaizumi et al., 1985).

Anatomically, A1 neurons of the CVL have a totally different projection field than do neurons of the C1 group. A1 neurons do not project to the spinal cord as previously believed (Dahlström & Fuxe, 1965), but rather project rostrally to other areas including the hypothalamus - notably the periventricular and supraoptic nuclei, the NTS, and of greatest interest, the C1 area (Blessing et al., 1981, 1982; Sawchenko & Swanson, 1982). Thus any action of neurons of the A1 area of CVL upon sympathetic discharge must be via indirect pathways to the intermediolateral column of the spinal cord.

Excitation of neurons in the A1 area by low-frequency electrical stimulation or by the local application of excitatory amino acids lowers AP and heart rate (Blessing & Reis, 1982, 1983). In contrast, interference of the activity of A1 neurons by electrolytic lesions or by local injection of excitatory-neurotoxins elevates AP (Blessing, West & Chalmers, 1981; Blessing, Sved & Reis, 1982; Imaizumi et al., 1985). In rabbit and rat the elevation is often fulminating. The elevation in AP in part represents augmented sympathetic discharge (Granata et al., 1985a). However, an augmented release of AVP contributes to the hypertension (Blessing et al., 1982; Imaizumi et al., 1985). GABA applied to the CVL elevates while bicucculine lowers AP (Blessing & Reis, 1983), a finding completely opposite to the effects of these agents within the C1 area (Ross et al., 1984a; Benarroch et al. 1986).

Since neurons of the A1 area do not project to the spinal cord (Blessing et al., 1981, 1982; Sawchenko & Swanson, 1982), the question is posed as to the network through which neurons of the A1 area mediate their

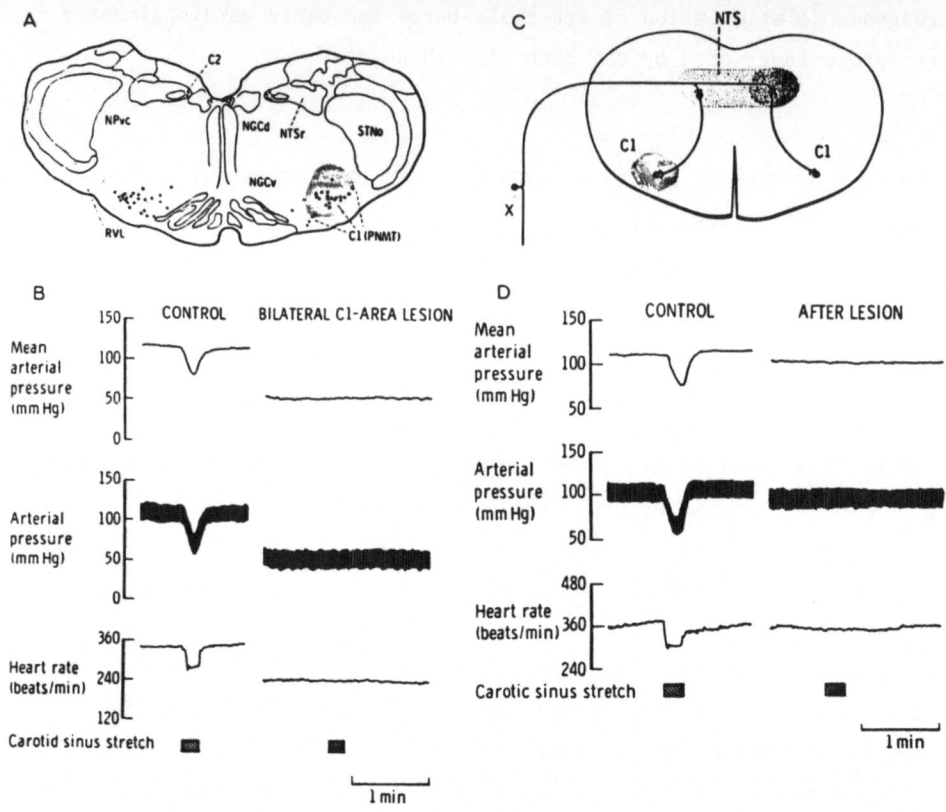

Fig. 4. (a) Representative case with bilateral lesions of C1 area. (b) Effect of bilateral C1 lesions on AP and baroreflex. Note that AP falls to approximately 55 mmHg and below maximum fall of AP with baroreceptor stimulation. (c) Schematic Representation. (d) Effect of crossed NTS-C1 lesion on AP, heart rate and baroreflex response. Note disappearance of baroreflex with preservation of arterial pressure. (From Granata et al., 1985a.)

sympathoinhibition. That they do so by a projection onto the C1 area has recently been demonstrated by us (Ross et al., 1985). Granata et al. (1986) have demonstrated that hypertension elicited by the local injection of the excitotoxin kainic acid (KA) into the A1 area bilaterally fails to be modified by destruction of projection targets including the NTS or hypothalamus, but is completely eliminated by bilateral lesions of the C1 area. The results therefore indicate that an A1-C1 projection is inhibitory upon sympathetic outflow. The question presently being pursued is whether or not the neurons of the Al area which are tonically inhibiting Cl area neurons are noradrenergic. Pharmacological data are consistent with the hypothesis, but further evidence is required before this point can be established.

A NEW MAP OF THE MEDULLA IN CIRCULATORY CONTROL

In summary, recent studies have helped to solve a classic problem in the neurophysiology of the circulation, namely the identification of the medullary neurons governing tonic levels of AP and mediating the baroreceptor reflex arc (Fig. 5) (see Reis, 1984). It is now clear that neurons in the rostral ventrolateral medulla, probably adrenaline neurons of the Cl group which directly innervate the autonomic column of the cord, represent the tonic sympathoexcitatory units. These neurons would be predicted, on the basis of their physiology, to be tonically active and inhibited by baroreceptor input. Such neurons have recently been identified electrophysiologically (Brown & Guyenet, 1984), although the demonstration that these are indeed Cl neurons remains to be established. Neurons of the Cl area are also sensitive to cerebral ischemia (Dampney & Moon, 1980) and hence may have a role as some sort of oxygen sensor. The importance of neurons of the Cl area in the life process cannot be underestimated. Loss of sympathetic tone and the abolition of orthostatic reflexes in veterbrates rapidly leads to death.

The Cl area neurons comprising tonic vasomotor cells themselves are held under tight inhibitory control. Three major inhibitory inputs have been identified: one from the NTS (Ross et al., 1985) mediates the tonic inhibition from baroreceptors (Granata et al., 1985a); a second is from neurons in the caudal ventrolateral medulla located within an area surrounding Al noradrenergic neurons (Granata et al., 1986). These inputs are tonically active. A third, inhibitory network is organized locally and appears to be exerted by GABAergic neurons within the Cl area (Meeley et al., 1985; Ruggiero et al., 1985a). The inhibitory inputs appear to function as tonic vasodepressor networks.

This information has now provided a map identifying the principal pathway mediating the cardiovascular responses to stimulation of arterial baroreceptors (Fig. 5) as well as potential transmitter candidates in the network. Afferent fibers from the IXth and Xth nerves innervating

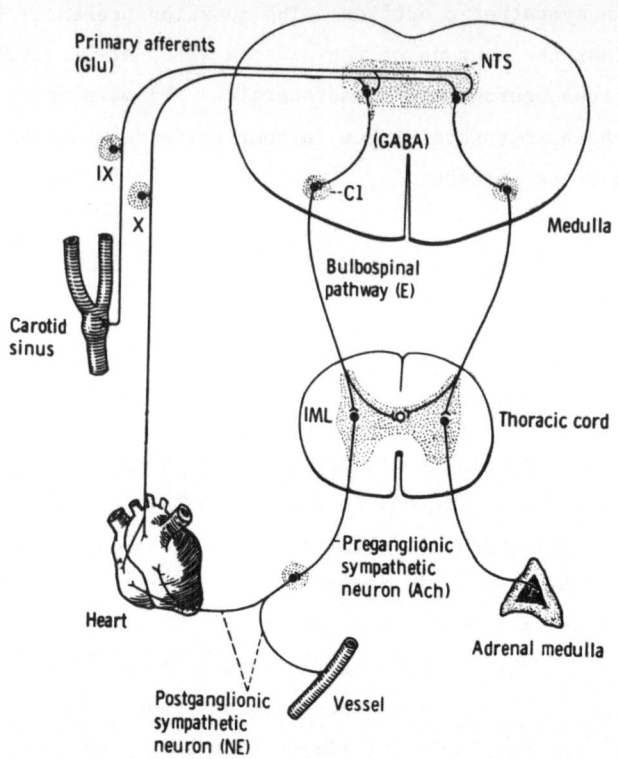

Fig. 5. Diagrammatic representation of proposed pathways and
         neurotransmitters that mediate baroreceptor reflex activity.
         Afferent fibers of the ninth (IX) and tenth (X) nerves terminate
         in the nucleus of the tractus solitarius (NTS) with a proposed
         neurotransmitter being L-glutamate (Glu). Here they synapse on
         neurons that relay into the C1 area. The transmitter of these
         neurons may be gamma-aminobutyric acid (GABA). These neurons,
         which synthesize epinephrine (E), innervate cholinergic neurons
         in the intermediolateral (IML) and intermediomedial levels of the
         spinal cord, which then relay either into the adrenal medulla or
         into sympathetic ganglion where they contact noradrenergic
         postganglionic neurons, which in turn innervate noradrenergic
         sympathetic neurons (NE), which innervate blood vessels, the
         heart, or both. Also shown are neurons in the dorsal motor
         nucleus (DMX) of Nosaka (N), which provide efferent innervation
         of the heart and nucleus ambiguous (NA). (From Reis, 1984.)

receptors in the cardiopulmonary tree and arterial baroreceptors terminate
in NTS and also dorsal motor nucleus. The transmitter in this limb may be

L-glutamate (e.g. Talman, Granata & Reis, 1984). Excitation of this
pathway excites intrinsic neurons of NTS, since excitation of NTS neurons
decreases while chemical blockade or lesions elevates it (see Reis, 1984).
The neurotransmitter in the pathway from the NTS to the C1 area is
unknown. Initially we believed that GABA might be a transmitter candidate
in this direct pathway since topical GABA applied to the C1 area simulates
baroreceptor stimulation while the GABA antagonist elicits the equivalent
of baroreceptor denervation. However, we have recently demonstrated that
destruction of NTS neurons does not deplete local GABA stores, indicating
that GABA is primarily contained in local neurons in the C1 area (Meeley
et al., 1985).

The projection from the C1 area to the intermediolateral column is
the next link in the arc. While the evidence strongly suggests that C1
area neurons represent this linkage, it is not certain whether the
transmitter released to produce sympathetic excitation is adrenaline.
Other transmitter candidates co-localized to the C1 neurons might be
critical.

The remainder of the loop consists of cholinergic preganglionic
sympathetic neurons innervating the adrenal medulla or nonadrenergic
neurons in sympathetic ganglia which, in turn, innervate blood vessels and
the heart.

The identification of the tonic vasomotor region and its major
pathways can only be viewed, however, as a start in our understanding of
the neurobiology of the brainstem control. The clarification on the micro
scale of the nature of the synaptic interactions between the various
substations upon neurons in the C1 area, the identification of the
transmitters and receptors therein, and the details of the functions of
the local circuits all remain to be elaborated. Moreover, the role of
these important cells in visceral disease, including foremost
hypertension, remains an intriguing challenge for the future.

REFERENCES

Alexander, R. S., 1946, Tonic and reflex functions of medullary
        sympathetic cardiovascular centers., J. Neurophysiol., 9:105-217.
Armstrong, D. M., Ross, C. A., Pickel, V. M., Joh, T. H., and Reis, D. J.,
        1982, Distribution of dopamine, noradrenaline-containing cell
        bodies in the rat medulla oblongata: Demonstrated by the

immunocytochemical localization of catecholamine biosynthetic enzymes, J. Comp. Neurol., 212:173-187.

Benarroch, E. E., Granata, A. R., Ruggiero, D. A., Park, D. H., and Reis, D. J., 1986 in press, Neurons of the C1 area mediate cardiovascular responses from the ventral medullary surface, Amer. J. Physiol.

Blessing, W. W., Costa, M., Furness, J. B., West, M. J., and Chalmers, J. P., 1981, Projection from A1 neurons towards the nucleus tractus solitarius in rabbit, Cell Tiss. Res., 220:27-40.

Blessing, W. W., Jaeger, C. B., Ruggiero, D. A., and Reis, D. J., 1982, Hypothalamic projections of medullary catecholamine neurons in the rabbit: A combined catecholamine fluorescence and HRP transport study, Brain Res. Bull, 9:279-286.

Blessing, W. W., and Reis, D. J., 1982, Inhibitory cardiovascular function of neurons in the caudal ventrolateral medulla of the rabbit: Relationship to the area containing A1 noradrenergic neurons, Brain Res., 253:161-171.

Blessing, W. W. and Reis, D. J., 1983, Evidence that GABA and glycine-like inputs inhibit vasodepressor neurons in the caudal ventrolateral medulla of the rabbit, Neurosci. Lett., 37:57-62.

Blessing, W. W., Sved, A. F., and Reis, D. J., 1982, Destruction of noradrenergic neurons in rabbit brainstem elevates plasma vasopressin, causing hypertension, Science, 217:661-663.

Blessing, W. W., West, M. J., and Chalmers, J., 1981, Hypertension, bradycardia, and pulmonary edema in the conscious rabbit after brainstem lesions coinciding with the A1 group of catecholamine neurons, Circ. Res. 49:949-958.

Brown, D. L., and Guyenet, P. G., 1984, Cardiovascular neurons of brain stem with projections to spinal cord, Am. J. Physiol. 247: (Regulatory Integrative Comp. Physiol. 16: R1009-R1016.

Dahlström, A., and Fuxe, K., 1964, Evidence for the existence of monoamine containing neurons in the central nervous system. I. Demonstration of monoamines in the cell bodies of brain stem neurons, Acta. Physiol. Scand., 62 (Suppl. 232):1-55.

Dahlström, A., and Fuxe, K., 1965, Evidence for the existence of monoamine neurons in the central nervous system. II. Experimentally induced changes in the intra-neuronal levels of bulbospinal neuron systems, Acta. Physiol. Scand., (Suppl. 247):1-36.

Dampney, R.A.L., Goodchild, A. K., Robertson, I. G., and Montgomery, W., 1982, Role of ventrolateral medulla in vasomotor regulation: A correlative anatomical and physiological study, Brain Res. 249:223-235.

Dampney, R.A.L., and Moon, E. A., 1980, Role of ventrolateral medulla in
    vasomotor response to cerebral ischemia, Am. J. Physiol.
    239:H349-H358.

Feldberg, W., and Guertzenstein, P. G., 1972, A vasodepressor effect of
    pentobarbitone sodium, J. Physiol. (London) 224:83-103.

Feldberg, W., and Guertzenstein, P. G., 1976), Vasodepressor effect
    obtained by drugs acting on the ventral surface of the brain stem,
    J. Physiol. (London) 258:337-355.

Granata, A. R., Numao, Y., Kumada, M., and Reis, D. J., 1986, Tonic
    sympathoinhibition of caudal ventrolateral medulla is mediated by
    brainstem area with C1 epinephrine neurons, Brain Res., in press.

Granata, A. R., Ruggiero, D. A., Park, D. H., Joh, T. H., and Reis, D. J.,
    1985a, Brain stem area with epinephrine neurons mediates baroreflex
    vasodepressor responses, Am. J. Physiol. (HCP) 17:H547-H567.

Guertzenstein, P. G., Hilton, S. M., Marshall, J. M., and Timms, R. J.,
    1978, Experiments on the origin of vasomotor tone, J. Physiol.
    (London) 275:78-79.

Guertzenstein, P. G., and Silver, A., 1974, Fall in blood pressure from
    discrete regions of the ventral surface of the medulla by glycine
    and lesions, J. Physiol. (London) 242:489-503.

Hilton, S. M., 1975, Ways of viewing the central nervous control of the
    circulation - old and new. Brain Res. 87:213-219.

Hökfelt, T., Fuxe, K., Goldstein, M., and Johansson, O., 1974,
    Immunohistochemical evidence for the existence of adrenaline
    neurons in rat brain, Brain Res. 66:235-251.

Imaizumi, T., Granata, A. R., Benarroch, E. E., Sved, A. E., and Reis, D.
    J., 1985, Contributions of arginine vasopressin and the sympathetic
    nervous system to the fulminating hypertension after destruction of
    neurons of caudal ventrolateral medulla of the rat, J. Hypertens.,
    in press.

Kumada, M., Dampney, R.A.L., and Reis, D. J., 1979, Profound hypotension
    and abolition of the vasomotor component of the cerebral ischemic
    response produced by restricted lesions of medulla oblongata:
    Relationship to the so-called tonic vasomotor center, Circ. Res.
    45:63-70.

Meeley, M. P., Ruggiero, D. A., Ishitsuka, T., and Reis, D. J., 1985,
    Intrinsic GABA neurons in the nucleus solitarii and the rostral
    ventrolateral medulla of the rat: an immunocytochemical and
    biochemical study, Neurosci. Lett., in press.

Reis, D. J., 1984, The brain and hypertension: Reflections on 35 years of inquiry into the neurobiology of the circulation, Circulation 70 (Suppl. III-31 - III-45.

Ross, C. A., Armstrong, D. M., Ruggiero, D. A., Pickel, V. M., Joh, T. H., and Reis, D. J., 1981, Adrenaline neurons in the rostral ventrolateral medulla innervate thoracic spinal cord: A combined immunocytochemical and retrograde transport demonstration, Neurosci. Lett. 25:257-262.

Ross, C. A., Ruggiero, D. A., Joh, T. H., Park, D. H., and Reis, D. J., 1984a, Rostral ventrolateral medulla: Selective projections to the thoracic autonomic cell column from the region containing C1 adrenaline neurons, J. Comp. Neurol. 223:168-185.

Ross, C. A., Ruggiero, D. A., Park, D. H., Joh, T. H., Sved, A. F., Fernandez-Pardal, J., Saavedra, J. M., and Reis, D. J., 1984b, Tonic vasomotor control by the rostral ventrolateral medulla: Effect of electrical or chemical stimulation of the area containing C1 adrenalin neurons on arterial pressure, heart rate and plasma catecholamines and vasopressin, J. Neurosci. 4:479-494.

Ross, C. A., Ruggiero, D. A., and Reis, D. J., 1985, Connections between the nucleus of the tractus solitarius and the rostral ventrolatral medulla: A possible anatomical substrate of baroreceptor and other visceral reflexes. J. Comp. Neurol., in press.

Ruggiero, D. A., Meeley, M. P., Anwar, M., and Reis, D. J., 1985a, Newly identified GABAergic neurons in regions of the ventrolateral medulla which regulate blood pressure, Brain Res., 339:171-177.

Ruggiero, D. A., Ross, C. A., Anwar, M., Park, D. H., Joh, T. H., and Reis, D. J., 1985b, Distribution of neurons containing phenylethanolamine N-methyltransferase in medulla and hypothalamus of rat, J. Comp. Neurol., in press.

Sawchenko, P. E., and Swanson, L. W., 1982, The organization of noradrenergic pathways from the brain stem to the paraventricular and supraoptic nuclei in the rat, Brain Res. Rev. 4:275-325.

Schläfke, M., and Löschke, H. H., 1967, Lokalization eines an der regulation von atmung und kreislauf beteiligten gebiets an der ventralen oberflaeche der medulla oblongata durch kalteblockade, Pflügers Arch. 297:201-220.

Talman, W. T., Granata, A. R., and Reis, D. J., 1984, Glutamatergic mechanisms in the nucleus tractus solitarius in blood pressure control, Fed. Proc. 43(1):39-44.

ASPECTS ON THE ROLE OF NEUROPEPTIDE Y AND ATRIAL PEPTIDES IN CONTROL OF

VASCULAR RESISTANCE

Luigi F. Agnati, Kjell Fuxe*, Roberta Grimaldi, Anders
Harfstrand*, Michele Zoli, Isabella Zini, Detlev Ganten+,

and Pasquale Bernardi
Department of Human Physiology, University of Modena
Modena, Italy *Department of Histology, Karolinska
Institute, Stockholm, Sweden, +Institute of Pharmacology
University of Heidelberg, Heidelberg, West Germany

INTRODUCTION

In the last years evidence has been gathered for a role of central
monoamine and peptide systems in the cardiovascular control (Fuxe et al.
1981a, Ganten et al. 1981, Reis et al. 1984). Our group has particularly
studied the modulation by central neuropeptides of the catecholamine (CA)
control of central cardiovascular function. Thus, in the medulla
oblongata there may exist neuronal networks, in which adrenaline (A) and
noradrenaline (NA) synapses can mediate hypotensive and hypertensive
responses, respectively (Fuxe et al. 1982). In the same brain region e.g.
substance P (SP), somatostatin (SS) and opioid peptides can modulate
cardiovascular responses (Bolme et al. 1978, Fuxe et al. 1981b) (Fig. 1).
Recently, we have obtained indications that neuropeptide Y (NPY), a brain
and pancreatic peptide (Tatemoto, Carlquist & Mutt 1982) present in A
(Hokfelt et al. 1983) neurons in the area of the nucleus tractus
solitarius has a vasodepressor effect on arterial blood pressure (ABP) and
lowers heart rate (HR) (Hokfelt et al. 1983, Fuxe et al. 1983). This
action may involve a modulation of A synaptic function. This paper will
briefly review these studies on NPY/CA interactions. New findings on the
peripheral actions of novel peptides, isolated from mammalian atria (Tang
et al. 1984) will also be reported. Atriopeptin III (APIII) was chosen
for these experiments.

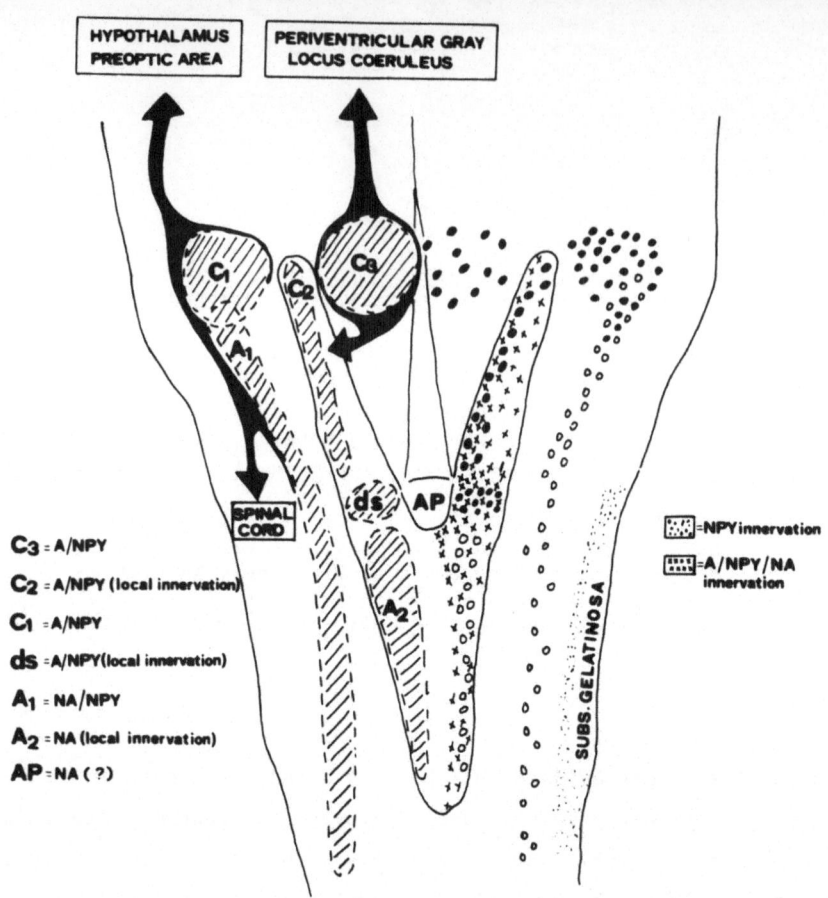

Fig. 1. Schematic representation of the NA, A and NPY immunoreactive cell body systems and their projections and terminal areas, present in the medulla oblongata, especially in the area of the n. tractus solitarius. Local NPY neurons innervate the substantia gelatinosa. Abbreviations: AP= area postrema; ds= dorsal strip.

MATERIALS AND METHODS

Male specific pathogen free Sprague Dawley rats (b.w. 200-250 g) were used. The animals were kept under standardized temperature, humidity and lighting conditions. They had free access to water and food pellets. APIII and NPY were obtained from Peninsula Laboratories (San Carlos, California, U.S.A.). ABP was recorded both in α-chloralose anesthetized rats (see Fig. 9) and in freely moving animals. In the latter a catheter had been inserted under halothane ansesthesia into the left common carotid artery and another one into the right jugular vein 2 - 3 days before the experiment (see Fig. 2).

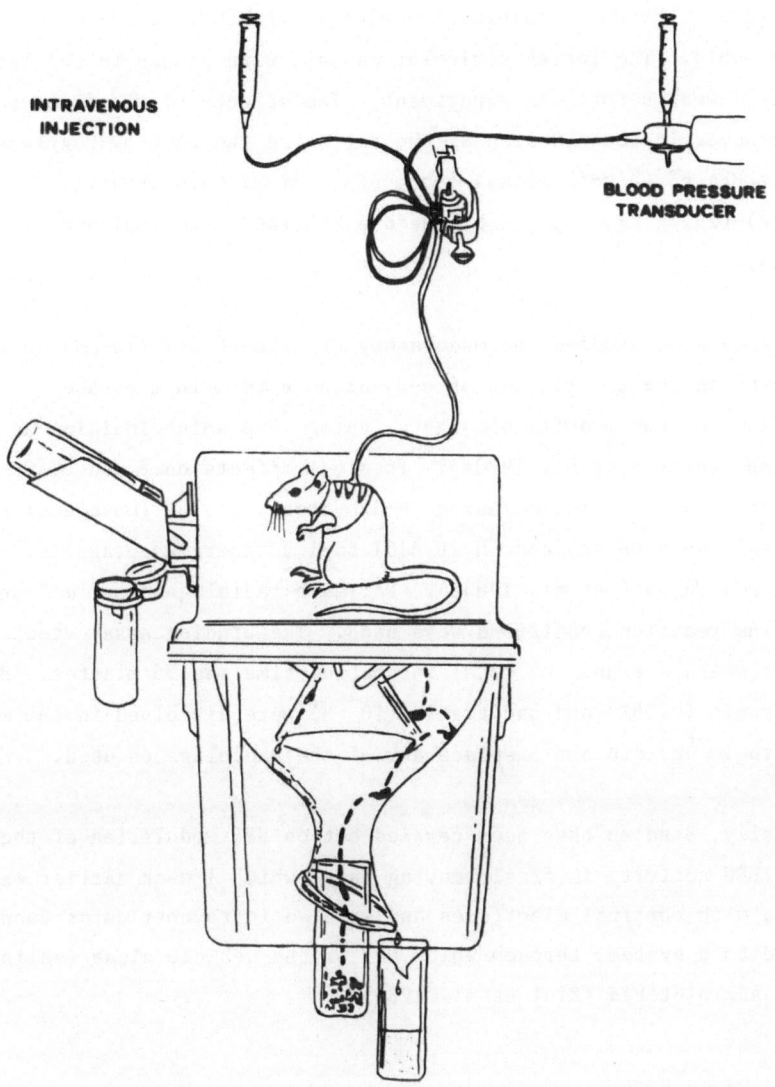

INTRAVENOUS
INJECTION

BLOOD PRESSURE
TRANSDUCER

Fig. 2. Schematic illustration of the animal model used to evaluate the
effects of rat atriopeptin (APIII) in awake and freely moving
rat. The animal was operated 3 days earlier. One catheter was
then inserted into the right jugular vein and another into the
left common carotid artery. These catheters remained open by
means of heparine washings. The catheters were connected with a
transducer, which led into a polygraph, to have a continuous
recording of the mean arterial blood pressure (ABP). The animal
is shown to move in a metabolic cage, which allows the collection
of urine.

The effects of intraventricular NPY injection on NA and A levels and turnover have been evaluated by means of high pressure liquid chromatography (HPLC) in combination with electrochemical detection (Mefford 1981). The intraventricular cannula were placed in the lateral ventricle 1 week before the experiment. The effects of NPY (1.6 nmol/rat) on CA turnover in the NTS area of the rat using the DA β-hydroxylase inhibitor FLA 63 (bis-(L-methyl-4-homopiperazinyl-thiocarbonyl)-disulfide) (25 mg/kg, i.p., 2 hrs before killing) were evaluated (Fuxe et al. 1981a).

We have also studied the modulatory effects of NPY (10 nM) in vitro on the binding characteristics of α-2 adrenoceptors in membrane preparations of the medulla oblongata, using ^{3}H-p-aminoclonidine as a radioligand (Agnati et al. 1983a). Possible effects on β and α-1 adrenoceptors were evaluated using ^{3}H-dihydroalprenolol (unselective β-adrenergic antagonist) and ^{3}H-WB 4101 (α-1 adrenergic antagonist) as radioligands (Agnati et al. 1983a). For each radioligand optimal and equilibrium reaction conditions were used. The binding assays took place at room temperature and the total incubation time was 30 minutes. Bovine serum albumin (0.05%) and bacitracin (10^{-6}M) were dissolved in the medium. NPY (up to 10^{-6}M) did not displace any of the radioligands used.

Finally, studies have been carried out on NPY modulation of the cortical EEG activity in freely moving rats, which 1 week earlier were implanted with cortical electrodes and with an intraventricular cannula, connected to a system, through which NPY or the vehicle alone (saline) could be administered (Zini et al. 1984).

POSSIBLE NPY/A INTERACTIONS AT THE LEVEL OF THE MEDULLA OBLONGATA

Morphological studies

The location of the A/NPY, NA/NPY and NA cell groups in the medulla oblongata is summarized in Figs. 1, 3 and 4 together with the A/NA/NPY innervation of the dorsal medulla, especially the area of the n. tractus solitarius (Hokfelt et al. 1984; Kalia et al. 1985a; and Kalia et al. 1985b). The vast majority of the adrenaline cells of C1, C2 and C3 as well as of the dorsal strip, contains NPY immunoreactivity as demonstrated after colchicine treatment (120 µg/rat, i.c. 24 h before decapitation).

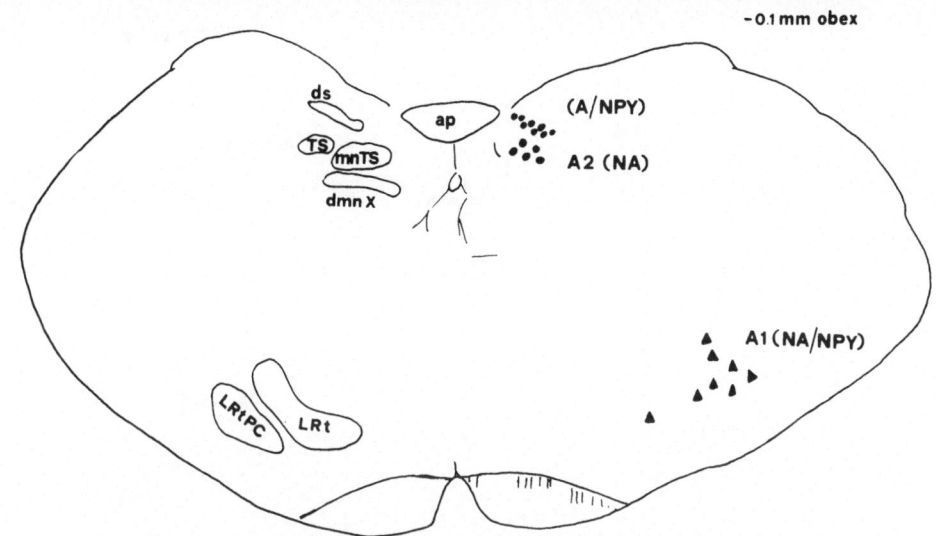

Fig. 3. Schematic representation of CA and NPY positive cells in a
coronal section −0.1 mm from the obex. Abbreviations: ap= area
postrema; TS= tractus solitarius; mnTS= medial part of nucleus
tractus solitarius; dmnX= dorsal motor nucleus of the vagus; LRn=
lateral reticular nucleus; LRnPC= lateral reticular nucleus,
parvocellular part.

Also the Al but not the A2 NA cell group contains NPY-like
immunoreactivity.

Biochemical Studies

    a) Effects of i.v.t. administered NPY on A and NA levels and
utilization. NPY (1.6 nmol/rat) dissolved in saline or saline alone was
intraventricularly injected in awake unrestrained male rats. A group of
rats was pretreated with saline i.p. or with FLA 63 (25 mg/kg, i.p.) two
hours before killing. The evaluation of A and NA contents in the NTS area
was performed by means of HPLC in combination with electrochemical
detection. No effects of NPY on A or NA utilization could be detected
(see Fig. 5), since depletion of the NA and A stores was similar in the
two groups after DA β-hydroxylase inhibition.

    b) Effects of NPY on α-2 adrenergic recognition sites. NPY (10 nM)
can modulate ^{3}H-p-aminoclonidine and ^{3}H-rauwolscine (see Fig. 6), but not
^{3}H-dihydrolalprenolol and ^{3}H-WB 4101 binding sites in membranes of the

507

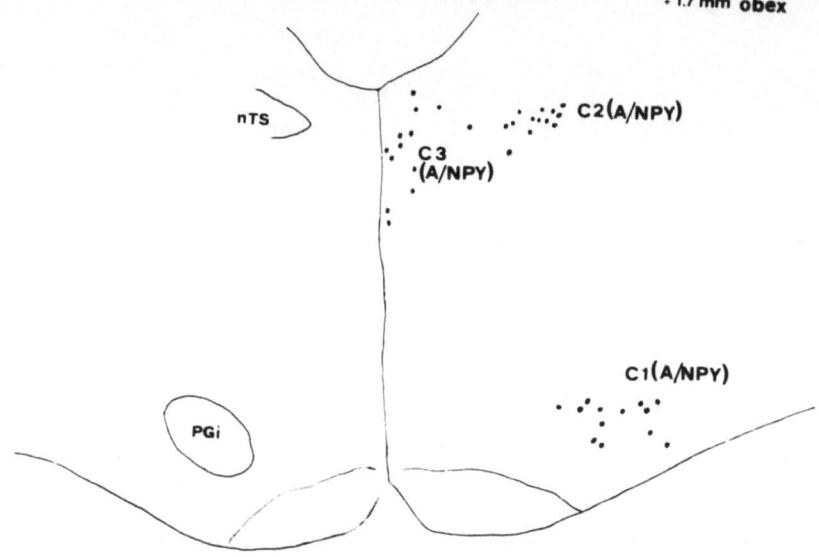

Fig. 4. Schematic representation of adrenaline and NPY positive cells in
a coronal section +1.7 mm rostral to the obex. Abbreviations:
nTS= nucleus tractus solitarius; PGi= paragigantocellular
reticular nucleus.

medulla oblongata (Agnati et al. 1983a; Fuxe et al. 1984b). The effects
of NPY are dose-dependent. In fact, it causes in a concentration of 100
nM a 178$\pm$20 percent change of the K_D values instead of a 15$\pm$5% increase as
seen with 10 nM. However, the increase in the B_{max} values was about 20$\pm$5%
following incubation with either 10 or 100 nM of NPY (Fig. 6).

Functional Studies

a) Effects of NPY on cardiovascular functions. NPY, when given
i.c., has been demonstrated to reduce arterial blood pressure, to reduce
heart rate, and to reduce respiratory frequency in the α-chloralose
anesthetized rat (Fig. 7). (Fuxe et al. 1983; Harfstrand et al. 1984).
Thus, the cardiovascular effects of NPY mimic those of the α-2 adrenergic
agonist clonidine. According to our hypothesis (Fuxe et al. 1981a)
clonidine produces its cardiovascular actions via activation of α-2
adrenergic receptors linked to adrenaline synapses in the brain stem. On
the basis of these physiological findings we have postulated that NPY may

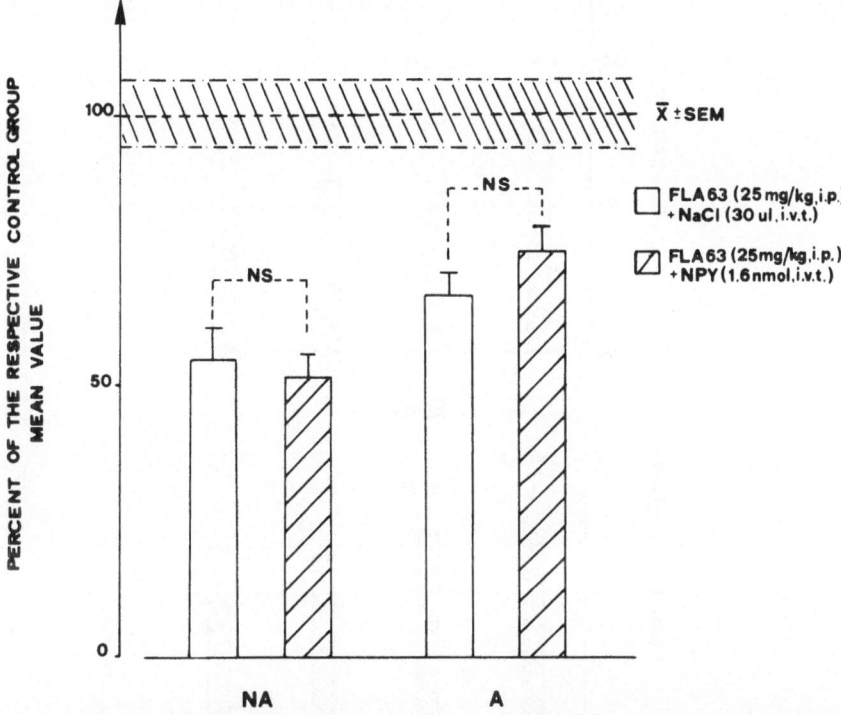

Fig. 5.   Effects of intraventricular injections of NPY (1.6 nmol/rat) on
          noradrenaline (NA) and adrenaline (A) utilization in the NTS
          area of the rat, using the DA-β-hydroxylase inhibitor FLA 63 (25
          mg/kg, i.p., 2 h before killing). NPY was given immediately
          before FLA 63 injection. The area sampled was the mediodorsal
          and caudal part of the medulla oblongata. HPLC in combination
          with electrochemical detection was used. The results are given
          as means ± s.e.m.. The sample size was equal to 10.
          Statistical analysis according to Mann Whitney U-test. The
          absolute levels of NA and A was in the control group: 1629±29
          (NA) and 41±7 ng/g (A).

produce these cardiovascular actions, at least in part, by increasing the
α-2 adrenergic transmission line via a receptor-receptor interaction in
the NPY adrenaline costoring synapses of the brain stem. The enhancement
of the α-2 adrenergic transmission line may mainly be brought about by
activation of the coupling unit (G protein, Ni) inhibiting the biological
effector, probably adenylate cyclase, linked to the α-2 adrenergic
recognition sites. Such a hypothesis is supported by the demonstration
(Harfstrand et al. 1984) that NPY can reduce arterial blood pressure in
the presence of blocked α-2 adrenergic receptors. RX 781094 (208 nmol)

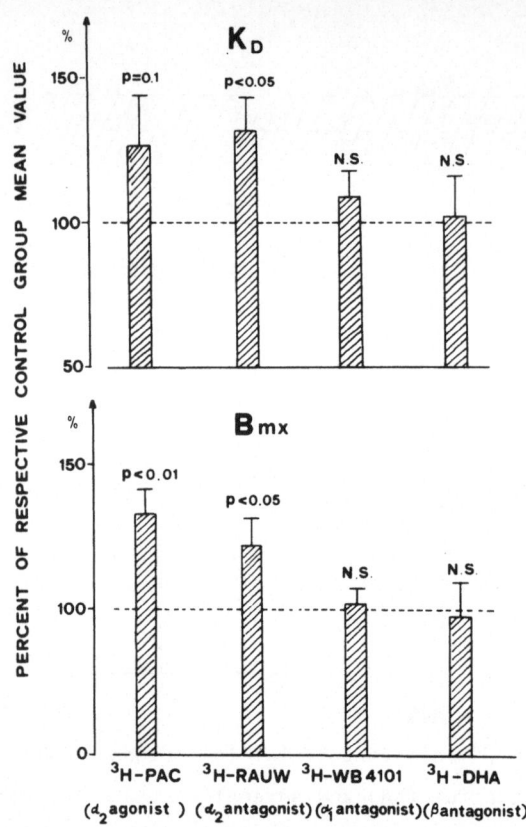

Fig. 6. A summary of the effects of NPY (10 nM) on the ^{3}H-paraaminoclonidine (^{3}H-PAC), ^{3}H-Rauwolscine (^{3}H-RAUW), ^{3}H-(2-(2'-6'-dimethoxy-phenoxyethylamino)methyl-benzodioxan) (^{3}H-WB4101) and ^{3}H-dihydroalprenolol (^{3}H-DHA) binding sites in membrane preparations of the medulla oblongata. For details on the binding procedures, see Agnati et al. (1983). Means $\pm$ s.e.m. are shown in per cent of solvent group mean value. The mean values of eight replications observed for the control experiments (no NPY in the incubation medium) were: for ^{3}H-PAC, $K_D = 1.28$ nM, $B_{max} = 0.109$ pmol/mg prot.; for ^{3}H-RAUW, $K_D = 4.15$ nM, $B_{max} = 0.93$ pmol/mg prot; for ^{3}H-WB4101, $K_D = 0.26$ nM, $B_{max} = 0.108$ pmol/mg prot.; for ^{3}H-DHA $K_D = 2.54$ nM, $B_{max} = 0.128$ pmol/mg prot. The statistical analysis was carried out by means of Student's paired t-test.

was given i.c. prior to the NPY (0.25 nmol) injection. It may be speculated that the putative NPY receptor on activation can undergo a

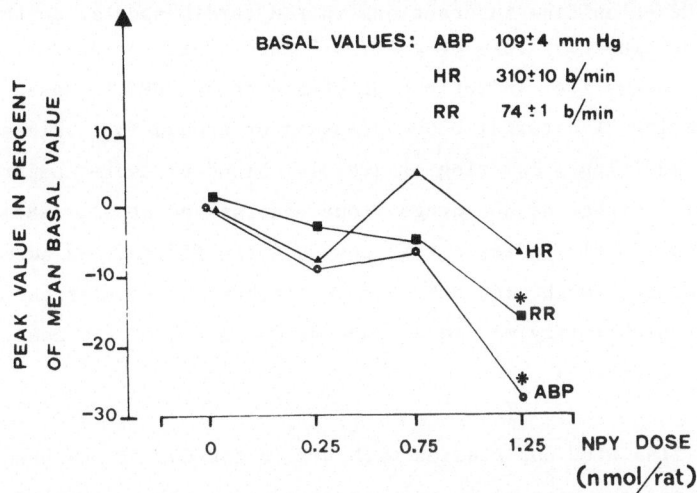

Fig. 7. Effects of intracisternal administration of NPY on cardiovascular and respiratory parameters. The experiments were performed in α-chloralose (100 mg/kg, i.v.) anaesthetized rats. ABP and HR recordings were obtained by means of a catheter in the common carotid artery, which was coupled to a polygraph via a Statham blood pressure transducer. Respiration rate was recorded via a cannula in the esophagus, and also connected to the polygraph via a pressure transducer. In the analysis the peak effect during the first 30 min of recording was selected. The results were analyzed by means of treatments versus control non parametric procedures. *= $p < 0.05$.

conformational change which leads to the activation of the inhibitory guanyl nucleotide regulated protein (Ni) of the α-2 adrenergic transmission line. In this way α-2 adrenergic dependent transmission can be activated and adenylate cyclase inhibited in spite of the blockade of α-2 adrenergic receptors. As discussed previously, this action at the level of the coupling device can then help to explain the reduced affinity in the α-2 adrenergic agonist binding sites found in the in vitro experiments. Obviously, the increase in the number of α-2 adrenergic receptors may also contribute to the enhancement of α-2 adrenergic receptor function and may be produced by another component of the receptor-receptor interaction. The present results also open up the possibility that NPY can produce its cardiovascular effects, at least in part, independently of α-2 adrenergic receptor function. However, studies on NPY-clonidine interactions in the cardiovascular control do underline the important role of a NPY/α-2 adrenergic receptor interaction. Thus, simultaneous i.c. injections of clonidine and NPY in doses giving an optimal lowering of arterial blood pressure or around ED_{50} values, do not lead to any additional lowering of arterial blood pressure compared to that seen with either of the drugs alone (Harfstrand et al. 1984). Thus, these studies clearly indicate that postsynaptic NPY and α-2 adrenergic receptors interact with each other in central cardiovascular control, an interaction which mainly may take place at the level of the coupling device, the Ni protein.

In the line with our studies with NPY in the CNS it has been found that NPY cooperates with NA in perivascular nerve fibres (Ekblad et al. 1984). Furthermore, the potentiating effect of NPY on the response of electrical stimulation is probably located at the postsynaptic level, since NPY does not affect the spontaneous nor the electrically evoked release of ^{3}H-NA. Also no effects of NPY, given i.v.t., have been observed on NA and A levels and turnover in doses producing cardiovascular actions (see Fig. 5). On the basis of these observations (Fuxe et al. 1984a; Fuxe et al. 1985a, Agnati & Fuxe 1983), we have postulated that receptor-receptor interactions make possible the existence of heterostatic mechanisms in central synapses and, in this case in central A synapses. Thus, by means of the receptor-receptor interactions the main transmission line, in this case the α-2 adrenergic transmission line, can be modulated by NPY without introducing compensatory changes in A levels, turnover and release. Thus, the set point of the α-2 adrenergic transmission line can be changed by means of the receptor-receptor interaction without interfering with homeostasis, opening up a new way to understand functional plasticity in synapses of the nervous system and to treat hypertensive diseases.

In agreement with our observations that α-2 adrenergic receptor blockade cannot counteract the central cardiovascular actions of NPY it has been found that NPY induced vasoconstriction (Lundberg & Tatemoto 1982) and contraction in cerebral arteries in vitro are not blocked by the α-2 adrenoreceptor antagonist rauwolscine. It is also of substantial interest that they were blocked by calcium removal or by calcium channel antagonists (Edvinsson et al. 1983). When focusing our interest on the postsynaptic actions of NPY it must, however, be realized that in certain organs evidence for presynaptic actions of the NPY has been noted. Thus, in the rat vas deferens a NPY receptor seems to exist presynaptically, which on stimulation by NPY can inhibit release of NA from noradrenergic nerve terminals (Ohhashi & Jacobowitz 1983). A similar phenomenon has also been demonstrated centrally, since NPY has been found to enhance the inhibitory feedback action of clonidine on CA release in slices of medulla oblongata (Fuxe, Cerrito, Martire and Agnati, unpublished data).

b) Effects of NPY on cortical EEG activity. It has been found that i.v.t. injections of NPY in nmol doses into normotensive Spraque-Dawley rats can produce a reduction of desynchronized and mixed EEG activity and an increase in the synchronized EEG activity during the first hours after injection (Fig. 8) (Fuxe et al. 1983; Zini et al. 1984). Thus, also with regard to the effects on the sleep-wakefulness cycle does NPY mimic the action of clonidine, the α-2 adrenoreceptor agonist. It is known that the locus coeruleus NA system is involved in maintenance of tonic arousal (Fuxe & Lidbrink 1973) and that it is being controlled by an inhibitory adrenergic innervation (Fuxe et al. 1974; Hokfelt et al. 1980). Thus, locus coeruleus is innervated by A nerve terminals. We have recently been able to demonstrate and to quantitate the existence of NPY-like immunoreactivity within large numbers of the A nerve terminal networks innervating the locus coeruleus (Agnati et al. 1985; Fuxe et al. 1985b). In addition, NPY-like immunoreactivity has been demonstrated in several of the NA cell bodies of the locus coeruleus (Everitt et al. 1984). It is therefore possible that part of the action of NPY on cortical EEG activity could be related to its ability to enhance the postsynaptic effects of A released onto the NA cell bodies of the locus coeruleus. Thus, it is assumed that NPY may enhance the transmission of the α-2 adrenergic transmission line also in the NPY / A coexisting synapses of the locus coeruleus as indicated in the NPY / A synapses controlling cardiovascular functions in the lower brain stem.

Possible Physiopathological Relevance of the Interactions Between NPY and α-2 Adrenergic Receptors

There probably exists a disturbance in the adrenergic vasodepressor mechanism in the spontaneously hypertensive rat. Thus, a reduction of the A turnover exists in the cardiovascular centers of the dorsal medulla oblongata in relation to the development of spontaneous hypertension. Also stress selectively depletes A stores in this region of the spontaneous hypertensive rat. It was therefore of interest to evaluate

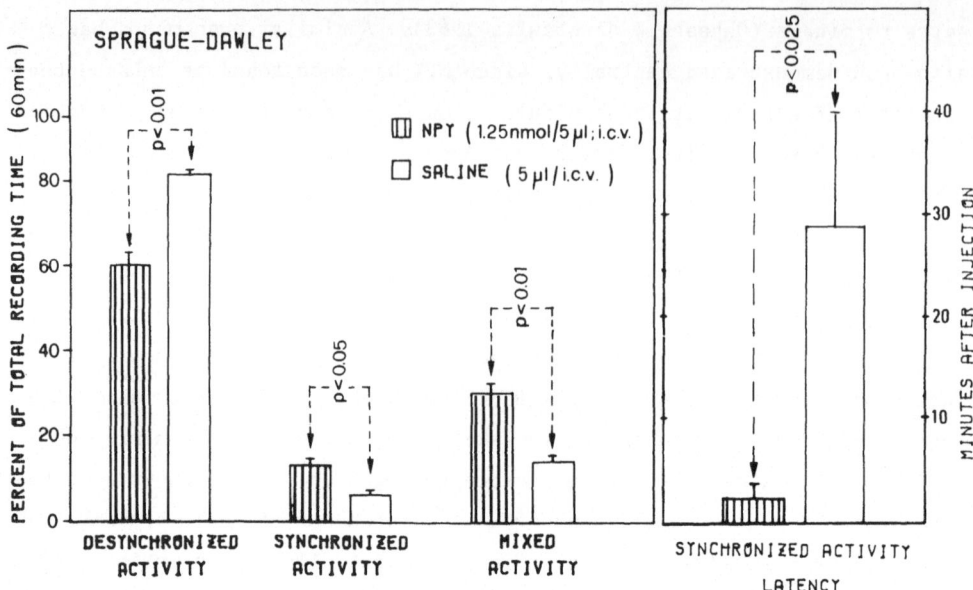

Fig. 8.   Effects of intraventricular injection of NPY on electrocortical
          activity in freely moving rats. Six rats per group have been
          studied. The rats were implanted with four cortical electrodes
          located in the anterior A = 12.5, L = + 2.0 and in the posterior
          A = 4.5, L = + 3.0 (Konig and Klippel Atlas) part of the skull.
          The anterior and posterior electrodes were connected in a bipolar
          fashion. The EMG electrodes were implanted in the neck muscles.
          The coordinates of the intraventricular cannula were: A = 9.5, L
          = 1.6, V = 4.7. The experiments started 7 days after operation
          and were run between 10 a.m. and 1 p.m. The EEG recording
          started 30 min after the animals entered the box. After 30 min
          of basal recording either saline of NPY was injected via a
          catheter without opening the box. Left panel: The recordings
          were divided into periods of 20 sec and the EEG patterns during
          these periods classified either as desynchronized, mixed or
          synchronized activity. The data obtained were expressed as
          percent of total recording time (60 min). Statistical analysis
          according to Mann-Whitney U test. Right panel: The latency
          between the saline of NPY administration and the appearance of
          the synchronized activity has been recorded. For each animal
          either saline or NPY were administered in a random fashion, with
          a wash-out period of at least 24 hours. Statistical analysis
          according to Wilcoxon test for paired samples.

the ability of NPY in vitro to modulate the high affinity α-2 agonist adrenergic binding sites in membranes of the medulla oblongata of the spontaneous hypertensive (SH) rat (adult rats).

In these experiments the SH rats were obtained from Charles River together with normotensive Wistar-Kyoto control rats of the same age (3 months old). The concentration of NPY was 10 nM as in the previous experiments on the Sprague-Dawley animals. NPY can no longer increase the number of α-2 adrenergic agonist binding sites in membranes of the medulla oblongata of the SH rat (Agnati et al. 1983b). In the membranes from the medulla oblongata of the normotensive control rat there was instead a trend for a reduction in the number of ^{3}H-aminoclonidine binding sites in the membrane preparation (Agnati et al. 1983b). The changes observed in the NPY/α-2 adrenergic receptor interaction in SH rats compared with normotensive animals are of special interest in view of the possibility to have an in vivo correlate by comparing the cardiovascular actions of NPY in normotensive and SH animals (Harfstrand et al. 1984).

In cardiovascular experiments (Harfstrand et al. 1984) using α-chloralose anaesthetized rats it was shown that the threshold dose of NPY (i.c. administered) to reduce arterial blood pressure is substantially increased in the SH rats in comparison with the Wistar-Kyoto controls. The ED_{50} value increased about 10 times from 12 pmol/rat to 120 pmol/rat given i.c.. However, the maximal lowering of arterial blood pressure by NPY is the same in the SH and normotensive rats. These cardiovascular results indicate an in vivo correlate to the in vitro experiments on NPY/ α-2 adrenergic receptor interactions in membranes of the medulla oblongata. Thus, the high ED_{50}-value for the cardiovascular actions of NPY in SH-rats correlates with the failure of NPY in low concentrations in vitro to modulate the characteristics of α-2 adrenergic receptors using ^{3}H-p-aminoclonidine as a radioligand. Instead, the central cardiovascular actions of NPY in high concentrations are the same in SH rats and normotensive rats. High concentrations of NPY in vitro can produce a similar increase of the K_D-values of the α-2 high affinity adrenergic agonist binding sites in membranes from young normotensive and SH rats (unpublished data). These results taken together indicate that the regulation of α-2 adrenergic receptors in membranes of the medulla oblongata is pathologically disturbed in the SH rat compared with the normotensive rat. It seems as if the NPY receptor in the high affinity range can no longer be sufficiently activated by NPY to induce its cardiovascular actions as well as its modulatory effects on the K_D- and B-max values of the α-2 adrenergic receptors of the high affinity type. In conclusion, the work indicates that integrative mechanisms in adrenergic synapses in cardiovascular centers of the brain stem are pathologically altered in SH animals.

It should also be underlined that the receptor-receptor interaction makes possible a very marked miniaturization of the circuit (Fig. 9).

THE CONCEPT OF HOMEOSTATIC AND HETEROSTATIC MECHANISMS AT SYNAPTIC LEVEL

There is a large body of evidence that at synaptic level powerful negative feedbacks are operative to keep the possible fluctuations of the efficiency of the transmission line under control (Fig. 10). Thus, e.g., treatment with DA receptor blocking agents leads to increase synthesis and release of DA and also to the development of a clearcut DA receptor supersensitivity caused by an increased receptor density and an increased coupling to their biological effector (Agnati & Fuxe 1980; Fuxe et al. 1981c). DA receptor supersensitivity in striatum can also be observed after lesion induced degeneration of the ascending DA nigrostriatal pathway. However, the mechanisms underlying lesion and neuroleptic induced DA supersensitivity development are at least partly different, since additive changes in DA dependent behaviors have been observed following neuroleptic treatment of animals having supersensitive DA receptors due to lesion of ascending DA pathways (Agnati & Fuxe 1980; Fuxe et al. 1981c).

Thus, in general, any action on transmitter recognition/decoding system (e.g. receptor blockade) triggers a compensatory response both at the postsynaptic (receptor sensitivity) and presynaptic (transmitter synthesis and release) level which tends to maintain the constancy of the efficacy of the transmission lines. Our findings on receptor-receptor interactions show that it is possible to change the efficacy of a transmission line (e.g. NPY effects on α-2 recognition and transduction system) without triggering compensatory responses (no effects of NPY on A release).

Thus, the synapse is still under the control of the negative feedbacks, which allow a high fidelity of the signal transmission, but these feedbacks have now a different set-point. This is obviously a very efficient and economic mechanism for synaptic plasticity.

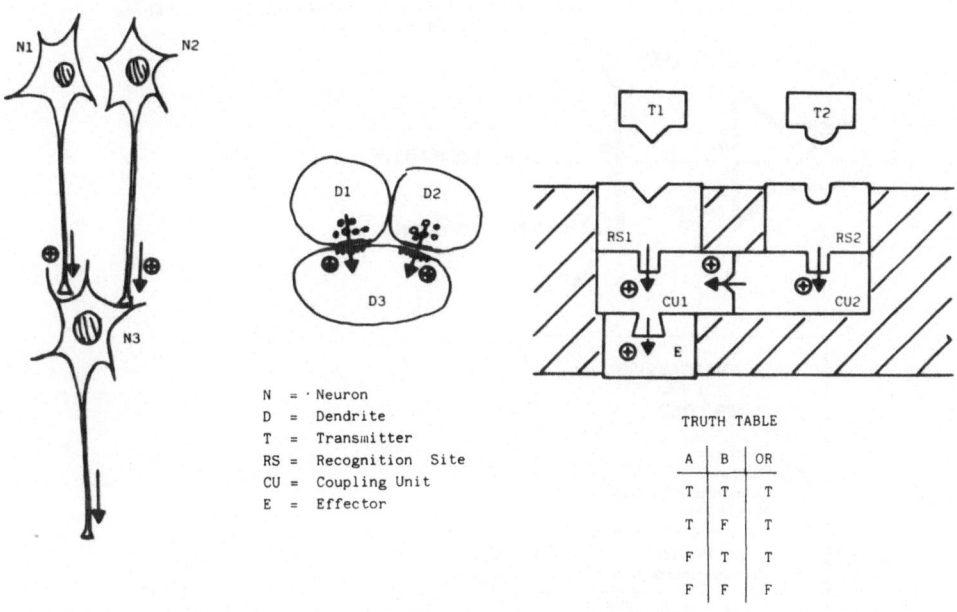

NETWORK LEVEL   LOCAL CIRCUIT LEVEL   MEMBRANE LEVEL

N = · Neuron
D = Dendrite
T = Transmitter
RS = Recognition Site
CU = Coupling Unit
E = Effector

TRUTH TABLE

A	B	OR
T	T	T
T	F	T
F	T	T
F	F	F

Fig. 9. Schematic representation of the possible logic model underlying the receptor-receptor interaction as exemplified for NPY receptors and α-2 receptors. It should be observed that this mechanism in the present case works as an "OR circuit," since the hypotensive action can be obtained after NPY or after clonidine but NPY cannot cause a clearcut increase of the hypotensive action of clonidine. Furthermore, NPY causes hypotension also after α-2 receptor blockade with R781094 (Harfstrand et al. 1984). Thus, as suggested previously (Agnati et al. 1982), by means of receptor-receptor interactions a miniaturization can be achieved (note the successive steps in miniaturization from the left to the right of neural circuits capable of an "OR" operation). The truth tables gives the logical value for the output (T= true= active; F= false= inactive) when the inputs (A,B) take different logical values.

This heterostatic mechanism appears to be energetically very convenient, since it can take place in isolated synaptic membranes. This means that the intracellular biochemical machinery is not directly

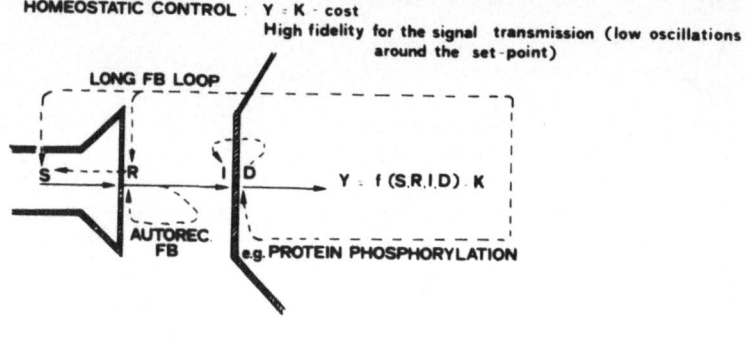

HOMEOSTATIC CONTROL : Y = K · cost

High fidelity for the signal transmission (low oscillations around the set-point)

LONG FB LOOP

$Y = f(S,R,I,D) \cdot K$

AUTOREC. FB

e.g. PROTEIN PHOSPHORYLATION

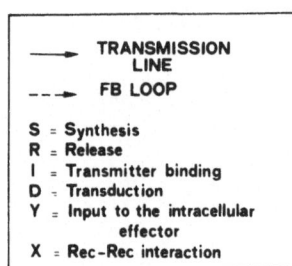

	TRANSMISSION LINE
	FB LOOP

S = Synthesis
R = Release
I = Transmitter binding
D = Transduction
Y = Input to the intracellular effector
X = Rec-Rec interaction

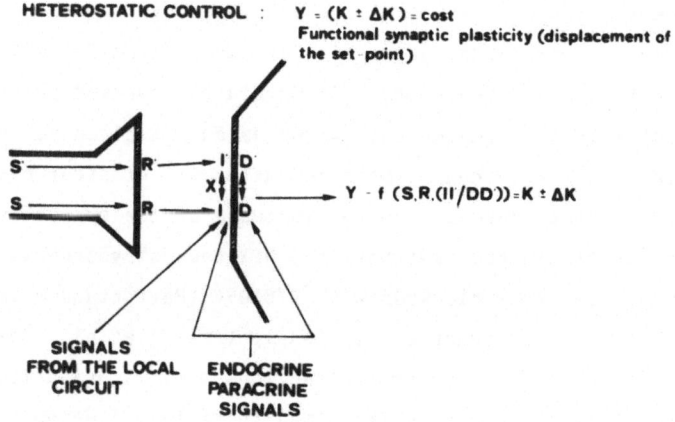

HETEROSTATIC CONTROL : $Y = (K \pm \Delta K) =$ cost

Functional synaptic plasticity (displacement of the set-point)

$Y = f(S,R,(II/DD)) = K \pm \Delta K$

SIGNALS FROM THE LOCAL CIRCUIT

ENDOCRINE PARACRINE SIGNALS

Fig. 10. Schematic representation of the mechanisms which may contribute to the two types of basic behaviors of the synapses: the constancy of the efficacy of the transmission line (synaptic homeostasis) and the change of the level at which this constancy is maintained (synaptic heterostasis).

involved. Obviously, other types of plastic synaptic adjustments are possible, some of these which certainly involve changes in protein phosphorylation. However, the receptor-receptor interaction mechanism being located at membrane level can integrate and filter signals before these act as inputs to the intracellular effectors.

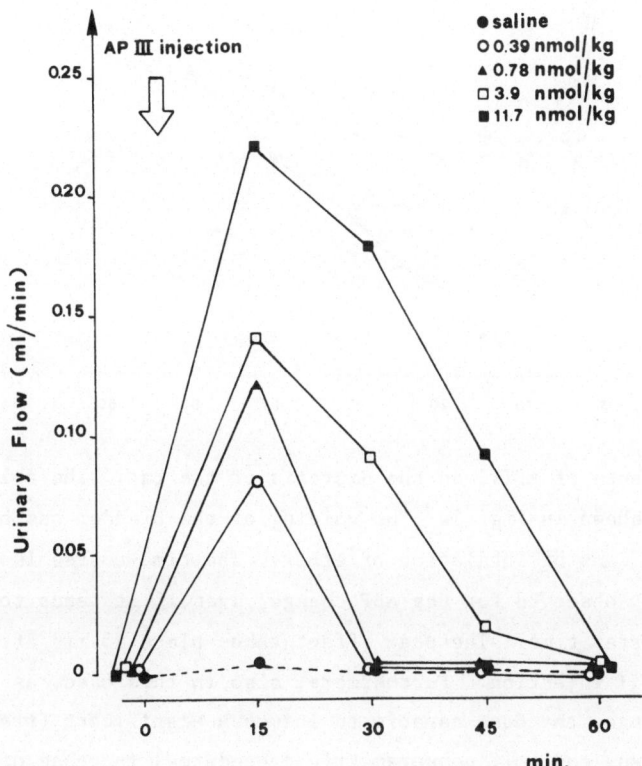

Fig. 11. Effects of intravenous injections of APIII on ABP. For the
animal model, see Fig. 2. The time-course of the hypotensive
effects of the peptide is given. The threshold dose for a
significant hypotension (treatment versus control, non
parametric procedures) is between 0.39 and 0.78 nmol/kg. Thus,
the hypotensive effects takes place within the first 10 min and
lasts for about 30 min. All percent changes larger than 10%
were found significant.

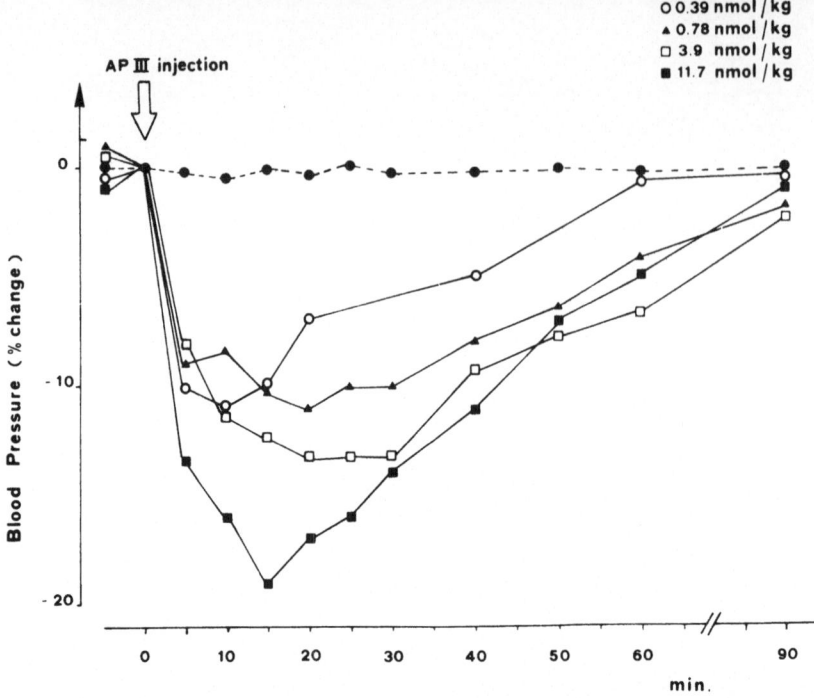

Fig. 12. Effects of APIII on the diuresis of the rat. The animal model
is shown in Fig. 2. The voiding of the bladder has been induced
by means of inhalation of ether. The time-course is similar to
that observed for the ABP change, even if it tends to last for a
shorter time. The peak effect takes place 15 min after the
APIII injection. Furthermore, also in this case, as for the ABP
change, the dose capable to induce a significant (treatments
versus control, nonparametric procedures) increase of diuresis
is between 0.39 and 0.78 nmol/kg. All values larger than 0.10
ml/min were found significant.

From a general standpoint we also suggest that any kind of functional
synaptic plasticity should be based on an heterostatic regulation of the
synapse.

EVIDENCE FOR A ROLE OF ATRIAL PEPTIDES IN THE CONTROL OF CARDIOVASCULAR
PARAMETERS

It has been shown that mammalian atria contain peptides that have
depressor vasodilator and natriuretic activities (Tange et al. 1984).
Depressor, diuretic and natriuretic effects of atriopeptin III (APIII) can

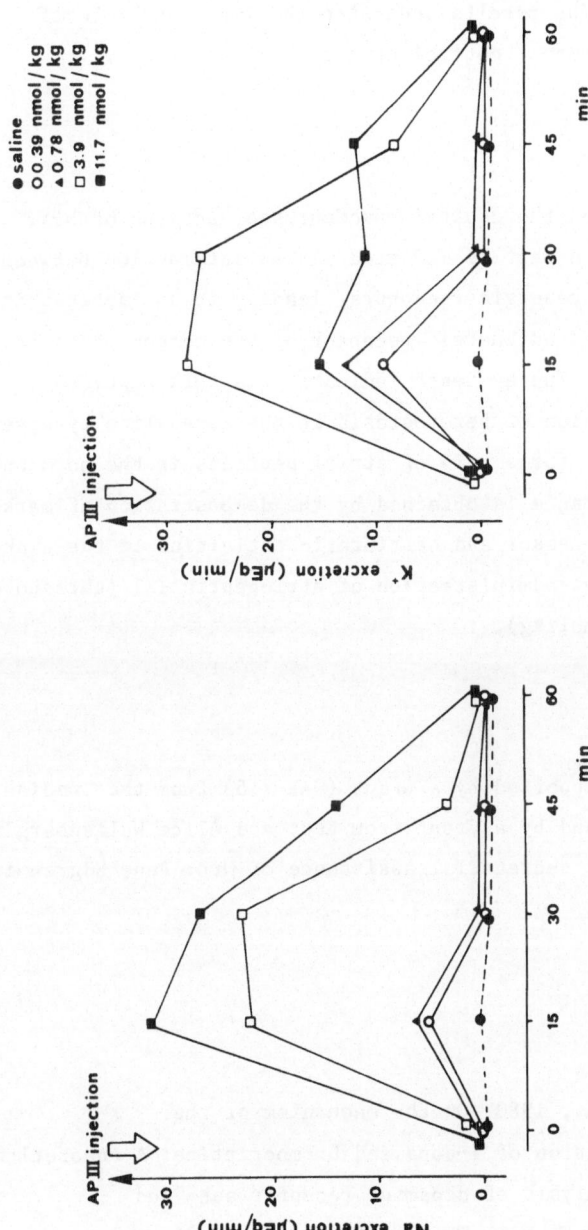

Fig. 13. Effects of APIII on Na$^+$ and K$^+$ excretion. For the animal model used see Fig. 2. The time-courses of Na$^+$ and K$^+$ excretions roughly resemble those found for urine excretion. However, it should be noted that for K$^+$ excretion there is no dose-response relationship and furthermore only the two higher doses produced a significant (treatments versus control, nonparametric procedures) increase in Na$^+$ and K$^+$ excretion.

occur in awake freely moving male rats (Fig. 11-13). APIII dose-dependently reduces with a high potency arterial blood pressure and increases urinary outflow and Na^+ and K^+ excretion in the awake unrestrained male rat. Instead, APIII in the same dose range lacks actions on cardiovascular parameters when given i.c. in the α-chloralose anaesthetized male rat. The results underline the important role of atrial peptides in the hormonal control of vascular resistance.

SUMMARY

This review focuses on the central vasodepressor actions of NPY. This action may at least in part be mediated via an interaction between the central NPY and α-2 adrenergic receptors, leading to an increase in the transduction of the α-2 adrenergic receptor of the adrenaline synapses in the medulla oblongata. The evidence indicates that NPY peptides participate in the regulation of heterostasis in the adrenaline synapses. Moreover, further evidence for a role of atrial peptides in the hormonal control of vascular resistance is obtained by the demonstration of marked and dose-dependent vasodepressor and natriuretic activities in the awake anaesthetized rat upon i.v. administration of atriopeptin III (threshold dose between 0.4 to 0.8 nmol/kg).

ACKNOWLEDGMENTS

This work has been supported by a Grant (04X-715) from the Swedish Medical Research Council and by a Grant from Knut and Alice Wallenberg's Foundation. The excellent secretarial assistance of Mrs. Anne Edgren is gratefully acknowledged.

REFERENCES

Agnati, L. F. and, Fuxe, K., 1980, On the mechanism of the anti-parkinsonian action of l-dopa and bromocriptine: A theoretical and experimental analysis of dopamine receptor sub- and supersensitivity, J. Neural. Transm. Suppl. 16:69-81.

Agnati, L. F., Fuxe, K., Zoli, M., Rondanini, C., and Ogren, S.O., 1982, New vistas on synaptic plasticity: mosaic hypothesis on the engram, Med. Biol. 60:183-190.

Agnati, L. F., and Fuxe, K., 1983, Subcortical limbic ^{3}H-N-propylnorapomorphine binding sites are markedly modulated by cholecystokinin-8 in vitro, Biosci. Reports 3:1101-1105.

Agnati, L. F., Fuxe, K., Benfenati, F., Battistini, N., Harfstrand, A., Tatemoto, K., Hokfelt, T., and Mutt, V., 1983a, Neuropeptide Y in vitro selectively increases the number of α-2 adrenergic binding sites in membranes of the medulla oblongata of the rat, Acta Physiol. Scand. 118:293-295.

Agnati, L. F., Fuxe, K., Benfenati, F., Battistini, N., Harfstrand, A., Hokfelt, T., Cavicchioli, L., Tatemoto, K., and Mutt, V., 1983b, Failure of neuropeptide Y in vitro to increase the number of α-2 adrenergic binding sites in membranes of medulla oblongata of the spontaneous hypertensive rat, Acta Physiol. Scand. 119:309-312.

Agnati, L. F., Fuxe, K., Calza, L., Giardino, L., Zini, I., Toffano, G., Goldstein, M., Marrama, P., Gustafsson, J.-A., Yu Z.-Y., Cuello, A. C., Terenius, L., Lang, R., and Ganten, D., 1985, Morphometrical and microdensitometrical studies on monoaminergic and peptidergic neurons in the aging brain, in: "Quantitative Neuroanatomy in Transmitter Research," K. Fuxe and L. F. Agnati, eds., pp. 91-112, MacMillan Press, England, in press.

Bolme, P., Fuxe, K., Agnati, L. F., Bradley, R., and Smythies, L., 1978, Cardiovascular effects of morphine and opioid peptides following intracisternal administration in chloralose-anaesthetized rats, Eur. J. Pharmacol. 48:319-324.

Edvinsson, L., Emson, P., McCulloch, J., Tatemoto, K., and Uddman, R., 1983, Neuropeptide Y: cerebrovascular innervation and vasomotor effects in the cat, Neurosci. Lett. 43:79-84.

Ekblad, E., Edvinsson, L., Wahlestedt, C., Uddman, R., Hakanson, R., and Sundler, F., 1984, Neuropeptide Y co-exists and co-operates with noradrenaline in perivascular nerve fibers, Regul. Pept. 8:225.

Everitt, B. J., Hokfelt, T., Terenius, L., Tatemoto, K., Mutt, V., and Goldstein, M., 1984, Differential co-existence of neuropeptide Y (NPY)-like immunoreactivity with catecholamines in the central nervous system of the rat, Neurosci 11:443-462.

Fuxe, K., and Lidbrink, P., 1973, Biogenic amine aspects of sleep and waking, in: "Sleep: Physiology, Biochemistry, Psychology, Pharmacology, Clinical Implications," W. P. Koella, and P. Levin, eds., pp. 12-26, Karger, Basel.

Fuxe, K., Lidbrink, P., Hokfelt, T., Bolme, P., and Goldstein, M., 1974, Effects of piperoxane on sleep and waking in the rat. Evidence for increased waking by blocking inhibitory adrenaline receptors on the locus coeruleus, Acta Physiol. Scand. 91:566-567.

Fuxe, K., Agnati, L. F., Ganten, D., Goldstein, M., Yukimura, T., Jonsson

G., Bolme, P., Hokfelt, T., Andersson, K., Harfstrand, A., Unger, T., and Rascher, W., 1981a, The role of noradrenaline and adrenaline neuron systems and substance P in the control of central cardiovascular functions, in: "Central Nervous System Mechanisms in Hypertension," J. P. Buckley, and C. M. Ferrario, eds., pp. 89-113, Raven Press, New York.

Fuxe, K., Agnati, L. F., Rosell, S., Harfstrand, A., Lundberg, J., Hokfelt, T., and Bernardi, P., 1981b, Vasopressor effects of substance P and its C-terminal fragments following intracisternal injection to α-chloralose anaesthetized rats: Blockade by a substance P antagonist, Eur. J. Pharmacol. 15:171-176.

Fuxe, K., Agnati, L. F., Kohler, C., Kuonen, D., Ogrens, S.-O., Andersson, K., and Hokfelt, T., 1981c, Characterization of normal and supersensitive dopamine receptors: Effects of ergot drugs and neuropeptides, J. Neural Transm. 51:3-37.

Fuxe, K., Agnati, L. F., Ganten, D., Andersson, K., Calza, L., Vincent, M., Sassard, J., Yukimura, T., Eneroth, P., Goldstein, M., Hokfelt, T., Rosell, S., Harfstrand, A., Vale, W., Brown, M., and Rivier, J., 1982, Part 1: Catecholamines and renin angiotensin-system. Central and peripheral hormones and peptides: Focus on the involvement of noradrenaline and adrenaline neurons, opioid peptides, substance P and somatostatin in central cardiovascular regulation of the rat, in: "Hypertensive Mechanisms: The Spontaneously Hypertensive Rat as a Model to Study Human Hypertension," W. Rascher, D. Clough, and D. Ganten, eds., pp. 417-441, Schattauer Verlag, Stuttgart.

Fuxe, K., Agnati, L. F., Harfstrand, A., Zini, I., Tatemoto, K., Merlo Pich, E., Hokfelt, T., Mutt, V., and Terenius, L., 1983, Central administration of neuropeptide Y induced hypotension, bradypnea and EEG synchronization in the rat, Acta Physiol. Scand. 118:189-192.

Fuxe, K., Agnati, L. F., Harfstrand, A., Zoli, M., Benfenati, F., Toffano, G., and Merlo Pich, E., 1984a, Evidence for functional synaptic plasticity in the central nervous system and its alteration during aging and in physiopathological conditions, in: "Clinical Neuropharmacology," G. Racagni, R. Paoletti, and P. Kielholz, eds., Vol. 7, Suppl. 1, pp. 590-591, Raven Press, N.Y.

Fuxe, K., Agnati, L. F., Harfstrand, A., Martire, M., Goldstein, M., Grimaldi, R., Bernardi, P., Zini, I., Tatemoto, K., and Mutt, V., 1984b, Evidence for a modulation by neuropeptide Y of the α-2 adrenergic transmission line in central adrenaline synapses. New possibilities for treatment of hypertensive disorders, Clin. Exp. Hypertension - Theory and Practice, A6 (10 & 11):1951-1956.

Fuxe, K.,Agnati, L. F., Harfstrand, A., Andersson, K., Mascagni, F., Zoli, M., Kalia, M., Battistini, N., Benfenati, F., Hokfelt, T., and Goldstein, M., 1985a, Studies on the peptide comodulator transmission line in central monoamine synapses. New openings for the treatment of disorders of the central nervous system, Progress in Brain Res., in press.

Fuxe, K., Agnati, L. F., Zoli, M., Harfstrand, A., Grimaldi, R., Bernardi, P., Camurri, M., Tucci, F., and Goldstein, M., 1985b, Development of quantitative methods for the evaluation of the entity of coexistence of neuroactive substances in nerve terminal populations in discrete areas of the central nervous system, in: "Quantitative Neuroanatomy in Transmitter Research," K. Fuxe, and L. F. Agnati, eds., pp. 157-174, MacMillan Press, England.

Ganten, D., Unger, T., Simon, W., Schaz, K., Scholkens, B., Mann, J., Speck, G., Lang, R., and Rascher, W., 1981, Central peptidergic stimulation: Focus on cardiovascular actions of angiotensin and opioid peptides, in: "Central Nervous System Mechanisms in Hypertension," J. P. Buckley and C. M. Ferrario, eds., pp. 265-282, Raven Press, New York.

Harfstrand, A., Fuxe, K., Agnati, L. F., Ganten, D., Eneroth, P., Tatemoto, K., and Mutt, V., 1984, Studies on neuropeptide Y catecholamine interactions in central cardiovascular regulation in the α-chloralose anesthetized rat. Evidence for a possible new way of activating the α-2 adrenergic transmission line, Clin. Exp. Hypertension - Theory and Practice, A6 (10 & 11):1947-1950.

Hokfelt, T, Goldstein, M., Fuxe, K., Johansson, O., Verhofstad, A., Steinbusch, H., Penke, B., and Vargas, J., 1980, Histochemical identification of adrenaline containing cells with special reference to neurons, in: "Central Adrenaline Neurons: Basic Aspects and their Role in Cardiovascular Functions," K. Fuxe, M. Goldstein, B. Hokfelt, and T. Hokfelt eds., pp. 19-47, Pergamon Press, Oxford.

Hokfelt, T., Lundberg, J. M., Tatemoto, K., Mutt, V., Terenius, L., Polak, J., Elde, R., and Goldstein, M., 1983, Neuropeptide Y (NPY)-and FMRF amide neuropeptide-like immunoreactivities in catecholamine neurons of the medulla oblongata, Acta Physiol. Scand. 117:315-318.

Hokfelt, T., Everitt, B. J., Fuxe, K., Kalia, M., Agnati, L. F., Johansson, O., Theordorsson-Norheim, E., and Goldstein, M., 1984, Transmitter and peptide systems in areas involved in the control of blood pressure, Clin. Exp. Hyper.-Theory and Practice A6 (1&2):23-41.

Kalia, M., Fuxe, K., and Goldstein, M., 1985a, Rat medulla oblongata. II.

Dopaminergic, noradrenergic (A1 and A2) and adrenergic neurons, nerve fibers, and presumptive terminal processes, J. Comp. Neurology 233:308-332.

Kalia, M., Fuxe, K., and Goldstein, M., 1985b, Rat medulla oblongata. III. adrenergic (C1 and C2) neurons, nerve fibers and presumptive terminal processes, J. Comp. Neurology 233:333-349.

Lundberg, J. M., and Tatemoto, K., 1982, Pancreatic polypeptide family (APP, BPP, NPY and PYY) in relation to sympathetic vasoconstriction resistant to α-adrenoceptor blockade, Acta Physiol. Scand. 116:393-402.

Mefford, I. N., 1981, Application of high performance liquid chromatography with electrochemical detection to neurochemical analysis: measurement of catecholamines, serotonin and metabolites in rat brain, J. Neurosci. Methods 3:207-224.

Ohhashi, T., and Jacobowitz, D. M., 1983, The effects of pancreatic polypeptides and neuropeptide Y on the rat vas deferens, Peptides 4:381-386.

Reis, D., Granata, A., Joh, T., Ross, C., Ruggiero, D., and Park, D., 1984, Brain stem catecholamine mechanisms in tonic and reflex control of blood pressure, in: "Hypertension. Central α-Adrenergic Blood Pressure Regulating Mechanisms," W. Abrams, and C. Sweet, eds., part II, Vol. 6, 5:7-15, American Heart Ass., Dallas.

Tang, J., Webber, R. J., Chang, D., Chang, J. K., Kiang, J., and Wei E. T., 1984, Depressor and natriuretic activities of several atrial peptides, Regulatory Peptides 9:53-59.

Tatemoto, K., Carlquist, M., and Mutt, V., 1982, Neuropeptide Y-a novel brain peptide with structural similarities to peptide YY and pancreatic polypeptide, Nature 296:659-660.

Zini, I, Merlo Pich, E., Fuxe, K., Lenzi, P. L., Agnati, L. F., Harfstrand, A., Mutt, V., Tatemoto, K., and Moscara, M., 1984, Actions of centrally administered neuropeptide Y on EEG activity in different rat strains and in different phases of their circadian cycle, Acta Physiol. Scand. 122:71-77.

NEUROCHEMICAL BASIS FOR THE VASOACTIVE ACTION OF PROLACTIN

Filippo Drago and Umberto Scapagnini

Institute of Pharmacology
University of Catania Medical School
Viale A. Doria 6
95125 Catania (Italy)

INTRODUCTION

The ancient hormone prolactin (PRL) has undergone a modern
renaissance with the discovery of its broad spectrum of action. There is
evidence that all adenohypophyseal hormones, except PRL, were committed
early in vertebrate philogenesis to the control of a single or at most a
few physiological processes (Nicoll, 1974). On the contrary, PRL did not
undergo this fate of early specialization but was bound to the control of
a wide variety of different physiological mechanisms. In fact, Nicoll and
Bern (1972) have listed 85 distinct effects of the hormone. These are far
in excess of the reported actions of all other adenohypophyseal hormones
together. The number to-date of the recognized effects of PRL is even
higher due to the recent discovery of specific actions of the hormone on
organs other than its classical target tissues. These recently discovered
actions are both central and peripheral. Central actions are those on the
brain, i.e. neurochemical and behavioral actions (Drago, 1982; Drago et
al., 1984). Among the peripheral actions of PRL that have newly been
recognized, the most interesting is probably that concerning the
cardiovascular system. In this respect, PRL may be a factor
physiologically involved in the regulation of heart function and blood
pressure.

This chapter will review the major findings in this field, focusing
particularly on the possible neurochemical mechanisms underlying the
vasoactive action of PRL.

PRL is evidently involved in water and electrolyte balance in virtually all classes of vertebrates (Nicoll, 1974). Research in recent years indicates that this hormone is of prime importance in regulating osmolar balance through the gill and the kidney in certain teleosts (for a review, see: Ball, 1969). In nonmammalian tetrapods, such as amphibians, reptiles and birds, PRL also plays an important role in osmolar balance (Nicoll, 1974).

PRL has been reported to exert antidiuretic effects similar to those of vasopressin in different mammalian species (Horrobin et al., 1971; Labella et al., 1975). The hormone may be involved in the physiological regulation of osmolar balance also in humans, and the kidney may be considered as an important target organ for PRL (Buckman and Peake, 1973). Indeed, the existence of specific receptors for PRL in the rat kidney (Turkington, 1972) and the localization of the hormone in this organ by immunohistochemical studies provide further evidence of a possible action of PRL on the kidney (Dube et al., 1978).

The intimate mechanism of the osmoregulatory action of PRL may be the promotion of sodium retention (Nicoll, 1974). If this is the case, blood PRL levels would be inversely related to plasma osmolality. However, a direct relationship between plasma osmolality and PRL is also possible, suggesting that the hormone may be more intimately associated with water retention than it is with conserving sodium.

In spite of the great number of findings concerning the physiological action of PRL on osmolar balance under normal conditions, there is no clear evidence that PRL may exert antidiuretic effects in pathological changes of urinary output, such as diabetes insipidus. Rats with hereditary diabetes insipidus of the Brattleboro strain represent a valid model for such a condition. The administration of exogenous rat PRL reduces urine excretion and water intake of these rats (Sartani et al., 1978). However, other studies have shown that endogenous PRL has no intrinsic antidiuretic activity in rats with hereditary diabetes insipidus made hyperprolactinaemic by pituitary homografts under the kidney capsule (Adler et al., 1978). It has been suggested that the antidiuretic activity associated with different PRL preparations could be entirely due to their contamination with vasopressin (North et al., 1979). This is in

agreement with recent experiments, where the effects of hyperprolactinaemia on the water intake and urine output were studied in Brattleboro
rats (Drago and Bohus, in preparation). The mean water intake and urine
output of homozygous (HOSH), heterozygous (HESH) and normal (NOSH)
sham-operated rats, and those of homozygous (HOHO), heterozygous (HEHO)
and normal (NOHO) homografted animals are shown in Table 1. Both water
intake and urine output of HOSH rats appeared to be higher than those of
the other sham-operated groups. Hyperprolactinaemia induced by pituitary
homografts did not change the water intake and urine output in HOHO, HEHO
and NOHO rats, as compared to those of sham-operated groups.

TABLE 1

Water Intake and Urinary Output in Homozygous, Heterozygous and Normal
Rats of Brattleboro Rats with Pituitary Homografts under the Kidney
Capsule (HOHO, HEHO, and NOHO, respectively) and Sham-Operated (HOSH,
HESH, and NOSH, respectively).

GROUPS	WATER INTAKE*	URINARY OUTPUT
HOSH	74.5+4.1**	71.3+4.5**
HOHO	70.8+3.6**	70.8+1.4**
HESH	19.6+2.1	17.3+1.9
HEHO	17.4+1.5	16.5+0.9
NOSH	18.6+2.3	16.2+1.7
NOHO	20.3+1.6	18.4+2.0

*Values are mean + SEM of ml/100 g body weight of 3 days observations.
**Significantly different as compared to NOSH or NOHO rats (p<0.05,
Dunnett's test or multiple comparisons).

Although the evidence to-date is equivocal whether PRL regulates
aldosterone secretion directly, there are consistent findings concerning
an influence of PRL on aldosterone function. Human placental prolactin is
reported to increase urinary excretion of a metabolite of aldosterone
(tetrahydroaldosterone) in humans (Melby et al., 1966). Furthermore, it

is clear that PRL and ACTH act synergistically causing an increase in adrenocortical steroidogenesis under various conditions (Nicoll, 1974; McMurtry and Wexler, 1981b; Drago and Continella, in press).

EFFECTS OF PROLACTIN ON BLOOD PRESSURE

In light of the findings on the effects of PRL on osmolar balance, a number of studies have been carried out on the possible role of this hormone in the regulation of blood pressure. First, Bryant et al. (1973) reported that PRL injections in female rats increase blood volume while decreasing blood pressure and the hypertensive action of angiotensin. In contrast, Horrobin et al. (1973) claimed that infusion of ovine PRL into decerebrate rabbits increases arterial blood pressure. Other data indicate the PRL effects on blood pressure may be rather complex (Nicoll, 1974).

Interest in the possible role of PRL in the pathogenesis of hypertension was heightened when Stumpe et al. (1977) found increased plasma PRL levels in patients with essential hypertension. These authors suggested that the increased production of this hormone in hypertensive patients may be due to a defective control of PRL secretion by hypothalamic dopamine, which is considered as the most important PRL-inhibiting factor (PIF). Other authors found that the spontaneously hypertensive rats (SHR), a close animal model of essential hypertension in humans, is hyperprolactinaemic and suggested that a similar defect in central dopaminergic control of PRL release may be operative in SHR (Sowers et al., 1979). Although SHR have not been found to be hyperprolactinaemic by McMurtry et al. (1980), these authors have shown that these animals secrete extra quantities of PRL in response to stress (McMurtry and Wexler, 1981a) and that their high blood pressure may be reduced effectively by the administration of a dopaminergic drug, e.g. bromocriptine, that lowers plasma PRL levels (McMurtry et al., 1979). Interestingly, genetically normotensive Sprague-Dawley pups nursed by SHR dams can develop high blood pressure, suggesting the transmission of some hypertensinogen factor through the mother's milk (McMurtry et al., 1981). This factor can be PRL, as this hormone has been detected in high concentrations in the milk.

In a recent experiment, a group of normotensive male rats of Wistar strain was made hyperprolactinaemic by pituitary homografts under the kidney capsule and their blood pressure was checked 10 days after the beginning of hyperprolactinaemia. Plasma PRL and corticosterone levels were measured at the same times by radioimmunoassay. Table 2 shows that hyperprolactinaemia (plasma PRL levels are not reported) was accompanied by an increase in corticosterone levels, but no change was found as far as the systolic blood pressure was concerned. These results agree with those found by McMurtry and Wexler (1981b) and suggest that the failure of hyperprolactinaemia to develop hypertension in normotensive Wistar rats may depend on the absence in this strain of a genetically mediated hypertensinogen factor which is activated by hyperprolactinaemia.

TABLE 2

Effects of Endogenous Hyperprolactinaemia (HPRL) as Induced by Pituitary Homografts Under the Kidney Capsule on the Systolic Blood Pressure and Corticosterone Plasma Levels of Wistar Male Rats.

GROUPS	SYSTOLIC BLOOD PRESSURE (mmHg)*	PLASMA CORTICOSTERONE LEVELS (ug/100 ml)
CONTROLS	96.0+7.8	8.6+0.5
HPRL	94.5+7.5	19.5+1.7**

*Values are mean + SEM

**Significantly different as compared to the control group (p<0.05, Student's t-test, two tailed).

EFFECTS OF PROLACTIN ON BLOOD VESSELS

Several investigations have been carried out on the role of PRL in the sensitivity of smooth muscles to different agonists. First, Manku et al. (1973) found that low levels of PRL may potentiate the vasoconstrictor action of norepinephrine and angiotensin. This effect has been confirmed in studies assessing the responses of the isolated mesenteric artery of the rat to norepinephrine (Muriuki et al., 1974). High concentrations of PRL appear to exert a transient potentiating effect followed by an inhibition of the action of norepinephrine and angiotensin (Manku et al., 1973). Pressor response to potassium which depend on extracellular calcium entry into the muscles remains unaffected by PRL at any concentration (Manku et al., 1979). Various vascular districts seem to be sensitive to PRL, as this effect has been described for the superior mesenteric vascular bed (Muriuki et al., 1974; Mtabaji et al., 1976), for the aortic smooth muscle preparation and for arterioles (Manku et al., 1973). Similar biphasic effects of PRL have also been shown on the contractile responses to cardiac muscle preparations (Nassar et al., 1974; Horrobin et al., 1974a).

A direct effect of PRL on smooth muscles has been widely demonstrated. This hormone can induce contractions of the guinea pig myometrium (Lipton et al., 1978) and ileum (Pillai et al., 1981), and of the gastro-intestinal tract in mice (Gopalakrishnan et al., 1981). Furthermore, PRL has no contractile effect in frog rectus abdominis muscle preparations indicating that only the smooth and not the skeletal muscles are sensitive to the contractile action of the hormone (Pillai et al., 1981).

NEUROCHEMICAL MECHANISMS OF THE VASOACTIVE ACTION OF PROLACTIN

It has been reported that the effects of PRL on vascular smooth muscles are generally accompanied by an increase in the output of prostaglandins (Horrobin et al., 1974b; Mtabaji et al., 1976). Inhibitors of prostaglandin synthesis, such as indomethacin and aspirin, block both the outflow of prostaglandins and the effects of PRL (Horrobin et al., 1978). Indeed, PRL can enhance the synthesis and release of prostaglandins (Manku et al., 1979), which are known to contract various smooth muscles (Bennett et al., 1975). Interestingly, a similar mechanism has been shown for PRL cytoprotective effect on stress-induced gastric ulcers (Drago et al., 1985). These ulcers can be prevented in hyperprolactinaemic rats or in animals injected intracerebroventricularly with PRL, and the administration of indomethacin blocks the cytoprotective effect of the hormone. Thus, it is possible that the vasoactive action of PRL may depend on an increased synthesis and release of prostaglandins.

Recently, it has been shown that PRL effects on vascular smooth muscles can also be blocked by cortisol or lithium (Manku et al., 1979). Di-homo-γ-linoleic acid (DHGL) and prostaglandin El had effects similar to those of PRL, in that they potentiated norepinephrine responses at low concentrations, inhibited at high ones, and had no effect on potassium responses. Arachidonic acid and prostaglandin E2 potentiated both norepinephrine and potassium responses. Furthermore, zinc had actions similar to those of PRL and DHGL, but which could be blocked only by lithium and not by cortisol (Manku et al., 1979). These data strongly support the concept that the vasoactive action of PRL is mediated by prostaglandins, probably of El type. In particular, the mobilization of DHGL by PRL seems to be an important mechanism in this respect.

A direct influence of PRL on vascular receptors is also possible. Interestingly, the administration of a specific prostaglandin blocker, SC-19220, failed to inhibit the contractile responses of guinea pig isolated ileum to PRL, suggesting that prostaglandins may not be involved in this action of PRL (Pillai et al., 1981). In contrast, the contractions induced by PRL were effectively inhibited by atropine and potentiated by neostigmine. This suggests the involvement of muscarinic receptors in the PRL-induced involvement of muscarinic receptors in the PRL-induced contractions of guinea pig ileum. Consistently, atropine abolished and physostigmine potentiated the PRL-enhanced gastro-intestinal propulsion in mice (Gopalakrishnan et al., 1981). Furthermore, this effect of PRL was inhibited by morphine (Pillai et al., 1981) or bromocriptine (Gopalakrishnan et al., 1981). However, apart from the difference in the mechanisms controlling the contraction of vascular and intestinal smooth muscles, it is of note that cholinergic transmission is far from relevant for the motility of blood vessels. Thus, it is possible that the involvement of cholinergic mechanisms in the PRL-induced intestinal contractions is not the same for blood vessels.

In a recent study, Katovich (1983) has demonstrated that chronic hyperprolactinaemia in male rats reduces peripheral β-adrenergic responsiveness, as assessed by three different parameters, i.e. isoproterenol-induced tachycardia, thirst and hyperthermia. However, this author failed to study the effect of PRL on isoproterenol-induced vasodilation and to give any explanation of this phenomenon. If this effect of PRL will be confirmed on β-adrenergic responsiveness of blood vessels, it would represent another possible mechanism of the vasoactive action of this hormone.

REFERENCES

Adler, R. A., Dolphin, S., and Sokol, H. W., 1978, Does endogenous prolactin have antidiuretic activity?, Endroc. Soc. Meet. abs. 552.

Ball, J. N., 1969, Prolactin and osmoregulation in teleost fishes: a review, Gen. Comp. Endocrinol. Suppl. 2:10-25.

Bennett, A., Eley, K. G., and Stockley, H. L., 1975, The effects of prostaglandins on guinea pig isolated intestine and their possible contribution to muscle activity and tone, Br. J. Pharmacol. 54:197-206.

Bryant, E. E., Douglas, B. H., and Ashburn, A. D., 1973, Circulatory changes following prolactin administration, Am. J. Obstet. Gynecol. 115:53-57.

Buckman, M. T. and Peake, G. T., 1973, Osmolar control of prolactin secretion in man, Science 181:755-757.

Drago, F., 1982, Prolactin and Behavior, Ph.D. Thesis, University of Utrecht Medical School, Utrecht.

Drago, F. and Continella, G., 1985, Somatic growth in hypophysectomized pituitary-homografted rats is promoted by prolactin, Experientia, in press.

Drago, F., Pennisi, G., and Scapagnini, U., 1984, Central mechanisms involved in hyperprolactinaemia-induced behavioral changes, in: "Pituitary Hyperfunction: Physiopathology and Clinical Aspects," F. Camanni and E. E. Muller, eds, pp 315-319, Raven Press, New York.

Drago, F., Continella, G., Conforto, G., and Scapagnini, U., 1985, Prolactin inhibits the development of stress-induced ulcers in the rat, Life Sci. 36:191-197.

Dube, D., Desy, L., Kelly, P. A., and Pelletier, G., 1978, Immunohistochemical localization of prolactin in the rat kidney, Endocr. Soc. Meet. abs. 749.

Gopalakrishnan, V., Ramaswamy, S., Pillai, N. P., and Ghosh, M. N., 1981, Efect of prolactin and bromocriptine on intestinal transit in mice, European J. Pharmacol. 74:369-372.

Horrobin, D. F., Burstyn, P. G., Lloyd, I. J., Durkin, N., Lipton, A., and Muriuki, K. L., 1971, Actions of prolactin on human renal function, Lancet 2:352-354.

Horrobin, D. F., Manku, M. S., and Burstyn, P. G., 1973, Effect of intravenous prolactin infusion on arterial blood pressure in rabbits, Cardiovascular Res. 7:585-587.

Horrobin, D. F., Manku, M. S., Nassar, B. A., and Davies, P. A., 1974a,

Aspirin, indomethacin, catecholamines and prostaglandin interactions on rat arterioles and rabbit hearts, Nature 250:425-426.

Horrobin, D. F., Manku, M. S., Karmali, R. A., and Greaves, M. N., 1974b, Prolactin and prostaglandin synthesis, Lancet 2:1154-1155.

Horrobin, D. F., Manku, M. S., Karmali, R. A., Ally, A. I., Karmazyn, N., Morgan, R. O., and Swift, R. A., 1978, Prostaglandins as second messengers of prolactin action, in: "Progress in Prolactin Physiology and Pathology, C. Robyn and M. Harter, eds, pp 189-198, Elsevier/North Holland, Amsterdam.

Katovich, M. J., 1983, Chronic hyperprolactinaemia reduces peripheral β-adrenergic responsiveness in male rats, Life Sci. 32:1213-1221.

Labella, F., Dular, R., Queen, G., and Vivian, S., 1975, Anterior pituitary hormone release in vitro inversely related to extracellular osmolarity, Endocrinology 96:1559-1567.

Lipton, A., Leah, J., and Parkington, H.C., 1978, Effects of ovarian steroids and prolactin on the contractility and sodium pump site density of guinea pig myometrium, J. Endocrinol. 77:43-49.

Manku, M. S., Nassar, B. A., and Horrobin, D. F., 1973, Effects of prolactin on the responses of rat aortic and arteriolar smooth muscle preparation to noradrenaline and angiotensin, Lancet 2:991-992.

Manku, M. S., Horrobin, D. F., Karmazyn, M., and Cunnane, S. C., 1979, Prolactin and zinc effects on rat vascular reactivity: possible relationship to dihomo-γ-linoleic acid and to protaglandin biosynthesis, Endocrinology 104:774-779.

McMurtry, J. P., and Wexler, B. C., 1981a, Hypersensitivity of spontaneously hypertensive rats (SHR) to heat, ether, and immobilization, Endocrinology 108:1730-1736.

McMurtry, J. P., and Wexler, B. C., 1981b, Hyperprolactinaemia and hyperadrenocorticism accompanied by normal blood pressure in Sprague-Dawley rats, Proc. Soc. Exp. Biol. Med. 168:114-118.

McMurty, J. P., Kazama, N., and Wexler, B. C., 1979, Effects of bromocryptine on hormone and blood pressure levels in the spontaneously hypertensive rat, Proc. Soc. Exp. Biol. Med. 161:186-191.

McMurtry, J. P., Kazama, N., and Wexler, B. C., 1980, Circadian rhythm of circulating and pituitary corticosterone and prolactin levels in spontaneously hypertensive rats, Life Sci. 26:2257-2263.

McMurty, J. P., Wright, G. L., and Wexler, B. C., 1981, Spontaneous hypertension in cross-suckled rats, Science 211:1173-1174.

Melby, J. C., Dale, S. L., Wilson, T. E., and Nichols, A. S., 1966,

Stimulation of aldosterone secretion by human placental lactogen, Clin. Res. 14:283.

Mtabaji, J. P., Manku, M. S., and Horrobin, D. F., 1976, Vascular actions of furosemide and bumetanide on the rat superior mesenteric vascular bed: interactions with prolactin and prostaglandins, Can. J. Physiol. Pharmacol. 54:357-362.

Muriuki, P. B., Mugambi, M., Thairu, K., Mathai, S., and Mati, J. K. C., 1974, Effects of prolactin on the responses of the isolated mesenteric artery of the rat to noradrenaline, J. Endorcinol. 63:249-256.

Nassar, B. A., Manku, M. S., Reed, J. D., Tynan, M., and Horrobin, D. F., 1974, Actions of prolactin and furosemide on heart rate and rhythm, Br. Med. J. 2:27-28.

Nicoll, C. S., 1974, Physiological actions of prolactin, in: "Handbook of Physiology," Section 7, Vol. 4, pp 253-292, American Physiological Society, Washington.

Nicoll, C. S., and Bern, H. A., 1972, On the actions of prolactin among vertebrates: is there a common denominator?, in: "Lactogenic Hormones," G. E. W. Wolstenholme and J. Knight, eds, pp 299-324, Churchill Livingstone, London.

North, W. G., Gellai, M., Sokol, H. W., and Adler, R., 1979, Rat prolactin has no intrinsic antidiuretic activity in the rat, Endocr. Soc. Meet. abs. 881.

Pillai, N. P., Ramaswamy, S., Gopalakrishnan, V., and Ghosh, M. N., 1981, Contractile effect of prolactin on guinea pig isolated ileum, European J. Phramacol. 72:11-16.

Sartani, A., Valiquette, G., and Motta, M., 1978, Antidiuretic effect of rat prolactin in Brattleboro rats, Endocr. Soc. Meet. abs. 547.

Sowers, J. R., Resch, G., Temple, G., Herzog, J., Colantino, M., 1979, Hyperprolactinaemia in the spontaneously hypertensive rat, Acta Endocrinol. 90:1-6.

Stumpe, K. O., Higuchi, M., Kolloch, R., Kruck, F., and Vetter, H., 1977, Increased prolactin production in patients with essential hypertension, Lancet 2:211.

Turkington, R. W., 1972, Molecular biological aspects of prolactin, in: "Lactogenic Hormones," G. E. W. Wolstenholme and J. Knight, eds, pp 111-136, Churchill Livingstone, London.

Vasoconstrictor
  effects of serotonin, 147
  nerve supply for vasomotor tone,
      465
Vasodepressor effect of
          neuropeptide Y, 503
Vasodilator
  antihypertensives, 210–212
    hydralazine, 211
    nifedipine, 211
    ninoxidil, 211
    nitrendipine, 211
    nitroprusside, 211
    verapamil, 211
  effects of serotonin, 147
Vasomotor centers, 467
Vasopressin, 444
  effects intrathecally, 453
  projections from PVN
          to spinal cord, 451
  vasoconstriction where
          baroreceptor reflex
          is impaired, 386
Ventral medulla, 487
Ventrolateral medulla, 354, 365
VMA, see methoxyhydroxymandelic
          acid
Vessel endothelium, 84
Voltage-sensitive ionic
          channels, 66